Fast Multipole Methods for the Helmholtz Equation in Three Dimensions

A Volume in the Elsevier Series in Electromagnetism

Elsevier Series in Electromagnetism

(Series formerly known as Academic Press Series in Electromagnetism)

Edited by

ISAAK MAYERGOYZ, University of Maryland, College Park, USA

Electromagnetism is a classical area of physics and engineering that still plays a very important role in the development of new technology. Electromagnetism often serves as a link between electrical engineers, material scientists, and applied physicists. This series presents volumes on those aspects of applied and theoretical electromagnetism that are becoming increasingly important in modern and rapidly development technology. Its objective is to meet the needs of researchers, students, and practicing engineers.

Books Published in the Series

Giorgio Bertotti, *Hysteresis in Magnetism: For Physicists, Material Scientists, and Engineers*
Scipione Bobbio, *Electrodynamics of Materials: Forces, Stresses, and Energies in Solids and Fluids*
Alain Bossavit, *Computational Electromagnetism: Variational Formulations, Complementarity, Edge Elements*
M.V.K. Chari and S.J. Salon, *Numerical Methods in Electromagnetism*
Goran Engdahl, *Handbook of Giant Magnetostrictive Materials*
Vadim Kuperman, *Magnetic Resonance Imaging: Physical Principles and Applications*
John C. Mallinson, *Magneto-Resistive Heads: Fundamentals and Applications*
Isaak Mayergoyz, *Nonlinear Diffusion of Electromagnetic Fields*
Giovanni Miano and Antonio Maffucci, *Transmission Lines and Lumped Circuits*
Shan X. Wang and Alexander M. Taratorin, *Magnetic Information Storage Technology*
Isaak Mayergoyz, *Mathematical Models of Hysteresis and Their Applications*
Fausto Fiorillo, *Measurements and Characterization of Magentic Materials*

Related Books

John C. Mallinson, *The Foundations of Magnetic Recording, Second Edition*
Reinaldo Perez, *Handbook of Electromagnetic Compatibility*

Related Journals

A sample journal issue is available online by visiting the homepage of the journal (homepage details at the top of this page). Free specimen copy gladly sent on request. Elsevier Science Ltd, The Boulevard, Langford Lane, Kidlington, Oxford, OX5 1GB, UK

Fast Multipole Methods for the Helmholtz Equation in Three Dimensions

NAIL A. GUMEROV
RAMANI DURAISWAMI

University of Maryland
Institute for Advanced Computer Studies
College Park, Maryland
USA

2004

ELSEVIER

Amsterdam • Boston • Heidelberg • London • New York • Oxford
Paris • San Diego • San Francisco • Singapore • Sydney • Tokyo

ELSEVIER B.V.
Radarweg 29
P.O. Box 211, 1000 AE Amsterdam
The Netherlands

ELSEVIER Inc.
525 B Street, Suite 1900
San Diego, CA 92101-4495
USA

ELSEVIER Ltd
The Boulevard, Langford Lane
Kidlington, Oxford OX5 1GB
UK

ELSEVIER Ltd
84 Theobalds Road
London WC1X 8RR
UK

First edition 2004

Library of Congress Cataloging in Publication Data
A catalog record is available from the Library of Congress.

British Library Cataloguing in Publication Data
A catalogue record is available from the British Library.

ISBN: 0-08-044371-0
ISSN: 1567-2131

⊗ The paper used in this publication meets the requirements of ANSI/NISO Z39.48-1992 (Permanence of Paper).
Printed and bound in Great Britain by CPI Antony Rowe, Eastbourne

To our wives
(Larisa Gumerov and Shashikala Duraiswami),
Children and Parents

Contents

Preface

Since Isaac Newton introduced a new descriptive method for the study of physics by using mathematical models for various physical phenomena, the solution of differential equations and interpretation of mathematical results have become one of the most important methods for scientific discovery in many branches of science and engineering. A century ago only mechanics and physics, and to a much smaller extent chemistry, enjoyed the use of the predictive and explanatory power of differential equations. At the end of the 20th century, mathematical models have become a commonplace in biology, economics, and many new interdisciplinary areas of science. The necessity for more accurate modeling and prediction, and the exponential growth and availability of computational capabilities has given rise to such disciplines as "computational physics", "computational chemistry", "computational biology", and more generally to "scientific computation." Contemporary engineering, physics, chemistry and biology actively use software for the solution of multi-dimensional problems. Material science, aerospace, chemical engineering, nuclear and environmental engineering, medical instrumentation—indeed, this list can be continued to include all sciences. Today, to a large extent modern technology depends on mathematical modeling and capabilities for the numerical solution of equations constituting these models.

The history of science knows many revolutions: the computational revolution at the end of the 20th century is closely related to the availability of cheap processing power through advances in electronics and materials science and improved algorithms and operating systems due to computer science and related disciplines. These have brought powerful desktop/laptop personal computers to researchers and engineers. These computers have sufficient speed and memory for the solution of such mathematical tasks as the three-dimensional boundary value problems for various partial differential equations. The availability of sophisticated front-end packages such as Matlab and Mathematica allows relatively naive users to access highly sophisticated algorithms, and makes simulation, and the analysis of simulation results a fundamental component of scientific discovery.

The computational capabilities of a modern computer, which children play with, is larger by several orders than the capabilities of huge mainframe computers and systems, exploited in the 1960s, 1970s, and even the 1980s. We should not forget that with such ancient computers humanity went to space, designed nuclear power stations, and experienced revolutions in science and technology of the 1960s. Mainframes and clusters of computers (supercomputers) of the end of the 20th century and the beginning of the 21st century have capabilities exceeding of those used just 10 years ago by orders of magnitude. Some limits for this growth are close enough today due to the limits of semiconductors and high-frequency electrical communications. However, new technologies based on new optical materials, optical switches, and optical analogs of semiconductor devices are under active research and development, which promise further growth of computational capabilities in the following decades. The exponential growth of computational power is captured in various "Moore's laws" named after the scientist Gordon Moore, a cofounder of Intel. In its original form [GM65], the law states that the number of components on a circuit doubles every 18 months. Today, this law is taken to mean that the capability of technology X doubles in Y months [K99].

Nevertheless, the evolution of computers (hardware) itself does not guarantee adequate growth of scientific knowledge or capabilities to solve applied problems unless appropriate algorithms (software) are also developed for the solution of the underlying mathematical problems. For example, for the solution of the most large-scale problems one needs to solve large systems of linear equations, which may consist of millions or billions of equations. Direct solution of a dense linear system for an $N \times N$ matrix requires $O(N^3)$ operations. Using this as a guideline we can say that the inversion of a million by million matrix would require about 10^{18} operations. The top computer in early 2004, the "Earth Simulator" in Yokohama, Japan, has a speed of about 36×10^{12} operations per second and would require about 8 h to solve this problem. If we were to consider a problem 10 times larger, this time would rise to about 1 year. It is impossible to conceive of using simulation as a means of discovery with direct algorithms even using such advanced computers.

Nevertheless, in many practical cases inversions of this type are routinely performed, since many matrices that arise in modeling have special structure. Using specially designed efficient methods for the solution of systems with such matrices, these systems can be solved in $O(N^2)$ or $O(N \log N)$ operations. This highlights the importance of research related to the development of fast and efficient methods for the solution of basic mathematical problems, particularly, multidimensional partial differential equations, since these solvers may be called many times during the solution of particular scientific or engineering design problems. In fact, improving the complexity of algorithms by an order of magnitude (decreasing the exponent by 1) can have a much more significant impact than even hardware advances. For a million variables, the improvement of the exponent can have the effect of skipping 16 generations of Moore's law!

It is interesting to observe how problems and methods of solution, which were formulated a century or two centuries ago, get a new life with advances in computational sciences and computational tools. One of the most famous examples here is related to the Fourier transform that appeared in the Fourier memoir and was submitted to public attention in 1807. This transform was first described in relation to a heat equation, but later it was found that the Fourier method is a powerful technique for the solution of the wave, Laplace, and other fundamental equations of mathematical physics. While used as a method to obtain analytical solutions for some geometries, it was not widely used as a computational method. A new life began for the Fourier transform only in 1965 after the publication of the paper by Cooley and Tukey [CT65], who described the Fast Fourier Transform (FFT) algorithm that enables multiplication of a vector by the $N \times N$ Fourier matrix for an expense of only $O(N \log N)$ operations as opposed to $O(N^2)$ operations. In practice, this meant that for the time spent for the Fourier transform of length, say, $N \sim 10^3$ with a straightforward $O(N^2)$ algorithm, one can perform the Fourier transform of a sequence of length $N \sim 10^5$, which is hundred times larger! Of course, this discovery caused methods based on the Fourier transform to be preferred over other methods, and revolutionized areas such as signal processing. This algorithm is described as one of the best ten algorithms of the 20th century [DS00].

Another example from these top ten algorithms is related to the subject of this book. This is an algorithm due to Rokhlin and Greengard [GR87] called the "Fast Multipole Method" (FMM). While it was first formulated for the solution of the Laplace equation in two and three dimensions, it was extended later for other equations, and more generally to the multiplication of $N \times N$ matrices with special structure by vectors of length N. This algorithm achieves approximate multiplication for expense of $O(\alpha N)$ operations, where α depends on the prescribed accuracy of the result, ε, and usually $\alpha \sim \log N + \log \varepsilon^{-1}$. For computations with large N, the significance of this algorithm is comparable with that of the FFT. While the algorithm itself is different from the FFT, we note that as the FFT did, it brings "new life" to some classical methods developed in the 19th century, which have not been used widely as general computational methods.

These are the methods of multipoles or multipole expansions, which, as the FFT, can be classified as spectral methods. Expansions over multipoles or some elementary factorized solutions for equations of mathematical physics were known since Fourier. However, they were used less frequently, say, for the solution of boundary value problems for complex-shaped domains. Perhaps, this happened because other methods such as the Boundary Element, Finite Element, or Finite Difference methods appeared to be more attractive from the computational point of view. Availability of a fast algorithm for solution of classical problems brought research related to multipole and local expansions to a new level. From an algorithmic point of view, the issues of fast and accurate translations, or conversions of expansions over different bases from one to the other have become of primary importance. For example, the issue of development of fast, computationally stable, translation methods and their relation to the structured matrices, for which fast matrix–vector multiplication is available, were not in the scope of 19th or 20th century researchers living in the era before the FMM. A more focused attention to some basic principles of multipole expansion theory is now needed with the birth of the FMM.

The latter sentence formulates the motivation behind the present book. When several years ago we started to work on the problems of fast solution of the Helmholtz equation in three dimensions we found a substantial lack in our knowledge on multipole expansions and translation theory for this equation. Some facts were well known, some scattered over many books and papers, and several things we had to rediscover by ourselves, since we did not find, at that time, the solution to our problems. A further motivation was from our desire to get a solution to some practically important problems such as scattering from multiple bodies and scattering from complex boundaries. Here again, despite many good papers from other researchers in the field, we could not find a direct answer to some of our problems, or find appropriate solutions (e.g. we were eager to have FFT-type algorithms for the translation and filtering of spherical harmonics, which are practically faster than our first $O(p^3)$ method based on a rotation–coaxial translation decomposition). We also found that despite a number of publications, some details and issues related to the error bounds and the complexity of the FMM were not worked out. In the present book we attempt to pay significant attention to these important issues. While future developments may make some of the results presented in this book less important, at the time of its writing, these issues are essential to the development of practical solvers for the Helmholtz equation using these fast algorithms.

Acknowledgments

We would like to thank the National Science Foundation for the support of much of the research on which the book is based through NSF awards 0086075 and 0219681. We would also like to acknowledge the support from the University of Maryland Institute for Advanced Computer Studies for the preparation of the book. In addition, the department of Computer Science and the Applied Mathematics and Scientific Computing Program at the University provided us the opportunity to develop some of this material while teaching a graduate course on the Fast Multipole Method. In particular we are very grateful to Profs Joseph F. JaJa (Director, UMIACS) and Larry S. Davis (Chair, Computer Science) for their support and encouragement. We would also like to thank Prof. Isaak Mayergoyz for his encouragement in writing this book.

Finally, writing this book required an investment of time, that necessarily reduced the time we could have spent otherwise. We would like to thank our families for their love, support, encouragement and forbearance.

Outline of the Book

The book is organized as follows.

Chapter 1: This is an introductory chapter whose main purpose is to present the scalar Helmholtz equation as a universal equation appearing in different areas of physics. Even though many problems are formulated in terms of systems of equations or are described by other well-known basic models, they can be reduced to the solution of the scalar Helmholtz equation using the scalar potentials and the Fourier or Laplace transforms. Here we also consider major types of boundary and transmission conditions and integral representation of solutions. Computation of the surface and volume integrals can be performed by discretization and reduction of the problem to summation of a large number of monopoles and dipoles. The rest of the book is dedicated to the solution of problems that arise from the scalar Helmholtz equation, whose solution can also be generalized to the summation of a large number of arbitrary multipoles.

Chapter 2: This chapter is dedicated to the fundamentals of the multipole and local expansions of the solutions of the Helmholtz equation. Most relations presented here are well known and one of the major goals of this chapter is to bring together in one place the necessary definitions and equations for easy reference. Another important goal is to establish the notation used in the book (because different authors use different functions under the same notation, e.g., spherical harmonics or "multipoles"). While the normalization factors to use may not seem important, our experience shows that one can spend substantial time to have a reliable analytical formula that can be used further. We introduce here the definition of the special functions used later in the book, and summarize useful relations for them.

Chapter 3: This is one of the key theoretical chapters. It introduces the concepts of reexpansion, translation, and rotation of solutions of the Helmholtz equation. Some equations and relations can be found in other sources while others are derived here for the first time. This chapter includes the basic concepts, structure of the reexpansion coefficients and special types and properties of these functions of vector argument. Since our major concern is the development and implementation of fast computational methods, we derive here some efficient methods for computations of the translation and rotation coefficients. While the explicit expressions for them via, say, Clebsch–Gordan coefficients, can be found elsewhere, these formulae are not practical for use in fast multipole methods. By designing and applying recursive methods, which allow one to compute all necessary coefficients spending not more than just a few operations for each of them, we achieve fast $O(p^4)$ and $O(p^3)$ translation methods, where p is the truncation number or bandwidth of functions used to approximate the solutions of the Helmholtz equation.

Chapter 4: The results of Chapter 3 can already be used for the solution of a number of problems of practical interest such as appearing in room acoustics and in scattering from multiple bodies. We identify the techniques used in this chapter as the "multipole reexpansion technique" or "multipole methods". In many cases this technique itself can substantially speed up solution of the problem compared to other methods (e.g. direct summation of sources or solution with boundary element methods). The purpose here is to show some problems of interest and provide the reader with some formulae that can be used for the solution of such complex problems as multiple scattering problem from arbitrarily shaped objects. This chapter comes before the chapters dedicated to fast multipole methods, and the methods presented can be speeded up further using the methods in subsequent chapters.

Chapter 5: In this chapter we introduce Fast Multipole Methods (FMM) in a general framework, which can be used for the solution of different multidimensional equations and problems, and where the solution of the Helmholtz equation in three dimensions is just a particular case. We start with some basic ideas related to factorization of solutions. We describe how rapid summation of functions can be performed. Next, we proceed to modifications of this basic idea, such as the "Single Level FMM", and the "Multilevel FMM", which is the FMM in its original form. While there exist a substantial number of papers in this area that may be familiar to the reader, we found that the presentation in these often obscure some important issues, which are important for the implementation of the method and for its understanding. This method is universal in a sense that it can be formalized and applied to problems arising not only in mathematical physics. One of the issues one faces is the data structures to be used and efficient implementation of algorithms operating with a large amount of data used in the FMM. This is one of the "hidden" secrets of the FMM that usually each developer must learn. We provide here several techniques based on spatial ordering and bit interleaving that enable fast "children" and "neighbor" search procedures in data organized in such structures as octrees. These techniques are known in areas which are not related to mathematical physics, and we tried to provide a detailed insight for the reader who may not be familiar with them.

Chapter 6: While one can consider the FMM for the Helmholtz equation as a particular case of a generalized FMM procedure, it has some very important peculiarities. In the form originally introduced by Rokhlin and Greengard for the solution of the Laplace equation, the FMM is practical only for the so-called "low-frequency" problems, where the size of the computational domain, D_0, and the wave number, k, are such that $kD_0 < A$, where A is some constant. While this class of problems is important, it prevents application of the FMM for "high-frequency" problems, which are equally important. The method to efficiently to solve these problems is to vary the truncation number with the level of hierarchical space subdivision. To illustrate this we introduce a model of the FMM for the Helmholtz equation, and derive several important theoretical complexity results. One of the basic parameters of this model is a parameter we call the "translation exponent" that characterizes the complexity of translations for some given truncation number. We also introduce some concepts such as the "critical translation exponent", which separates the complexity of the method for higher frequencies from one type to the other. The critical value of the exponent depends on the dimensionality and "effective" dimensionality of the problem, which is determined by the non uniformity of the spatial distributions of the sources and receivers. We also provide some optimization results and suggest a fully adaptive FMM procedure based on tree-structures, opposed to the pyramid data structures used in the regular FMM. This method was found to be useful for the solution of some "low-frequency" problems, while additional research is needed for other problems.

Chapter 7: This chapter is dedicated to the theory which underlies fast translation methods, and serves as a guide for further developments in this field. While providing substantial background theory, we focus here on two translation methods of complexity $O(p^3)$ which are based on rotation–coaxial decomposition of the translation operator and on sparse matrix decompositions of the operators. While the first method is known in the literature and can be applied to the decomposition of any translation for any space-invariant equation (which follows from the group theory), the second method is presented here, to the best of our knowledge, for the first time. This method can be derived from the commutativity properties of the sparse matrices representing differential operators and dense matrices representing translation operators. We implemented and tested both the methods and found them to be reliable and fast. While the first method seems to have smaller asymptotic constants and, so is faster, we believe that new research opportunities for fast translation methods are uncovered by the second method.

Chapter 8: In this chapter we consider both new and existing translation methods that bring the complexity of translations to $O(p^2 \log^\alpha p)$ with some α ranging from 0 to 2. They are based on the use of properties of structured matrices, such as Toeplitz or Cauchy matrices or on the diagonal forms of the translation and rotation operators. While some techniques developed over the last decade have been implemented and studied, this is still an active area for research. We have attempted to summarize and advance the knowledge in this area, though we are sure new fast techniques, filters, or transforms, will continue to be developed. We provide a link between the methods operating in the functional space of expansion coefficients and the methods operating in the space of samples of surface functions, where the transform from one space to the other can be done theoretically with $O(p^2 \log^\alpha p)$ complexity. We also present here some asymptotic results that can be used for the development of fast translation methods at low and high frequencies.

Chapter 9: One of the most important issues in any numerical method is connected with the sources of errors in the method, and bounds for these errors. This particularly relates to the FMM, where the error control is performed based on theory. There are several studies in the literature related to this issue for the Helmholtz equation, which are mostly concerned with proper selection of the truncation number for expansion of monopoles. Here we present some results from our study of the error bounds, which we extend to the case of arbitrary multipoles, and in addition establish the error bounds for the truncation of translation operators represented by infinite matrices. The theoretical formulae derived were tested numerically on some example problems for the expansion of single monopoles and while running the FMM for many sources. The latter results bring interesting findings, which should be theoretically explained by further studies. This includes, e.g. the error decay exponent at low frequencies, that shows that evaluations based on the "worst" case analysis substantially overestimate actual errors.

Chapter 10: In the final chapter we demonstrate the application of the FMM to the solution of the multiple scattering problem. We discuss this in details as well as some issues concerned with the iterative techniques combined with the FMM. Also we show how the FMM can be applied to imaging of the three-dimensional fields that are described by the Helmholtz equation. Finally, we present some results of numerical study of these problems including convergence of the iterative methods and overall method performance.

The book is written in an almost "self-contained" manner so that a reader with appropriate background in mathematics and computational methods, who, for the first time faces the problem of fast solution of the Helmholtz equation in three dimensions, can learn everything from scratch and can implement a working FMM algorithm. Chapter 8 is an exception, since there we refer to algorithms such as Fast Legendre Transform or Fast Spherical Filters, whose detailed presentation is not given, since it would require a special book chapter. As we mentioned, these algorithms are under active research, and so, if a beginning reader reaches this stage, we hope that he or she will be able to read and understand the appropriate papers from the literature that in any case may be substantially updated by that time.

An advanced reader can go directly to chapters or sections of interest and use the other chapters as reference for necessary formulae, definitions and explanations. We need to emphasize that while we have tried to use notations and definitions consistent with those used in the field, we found that different authors often define similar functions differently. As in any new work, at times we have had to introduce some of our own notations for functions and symbols, which are still not in common use. In any case we recommend that the reader be careful, especially if the formulae are intended to be used for numerical work, and follow the derivations and definitions presented carefully to avoid inconsistency with definitions in other literature.

CHAPTER 1

Introduction

We begin the book with a review of the basic physical problems that lead to the various equations we wish to solve.

1.1 HELMHOLTZ EQUATION

The *scalar* Helmholtz equation

$$\nabla^2 \psi(\mathbf{r}) + k^2 \psi(\mathbf{r}) = 0, \qquad \nabla^2 = \frac{\partial^2}{\partial x^2} + \frac{\partial^2}{\partial y^2} + \frac{\partial^2}{\partial z^2}, \tag{1.1.1}$$

where $\psi(\mathbf{r})$ is a complex scalar function (potential) defined at a spatial point $\mathbf{r} = (x, y, z) \in \mathbb{R}^3$ and k is some real or complex constant, takes its name from Hermann von Helmholtz (1821–1894), the famous German scientist, whose impact on acoustics, hydrodynamics, and electromagnetics is hard to overestimate. This equation naturally appears from general conservation laws of physics and can be interpreted as a wave equation for monochromatic waves (wave equation in the frequency domain). The Helmholtz equation can also be derived from the heat conduction equation, Schrödinger equation, telegraph and other wave-type, or evolutionary, equations. From a mathematical point of view it appears also as an eigenvalue problem for the Laplace operator ∇^2. Below we show the derivation of this equation in several cases.

1.1.1 Acoustic waves

1.1.1.1 *Barotropic fluids*

The usual assumptions for acoustic problems are that acoustic waves are perturbations of the medium density $\rho(\mathbf{r}, t)$, pressure $p(\mathbf{r}, t)$, and velocity, $v(\mathbf{r}, t)$, where t is time. It is also assumed that the medium is inviscid, and that perturbations are small, so that:

$$\rho = \rho_0 + \rho', \qquad p = p_0 + p', \qquad \rho' \ll \rho_0, \qquad p' \ll p_0,$$
$$|\mathbf{v}'| \ll c \sim \sqrt{\frac{p_0}{\rho_0}}. \tag{1.1.2}$$

Here the perturbations are about an initial spatially uniform state (ρ_0, p_0) of the fluid at rest ($\mathbf{v}_0 = \mathbf{0}$) and are denoted by primes. The latter equation states that the velocity of the fluid is much smaller than the speed of sound c in that medium. In this case the linearized continuity (mass conservation) and momentum conservation equations can be written as:

$$\frac{\partial \rho'}{\partial t} + \nabla \cdot (\rho_0 \mathbf{v}') = 0, \qquad \rho_0 \frac{\partial \mathbf{v}'}{\partial t} + \nabla p' = 0, \tag{1.1.3}$$

where

$$\nabla = \mathbf{i}_x \frac{\partial}{\partial x} + \mathbf{i}_y \frac{\partial}{\partial y} + \mathbf{i}_z \frac{\partial}{\partial z}, \tag{1.1.4}$$

is the invariant "nabla" operator, represented by formula (1.1.4) in Cartesian coordinates, where $(\mathbf{i}_x, \mathbf{i}_y, \mathbf{i}_z)$ are the Cartesian basis vectors.

Differentiating the former equation with respect to t and excluding from the obtained expression $\partial \mathbf{v}'/\partial t$ due to the latter equation, we obtain:

$$\frac{\partial^2 \rho'}{\partial t^2} = \nabla^2 p'. \tag{1.1.5}$$

Note now that system (1.1.3) is not closed since the number of variables (three components of velocity, pressure, and density) is larger than the number of equations. The relation needed to close the system is the equation of state, which relates perturbations of the pressure and density. The simplest form of this relation is provided by *barotropic* fluids, where the pressure is a function of density alone:

$$p = p(\rho). \tag{1.1.6}$$

We can expand this in the Taylor series near the unperturbed state:

$$p = p(\rho_0) + \left.\frac{dp}{d\rho}\right|_{\rho=\rho_0} (\rho - \rho_0) + O((\rho - \rho_0)^2). \tag{1.1.7}$$

Taking into account that $p(\rho_0) = p_0$ we, obtain, neglecting the second-order nonlinear term:

$$p' = c^2 \rho', \qquad c^2 = \left.\frac{dp}{d\rho}\right|_{\rho=\rho_0}, \tag{1.1.8}$$

where we used the definition of the speed of sound in the unperturbed fluid, which is a real positive constant (property of the fluid). Substitution of expression (1.1.8) into relation (1.1.5) yields the *wave equation* for pressure perturbations:

$$\frac{1}{c^2}\frac{\partial^2 p'}{\partial t^2} = \nabla^2 p'. \tag{1.1.9}$$

Obviously, the density perturbations satisfy the same equation. The velocity is a vector and satisfies the *vector* wave equation:

$$\frac{1}{c^2}\frac{\partial^2 \mathbf{v}'}{\partial t^2} = \nabla^2 \mathbf{v}'. \tag{1.1.10}$$

This also means that each of the components of the velocity $\mathbf{v}' = (v'_x, v'_y, v'_z)$ satisfies the *scalar* wave equation (1.1.9). Note that these components are not independent. The momentum equation (1.1.3) shows that there exists some scalar function ϕ', which is called the *velocity potential*, such that

$$\mathbf{v}' = \nabla\phi, \qquad \frac{1}{c^2}\frac{\partial^2 \phi}{\partial t^2} = \nabla^2\phi \qquad \left(\rho_0\frac{\partial\phi}{\partial t} = -p'\right). \tag{1.1.11}$$

So the problem can be solved for the potential and then the velocity field can be found as the gradient of this scalar field.

1.1.1.2 *Fourier and Laplace transforms*

The wave equation derived above is linear and has particular solutions that are periodic in time. In particular, if the time dependence is a harmonic function of *circular frequency* ω, we can write

$$\phi(\mathbf{r}, t) = \mathrm{Re}(\mathrm{e}^{-\mathrm{i}\omega t}\psi(\mathbf{r})), \qquad \mathrm{i}^2 = -1, \tag{1.1.12}$$

where $\psi(\mathbf{r})$ is some complex valued scalar function and the real part is taken, since $\phi(\mathbf{r}, t)$ is real. Substituting expression (1.1.12) into the wave equation (1.1.11), we see that the latter is satisfied if $\psi(\mathbf{r})$ is a solution of the Helmholtz equation:

$$\nabla^2\psi(\mathbf{r}) + k^2\psi(\mathbf{r}) = 0, \qquad k = \frac{\omega}{c}. \tag{1.1.13}$$

The constant k is called the *wavenumber* and is real for real ω. The name is related to the case of plane wave propagating in the fluid, where the *wavelength* is $\lambda = 2\pi/k$ and so k is the *number* of waves per 2π units of length.

The Helmholtz equation, therefore, stands for monochromatic waves, or waves of some given frequency ω. For *polychromatic* waves, or sums of waves of different frequencies, we can sum up solutions with different ω. More generally, we can perform the *inverse Fourier transform* of the potential $\phi(\mathbf{r}, t)$ with respect to the temporal variable:

$$\psi(\mathbf{r}, \omega) = \int_{-\infty}^{\infty} e^{i\omega t}\phi(\mathbf{r}, t)dt. \tag{1.1.14}$$

In this case $\psi(\mathbf{r}, \omega)$ satisfies the Helmholtz equation (1.1.13). Solving this equation we can determine the solution of the wave equation using the *forward Fourier transform*:

$$\phi(\mathbf{r}, t) = \frac{1}{2\pi}\int_{-\infty}^{\infty} e^{-i\omega t}\psi(\mathbf{r}, \omega)d\omega, \qquad \omega = ck. \tag{1.1.15}$$

We note that in the Fourier transform the frequency ω can either be negative or positive. This results in either negative or positive values of the wavenumber. However, the Helmholtz equation depends on k^2 and is *invariant* with respect to a change of sign in k. This phenomenon, in fact, has a deep physical and mathematical origin, and appears from the property that the wave equation is a *two-wave equation*. It describes solutions which are a superposition of two waves propagating with the same velocity in *opposite directions*. We will consider this property and rules for proper selection of sign in Section 1.2.

While monochromatic waves are important solutions with physical meaning, we note that mathematically we can also consider solutions of the wave equation of the type:

$$\phi(\mathbf{r}, t) = \mathrm{Re}(e^{st}\psi(\mathbf{r})), \qquad s \in \mathbb{C}, \tag{1.1.16}$$

where s is an arbitrary complex constant. In this case, as follows from the wave equation, $\psi(\mathbf{r})$ satisfies the following Helmholtz equation

$$\nabla^2\psi(\mathbf{r}) + k^2\psi(\mathbf{r}) = 0, \qquad k^2 = -\frac{s}{c}. \tag{1.1.17}$$

Here the constant k^2 can be an arbitrary complex number. This type of solution also has physical meaning and can be applied to solve initial value problems for the wave equation. Indeed, if we consider solutions of the wave equation, such that the fluid was unperturbed for $t \leq 0$ ($\phi(\mathbf{r}, t) = 0, t \leq 0$) while for $t > 0$ we have a non-trivial solution, then we can use the *Laplace transform*:

$$\psi(\mathbf{r}, s) = \int_{-\infty}^{\infty} e^{-st}\phi(\mathbf{r}, t)dt, \qquad \mathrm{Re}(s) > 0, \tag{1.1.18}$$

which converts the wave equation into the Helmholtz equation with complex k, (Eq. (1.1.17)). If an appropriate solution of the Helmholtz equation is available, then we can determine the solution of the wave equation using the *inverse* Laplace transform:

$$\phi(\mathbf{r}, t) = \frac{1}{2\pi i} \int_{\sigma - i\infty}^{\sigma + i\infty} e^{st} \psi(\mathbf{r}, s) ds, \qquad \sigma > 0. \tag{1.1.19}$$

The above examples show that integral transforms with exponential kernels convert the wave equation into the Helmholtz equation. In the case of the Fourier transform we can state that the Helmholtz equation is the wave equation in the *frequency domain*. Since methods for fast Fourier transform are widely available, conversion from time to frequency domain and back are computationally efficient, and so the problem of the solution of the wave equation can be reduced to the solution of the Helmholtz equation, which is an equation of lower dimensionality (3 instead of 4) than the wave equation.

1.1.2 Scalar Helmholtz equations with complex k

1.1.2.1 Acoustic waves in complex media

Despite the fact that the barotropic fluid model is a good idealization for real fluids in certain frequency ranges, it may not be adequate for complex fluids, where internal processes occur under external action. Such processes may happen at very high frequencies due to molecular relaxation and chemical reactions or at lower frequencies if some inclusions in the form of solid particles or bubbles are present. One typical example of a medium with internal relaxation processes is plasma.

To model a medium with relaxation one can use the mass and momentum conservation equations (1.1.3) or a consequence of these. The difference between the models of barotropic fluid and relaxating medium occurs in the equation of state, which can sometimes be written in the form:

$$p = p(\rho, \dot{\rho}), \tag{1.1.20}$$

where the dot denotes the substantial derivative with respect to time. Being perturbed, the density of such a medium does not immediately follow the pressure perturbations, but rather returns to the equilibrium state with some dynamics. In the case of small perturbations, linearization of this equation yields:

$$p' = c^2\left(\rho' + \tau_\rho \frac{\partial \rho'}{\partial t}\right). \tag{1.1.21}$$

Here τ_ρ is a constant having dimensions of time, and can be called the density relaxation time.

Equations (1.1.5) and (1.1.21) form a closed system, which has a particular solution oscillating with time so solutions of type Eq. (1.1.12) can be considered. To obtain corresponding Helmholtz equations for the wave equations considered, note that for a harmonic function we can simply replace the time derivative symbols:

$$\frac{\partial}{\partial t} \rightarrow -i\omega. \tag{1.1.22}$$

This replacement of the time derivative with $-i\omega$ can be interpreted as a transform of the equations from the time to the frequency domain. With this remark, we can derive from relations (1.1.5) and (1.1.21) a single equation for the density perturbation in the frequency domain (denoted by symbol ψ):

$$\nabla^2 \psi(\mathbf{r}) + k^2 \psi(\mathbf{r}) = 0, \qquad k^2 = \frac{\omega^2}{c^2(1 - i\omega\tau_\rho)}. \tag{1.1.23}$$

Thus we obtained the Helmholtz equation with a complex wavenumber. At low frequencies, $\omega \ll \tau_\rho^{-1}$, the relaxation term is not important and we obtain the Helmholtz equation with real k. For high frequencies, the character of the dependency $k(\omega)$ is different, plus k appears to be complex.

A more general dependence of the pressure on the density and its derivatives can be considered for waves in complex media, e.g.

$$p = p(\rho, \dot{\rho}, \ddot{\rho}, \ldots). \tag{1.1.24}$$

Dependences of this type may also include integrals over time with kernels (media with memory). It is not difficult to see that various linearized equations of state result in equations of type:

$$\nabla^2 \psi(\mathbf{r}) + k^2 \psi(\mathbf{r}) = 0, \qquad k^2 = k^2(\omega) \tag{1.1.25}$$

in the frequency domain. The function $k^2(\omega)$ depends on the particular medium considered and is related to its equation of state. The frequency dependence of the wavenumber is also called the *dispersion relationship*. If ω/k is characterized as the speed of sound, we can see that, in contrast to barotropic fluids in complex media, the speed of sound is now a function of frequency, and moreover, can be complex. In fact, the quantity

$$c_p = \frac{\omega}{\mathrm{Re}(k)} \tag{1.1.26}$$

is called the *phase velocity*, and characterizes the velocity of propagation of lines of constant phase, and the imaginary part of ω/k characterizes the attenuation of waves in the medium. The dependence of the real part on frequency is also known as the dispersion of the phase velocity (or simply, dispersion), which means that waves of different frequencies propagate with different velocities. There is a special case of the dispersion relationship, when $k(\omega)$ is real (e.g. this appears in some models of waves in plasma, and gravitation waves on a liquid free surface). In this case the medium is characterized as a medium with dispersion and without dissipation. As we can see from examples of medium with relaxation, the dispersion relationship (1.1.23) contains both real and imaginary parts, and so a medium with relaxation can also be characterized as a medium with dispersion and dissipation.

1.1.2.2 Telegraph equation

Another example of an equation that can be reduced to the Helmholtz equation is the telegraph equation. It is closely related to the equation for waves in a relaxating medium. The one-dimensional version of this equation first appeared in the description of signal transmission through a cable. It can be interpreted as a general wave equation with attenuation and extended to two and three dimensions, and is used by some researchers for modeling media with relaxation and dissipation, extremely low frequency electromagnetic wave propagation in the ionosphere, etc. This equation can be written in the time domain as:

$$\frac{1}{c^2}\frac{\partial^2 \Phi}{\partial t^2} + a\frac{\partial \Phi}{\partial t} + b\Phi = \nabla^2 \Phi, \qquad (1.1.27)$$

where a, b, and c are some constants. In the frequency domain we obtain the following Helmholtz equation (see Eq. (1.1.22)):

$$\nabla^2 \psi(\mathbf{r}) + k^2 \psi(\mathbf{r}) = 0, \qquad k^2 = \frac{\omega^2}{c^2} + \mathrm{i}\omega a - b. \qquad (1.1.28)$$

This is a special case of Eq. (1.1.25) where k is complex.

It is interesting to note that for $a = b = 0$ the telegraph equation turns into the usual wave equation, while for $c = \infty$, $b = 0$ it turns into the diffusion equation, discussed in Section 1.1.2.3, and for $c = \infty$, $a = 0$ it reduces to the Helmholtz equation in the time or frequency domains, with $k^2 = -b$. For real positive b both roots of the latter equation appear to be purely imaginary and this corresponds to two types of waves: exponentially growing and exponentially decaying. If there are no energy sources in the medium only decaying waves should be considered.

1.1.2.3 *Diffusion*

The heat conduction equation in solids can be written in the form:

$$\frac{\partial T}{\partial t} = \kappa \nabla^2 T, \qquad (1.1.29)$$

where T is the perturbation of the temperature and κ is the thermal diffusivity. This equation also describes heat conduction in incompressible liquids if the convective term is negligibly small compared to the conductive term and is the case when the liquid is at rest or the temperature of the liquid changes much faster than the liquid flows.

The heat conduction equation is universal and appears in many other problems, e.g. for description of mass diffusion. In this case T should be interpreted as the perturbation of mass concentration and κ as the mass diffusivity. Another example interprets Eq. (1.1.29) as one describing diffusion of vorticity in viscous fluids. In this case T is a component of the vorticity vector and κ is a kinematic viscosity of the fluid.

In any case the heat conduction equation can be considered in the frequency domain using the Fourier transform. In that domain a corresponding equation can be easily found using rule (1.1.22):

$$\nabla^2 \psi(\mathbf{r}) + k^2 \psi(\mathbf{r}) = 0, \qquad k^2 = \frac{i\omega}{\kappa}. \qquad (1.1.30)$$

This is the Helmholtz equation with purely imaginary k^2. Therefore, the wavenumber in this case will have both real and imaginary parts. Since both k and $-k$ provide solutions of the Helmholtz equation then either two "thermal waves" can be considered or one solution can be set to zero, based on the problem. For example, for heat propagation from a body in an infinite medium, one should select solutions which decay at infinity.

It is also interesting to note that at high frequencies, $\omega \gg \tau_p^{-1}$, propagation of waves in media with relaxation is described by the same type of dispersion relationship as for the heat conduction equation (see Eq. (1.1.23)).

1.1.2.4 *Schrödinger equation*

Having its origin in quantum mechanics, the Schrödinger equation appears as a universal equation for modulations of quasi-monochromatic acoustic and electromagnetic waves in complex media, e.g. in plasma. Modulation waves in weakly nonlinear approximation are described by the nonlinear Schrödinger equation and in linear approximation by the linear Schrödinger equation:

$$-\mathrm{i}\frac{\partial \Psi}{\partial t} = \nabla^2 \Psi. \tag{1.1.31}$$

Here Ψ is the wavefunction or the complex amplitude of a modulation wave. Transforming this equation into the frequency domain by using rule (1.1.22), we get:

$$\nabla^2 \psi(\mathbf{r}) + k^2 \psi(\mathbf{r}) = 0, \qquad k^2 = \omega. \tag{1.1.32}$$

Therefore, as in the case of wave equation (1.1.13), we obtained the Helmholtz equation with real k^2. Despite a similarity to the Helmholtz equations, we note the following important differences from the physical view point. First, in the case of the Schrödinger equation we have a dispersion of the phase velocity (Eq. (1.1.26)), $c_p = c_p(\omega)$ while for the wave equation $c_p = c_0 = \text{const}$. Second, since ω in the Fourier transform changes from $-\infty$ to ∞, k^2 appears to be negative for $\omega < 0$ and so k in this case is purely imaginary.

1.1.2.5 Klein–Gordan equation

The Schrödinger equation is a quantum mechanical equation for non-relativistic mechanics. For relativistic quantum mechanics the corresponding equation, which describes a free particle with zero spin, is called the Klein–Gordan equation and can be written in the form:

$$\frac{\partial^2 \Psi}{\partial t^2} = \nabla^2 \Psi - m^2 \Psi, \tag{1.1.33}$$

where Ψ is the wavefunction and m is the normalized particle mass. In the frequency domain, this corresponds to the following Helmholtz equation:

$$\nabla^2 \psi(\mathbf{r}) + k^2 \psi(\mathbf{r}) = 0, \qquad k^2 = -(m^2 - \omega^2). \tag{1.1.34}$$

Let us look for time-independent solutions ($\partial \Psi / \partial t = 0$ or $\omega = 0$) of the Klein–Gordan equation. In this case it reduces to the Helmholtz equation with purely imaginary $k = \mathrm{i}m$. The spherically symmetrical solution of this equation decaying at infinity is

$$\Psi = \frac{C}{r} \mathrm{e}^{-mr}, \qquad r = |\mathbf{r}|, \tag{1.1.35}$$

where C is some constant. The case $m = 0$ corresponds to non-relativistic approximation and in this case the potential is a fundamental solution of the Laplace equation $\sim r^{-1}$, which is known from electrostatics or gravitation theory. For non-zero m this potential is known as the *Yukawa potential*. The Yukawa potential can be used in the theory of relativistic gravitation or as an analog of the electrostatic potential for description of intermolecular forces and interactions.

1.1.3 Electromagnetic waves

1.1.3.1 Maxwell's equations

Consider now the appearance of the scalar Helmholtz equation in the context of Maxwell equations describing propagation of electromagnetic waves. For a medium free of charges and imposed currents, these equations can be written as:

$$\nabla\times\mathbf{E}=-\mu\frac{\partial\mathbf{H}}{\partial t},\quad \nabla\times\mathbf{H}=\varepsilon\frac{\partial\mathbf{E}}{\partial t},\quad \nabla\cdot\mathbf{E}=0,\quad \nabla\cdot\mathbf{H}=0, \tag{1.1.36}$$

where $\mathbf{E}$ and $\mathbf{H}$ are the electric and magnetic field vectors, and μ and ε are the magnetic permeability and electric permittivity in the medium, respectively. In the case of a vacuum we have

$$\mu=\mu_0,\quad \varepsilon=\varepsilon_0,\quad c=(\mu_0\varepsilon_0)^{-1/2},$$

where c is the speed of light in a vacuum and $c\approx 3\times 10^8$ m/s.

Taking the curl of the first equation and using the second equation, we have

$$\nabla\times\nabla\times\mathbf{E}=-\mu\frac{\partial}{\partial t}\nabla\times\mathbf{H}=-\mu\varepsilon\frac{\partial^2\mathbf{E}}{\partial t^2}=-\frac{1}{c^2}\frac{\partial^2\mathbf{E}}{\partial t^2}, \tag{1.1.37}$$

Due to the following well-known identity for the ∇ operator and the third equation in Eq. (1.1.36):

$$\nabla^2\mathbf{E}=\nabla(\nabla\cdot\mathbf{E})-\nabla\times\nabla\times\mathbf{E}=-\nabla\times\nabla\times\mathbf{E}, \tag{1.1.38}$$

we obtain

$$\frac{1}{c^2}\frac{\partial^2\mathbf{E}}{\partial t^2}=\nabla^2\mathbf{E}. \tag{1.1.39}$$

Similarly,

$$\frac{1}{c^2}\frac{\partial^2\mathbf{H}}{\partial t^2}=\nabla^2\mathbf{H}. \tag{1.1.40}$$

Thus, both the electric and magnetic field vectors satisfy the vector wave equation. Transformation of this equation into the frequency domain yields

$$\begin{aligned}(\nabla^2+k^2)\hat{\mathbf{E}}&=0,\qquad k=\frac{\omega}{c}\\(\nabla^2+k^2)\hat{\mathbf{H}}&=0,\end{aligned} \tag{1.1.41}$$

where we use circumflex to denote that we are in the frequency domain and $\hat{\mathbf{E}}$ and $\hat{\mathbf{H}}$ are the complex amplitudes of **E** and **H** for harmonic oscillations with frequency ω. These complex functions are also known as *phasors*.[1]

Note that the number of scalar equations (1.1.41) in three dimensions is six (each Cartesian component of $\hat{\mathbf{E}}$ or $\hat{\mathbf{H}}$ satisfies the scalar Helmholtz equation), while the original formulation (1.1.36) provide eight relations for the same quantities. The missing relations are equations stating that the divergence of the electric and magnetic fields is zero. This imposes limitations on the components of the electric and magnetic field vectors, since they should be constrained to satisfy these additional equations. So any of these fields is described by the following system of equations:

$$(\nabla^2 + k^2)\hat{\mathbf{E}} = 0, \qquad \nabla\cdot\hat{\mathbf{E}} = 0, \tag{1.1.42}$$

where $\hat{\mathbf{E}}$ can be replaced by $\hat{\mathbf{H}}$. It is interesting to note that, in the equation describing propagation of acoustic waves (Eq. (1.1.10)), the velocity field also satisfies the vector wave equation (so its phasor satisfies the vector Helmholtz equation) with an additional condition $\nabla \times \mathbf{v}' = 0$. This is equivalent to the existence of a scalar potential which satisfies the scalar wave equation, $\mathbf{v}' = \nabla\phi$. Below we will show that the second equation in Eq. (1.1.42) enables the introduction of two scalar potentials, which satisfy the scalar equations, and both vectors $\hat{\mathbf{E}}$ and $\hat{\mathbf{H}}$ can be expressed via these functions.

1.1.3.2 *Scalar potentials*

Since the components of $\hat{\mathbf{E}}$ are related via the divergence free condition, we can consider the representation of $\hat{\mathbf{E}}$ via two independent scalar functions ψ_1 and ψ_2, where each of these the functions satisfies the scalar Helmholtz equation:

$$(\nabla^2 + k^2)\psi_1 = 0, \qquad (\nabla^2 + k^2)\psi_2 = 0. \tag{1.1.43}$$

To do this we prove the following two theorems using vector algebra (these theorems can be found elsewhere).

THEOREM 1 *Let ψ be a scalar function that satisfies the Helmholtz equation (1.1.43). Then the function $\hat{\mathbf{E}} = \nabla\psi \times \mathbf{r}$ satisfies the constrained vector Helmholtz equations (1.1.42).*

[1] In general we can define the phasor as a Fourier-image of the function. For example, in Eq. (1.1.14) function Ω is the phasor of ϕ'.

PROOF. First we note that this function is a curl of some vector:

$$\hat{\mathbf{E}} = \nabla\psi \times \mathbf{r} = \nabla \times (\mathbf{r}\psi) \tag{1.1.44}$$

So it is a solenoidal (or divergence free) field:

$$\nabla\cdot\hat{\mathbf{E}} = \nabla\cdot[\nabla \times (\mathbf{r}\psi)] = 0. \tag{1.1.45}$$

Thus the second equation in Eq. (1.1.42) is satisfied. Consider now the first equation in Eq. (1.1.42). We have

$$(\nabla^2 + k^2)\hat{\mathbf{E}} = -\nabla \times \nabla \times \nabla \times (\mathbf{r}\psi) + k^2\nabla \times (\mathbf{r}\psi)$$
$$= \nabla \times [-\nabla \times \nabla \times (\mathbf{r}\psi) + k^2\mathbf{r}\psi] = \nabla \times [\nabla^2(\mathbf{r}\psi) + k^2\mathbf{r}\psi]. \tag{1.1.46}$$

The last equality holds due to:

$$-\nabla \times \nabla \times (\mathbf{r}\psi) = \nabla^2(\mathbf{r}\psi) - \nabla[\nabla\cdot(\mathbf{r}\psi)], \tag{1.1.47}$$

and the curl of of the last term in Eq. (1.1.47) is zero (the curl of gradient). Consider now:

$$\nabla^2(\mathbf{r}\psi) = \nabla^2(\mathbf{i}_x x\psi) + \nabla^2(\mathbf{i}_y y\psi) + \nabla^2(\mathbf{i}_z z\psi)$$
$$= \mathbf{i}_x\nabla^2(x\psi) + \mathbf{i}_y\nabla^2(y\psi) + \mathbf{i}_z\nabla^2(z\psi)$$
$$= (\mathbf{i}_x x + \mathbf{i}_y y + \mathbf{i}_z y)\nabla^2\psi + 2(\mathbf{i}_x\nabla x\cdot\nabla\psi + \mathbf{i}_y\nabla y\cdot\nabla\psi + \mathbf{i}_z\nabla z\cdot\nabla\psi)$$
$$= \mathbf{r}\nabla^2\psi + 2\nabla\psi. \tag{1.1.48}$$

So from Eqs. (1.1.43) and (1.1.46) we have:

$$(\nabla^2 + k^2)\hat{\mathbf{E}} = \nabla \times [\nabla^2(\mathbf{r}\psi) + k^2\mathbf{r}\psi] = \nabla \times [\mathbf{r}(\nabla^2 + k^2)\psi + 2\nabla\psi]$$
$$= 2\nabla \times \nabla\psi = 0. \tag{1.1.49}$$

This proves the theorem. □

COROLLARY 1 *Let ψ be a scalar function that satisfies the Helmholtz equation (1.1.43). Then the function $\hat{\mathbf{E}} = \nabla\psi \times \mathbf{r}_*$, where $\mathbf{r}_*$ is an arbitrary constant vector (also called a "pilot vector"), satisfies Eqs. (1.1.42).*

PROOF. To prove this statement it is sufficient to see that the Maxwell (or corresponding Helmholtz) equations are invariant with respect to selection of the origin of the reference frame. Therefore, the function $\hat{\mathbf{E}}_1 = \nabla\psi \times (\mathbf{r} - \mathbf{r}_*)$ satisfies Eqs. (1.1.42). Because of the linearity of equations the difference $\hat{\mathbf{E}} = \nabla\psi\times\mathbf{r} - \nabla\psi\times(\mathbf{r}-\mathbf{r}_*) = \nabla\psi\times\mathbf{r}_*$ also satisfies Eqs. (1.1.42). □

THEOREM 2 *Let ψ be a scalar function that satisfies the Helmholtz equation (1.1.43). Then $\hat{\mathbf{E}} = \nabla\times(\nabla\psi\times\mathbf{r})$ satisfies Eqs. (1.1.42).*

PROOF. Since $\hat{\mathbf{E}}$ is the curl of some vector we immediately have $\nabla\cdot\hat{\mathbf{E}} = 0$. We also have

$$(\nabla^2 + k^2)\hat{\mathbf{E}} = \nabla\times\nabla\times[\nabla^2(\mathbf{r}\psi) + k^2\mathbf{r}\psi] = 0, \qquad (1.1.50)$$

which follows from identities (1.1.46) and (1.1.49). This proves the theorem. □

COROLLARY 2 *Let ψ be a scalar function that satisfies the Helmholtz equation (1.1.43). Then the function $\hat{\mathbf{E}} = \nabla\times(\nabla\psi\times\mathbf{r}_*)$, where $\mathbf{r}_*$ is an arbitrary constant (pilot) vector, satisfies Eqs. (1.1.42).*

PROOF. The proof is similar to the proof of the corollary for Theorem 2. □

One can then think that, by application of the curl, more linearly independent solutions can be generated. This is not true since operator $\nabla\times\nabla\times$ can be expressed via the Laplacian as:

$$\nabla\times\nabla\times(\nabla\psi\times\mathbf{r}) = -\nabla^2(\nabla\psi\times\mathbf{r}) = k^2(\nabla\psi\times\mathbf{r}). \qquad (1.1.51)$$

Thus the function $\hat{\mathbf{E}}^{(2)} = \nabla\times\nabla\times(\nabla\psi\times\mathbf{r}) = k^2\hat{\mathbf{E}}^{(0)}$ linearly depends on $\hat{\mathbf{E}}^{(0)}$, where $\hat{\mathbf{E}}^{(0)} = \nabla\psi\times\mathbf{r}$ and is a solution of the Maxwell equations. This shows that all solutions produced by multiple application of the curl operator to $\nabla\psi\times\mathbf{r}$ can be expressed via the two basic solutions $\nabla\psi\times\mathbf{r}$ and $\nabla\times(\nabla\psi\times\mathbf{r})$ and, generally, we can represent solutions of the Maxwell equations in the form:

$$\hat{\mathbf{E}} = \nabla\psi_1\times\mathbf{r} + \nabla\times(\nabla\psi_2\times\mathbf{r}), \qquad (1.1.52)$$

where ϕ and ψ are two arbitrary scalar functions that satisfy Eqs. (1.1.43). Owing to identity (1.1.44) this also can be rewritten as:

$$\hat{\mathbf{E}} = \nabla\times[\mathbf{r}\psi_1 + \nabla\times(\mathbf{r}\psi_2)]. \qquad (1.1.53)$$

Note that the above decomposition is *centered* at $\mathbf{r} = \mathbf{0}$ ($\mathbf{r} = \mathbf{0}$ is a special point). Obviously the center can be selected at an arbitrary point $\mathbf{r} = \mathbf{r}_*$. For some problems it is more convenient to use a constant vector $\mathbf{r}_*$ instead of $\mathbf{r}$ for the decomposition we used above. More generally, one can use a decomposition in the form:

$$\hat{\mathbf{E}} = \nabla \times \{(a_1\mathbf{r} + \mathbf{r}_{*1})\psi_1 + \nabla \times [(a_2\mathbf{r} + \mathbf{r}_{*2})\psi_2]\}, \qquad (1.1.54)$$

which is valid for arbitrary constants a_1 and a_2 (can be zero) and vectors $\mathbf{r}_{*1}$ and $\mathbf{r}_{*2}$, which can be selected as convenience dictates for the solution of a particular problem.

Consider now the phasor of the magnetic field vector. As follows from the first equation (1.1.36) it satisfies the equation:

$$\mathrm{i}\omega\mu\hat{\mathbf{H}} = \nabla \times \hat{\mathbf{E}}. \qquad (1.1.55)$$

Substituting here decomposition (1.1.43) and using identities (1.1.44) and (1.1.51), we obtain

$$\mathrm{i}c\mu k\hat{\mathbf{H}} = \nabla \times (\nabla\psi_1 \times \mathbf{r}) + k^2(\nabla\psi_2 \times \mathbf{r}) = \nabla \times [k^2\mathbf{r}\psi_2 + \nabla \times (\mathbf{r}\psi_1)]. \quad (1.1.56)$$

This form is similar to the representation of the phasor of the electric field (1.1.53) where functions ψ_1 and ψ_2 exchange their roles and some coefficients appear. In the case of more general decomposition (1.1.54) we have:

$$\mathrm{i}c\mu k\hat{\mathbf{H}} = \nabla \times \{k^2(a_2\mathbf{r} + \mathbf{r}_{*2})\psi_2 + \nabla \times [(a_1\mathbf{r} + \mathbf{r}_{*1})\psi_1]\}. \qquad (1.1.57)$$

This shows that solution of Maxwell equations in the frequency domain is equivalent to two scalar Helmholtz equations. These equations can be considered as independent, while their coupling occurs via the boundary conditions for particular problems.

Note that the wavenumber k in the scalar Helmholtz equations (1.1.43) is real. More complex models of the medium can be considered (say, owing to the presence of particles of sizes much smaller than the wavelength and for waves whose period is comparable with the periods of molecular relaxation or resonances once we consider waves in some media, not vacuum). In such a medium one can expect effects of dispersion and dissipation such as we considered for acoustic wave propagation in complex media. This will introduce a dispersion relationship $k = k(\omega)$ and complex k.

1.2 BOUNDARY CONDITIONS

The Helmholtz equation is an equation of the elliptic type, for which it is usual to consider boundary value problems. Boundary conditions follow from particular physical laws (conservation equations) formulated on the boundaries of the domain in which a solution is required. This domain can be finite (internal problems) or infinite (external problems). For infinite domains, the solutions should satisfy some conditions at the infinity. These conditions also have a physical origin. For the Helmholtz equation that arises as a transform of the wave equation into the frequency domain, the boundary conditions should be understood in the context of the original wave equation.

1.2.1 Conditions at infinity

1.2.1.1 Spherically symmetrical solutions

To understand the conditions which should be imposed on solutions of the Helmholtz equation in infinite domains, we start with the consideration of spherically symmetrical solutions of the scalar wave equation. In this case a function ϕ, which satisfies the wave equation (1.1.11), depends on the distance $r = |\mathbf{r}|$ only. It is well known that a solution of this equation can be written in the following D'Alembert form:

$$\phi(r,t) = \frac{1}{r}[f(t + r/c) + g(t - r/c)], \qquad (1.2.1)$$

where f and g are two arbitrary differentiable functions. The former function describes *incoming waves* towards the center $r = 0$ and the latter function describes *outgoing waves* from the center $r = 0$. Indeed the incoming wave phase can be characterized by some constant value of f, which is realized at $r = -ct + \text{const}$, and so the wavefronts converge towards the center as t is growing. Inversely, the outgoing wave phase is characterized by some constant value of g, which is realized at $r = ct + \text{const}$ and so the wavefronts for the outgoing waves diverge from the center as increasing t.

Therefore, a spherically symmetrical solution of the scalar wave equation can be characterized by specification of two functions of time $f(t)$ and $g(t)$. Assume that these functions satisfy the necessary conditions to perform the Fourier transform. Then, in the frequency domain we have images of these functions according to Eq. (1.1.11):

$$\hat{f}(\omega) = \int_{-\infty}^{\infty} e^{i\omega t} f(t)dt, \qquad \hat{g}(\omega) = \int_{-\infty}^{\infty} e^{i\omega t} g(t)dt. \qquad (1.2.2)$$

With these definitions and solution (1.2.1) we can determine the image, or phasor $\psi(r, \omega)$ of $\phi(r, t)$ in the frequency domain as:

$$\begin{aligned}
\psi(r, \omega) &= \int_{-\infty}^{\infty} e^{i\omega t}\phi(r, t)dt \\
&= \frac{1}{r}\left[\int_{-\infty}^{\infty} e^{i\omega t} f(t + r/c)dt + \int_{-\infty}^{\infty} e^{i\omega t} g(t - r/c)dt\right] \\
&= \frac{1}{r}\left[\int_{-\infty}^{\infty} e^{i\omega(t' - r/c)} f(t')dt' + \int_{-\infty}^{\infty} e^{i\omega(t' + r/c)} g(t')dt\right] \\
&= \frac{1}{r}\hat{f}(\omega)e^{-ikr} + \frac{1}{r}\hat{g}(\omega)e^{ikr}, \qquad k = \frac{\omega}{c}.
\end{aligned} \tag{1.2.3}$$

Here we defined $k = \omega/c$ and so this quantity is negative for $\omega < 0$ and positive for $\omega > 0$. The function $\psi(r, \omega)$ is a solution of the spherically symmetrical Helmholtz equation (1.1.13). It is seen that solutions corresponding to the incoming waves are proportional to e^{-ikr} while solutions corresponding to the outgoing waves are proportional to e^{ikr}.

It is not difficult to see also that at large r we have:

$$r\left(\frac{\partial\psi}{\partial r} - ik\psi\right) = -2ik\hat{f}(\omega)e^{-ikr} + O\left(\frac{1}{r}\right), \tag{1.2.4}$$

$$r\left(\frac{\partial\psi}{\partial r} + ik\psi\right) = 2ik\hat{g}(\omega)e^{ikr} + O\left(\frac{1}{r}\right). \tag{1.2.5}$$

This means that if the condition

$$\lim_{r\to\infty}\left[r\left(\frac{\partial\psi}{\partial r} - ik\psi\right)\right] = 0 \tag{1.2.6}$$

holds then $\hat{f}(\omega) \equiv 0$. This results in $f(t) \equiv 0$ and in this case $\phi(r, t)$ consists only of outgoing waves. Similarly, in the case if the condition

$$\lim_{r\to\infty}\left[r\left(\frac{\partial\psi}{\partial r} + ik\psi\right)\right] = 0 \tag{1.2.7}$$

holds, the solution consists only of incoming waves and $g(t) \equiv 0$.

1.2.1.2 Sommerfeld radiation condition

The problems which are usually considered in relation to the wave equation in three-dimensional unbounded domains are scattering problems. In this case the wave function is specified as:

$$\phi(\mathbf{r}, t) = \phi_{\text{in}}(\mathbf{r}, t) + \phi_{\text{scat}}(\mathbf{r}, t), \tag{1.2.8}$$

where both functions $\phi_{in}(\mathbf{r}, t)$ and $\phi_{scat}(\mathbf{r}, t)$ satisfy the wave equation. The function $\phi_{in}(\mathbf{r}, t)$ is the potential of the incident field, while $\phi_{scat}(\mathbf{r}, t)$ is the potential of the scattered field, which arises due to the presence of one or several scatterers. In the absence of scatterers $\phi(\mathbf{r}, t) = \phi_{in}(\mathbf{r}, t)$ is some given function (e.g. the potential of a plane wave propagating along the z-direction, $\phi_{in}(\mathbf{r}, t) = F(t - z/c)$).

To understand the scattered field we may turn our attention to the *Huygens principle*, which represents wave propagation as an emission of secondary wave from the points located on the current wavefront. When the primary wave described by $\phi_{in}(\mathbf{r}, t)$ reaches the scatterer boundary the secondary waves are generated from the boundary points located at the intersection of the boundary and the wavefront. Owing to the finite speed of wave propagation, spatial points far from the boundary "do not know" about these secondary waves, so these waves can be thought of as waves *outgoing* from the boundary points. For each point we can then write the secondary wave potential in the form (1.2.1), where $f \equiv 0$ and, therefore, in the frequency domain condition (1.2.6) holds. Since the total scattered field, $\phi_{scat}(\mathbf{r}, t)$, can now be seen as a superposition of outgoing waves, the corresponding potential in the frequency domain should satisfy the condition:

$$\lim_{r \to \infty} \left[r \left(\frac{\partial \psi_{scat}}{\partial r} - ik\psi_{scat} \right) \right] = 0. \tag{1.2.9}$$

This condition is called the *Sommerfeld radiation condition* or just the *radiation condition*. It states that the scattered field consists of outgoing waves only. Solutions of the Helmholtz equation which satisfy the radiation condition are called *radiating solutions* or *radiating functions*.

In some wave problems considered in infinite domains all the wave sources and scatterers can be enclosed inside some sphere. Since in the absence of the wave sources the solution of the wave equation is trivial, $\phi(\mathbf{r}, t) \equiv 0$, then all perturbations for points located outside the sphere come only from some events inside the sphere. This means that in this case the total field in the frequency domain, $\psi(\mathbf{r}, \omega)$ is a radiating function.

We emphasize that the radiation condition (1.2.9) derived from consideration of point sources is applied to a set of sources, i.e. to the case $\psi_{scat} = \psi_{scat}(\mathbf{r}, k)$. Generally, the far-field asymptotics of ψ_{scat} is

$$\psi_{scat} \sim \frac{1}{r} \Psi(\theta, \varphi) e^{ikr}, \tag{1.2.10}$$

where $\Psi(\theta, \varphi)$ is the angular dependence on spherical angles θ and φ, and so condition (1.2.9) holds. Indeed, from a very remote point, a set of sources or scatterers is seen as one point (like we see galaxies consisting of many stars as one "star"). While for different angles there will be different values of of ψ_{scat} (so it is not spherically symmetrical), for given, or fixed, angles θ and φ there is no difference between the asymptotic behavior of a set of sources and an equivalent single source.

1.2.1.3 *Complex wavenumber*

As we showed above, the Helmholtz equation with complex k can appear in some models:

$$k = k_r + \mathrm{i}k_i, \qquad k_i \neq 0. \tag{1.2.11}$$

For any $k \neq 0$, the solution of the spherically symmetrical Helmholtz equation can be written in the form:

$$\begin{aligned}\psi(r,k) &= \frac{1}{r}\hat{f}(k)\mathrm{e}^{-\mathrm{i}kr} + \frac{1}{r}\hat{g}(k)\mathrm{e}^{\mathrm{i}kr} \\ &= \frac{1}{r}\hat{f}(k)\mathrm{e}^{k_i r}\,\mathrm{e}^{-\mathrm{i}k_r r} + \frac{1}{r}\hat{g}(k)\mathrm{e}^{-k_i r}\,\mathrm{e}^{\mathrm{i}k_r r},\end{aligned} \tag{1.2.12}$$

where $\hat{f}(k)$ and $\hat{g}(k)$ are some integration constants.

In the case $k_r = 0$, which is realized, e.g. for the Klein–Gordan equation, we have a sum of exponentially growing and decaying solutions. The decaying solution is nothing but the Yukawa potential (1.1.35) and so it should be selected if we request that solutions are bounded outside a sphere which contains the point of singularity $r = 0$. Hence in this case the boundary condition is:

$$\lim_{r\to\infty} \psi = 0. \tag{1.2.13}$$

The case $k_r \neq 0$ deserves a more detailed consideration. Assume that solution (1.2.12) represents the complex amplitude of a monochromatic wave propagating in complex medium (1.1.12):

$$\phi(r,t) = \frac{1}{r}\mathrm{Re}(\hat{f}(k)\mathrm{e}^{k_i r}\,\mathrm{e}^{-\mathrm{i}k_r(r+c_p t)} + \hat{g}(k)\mathrm{e}^{-k_i r}\,\mathrm{e}^{\mathrm{i}k_r(r-c_p t)}), \tag{1.2.14}$$

where c_p is the phase velocity (1.1.26). Since k appears in the Helmholtz equation as k^2 and the definition of the sign of k depends on our choice, we can define its sign, as in the case of real k, in such a way that the phase velocity is positive at positive frequencies, i.e. $k_r > 0$ at $\omega > 0$. This means that the first term in expression (1.2.14) describes the incoming wave, while the second term corresponds to the outgoing wave. In the case of wave scattering or propagation outside of waves generated in a finite spatial domain we need to select only the solution corresponding to outgoing waves—in other words to set $\hat{f}(k) = 0$. Therefore, as in the case of real k, we impose the following condition for asymptotic behavior as $r \to \infty$ of solutions of the Helmholtz equation:

$$\psi(\mathbf{r}, k) \sim \frac{1}{r} e^{ikr}. \tag{1.2.15}$$

Now we note that if $k_i > 0$ this solution is decaying at $r \to \infty$, so it can be replaced with condition (1.2.12). This is the case for dissipative media, which means that small perturbations should not grow as they propagate in the medium. For example, for relaxating media with a dispersion relationship (1.1.23) the root corresponding to $k_r > 0$ at $\omega > 0$ is:

$$k = \frac{\omega}{c(1 + \tau_\rho^2 \omega^2)^{1/2}} (1 + i\omega\tau_\rho)^{1/2}, \qquad k_i > 0. \tag{1.2.16}$$

Therefore, condition (1.2.13) can be used in this case. The same holds for the diffusion equation, where for positive ω, k_r, and diffusivity κ, Eq. (1.1.30) yields:

$$k = (1 + i)\left(\frac{\omega}{2\kappa}\right)^{1/2}, \qquad k_i > 0. \tag{1.2.17}$$

Despite the situation where $k_i < 0$ at $\omega/k_r > 0$ being rather unusual, it is not impossible. In this case we can see that the outgoing wave should grow in amplitude as it propagates in the medium. The unperturbed state of such media should be characterized as *linearly unstable*, since any small perturbation will exponentially (explosively) grow as it propagates. The examples of media of such type can be found in the theory of superheated liquids, active media, which can release energy under perturbations (explosives), etc. We can see that condition (1.2.12) is not applicable in this case, since the physical meaning requires to select a solution not decaying at infinity, but a growing solution. Our reasoning here is based on the *causality principle*.

1.2.1.4 Silver–Müller radiation condition

The Silver–Müller radiation condition is a condition that is imposed on the scattered electromagnetic field when solving the Maxwell equations. It can be stated as:

$$\begin{aligned}
&\lim_{r \to 0} (\mu^{1/2} \hat{\mathbf{H}}_{\mathrm{scat}} \times \mathbf{r} - r\varepsilon^{1/2} \hat{\mathbf{E}}_{\mathrm{scat}}) = \mathbf{0}, \\
&\lim_{r \to 0} (\varepsilon^{1/2} \hat{\mathbf{E}}_{\mathrm{scat}} \times \mathbf{r} + r\mu^{1/2} \hat{\mathbf{H}}_{\mathrm{scat}}) = \mathbf{0},
\end{aligned} \tag{1.2.18}$$

where $\hat{\mathbf{E}}_{\mathrm{scat}}$ and $\hat{\mathbf{H}}_{\mathrm{scat}}$ are the phasors of the scattered electric and magnetic fields arising from the decomposition of the total field in to the incident and scattered fields:

$$\hat{\mathbf{E}} = \hat{\mathbf{E}}_{\rm in} + \hat{\mathbf{E}}_{\rm scat}, \qquad \hat{\mathbf{H}} = \hat{\mathbf{H}}_{\rm in} + \hat{\mathbf{H}}_{\rm scat}.$$

The physical meaning of these conditions is similar to those for the scalar field—they simply state that the scattered field consists of outgoing waves only. Indeed, if we consider a point $\mathbf{r} = (r, \theta, \varphi)$ located on a very large sphere enclosing all the scatterers, we can see that quantities $\hat{\mathbf{E}}_{\rm scat}$ and $\hat{\mathbf{H}}_{\rm scat}$ represent plane waves propagating in the radial direction from the center of the sphere, whose amplitude decays as r^{-1} due to the sphericity.

Consider now a plane wave solution of Maxwell equations. Let unit vector $\mathbf{s}$, $|\mathbf{s}| = 1$, characterize the direction of the plane wave propagation. In this case

$$\hat{\mathbf{E}} = \mathbf{c}\,e^{ik\mathbf{s}\cdot\mathbf{r}}, \tag{1.2.19}$$

satisfies the vector Helmholtz equation (1.1.42) and vector $\mathbf{c}$ should be orthogonal to $\mathbf{s}$ to satisfy the divergence free condition, $\mathbf{c}\cdot\mathbf{s} = 0$. This condition can be enforced if we take $\mathbf{c} = \mathbf{s} \times \mathbf{q}$, where $\mathbf{q}$ is an arbitrary vector. So the plane wave solution of Maxwell equations can be written in the form:

$$\hat{\mathbf{E}} = (\mathbf{s} \times \mathbf{q})e^{ik\mathbf{s}\cdot\mathbf{r}}. \tag{1.2.20}$$

From relation (1.1.55) we can then determine the phasor of the magnetic field vector:

$$i\omega\mu\hat{\mathbf{H}} = \nabla \times [(\mathbf{s} \times \mathbf{q})e^{ik\mathbf{s}\cdot\mathbf{r}}] = ik[\mathbf{s} \times (\mathbf{s} \times \mathbf{q})]e^{ik\mathbf{s}\cdot\mathbf{r}} = ik\mathbf{s} \times \hat{\mathbf{E}}. \tag{1.2.21}$$

Due to the vector identity

$$[\mathbf{s} \times (\mathbf{s} \times \mathbf{q})] \times \mathbf{s} = [(\mathbf{s}\cdot\mathbf{q})\mathbf{s} - \mathbf{q}] \times \mathbf{s} = (\mathbf{s} \times \mathbf{q}), \tag{1.2.22}$$

we can see that

$$i\omega\mu\hat{\mathbf{H}} \times \mathbf{s} = ik\{[\mathbf{s} \times (\mathbf{s} \times \mathbf{q})] \times \mathbf{s}\}e^{ik\mathbf{s}\cdot\mathbf{r}} = ik(\mathbf{s} \times \mathbf{q})e^{ik\mathbf{s}\cdot\mathbf{r}} = ik\hat{\mathbf{E}}. \tag{1.2.23}$$

This shows that for plane waves propagating in direction $\mathbf{s}$ we have:

$$\mu^{1/2}\hat{\mathbf{H}} \times \mathbf{s} - \varepsilon^{1/2}\hat{\mathbf{E}} = \mathbf{0}, \qquad \varepsilon^{1/2}\hat{\mathbf{E}} \times \mathbf{s} + \mu^{1/2}\hat{\mathbf{H}} = \mathbf{0},$$
$$c = \frac{\omega}{k} = \frac{1}{\sqrt{\mu\varepsilon}}. \tag{1.2.24}$$

From the above arguments as $r \to \infty$, we can see that if we replace $\mathbf{s} = \mathbf{r}/r$ in relations (1.2.24) and take into account that for a given direction (θ, φ)

the principal term of the scattered field is proportional to r^{-1} then we obtain the Silver–Müller conditions (1.2.18).

Note that these conditions are not necessary when considering the reduction of Maxwell equations to the scalar Helmholtz equations. If the scalar potentials satisfy the Sommerfield radiation conditions, the Silver–Müller conditions hold automatically, since both state the same physical fact.

1.2.2 Transmission conditions

Real systems can be considered as a unity of domains occupied by relatively homogeneous media. While the physical properties of different substances can differ substantially (say, air and rigid particles), one should keep in mind that waves of different nature can propagate in any substance (e.g. acoustic waves) and, therefore, wave-type equations can be used for their description. Owing to the difference in properties, the speed of propagation of perturbations is different for different media. Therefore, for descriptions of waves in each domain we can apply the wave equation with the speed of sound or light corresponding to the medium that occupies that domain. The problem then is to provide sufficient conditions on the domain boundaries that enable us to match solutions in different domains and build solutions for the corresponding wave equation. These conditions are known as *transmission* conditions, which can also be interpreted as *jump conditions* or conditions for discontinuities, since the wave function and/or its derivatives may jump on the contact boundaries. In general, the jump conditions can be derived from the same conservation equations that lead to the governing equations. The form of these conservation laws should be written in integral form to allow discontinuities and then the conditions arise after shrinking the domain to the contact surfaces. Below we provide examples of transmission conditions for acoustic and electromagnetic waves.

1.2.2.1 Acoustic waves

In acoustics we usually consider problems when the boundaries of the domains are either immovable or move with velocities much smaller than the speed of sound. We also consider the case when the amplitude of pressure perturbations is small and perturbations of the mass velocity are small as well. In the linear approximation, this results in the following two conditions on a contact surface S with normal n separating two media marked as 1 and 2:

$$\mathbf{v}_1'\cdot\mathbf{n} = \mathbf{v}_2'\cdot\mathbf{n}, \qquad p_1' = p_2'. \tag{1.2.25}$$

The first condition states that the velocities normal to the surface are the same. In fluid mechanics this is known as kinematic condition. In fact, it follows from the mass conservation equation in its assumption that there is no mass transfer through the surface S. The second condition, sometimes called dynamic conditions, follows from the momentum conservation equation and is valid if there are no surface forces. As follows from this description, these conditions should be modified if mass is transferred through the surface (say, owing to phase transitions), and if there are some appreciable surface forces (for example, surface tension). These conditions are sufficient to match solutions of the wave or Helmholtz equation in two domains.

Depending on the problem to be solved (wave equation for pressure or for the velocity potential), conditions (1.2.25) can be written in terms of pressure or velocity potential and their derivatives only. Consider first the pressure equations. As follows from the momentum conservation equation (1.1.3) written in the frequency domain, the phasors of pressure and velocity perturbations, $\hat{p}'$ and $\hat{\mathbf{v}}'$, satisfy equations

$$-i\omega\rho_1\hat{\mathbf{v}}'_1 + \nabla\hat{p}'_1 = \mathbf{0}, \qquad -i\omega\rho_2\hat{\mathbf{v}}'_2 + \nabla\hat{p}'_2 = \mathbf{0}. \tag{1.2.26}$$

Taking the scalar product of these equations with normal $\mathbf{n}$, denoting the normal derivative $\partial/\partial n = \mathbf{n}\cdot\nabla$, and noticing that relations (1.2.25) also hold in the frequency domain (remember our assumption that the speed of the boundary is much smaller than the speed of sound!), we obtain the following transmission conditions for pressure perturbations applicable to matching solutions of the Helmholtz equation in domains 1 and 2:

$$\frac{1}{\rho_1}\frac{\partial\hat{p}'_1}{\partial n} = \frac{1}{\rho_2}\frac{\partial\hat{p}'_2}{\partial n}, \qquad \hat{p}'_1 = \hat{p}'_2. \tag{1.2.27}$$

Here ρ_1 and ρ_2 are the respective medium densities.

Now consider the problem formulation for the Helmholtz equation written in terms of the velocity potential (1.1.11). The integral of the momentum equation (the expression in parentheses in equation (1.1.11)) can be written in phasor space, where we use notation ψ for the phasor of ϕ (see Eq. (1.1.14)) as:

$$i\omega\rho_1\psi_1 = p'_1, \qquad i\omega\rho_2\psi_2 = p'_2. \tag{1.2.28}$$

Hence, relations (1.2.25) lead to the following transmission conditions:

$$\frac{\partial\psi_1}{\partial n} = \frac{\partial\psi_2}{\partial n}, \qquad \rho_1\psi_1 = \rho_2\psi_2. \tag{1.2.29}$$

Comparing these conditions with relation (1.2.27) we can see that, in the case of the pressure formulation, the function which satisfies Helmholtz equations in two different domains is continuous, while its normal derivatives have a discontinuity. The opposite situation, when the normal derivative is continuous while the wave function itself has a jump on the boundary, is the case in terms of the velocity potential.

Note that, for acoustic waves in complex media (dispersion, dissipation, relaxation), conditions (1.2.27) and (1.2.29) should be modified according to the model of the media. Proper transmission conditions in this case can be obtained from general mass and momentum conservation relations (1.2.25) and specific equations of state for the medium, such as Eq. (1.1.21), written in the frequency domain.

1.2.2.2 Electromagnetic waves

Here we consider transmission conditions in the case when two dielectrics with electric permittivities ε_1 and ε_2 and magnetic permeabilities μ_1 and μ_2 are in contact over a surface S with normal $\mathbf{n}$. To match two solutions of the Maxwell equation we require that the tangential components of the vectors of electric and magnetic fields are continuous. These components can be found by taking the cross product of the respective vectors with the normal, and therefore, can be written in the phasor space as:

$$\mathbf{n} \times \hat{\mathbf{E}}_1 = \mathbf{n} \times \hat{\mathbf{E}}_2, \qquad \mathbf{n} \times \hat{\mathbf{H}}_1 = \mathbf{n} \times \hat{\mathbf{H}}_2. \tag{1.2.30}$$

1.2.3 Conditions on the boundaries

Conditions on the boundaries of domain 1 are used when either the properties of the boundary material (medium 2) are very different from the properties of medium 1 or can be modeled or assumed. In the former case the transmission conditions can be simplified and provide sufficient conditions for solution of the Helmholtz equation. In the latter case, simplification of the general problem usually follows from consideration of some model problem by applying the results to a more general case. Since such modeling is outside the scope of this book, we mention here the following basic types of boundary conditions for scalar wave equation and Maxwell equations. Here we assume that the domain of consideration is medium 1 (we also call it the host, carrier, or just a medium with no index), and the material of the boundary has the properties of medium 2 (we will drop the indexing if it is clear from the context). The normal derivative in each case is taken inward in to the domain of the carrier medium (direction from medium 2 to medium 1).

1.2.3.1 *Scalar Helmholtz equation*

- The Dirichlet boundary condition

$$\psi|_S = 0 \tag{1.2.31}$$

  appears, e.g. for complex amplitude pressure in acoustics, when the surface material has a very low acoustic impedance compared to the acoustic impedance of the carrier medium ($\rho_2 c_2 \ll \rho_1 c_1$). In this case the surface is called *sound soft*.
- The Neumann boundary condition

$$\left.\frac{\partial \psi}{\partial n}\right|_S = 0 \tag{1.2.32}$$

  in acoustics holds for complex amplitude of pressure, when the surface material has a much higher acoustic impedance than the acoustic impedance of the host medium ($\rho_2 c_2 \gg \rho_1 c_1$). In this case the surface is called *sound hard*.
- The Robin (or mixed, or impedance) boundary condition

$$\left.\left(\frac{\partial \psi}{\partial n} + i\sigma\psi\right)\right|_S = 0 \tag{1.2.33}$$

  in acoustics is used to model the finite acoustic impedance of the boundary. In this case σ is the admittance of the surface. Solutions of the Helmholtz equation with the Robin boundary condition in limiting cases $\sigma \to 0$ and $\sigma \to \infty$ turn into solutions of the same equation with the Neumann and Dirichlet boundary conditions, respectively.

The boundary value problems with those conditions are called the Dirichlet, Neumann, and Robin problems, respectively.

1.2.3.2 *Maxwell equations*

We mention here two cases important for wave scattering problems:

- Perfect conductor boundary condition

$$\mathbf{n} \times \hat{\mathbf{E}}|_S = 0. \tag{1.2.34}$$

  When we express the electric field vector via scalar potentials (1.1.53), this condition turns into

$$\mathbf{n} \times \nabla \times [\mathbf{r}\psi_1 + \nabla \times (\mathbf{r}\psi_2)]|_S = 0. \tag{1.2.35}$$

  This can also be modified for the more general representation (1.1.54).

- The Leontovich (or impedance) boundary condition

$$[\mathbf{n} \times \hat{\mathbf{H}} - \lambda(\mathbf{n} \times \hat{\mathbf{E}}) \times \mathbf{n}]|_S = 0. \tag{1.2.36}$$

Here λ is a constant called boundary impedance. In terms of scalar potentials (1.1.53) and (1.1.56) this can be written as:

$$\mathbf{n} \times \nabla \times [(i\lambda' + \nabla\times)(\mathbf{r}\psi_1) + (k^2 + i\lambda'\nabla\times)(\mathbf{r}\psi_2)]|_S = \mathbf{0},$$
$$\lambda' = \mu ck\lambda. \tag{1.2.37}$$

Modification for more general forms can be achieved by substituting equations (1.1.54) and (1.1.57) into relation (1.2.36). We can also see that, in limiting case $\lambda' \to \infty$, condition (1.2.37) transforms to condition (1.2.35).

1.3 INTEGRAL THEOREMS

Integral equation approaches are fundamental tools in the numerical solution of the Helmholtz equation and equations related to it. These approaches have significant advantages for solving both external and internal problems. They also have a few disadvantages and we discuss both below.

A major advantage of these methods is that they effectively reduce the dimensionality of the domain over which the problem has to be solved. The integral equation statement reduces the problem to one of an integral over the surface of the boundary. Thus, instead of the discretization of a volume (or a region in 2D), we need only discretize surfaces (or curves in 2D). The problem of creating discretizations ("meshing") is well known to be a difficult task—almost an art—and the simplicity achieved by a reduction in dimensionality must not be underestimated. Further, the number of variables required to resolve a solution is also significantly reduced.

Another major advantage of the integral equation representation for external problems is that these ensure that the far-field Sommerfeld (or Silver–Müller conditions) are automatically exactly satisfied. Often volumetric discretizations must be truncated artificially and effective boundary conditions imposed on the artificial boundaries. While considerable progress has been made in developing so called perfectly matched layers to imitate the properties of the far-field, these are relatively difficult to implement.

Despite these advantages the integral equation approaches have some minor disadvantages. The first is that their formulation is usually more complex mathematically. However, this need not be an obstacle to their understanding and implementation since there are many clear expositions of the integral equation approaches.

A second disadvantage of the integral equation approach is that it leads to linear systems with dense matrices. These dense matrices are expensive with which to perform computations. Many modern calculations require characterization of the scattering off complex shaped objects. While integral equation methods may allow such calculations, they can be relatively slow. The fast multipole methods discussed in this book alleviate this difficulty. They allow extremely rapid computation of the product of a vector with a dense matrix of the kind that arises upon discretization of the integral equation and go a long way towards alleviating this disadvantage.

Below we provide a brief introduction to the integral theorems and identities which serve as a basis for methods using integral formulations.

1.3.1 Scalar Helmholtz equation

1.3.1.1 Green's identity and formulae

Green's function

The free-space Green's function G for the scalar Helmholtz equation in three dimensions is defined as:

$$G(\mathbf{x},\mathbf{y}) = \frac{\exp(ik|\mathbf{x}-\mathbf{y}|)}{4\pi|\mathbf{x}-\mathbf{y}|}, \qquad \mathbf{x},\mathbf{y} \in \mathbb{R}^3. \tag{1.3.1}$$

As follows from the definition, this is a symmetric function of two spatial points $\mathbf{x}$ and $\mathbf{y}$:

$$G(\mathbf{x},\mathbf{y}) = G(\mathbf{y},\mathbf{x}), \tag{1.3.2}$$

and is a distance function between points $\mathbf{x}$ and $\mathbf{y}$. This function satisfies the equation:

$$\nabla^2 G(\mathbf{x},\mathbf{y}) + k^2 G(\mathbf{x},\mathbf{y}) = -\delta(\mathbf{x}-\mathbf{y}), \tag{1.3.3}$$

where $\delta(x-y)$ refers to the Dirac delta function (distribution) which is defined as:

$$\int_{\mathbb{R}^3} f(\mathbf{x})\delta(\mathbf{x}-\mathbf{y})dV(\mathbf{x}) = \begin{cases} f(\mathbf{y}), & \text{for } \mathbf{y} \in \mathbb{R}^3, \\ 0, & \text{otherwise.} \end{cases} \tag{1.3.4}$$

Here $f(\mathbf{x})$ is an arbitrary function and integration is taken over the entire space. The Green's function is thus a solution of the Helmholtz equation in the domain $\mathbf{x} \in \mathbb{R}^3 \backslash \mathbf{y}$ or $\mathbf{y} \in \mathbb{R}^3 \backslash \mathbf{x}$. Note that, in the entire space $\mathbb{R}^3$, the Green's function does not satisfy the Helmholtz equation, since the right-hand side of this equation is not zero everywhere. The equation which it satisfies is a *non-uniform Helmholtz equation*. Generally written as

$$\nabla^2 \psi(\mathbf{r}) + k^2 \psi(\mathbf{r}) = -f(\mathbf{r}), \tag{1.3.5}$$

it is a wave analog (in the frequency domain) of the Poisson equation (the case $k = 0$), which has in the right-hand side some function $f(\mathbf{r})$ responsible for the spatial distribution of charges (or sources).

This "impulse response" of the Helmholtz equation is a fundamental tool for studying the Helmholtz equation. It is also referred to as the point source solution or the fundamental solution.

Divergence theorem

The following theorem from Gauss relates an integral over a domain $\Omega \subset \mathbb{R}^3$ to the surface integral over the boundary S of this domain:

$$\int_\Omega (\nabla \cdot \mathbf{u}) \mathrm{d}V = \int_S (\mathbf{n} \cdot \mathbf{u}) \mathrm{d}S, \tag{1.3.6}$$

where $\mathbf{n}$ is the normal vector on the surface S that is outward to the domain Ω. This theorem holds for finite or infinite domains assuming that the integrals converge. In generalized informal form for n-dimensional space, the divergence theorem can be written as:

$$\int_{\Omega^n} (\nabla \circ \mathbf{A}) \mathrm{d}V = \int_{\partial\Omega^n} (\mathbf{n} \circ \mathbf{A}) \mathrm{d}S, \qquad \Omega^n \subset \mathbb{R}^n, \tag{1.3.7}$$

where $\circ$ is any operator and $\mathbf{A}$ is a scalar or vector quantity for which $\circ$ is defined.

Green's integral theorems

These theorems play a role analogous to the familiar "integration by parts" in the case of integration over the line. Recall that, for integrals over a line, we can write:

$$\int_a^b u \, \mathrm{d}v = -\int_a^b v \, \mathrm{d}u + (uv)|_a^b. \tag{1.3.8}$$

Green's first integral theorem states that for a domain Ω with boundary S, given two functions $u(\mathbf{x})$ and $v(\mathbf{x})$, we can write:

$$\int_\Omega (u\nabla^2 v + \nabla u \cdot \nabla v) \mathrm{d}V = \int_\Omega \nabla \cdot (u \nabla v) \mathrm{d}V = \int_S \mathbf{n} \cdot (u \nabla v) \mathrm{d}S, \tag{1.3.9}$$

where we have used the divergence theorem on the quantity $u\nabla v$. This formula may be put into a form that is reminiscent of the formula of integration by parts by rearranging terms

$$\int_{\Omega} u\nabla\cdot(\nabla v)\mathrm{d}V = -\int_{\Omega} \nabla u\cdot\nabla v \,\mathrm{d}V + \int_{S} u(\mathbf{n}\cdot\nabla v)\mathrm{d}S, \tag{1.3.10}$$

where we observe that the derivative operator has been exchanged from the function v to the function u, and that the boundary term has appeared.

To derive *Green's second integral theorem*, we write Eq. (1.3.9) by exchanging the roles of u and v, as:

$$\int_{\Omega} (v\nabla^2 u + \nabla u\cdot\nabla v)\mathrm{d}V = \int_{S} v(\mathbf{n}\cdot\nabla u)\mathrm{d}S, \tag{1.3.11}$$

and subtract it from Eq. (1.3.9). This yields:

$$\int_{\Omega} (u\nabla^2 v - v\nabla^2 u)\mathrm{d}V = \int_{S} \mathbf{n}\cdot(u\nabla v - v\nabla u)\mathrm{d}S. \tag{1.3.12}$$

This equation can also be written as:

$$\int_{\Omega} u\nabla^2 v \,\mathrm{d}V = \int_{\Omega} v\nabla^2 u \,\mathrm{d}V + \int_{S} \mathbf{n}\cdot(u\nabla v - v\nabla u)\mathrm{d}S. \tag{1.3.13}$$

Green's formula

Let us consider a domain Ω with boundary S. Using the sifting property of the delta function (1.3.4) we may write for a given function ψ at a point $\mathbf{y} \in \Omega$:

$$\int_{\Omega} \psi(\mathbf{x})\delta(\mathbf{x} - \mathbf{y})\mathrm{d}V(\mathbf{x}) = \psi(\mathbf{y}), \qquad \mathbf{y} \in \Omega. \tag{1.3.14}$$

Using Eq. (1.3.3) the function may be written as:

$$\psi(\mathbf{y}) = -\int_{\Omega} \psi(\mathbf{x})[\nabla_{\mathbf{x}}^2 G(\mathbf{x}, \mathbf{y}) + k^2 G(\mathbf{x}, \mathbf{y})]\mathrm{d}V(\mathbf{x}), \tag{1.3.15}$$

where $\nabla_{\mathbf{x}}$ is the nabla operator with respect to variable $\mathbf{x}$. Using Green's second integral theorem (1.3.13), where we set $u = \psi$ and $v = G$, we can write the above as:

$$\psi(\mathbf{y}) = -\int_{\Omega} k^2\psi(\mathbf{x})G(\mathbf{x},\mathbf{y})\mathrm{d}V(\mathbf{x}) - \int_{\Omega} G(\mathbf{x},\mathbf{y})(\nabla_{\mathbf{x}}^2\psi)\mathrm{d}V(\mathbf{x})$$
$$- \int_S \mathbf{n}\cdot[\psi(\mathbf{x})\nabla_{\mathbf{x}}G(\mathbf{x},\mathbf{y}) - G(\mathbf{x},\mathbf{y})\nabla_{\mathbf{x}}\psi(\mathbf{x})]\mathrm{d}S(\mathbf{x})$$
$$= -\int_{\Omega} [\nabla_{\mathbf{x}}^2\psi(\mathbf{x}) + k^2\psi(\mathbf{x})]G(\mathbf{x},\mathbf{y})\mathrm{d}V(\mathbf{x})$$
$$- \int_S \mathbf{n}\cdot[\psi(\mathbf{x})\nabla_{\mathbf{x}}G(\mathbf{x},\mathbf{y}) - G(\mathbf{x},\mathbf{y})\nabla_{\mathbf{x}}\psi(\mathbf{x})]\mathrm{d}S(\mathbf{x}). \tag{1.3.16}$$

Let us assume now that the function $\psi(\mathbf{x})$ satisfies the non-uniform Helmholtz equation (1.3.5). Then we see that the solution to this equation can be written as:

$$\psi(\mathbf{y}) = \int_{\Omega} f(\mathbf{x})G(\mathbf{x},\mathbf{y})\mathrm{d}V(\mathbf{x}) - \int_S \mathbf{n}\cdot[\psi(\mathbf{x})\nabla_{\mathbf{x}}G(\mathbf{x},\mathbf{y})$$
$$- G(\mathbf{x},\mathbf{y})\nabla_{\mathbf{x}}\psi(\mathbf{x})]\mathrm{d}S(\mathbf{x}). \tag{1.3.17}$$

If the domain has no boundaries, we see that the solution to the problem is obtained as a convolution of the right-hand side with the impulse response:

$$\psi(\mathbf{y}) = \int_{\Omega} f(\mathbf{x})G(\mathbf{x},\mathbf{y})\mathrm{d}V(\mathbf{x}). \tag{1.3.18}$$

Let us consider the case when ψ in domain Ω satisfies the Helmholtz equation, or Eq. (1.3.5) with $f = 0$. Then relation (1.3.17) provides us with the solution for ψ in the domain from its boundary values:

$$\psi(\mathbf{y}) = \int_S \left[G(\mathbf{x},\mathbf{y})\frac{\partial\psi(\mathbf{x})}{\partial n(\mathbf{x})} - \psi(\mathbf{x})\frac{\partial G(\mathbf{x},\mathbf{y})}{\partial n(\mathbf{x})}\right]\mathrm{d}S(\mathbf{x}), \tag{1.3.19}$$
$$\mathbf{y}\in\Omega, \quad \mathbf{n} \text{ directed outside } \Omega,$$

where we denoted $\partial/\partial n(\mathbf{x}) = \mathbf{n}\cdot\nabla_{\mathbf{x}}$.

The obtained equation is valid for the case when $\mathbf{y}$ is in the domain (and not on the boundary). In the case of infinite domains, function $\psi(\mathbf{y})$ satisfies the Sommerfeld condition as $|\mathbf{y}| \to \infty$. This equation is also called the Helmholtz integral equation or the Kirchhoff integral equation. Note that we derived this equation assuming that $\mathbf{n}$ is the normal directed outward the domain Ω. In the case of infinite domains, when S is the surface of some body (scatterer), usually the opposite direction of $\mathbf{n}$ is used, since it is defined as a normal *outer to the body*. In the case of this definition of the normal, which we also accept for the solution of scattering problems, Green's formula is:

$$\psi(\mathbf{y}) = \int_S \left[\psi(\mathbf{x}) \frac{\partial G(\mathbf{x},\mathbf{y})}{\partial n(\mathbf{x})} - G(\mathbf{x},\mathbf{y}) \frac{\partial \psi(\mathbf{x})}{\partial n(\mathbf{x})} \right] dS(\mathbf{x}), \tag{1.3.20}$$

$$\mathbf{y} \in \Omega, \quad \mathbf{n} \text{ directed inside } \Omega.$$

If ψ and $\partial\psi/\partial n$ vanish at the boundary, or more generally in a region, the above equation says that ψ vanishes identically.

1.3.1.2 Integral equation from Green's formula for ψ

In general a well-posed problem for ψ that satisfies Helmholtz equation will specify boundary conditions for ψ (Dirichlet boundary conditions (1.2.31)) or for its normal derivative $\partial\psi/\partial n$ (Neumann boundary conditions (1.2.32)) or for some combination of the two (Robin or "impedance" boundary condition (1.2.33)), but not both ψ and $\partial\psi/\partial n$. Thus at the outset we will only know either ψ or $\partial\psi/\partial n$ or a combination of them on the boundary, but not both. However, to compute ψ in the domain using Eq. (1.3.20), both ψ and $\partial\psi/\partial n$ are needed on the boundary.

To obtain both these quantities, we can take the ψ on the right-hand side to lie on the boundary. However, there are two issues with this. First, the Green's function G is singular when $\mathbf{x} \to \mathbf{y}$ so we need to consider the behavior of the integrals involving G and $\mathbf{n}\cdot\nabla G$ for $\mathbf{y}$ on the boundary and as $\mathbf{x} \to \mathbf{y}$. Second, we derived this formula using the definition of the δ function, where we assumed that the point $\mathbf{y}$ was in the domain.

Our intuition would be that, if the point $\mathbf{y}$ were on a smooth portion of the boundary, it would include half the effect of the δ function. If $\mathbf{y}$ were at a corner it would include a fraction of the local volume determined by the solid angle, γ, subtended in the domain by that point. In fact the analysis will mostly bear out this intuition, and the equation for ψ when $\mathbf{y}$ is on the boundary is:

$$\alpha\psi(\mathbf{y}) = \int_S \left(\psi(\mathbf{x}) \frac{\partial G(\mathbf{x},\mathbf{y})}{\partial n(\mathbf{x})} - G(\mathbf{x},\mathbf{y}) \frac{\partial \psi(\mathbf{x})}{\partial n(\mathbf{x})} \right) dS(\mathbf{x}),$$

$$\alpha = \begin{cases} 1/2 & y \text{ on a smooth part of the boundary} \\ \gamma/4\pi & y \text{ at a corner on the boundary} \\ 1 & y \text{ inside the domain} \end{cases} \tag{1.3.21}$$

Using the boundary condition for ψ or $\partial\psi/\partial n$, we can solve for the unknown component on the boundary. Once the boundary values are known we can obtain ψ elsewhere in the domain. The only caveat is that, for a given boundary condition, there are some wavenumbers k at which the integral on the boundary vanishes, even though the solution exists.

The theory of layer potentials provides a way to study these integrals, identify the problems associated with them, and avoid these problems.

1.3.1.3 Solution of the Helmholtz equation as distribution of sources and dipoles

Distribution of sources

The Green's function $G(x, y)$ can be interpreted in acoustics as the potential or free space field measured at point **y** and generated by a point source of unit intensity located at **x**. Due to symmetry of Green's function with respect to its arguments, the locations of the field point and the source can be exchanged. This gives rise to the so-called *reciprocity principle*, which can be written in more general terms, but we do not proceed with this issue here. If we are interested in solutions of the Helmholtz equation in some domain Ω to which **y** belongs, owing to the linearity of this equation we can decompose the solution to a sum of linearly independent functions, such that each function satisfies the Helmholtz equation in this domain. A set of Green's functions corresponding to sources located at various points *outside* Ω is a good candidate for this decomposition.

Some problems naturally provide a distribution of sources. For example, if one considers computation of a sound field generated by N speakers which emit sound in all directions more or less uniformly (omnidirectional speakers), and the size of the speakers is much smaller than the scale of the problem considered, then the field can be modeled as:

$$\psi(\mathbf{y}) = \sum_{j=1}^{N} Q_j G(\mathbf{x}_j, \mathbf{y}), \qquad \mathbf{y} \in \mathbb{R}^3 \backslash \{\mathbf{x}_j\}, \tag{1.3.22}$$

where Q_j and $\mathbf{x}_j$ are the intensity and location of the jth speaker, respectively.

In a more general case we can consider a continuous analog of these formulae and represent the solution in the form:

$$\psi(\mathbf{y}) = \int_{\bar{\Omega}} q(\mathbf{x}) G(\mathbf{x}, \mathbf{y}) dV(\mathbf{x}), \qquad \mathbf{y} \in \Omega, \ \ \bar{\Omega} \cap \Omega = \emptyset. \tag{1.3.23}$$

Here $q(\mathbf{x})$ is the distribution of source intensities, or *volume density* of sources and integration is taken over domain $\bar{\Omega}$, which is outer to Ω. In the case if $\bar{\Omega}$ is finite and Ω is infinite, the nice thing is that $\psi(\mathbf{y})$ satisfies the Sommerfield radiation conditions automatically. The problem then is to find an appropriate for particular problem distribution of sources $q(\mathbf{x})$. This can be done, say by solving appropriate integral equations.

Single layer potential

A particularly important case for construction of solutions of the Helmholtz equation is the case when all the sources are located on the surface S, which is the boundary of domain Ω. In this case, instead of integration over the volume, we can sum up all the sources over the surface:

$$\psi(\mathbf{y}) = \int_S q_\sigma(\mathbf{x})G(\mathbf{x},\mathbf{y})dS(\mathbf{x}), \qquad \mathbf{y} \in \Omega, \ S = \partial\Omega. \tag{1.3.24}$$

Function $q_\sigma(\mathbf{x})$ is defined on the surface points and is called the *surface density* of sources. Being represented in this form, function $\psi(\mathbf{y})$ is called the *single layer potential*. The term "single layer" is historical, and comes here to denote that we have only one "layer" of sources (one can imagine each source as a tiny ball and a surface covered by one layer of these balls).

Dipoles

Once we have two sources of intensities Q_1 and Q_2 located at $\mathbf{x}_1$ and $\mathbf{x}_2$, we can consider a field generated by this pair in the assumption that $\mathbf{x}_1$ and $\mathbf{x}_2$ are very close to each other. The field due to the pair is omnidirectional and we have:

$$\psi(\mathbf{y}) = \lim_{x_1 \to x_2} [Q_1 G(\mathbf{x}_1,\mathbf{y}) + Q_2 G(\mathbf{x}_2,\mathbf{y})] = (Q_1 + Q_2)G(\mathbf{x}_2,\mathbf{y}). \tag{1.3.25}$$

Assume now that $Q_1 = -Q_2 = 1$. The above equation shows that in this case $\psi(\mathbf{y}) \equiv 0$. Since $\psi(\mathbf{y})$ is not zero at $\mathbf{x}_1 \neq \mathbf{x}_2$ and zero otherwise, we can assume that it is proportional to the distance $|\mathbf{x}_1 - \mathbf{x}_2|$ (the validity of this assumption is clear from the further consideration):

$$\psi(\mathbf{y}) = |\mathbf{x}_2 - \mathbf{x}_1| M(\mathbf{x}_2,\mathbf{y}) + o(|\mathbf{x}_2 - \mathbf{x}_1|), \qquad |\mathbf{x}_2 - \mathbf{x}_1| \to 0 \tag{1.3.26}$$

Then we can determine the first order term as:

$$M^{(\mathbf{p})}(\mathbf{x}_2,\mathbf{y}) = \lim_{\mathbf{x}_1 \to \mathbf{x}_2} \frac{G(\mathbf{x}_1,\mathbf{y}) - G(\mathbf{x}_2,\mathbf{y})}{|\mathbf{x}_2 - \mathbf{x}_1|} = -\mathbf{p}\cdot\nabla_{\mathbf{x}} G(\mathbf{x}_2,\mathbf{y}), \qquad \mathbf{p} = \frac{\mathbf{x}_2 - \mathbf{x}_1}{|\mathbf{x}_2 - \mathbf{x}_1|}. \tag{1.3.27}$$

The obtained solution is called *dipole* ("two poles"). While this function satisfies the Helmholtz equation at $\mathbf{x} \neq \mathbf{y}$, we can see that, in contrast to the source, the field of the dipole is not omnidirectional, but has one preferred direction specified by vector $\mathbf{p}$, which is called the *dipole moment*. As we can see, this direction is determined by the relative location of the positive and negative sources generating the multipole.

Distribution of dipoles and double layer potential

The field of the dipole is different from the field of the monopole, so the dipole $M^{(\mathbf{p})}(\mathbf{x}, \mathbf{y})$ presents another solution of the Helmholtz equation, singular at $\mathbf{x} = \mathbf{y}$. As earlier, we can then construct a solution of the Helmholtz equation as a sum of dipoles with different intensities and moments distributed in space. As in the case with omnidirectional speakers, some problems can be solved immediately if the singularity is modeled as a dipole. By the way, in modeling of speakers, dipoles are also used to model the fact that the sound from the speaker comes in a certain direction. So the sound field generated by a set of N-directional speakers with intensities Q_j and dipole moments p_j will be:

$$\psi(\mathbf{y}) = \sum_{j=1}^{N} Q_j M^{(\mathbf{p}_j)}(\mathbf{x}_j, \mathbf{y}), \qquad \mathbf{y} \in \mathbb{R}^3 \backslash \{\mathbf{x}_j\}. \tag{1.3.28}$$

This can be generalized for continuous distributions. The case of particular interest is the field generated by a set of dipoles which are distributed over the boundary of the domain whose moments are directed as the normal to the surface. The potential of the field in this case is called the *double layer potential* and can be written according to definition (1.3.27) as:

$$\psi(\mathbf{y}) = \int_S q_\mu(\mathbf{x}) \frac{\partial G(\mathbf{x}, \mathbf{y})}{\partial n(\mathbf{x})} dS(\mathbf{x}), \qquad \mathbf{y} \in \Omega, \qquad S = \partial\Omega. \tag{1.3.29}$$

Here function $q_\mu(\mathbf{x})$ is a distributed strength of the dipoles or the surface density of dipoles. The term "double layer" is clear in the context of the representation of a dipole as a superposition of the fields due to negative and positive sources (so if one imagines each source as a tiny ball, then the surface should be covered by two layers, positive and negative, of these balls).

Connection to the Green's formula

Green's formula (1.3.20) provides an amazing finding that any solution $\psi(\mathbf{y})$ of the Helmholtz equation in an arbitrarily shaped domain can be represented as a sum of single and double layer potentials (1.3.24) and (1.3.29) with surface densities $q_\sigma(\mathbf{x}) = -\partial\psi/\partial n(\mathbf{x})$ and $q_\mu(\mathbf{x}) = \psi(\mathbf{x})$, respectively. The surface densities here are expressed via the values of the function itself.

1.3.2 Maxwell equations

In the case of Maxwell equations, which can be reduced to two vector Helmholtz equations for the phasors of the electric and magnetic field vectors with additional conditions that the fields should be solenoidal, the concept of Green's function can be generalized to handle the vector case and represent the field as a sum of corresponding vector (in fact, tensor) Green's functions.

To derive the Green's function for the Maxwell equation, we remind ourselves that the Green's function is not a solution of the Helmholtz or Maxwell equations in the entire space, since this function is singular at the location of a charge (or source). The Maxwell equations as written in form (1.1.36) do not have any terms which generate the electromagnetic field and describe the propagation of waves generated somewhere in the source/current free domain. These equations can be modified to include generators of the electromagnetic field. In fact, for homogeneous media ($\mu, \varepsilon = \text{const}$), we can modify only the second and the third equations in Eq. (1.1.36) as:

$$\nabla \times \mathbf{H} = \varepsilon \frac{\partial \mathbf{E}}{\partial t} + \mathbf{J}, \qquad \varepsilon \nabla \cdot \mathbf{E} = \rho, \tag{1.3.30}$$

where $\mathbf{J}$ is the current density and ρ is the charge density. These equations in the frequency domain take the form:

$$\nabla \times \hat{\mathbf{H}} = -\mathrm{i}\omega\varepsilon\hat{\mathbf{E}} + \hat{\mathbf{J}}, \qquad \varepsilon \nabla \cdot \hat{\mathbf{E}} = \hat{\rho}, \tag{1.3.31}$$

where $\hat{\mathbf{J}}$ and $\hat{\rho}$ are the phasors of $\mathbf{J}$ and ρ. Taking the divergence of the first equation and using the second equation, we can see that:

$$0 = -\mathrm{i}\omega\varepsilon\nabla \cdot \hat{\mathbf{E}} + \nabla \cdot \hat{\mathbf{J}} = -\mathrm{i}\omega\hat{\rho} + \nabla \cdot \hat{\mathbf{J}}. \tag{1.3.32}$$

To obtain a single equation for $\hat{\mathbf{E}}$ we substitute expression of the magnetic field phasor (1.1.55) via $\hat{\mathbf{E}}$ into the first equation (1.3.31). This yields

$$\nabla \times \nabla \times \hat{\mathbf{E}} - k^2\hat{\mathbf{E}} = \mathrm{i}\omega\mu\hat{\mathbf{J}}. \tag{1.3.33}$$

Owing to vector identity $\nabla \times \nabla \times = -\nabla^2 + \nabla(\nabla \cdot)$ we have from this equation and Eq. (1.3.32):

$$\nabla^2\hat{\mathbf{E}} + k^2\hat{\mathbf{E}} = \frac{1}{\varepsilon}\nabla\hat{\rho} - \mathrm{i}\omega\mu\hat{\mathbf{J}} = -\mathrm{i}\omega\mu\left[\hat{\mathbf{J}} + \frac{1}{\varepsilon\mu\omega^2}\nabla(\nabla \cdot \hat{\mathbf{J}})\right]$$

$$= -\mathrm{i}\omega\mu\left(\mathbf{I} + \frac{1}{k^2}\nabla\nabla\right) \cdot \hat{\mathbf{J}}. \tag{1.3.34}$$

Here, assuming that the reader is familiar with elements of tensor analysis (otherwise we recommend reading definitions from appropriate handbooks), we introduced notation **I** and $\nabla\nabla$ for *second rank tensors*, or *dyadics*, which are represented in the three-dimensional case by the following 3×3 symmetric matrices in the basis of Cartesian coordinates:

$$\mathbf{I} = \begin{pmatrix} 1 & 0 & 0 \\ 0 & 1 & 0 \\ 0 & 0 & 1 \end{pmatrix}, \qquad \nabla\nabla = \begin{pmatrix} \dfrac{\partial^2}{\partial x^2} & \dfrac{\partial^2}{\partial x\,\partial y} & \dfrac{\partial^2}{\partial x\,\partial z} \\ \dfrac{\partial^2}{\partial x\,\partial y} & \dfrac{\partial^2}{\partial y^2} & \dfrac{\partial^2}{\partial y\,\partial z} \\ \dfrac{\partial^2}{\partial x\,\partial z} & \dfrac{\partial^2}{\partial y\,\partial z} & \dfrac{\partial^2}{\partial z^2} \end{pmatrix}. \tag{1.3.35}$$

There are two ways to proceed with the representation of solutions of the *non-uniform* Maxwell equations. In fact, they lead to the same result, and the difference is from a methodological point of view. The first way is based simply on a notice that the vector non-uniform Helmholtz equation (1.3.34) is nothing but a set of three scalar equations for each Cartesian component of the phasor of the electric field vector. For example, taking a scalar product with $\mathbf{i}_x$ of each term of Eq. (1.3.34), we obtain the following scalar equation for the x component:

$$\nabla^2 \hat{E}_x + k^2 \hat{E}_x = -i\omega\mu\left[\left(\mathbf{I} + \frac{\nabla\nabla}{k^2}\right)\cdot\hat{\mathbf{j}}\right]\cdot\mathbf{i}_x = -i\omega\mu\left[\hat{j}_x + \frac{\partial}{\partial x}(\nabla\cdot\hat{\mathbf{j}})\right], \tag{1.3.36}$$
$$\hat{E}_x = \hat{\mathbf{E}}\cdot\mathbf{i}_x, \qquad \hat{j}_x = \hat{\mathbf{j}}\cdot\mathbf{i}_x.$$

The solution of this equation for free space, which is Eq. (1.3.5), can be obtained using the scalar Green's function (1.3.18):

$$\hat{E}_x(\mathbf{y}) = i\omega\mu \int_{\bar{\Omega}} \left[\hat{j}_x(\mathbf{x}) + \frac{\partial}{\partial x}(\nabla_x\cdot\hat{\mathbf{j}}(\mathbf{x}))\right] G(\mathbf{x}, \mathbf{y})\, dV(\mathbf{x}). \tag{1.3.37}$$

Here we assumed that, in domain Ω, which is outside some domain $\bar{\Omega}$, there are no imposed currents, $\mathbf{j}(\mathbf{x}) = 0$. Writing similar equations for the other two components of $\hat{\mathbf{E}}$, we can summarize the result in one vector formula:

$$\hat{\mathbf{E}}(\mathbf{y}) = i\omega\mu \int_{\bar{\Omega}} \left[\hat{\mathbf{j}}(\mathbf{x}) + \nabla_x(\nabla_x\cdot\hat{\mathbf{j}}(\mathbf{x}))\right] G(\mathbf{x}, \mathbf{y})\, dV(\mathbf{x}). \tag{1.3.38}$$

If $\mathbf{y}$ is from a domain free of imposed currents, $\mathbf{y} \in \Omega$, $\bar{\Omega} \cap \Omega = \emptyset$, then, as follows from Eqs. (1.3.32) and (1.3.34), the phasor of the electric field vector satisfies uniform Maxwell equations (1.1.42). This is a situation, similar to the scalar case (1.3.23), where we constructed a solution of the Helmholtz equation by placing sources outside the domain. In the case of Maxwell equations, instead of some scalar field $q(\mathbf{x})$ characterizing the source density distribution, we have a vector field $\hat{\mathbf{J}}(\mathbf{x})$ or *current density distribution*. As in the scalar case, the problem is to determine this unknown distribution and this can be done by different methods, including boundary integral equations. Indeed, we can derive these equations in the same way as we derived Eq. (1.3.38) by considering scalar equations for each component of the electric field vector.

The second way is to introduce dyadic Green's function immediately as a solution of Maxwell equations (1.1.42) everywhere in free space except for one singular point, where the solution blows up. This is the way we introduced the scalar Green's function (1.3.3). The "impulse response" of Maxwell equations can be found by solving the following equation for dyadic Green's function:

$$-\nabla \times \nabla \times \mathbf{G}(\mathbf{x}, \mathbf{y}) + k^2 \mathbf{G}(\mathbf{x}, \mathbf{y}) = -\mathbf{I}\delta(\mathbf{x} - \mathbf{y}). \qquad (1.3.39)$$

This form is dictated first by Eq. (1.3.33), where we should assume that the source term is due to currents in the domain, which can be contracted into one point. Second, representing $\hat{\mathbf{E}}(\mathbf{y})$ as a convolution of $\hat{\mathbf{J}}(\mathbf{x})$ with the impulse response function

$$\hat{\mathbf{E}}(\mathbf{y}) = -\mathrm{i}\omega\mu \int_{\bar{\Omega}} \mathbf{G}(\mathbf{x}, \mathbf{y}) \cdot \hat{\mathbf{J}}(\mathbf{x}) \mathrm{d}V(\mathbf{x}), \qquad (1.3.40)$$

we can see, comparing this result with relation (1.3.38), that $\mathbf{G}(\mathbf{x}, \mathbf{y})$ should be a second rank tensor.

Function $\mathbf{G}(\mathbf{x}, \mathbf{y})$ defined by Eq. (1.3.39) can be related to scalar Green's function (1.3.1). The relation is:

$$\mathbf{G}(\mathbf{x}, \mathbf{y}) = \left(\mathbf{I} + \frac{1}{k^2} \nabla_x \nabla_x \right) G(\mathbf{x}, \mathbf{y}), \qquad (1.3.41)$$

and can be checked by substitution into Eq. (1.3.40) followed by integration by parts to obtain form (1.3.38). It is interesting to note that the dyadic Green's function for Maxwell equations involves not only $G(\mathbf{x}, \mathbf{y})$, but also the second derivatives of this function. As will be shown in Chapter 2, the second derivatives of a *monopole* ($G(\mathbf{x}, \mathbf{y})$) can be expressed in terms of dipoles and *quadrupoles*.

However, if the statement of the problem allows determination of $\hat{\mathbf{E}}$, then $\hat{\mathbf{H}}$ can be found simply from relation (1.1.55) and the problems for $\hat{\mathbf{H}}$ can be considered in terms of this vector only. This is due to an obvious symmetry between $\hat{\mathbf{E}}$ and $\hat{\mathbf{H}}$ in the free space Maxwell equations (one can replace these vectors taking care with constants and signs).

Consideration of the solution for $\hat{\mathbf{E}}$ given above is physics based, since it operates with such terms as "charges" and "currents". It is noteworthy that, despite there being no magnetic charges in the Maxwell equations, one can nonetheless introduce "fictitious" magnetic charges and currents *outside* the domain, where solution of the free-field Maxwell equations should be obtained, as a *mathematical* trick. Indeed, relation (1.3.38) provides integral representation for solenoidal solutions of the vector Helmholtz equation in the domain free of charges ($\mathbf{y} \in \Omega, \bar{\Omega} \cap \Omega = \emptyset$), which is the equation for solenoidal vector $\hat{\mathbf{H}}$.

1.4 WHAT IS COVERED IN THIS BOOK AND WHAT IS NOT

The research field of acoustics and electromagnetics is huge and every year hundreds of publications in the form of papers, technical reports, monographs, and text books extend the knowledge in this field. Thus the objective of the book is far from giving a review of all these materials or the state of art in the entire field. We also did not have as an objective to provide the basics of wave theory, for which we can refer the reader to several well-written books on the fundamentals of acoustics [FHLB99, LL75, MI68] and electromagnetics [Che90, Jac75, LL60]. The book is also not about the field of differential equations or pure mathematics dedicated to the theory of elliptic equations and, particularly, the Helmholtz equation.

Trying to present some mathematical theory which can be applied to a solution of the Helmholtz equation, we focus on some issues that are important from the computational point of view, and therefore, miss several cases of this equation. For example, in the book we consider only the three-dimensional case for the scalar Helmholtz equation. The two-dimensional Helmholtz equation has its own beauty and symmetries and, while the translation theory for this case appears to be simpler than in three dimensions, this case deserves separate consideration for fast computational methods.

Another item missed is the theory for Maxwell equations. While these equations can be reduced to a solution of scalar wave equations and solved with the methods described in this book, efficient application of multipole methods here seems to require a deeper study of operations with vector or tensor spherical harmonics. Despite the fact that the technique for fast operations with vector spherical harmonics and vector spherical basis functions is currently developed in many aspects, we considered that putting this technique in the present book will make it less transparent and, since it cannot be understood without the basics of translation theory for scalar spherical harmonics, that we should present this theory elsewhere.

As can be seen from the present chapter, the Helmholtz equation can appear in various physical problems with complex k. Most results from the translation theory described in this book are universal and can be applied for any $k \neq 0$. However, since our focus was on the case of real k, we do not provide results such as error bounds for the case when the imaginary part of k is not zero. It can be argued that, in the case of complex k due to exponential decay of solutions, these solutions should be "better" than in the case of real k, and the case of real k is, in fact, the more difficult case. In our view, this situation requires a separate study and again must be presented elsewhere.

CHAPTER 2

Elementary Solutions

We start the consideration of the solutions of the Helmholtz equation in three dimensions with *elementary* or *factored* solutions in spherical coordinates. While there are several advantages to using spherical coordinates, the major one is that they provide a countable basis of functions orthogonal on a sphere that can be used for the representation of any other solution of the Helmholtz equation.

2.1 SPHERICAL COORDINATES

The transformation between spherical coordinates and Cartesian coordinates with a common origin $(x, y, z) \rightarrow (r, \theta, \varphi)$ is given by

$$x = r \sin\theta \cos\varphi, \qquad y = r \sin\theta \sin\varphi, \qquad z = r\cos\theta, \tag{2.1.1}$$

where $r \geq 0$ is the distance to the point $\mathbf{r} = (x, y, z)$ from the origin $(0, 0, 0)$, while θ and φ are spherical polar angles. At fixed r these angles can be treated as the coordinates of the points on a sphere of radius r. Varying these angles in the range

$$0 \leq \varphi < 2\pi, \qquad 0 \leq \theta \leq \pi \tag{2.1.2}$$

covers the whole surface of this sphere. The range of φ and θ can be extended to the set of all real numbers, since

$$\begin{aligned} &\mathbf{r}(r, \theta, \varphi + 2\pi) = \mathbf{r}(r, \theta, \varphi), \qquad \mathbf{r}(r, \theta + 2\pi, \varphi) = \mathbf{r}(r, \theta, \varphi), \\ &\mathbf{r}(r, -\theta, \varphi + \pi) = \mathbf{r}(r, \theta, \varphi). \end{aligned} \tag{2.1.3}$$

So any function of $\mathbf{r}$ should be 2π-periodical with respect to φ and to θ, and further the half-period relation, expressed by the latter equation should hold.

The spherical coordinates defined by transform (2.1.1) are orthonormal. This means that the unit vectors $(\mathbf{i}_r, \mathbf{i}_\theta, \mathbf{i}_\varphi)$ form a local basis at each spatial point, where $\mathbf{i}_r$ is the outer normal to a sphere of radius r, while $\mathbf{i}_\theta$

and $\mathbf{i}_\varphi$ are tangential vectors to this sphere (see Fig. 2.1). Since the basis is right-handed we have

$$\mathbf{i}_r \times \mathbf{i}_\theta = \mathbf{i}_\varphi, \qquad \mathbf{i}_\theta \times \mathbf{i}_\varphi = \mathbf{i}_r, \qquad \mathbf{i}_\varphi \times \mathbf{i}_r = \mathbf{i}_\theta. \tag{2.1.4}$$

The gradient and Laplacian of a scalar function ψ in spherical coordinates can be written as

$$\begin{aligned}\nabla\psi &= \mathbf{i}_r \frac{\partial \psi}{\partial r} + \mathbf{i}_\theta \frac{1}{r}\frac{\partial \psi}{\partial \theta} + \mathbf{i}_\varphi \frac{1}{r\sin\theta}\frac{\partial \psi}{\partial \varphi},\\ \nabla^2\psi &= \nabla\cdot(\nabla\psi) = \frac{1}{r^2}\frac{\partial}{\partial r}\left(r^2\frac{\partial\psi}{\partial r}\right) + \frac{1}{r^2\sin\theta}\frac{\partial}{\partial\theta}\left(\sin\theta\frac{\partial\psi}{\partial\theta}\right)\\ &\quad + \frac{1}{r^2\sin^2\theta}\frac{\partial^2\psi}{\partial\varphi^2}.\end{aligned} \tag{2.1.5}$$

2.1.1 Separation of variables

Consider now complex-valued solutions $\psi(\mathbf{r})$ of the Helmholtz equation

$$\nabla^2\psi + k^2\psi = 0, \tag{2.1.6}$$

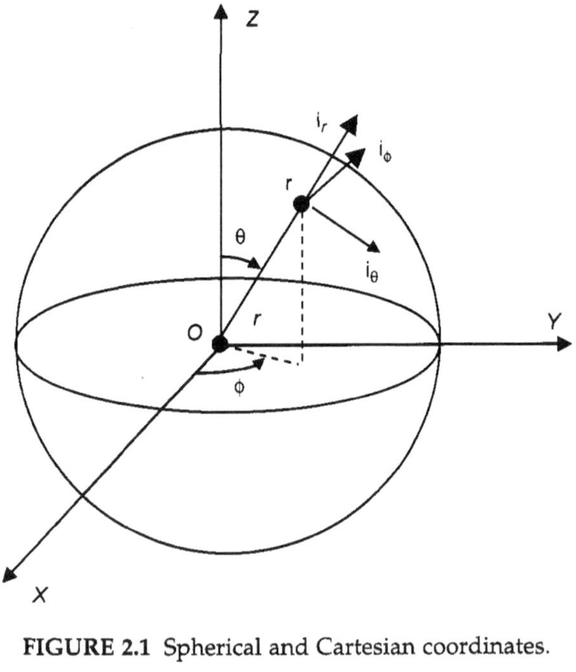

FIGURE 2.1 Spherical and Cartesian coordinates.

which can be expressed in factored form ("separation of variables")

$$\psi(r, \theta, \varphi) = \Pi(r)\Theta(\theta)\Phi(\varphi). \tag{2.1.7}$$

If we substitute this factorization into the Helmholtz equation and use the expression for the Laplacian in spherical coordinates given by formula (2.1.5), after multiplication by $r^2 \sin^2\theta/\psi$ we obtain

$$\frac{\sin^2\theta}{\Pi}\left[\frac{\mathrm{d}}{\mathrm{d}r}\left(r^2\frac{\mathrm{d}\Pi}{\mathrm{d}r}\right)+k^2r^2\Pi\right]+\frac{\sin\theta}{\Theta}\frac{\mathrm{d}}{\mathrm{d}\theta}\left(\sin\theta\frac{\mathrm{d}\Theta}{\mathrm{d}\theta}\right)+\frac{1}{\Phi}\frac{\mathrm{d}^2\Phi}{\mathrm{d}\varphi^2}=0. \tag{2.1.8}$$

2.1.1.1 *Equation with respect to the angle φ*

Since neither the first, nor the second term in the left-hand side of relation (2.1.8) depends on φ, the third term must also be independent of φ. Therefore, we have

$$\frac{\mathrm{d}^2\Phi}{\mathrm{d}\varphi^2} = \lambda\Phi, \tag{2.1.9}$$

where λ is a separation constant. Because Φ is periodic in 2π this separation constant must be

$$\lambda = -m^2, \qquad m = 0, \pm 1, \pm 2, \ldots \tag{2.1.10}$$

where m is an integer.

In this case the solution of Eq. (2.1.9) becomes

$$\Phi = C_1\,\mathrm{e}^{im\varphi} + C_2\,\mathrm{e}^{-im\varphi}, \tag{2.1.11}$$

or

$$\Phi = B_1 \sin m\varphi + B_2 \cos m\varphi, \tag{2.1.12}$$

where B_1, B_2, C_1 and C_2 are arbitrary constants of integration.

2.1.1.2 *Equation with respect to the angle θ*

Using Eqs. (2.1.9) and (2.1.10) we can now rewrite relation (2.1.8) in the form

$$\frac{1}{\Pi}\left[\frac{\mathrm{d}}{\mathrm{d}r}\left(r^2\frac{\mathrm{d}\Pi}{\mathrm{d}r}\right)+k^2r^2\Pi\right] = \frac{m^2}{\sin^2\theta} - \frac{1}{\Theta\sin\theta}\frac{\mathrm{d}}{\mathrm{d}\theta}\left(\sin\theta\frac{\mathrm{d}\Theta}{\mathrm{d}\theta}\right). \tag{2.1.13}$$

The left-hand side of this equation is a function of r only, while the right-hand side can depend only on θ. This may occur only if both the left-hand side and the right-hand side are constants. As before, let

λ denote the separation constant. So we have

$$\frac{m^2}{\sin^2\theta} - \frac{1}{\Theta \sin\theta}\frac{d}{d\theta}\left(\sin\theta\frac{d\Theta}{d\theta}\right) = \lambda, \tag{2.1.14}$$

$$\frac{1}{\Pi}\left[\frac{d}{dr}\left(r^2\frac{d\Pi}{dr}\right) + k^2r^2\Pi\right] = \lambda. \tag{2.1.15}$$

Further, we will frequently use the following angular variable μ instead of θ:

$$\mu = \cos\theta. \tag{2.1.16}$$

Derivatives with respect to θ can be expressed in terms of derivatives with respect to μ via

$$\frac{d\Theta}{d\theta} = \frac{d\Theta}{d\mu}\frac{d\mu}{d\theta} = -\frac{d\Theta}{d\mu}\sin\theta.$$

Accordingly

$$\frac{1}{\sin\theta}\frac{d}{d\theta}\left(\sin\theta\frac{d\Theta}{d\theta}\right) = \frac{1}{\sin\theta}\frac{d}{d\mu}\left(\sin\theta\frac{d\Theta}{d\theta}\right)\frac{d\mu}{d\theta} = \frac{d}{d\mu}\left(\sin^2\theta\frac{d\Theta}{d\mu}\right)$$
$$= \frac{d}{d\mu}\left[(1-\mu^2)\frac{d\Theta}{d\mu}\right].$$

Thus, Eq. (2.1.14) can be rewritten as

$$\frac{d}{d\mu}\left[(1-\mu^2)\frac{d\Theta}{d\mu}\right] + \left[\lambda - \frac{m^2}{1-\mu^2}\right]\Theta = 0. \tag{2.1.17}$$

The general solutions of this equation, which is known as the *associated Legendre differential equation* are superpositions of two special functions—the *associated Legendre functions of the first kind* and the *associated Legendre functions of the second kind*. The associated Legendre functions of the second kind are singular at $\mu = 1$ and can be excluded from consideration as soon as $\mu = 1$ is a regular point on the sphere and solution is regular at this point. Since we deal only with the associated Legendre functions of the first kind, we drop the words "of the first kind" for brevity. Regular solutions Θ which are regular 2π-periodic functions of θ are realized for

$$\lambda = n(n+1), \qquad n = 0, \pm 1, \pm 2, \ldots \tag{2.1.18}$$

where n is an integer. In this case we have

$$\Theta = CP_n^m(\mu), \tag{2.1.19}$$

where C is a constant, and $P_n^m(\mu)$ are the associated Legendre functions of degree n and order m.

For non-negative n and m they can be expressed via the *Legendre polynomials*, $P_n(\mu) = P_n^0(\mu)$, using the following *Rodrigues' formula*

$$P_n^m(\mu) = (-1)^m(1-\mu^2)^{m/2}\frac{d^m}{d\mu^m}P_n(\mu), \qquad n \geq 0, \qquad m \geq 0, \quad (2.1.20)$$

$$P_n(\mu) = \frac{1}{2^n n!}\frac{d^n}{d\mu^n}(\mu^2-1)^n, \qquad n \geq 0. \quad (2.1.21)$$

For negative m or n the associated Legendre functions can be expressed via those of positive n and m using symmetry relations, which are considered in detail below.

2.1.1.3 Equation with respect to the distance r

Using λ from relation (2.1.18) the equation for the radial function (2.1.15) becomes

$$\frac{d}{dr}\left(r^2\frac{d\Pi}{dr}\right) + \left[k^2r^2 - n(n+1)\right]\Pi = 0. \quad (2.1.22)$$

Denoting

$$\rho = kr, \qquad v(\rho) = \Pi(r), \quad (2.1.23)$$

we have the following differential equation for $v(\rho)$:

$$\rho^2 v'' + 2\rho v' + \left[\rho^2 - n(n+1)\right]v = 0. \quad (2.1.24)$$

This is a *spherical Bessel equation*. Solutions of this equation for different values of n are the *spherical Bessel functions of the first kind*

$$j_n(\rho) = \sqrt{\frac{\pi}{2\rho}}J_{n+1/2}(\rho), \quad (2.1.25)$$

and the *spherical Bessel functions of the second kind*

$$y_n(\rho) = \sqrt{\frac{\pi}{2\rho}}Y_{n+1/2}(\rho). \quad (2.1.26)$$

A third set of solutions called the *spherical Bessel functions of the third kind* or *spherical Hankel functions of the first and the second kinds* can be constructed by linear combination of the spherical Bessel functions of the first and second kinds as

$$h_n^{(1)}(\rho) = j_n(\rho) + iy_n(\rho) = \sqrt{\frac{\pi}{2\rho}}H_{n+1/2}^{(1)}(\rho), \quad (2.1.27)$$

$$h_n^{(2)}(\rho) = j_n(\rho) - \mathrm{i}y_n(\rho) = \sqrt{\frac{\pi}{2\rho}} H_{n+1/2}^{(2)}(\rho). \tag{2.1.28}$$

The functions $J_{n+1/2}(\rho)$ and $Y_{n+1/2}(\rho)$ are the *Bessel and Neumann functions of fractional order*, and $H_{n+1/2}^{(1)}(\rho)$ and $H_{n+1/2}^{(2)}(\rho)$ are the *Hankel functions of the first and the second kind of fractional order*. The pairs of solutions $j_n(\rho)$, $y_n(\rho)$ and $h_n^{(1)}(\rho)$, $h_n^{(2)}(\rho)$ are linearly independent solutions for every n.

The spherical Bessel functions can be expressed through elementary transcendental functions. This is clear from the following *Rayleigh's formulas*:

$$j_n(\rho) = \rho^n \left(-\frac{1}{\rho}\frac{\mathrm{d}}{\mathrm{d}\rho}\right)^n \frac{\sin\rho}{\rho}, \tag{2.1.29}$$

$$y_n(\rho) = -\rho^n \left(-\frac{1}{\rho}\frac{\mathrm{d}}{\mathrm{d}\rho}\right)^n \frac{\cos\rho}{\rho}, \qquad n = 0, 1, 2, \ldots \tag{2.1.30}$$

These formulas for spherical Bessel functions of the third kind become

$$h_n^{(1)}(\rho) = \rho^n \left(-\frac{1}{\rho}\frac{\mathrm{d}}{\mathrm{d}\rho}\right)^n \frac{\sin\rho - \mathrm{i}\cos\rho}{\rho} = -\mathrm{i}\rho^n \left(-\frac{1}{\rho}\frac{\mathrm{d}}{\mathrm{d}\rho}\right)^n \frac{\mathrm{e}^{\mathrm{i}\rho}}{\rho}, \tag{2.1.31}$$

$$h_n^{(2)}(\rho) = \rho^n \left(-\frac{1}{\rho}\frac{\mathrm{d}}{\mathrm{d}\rho}\right)^n \frac{\sin\rho + \mathrm{i}\cos\rho}{\rho} = \mathrm{i}\rho^n \left(-\frac{1}{\rho}\frac{\mathrm{d}}{\mathrm{d}\rho}\right)^n \frac{\mathrm{e}^{-\mathrm{i}\rho}}{\rho}. \tag{2.1.32}$$

Note that there are no formal limitations on k, except that $k \neq 0$. So this solution is valid for any complex-valued $k \neq 0$. The Helmholtz equation which appears from the linear wave equation corresponds to real $k^2 > 0$. This can also happen in the case of wave propagation in purely dispersive media. The Helmholtz equation with real $k^2 < 0$ appears, for example, as a field generated by Yukawa potentials (in molecular dynamics with screened Coulombic interactions). Cases with complex k^2 appear in Fourier transforms of the heat conduction equation (where k^2 is purely imaginary), and in the equation describing wave propagation in media with relaxation or dissipation (see Chapter 1).

2.1.2 Special functions and properties

2.1.2.1 Associated Legendre functions

Legendre polynomials

The simplest representatives of the associated Legendre functions are the Legendre polynomials, which are functions of zero order: $P_n(\mu) = P_n^0(\mu)$. The Legendre polynomials appear, for example, as expansion coefficients of the following generating function:

$$\frac{1}{(1-2\mu t+t^2)^{1/2}} = \sum_{n=0}^{\infty} P_n(\mu)t^n, \qquad |\mu| \leq 1, \qquad |t| < 1. \tag{2.1.33}$$

Explicit expressions for these polynomials are

$$P_n(\mu) = \frac{1}{2^n} \sum_{l=0}^{[n/2]} (-1)^l \binom{n}{l} \binom{2n-2l}{n} \mu^{n-2l}, \tag{2.1.34}$$

where [ ] denotes the integer part of the argument, and

$$\binom{n}{l} = \frac{n!}{l!(n-l)!} \tag{2.1.35}$$

are the *binomial coefficients*. The first three Legendre polynomials are

$$P_0(\mu) = 1, \qquad P_1(\mu) = \mu, \qquad P_2(\mu) = \tfrac{1}{2}(3\mu^2 - 1). \tag{2.1.36}$$

Figure 2.2 shows the first few Legendre polynomials in the range $[-1, 1]$.

The Legendre polynomials form a complete orthogonal basis on $L^2[-1, 1]$, which means that a scalar product in $L^2[-1, 1]$ of two

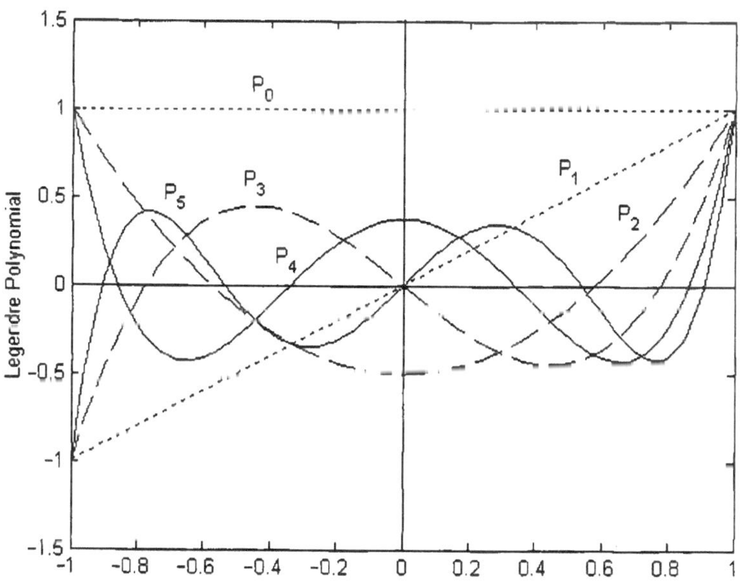

FIGURE 2.2 The Legendre polynomials.

polynomials of different degrees is zero, while the norm of each polynomial is not zero. More precisely,

$$(P_n, P_l) = \int_{-1}^{1} P_n(\mu)P_l(\mu)\mathrm{d}\mu = \frac{2}{2n+1}\delta_{nl}, \tag{2.1.37}$$

where δ_{nl} is the *Kronecker symbol,*

$$\delta_{nl} = \begin{cases} 1, & n = l, \\ 0, & n \neq l. \end{cases} \tag{2.1.38}$$

This property shows that any function $f \in L^2[-1, 1]$ can be expanded into series as

$$f(\mu) = \sum_{n=0}^{\infty} c_n P_n(\mu). \tag{2.1.39}$$

Taking scalar product of f with some P_l and using property (2.1.37), we find

$$c_l = \frac{2l+1}{2}(f, P_l) = \frac{2l+1}{2}\int_{-1}^{1} f(\mu)P_l(\mu)\mathrm{d}\mu. \tag{2.1.40}$$

Taking scalar product of f with f we obtain the *Parseval equality*:

$$(f, f) = \|f\|^2 = \int_{-1}^{1} f^2(\mu)\mathrm{d}\mu = \sum_{n=0}^{\infty} \frac{2}{2n+1} c_n^2. \tag{2.1.41}$$

In the range $\mu \in [-1, 1]$ all the Legendre polynomials are bounded as

$$|P_n(\mu)| \leq 1. \tag{2.1.42}$$

Moreover, the maximum of $|P_n(\mu)|$ is always reached at ± 1:

$$P_n(-1) = (-1)^n, \qquad P_n(1) = 1. \tag{2.1.43}$$

Functions of arbitrary order

The associated Legendre functions of integer order m and degree n satisfy the symmetry relations

$$P^m_{-n-1}(\mu) = P^m_n(\mu), \qquad P_n^{-m}(\mu) = \frac{(n-m)!}{(n+m)!} P^m_n(\mu), \tag{2.1.44}$$

$$n = 0, \pm 1, \ldots, \qquad m = 0, \pm 1, \ldots$$

and that is why only functions of non-negative n and m need be considered.

Moreover, the Rodrigues formula (2.1.20) shows that

$$P_n^m(\mu) = 0, \qquad m > n \geq 0. \tag{2.1.45}$$

So only orders $0 \leq m \leq n$ will be considered herein. Note also that these functions are even for even $n+m$ and odd for odd $n+m$

$$P_n^m(-\mu) = (-1)^{n+m} P_n^m(\mu). \tag{2.1.46}$$

The Rodrigues formula (2.1.20) shows that $P_n^m(\mu)$ of even order m are polynomials of degree n, while for odd m they are the products of the factor $(1-\mu^2)^{1/2} = \sin\theta$ and polynomials of degree $n-1$.

An explicit expression for the associated Legendre functions can be found using the following representation via the *hypergeometric function* $F(a, b; c; z)$

$$\begin{aligned} P_n^m(\mu) &= \frac{(-1)^m}{2^m} \frac{(n+m)!}{(n-m)!m!} \\ &\quad \times (1-\mu^2)^{m/2} F\left(m-n, m+n+1; m+1; \frac{1-\mu}{2}\right) \\ &= \frac{(-1)^m}{2^m} \frac{(n+m)!}{(n-m)!m!} (1-\mu^2)^{m/2} \\ &\quad \times \sum_{l=0}^{n-m} \frac{(-1)^l (n-m-l+1)_l (n+m+1)_l}{2^l l! (m+1)_l} (1-\mu)^l, \end{aligned} \tag{2.1.47}$$

where $(n)_l$ is the *Pochhammer symbol*.

$$(n)_0 = 1, \qquad (n)_l = \frac{(n+l-1)!}{(n-1)!}. \tag{2.1.48}$$

This formula yields the following particular functions:

$$\begin{aligned} &P_1^1(\mu) = -(1-\mu^2)^{1/2}, \qquad P_2^1(\mu) = -3\mu(1-\mu^2)^{1/2}, \\ &P_2^2(\mu) = 3(1-\mu^2). \end{aligned} \tag{2.1.49}$$

Figure 2.3 shows several first associated Legendre functions for $m > 0$.

Note also that

$$P_n^n(\mu) = (-1)^n (2n-1)!!(1-\mu^2)^{n/2}, \tag{2.1.50}$$

where we use the following definition for the *double factorial*

$$\begin{aligned} &(2n-1)!! = 1 \cdot 3 \cdots (2n-1), \qquad (2n)!! = 2 \cdot 4 \cdots (2n), \\ &(-1)!! = 0!! = 1, \qquad n = 0, 1, \ldots \end{aligned} \tag{2.1.51}$$

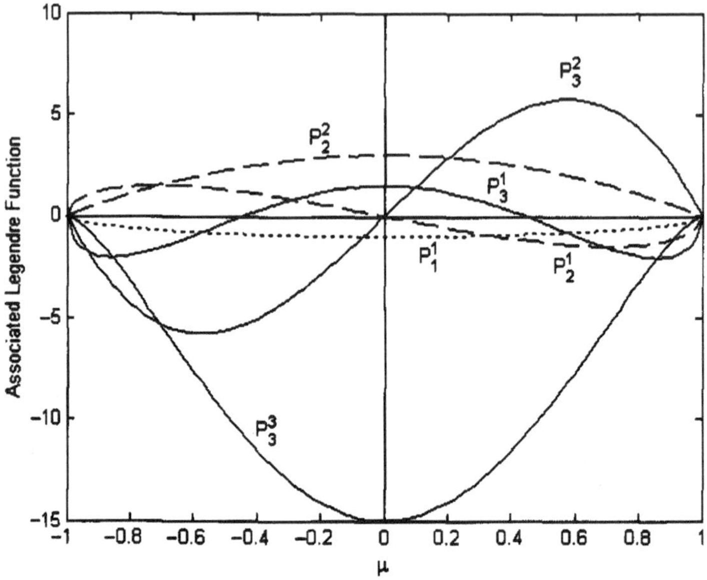

FIGURE 2.3 The associated Legendre functions.

The associated Legendre functions satisfy various recurrence relations. Some important relations, which are used in this book are listed below:

$$\mu P_n^m(\mu) = \frac{n+m}{2n+1} P_{n-1}^m(\mu) + \frac{n-m+1}{2n+1} P_{n+1}^m(\mu), \tag{2.1.52}$$

$$(1-\mu^2)\frac{\mathrm{d}P_n^m(\mu)}{\mathrm{d}\mu} = \frac{(n+1)(n+m)}{2n+1} P_{n-1}^m(\mu) - \frac{n(n-m+1)}{2n+1} P_{n+1}^m(\mu), \tag{2.1.53}$$

$0 \leq m \leq n:$

$$\sqrt{1-\mu^2}P_n^m(\mu) = \frac{1}{2n+1} P_{n-1}^{m+1}(\mu) - \frac{1}{2n+1} P_{n+1}^{m+1}(\mu), \tag{2.1.54}$$

$$\mu\sqrt{1-\mu^2}\frac{\mathrm{d}P_n^m(\mu)}{\mathrm{d}\mu} + \frac{mP_n^m(\mu)}{\sqrt{1-\mu^2}} = -\frac{n+1}{2n+1} P_{n-1}^{m+1}(\mu) - \frac{n}{2n+1} P_{n+1}^{m+1}(\mu), \tag{2.1.55}$$

$1 \le m \le n:$

$$\sqrt{1-\mu^2}P_n^m(\mu) = \frac{(n-m+1)(n-m+2)}{2n+1}P_{n+1}^{m-1}(\mu) - \frac{(n+m-1)(n+m)}{2n+1}P_{n-1}^{m-1}(\mu), \tag{2.1.56}$$

$$\mu\sqrt{1-\mu^2}\frac{dP_n^m(\mu)}{d\mu} - \frac{mP_n^m(\mu)}{\sqrt{1-\mu^2}} = \frac{(n+1)(n+m-1)(n+m)}{2n+1}P_{n-1}^{m-1}(\mu) + \frac{n(n-m+1)(n-m+2)}{2n+1}P_{n+1}^{m-1}(\mu). \tag{2.1.57}$$

One of the most important properties of the associated Legendre functions is that these functions of given order m are orthogonal in $L^2[-1,1]$:

$$(P_n^m, P_l^m) = \int_{-1}^{1} P_n^m(\mu)P_l^m(\mu)d\mu = \frac{2}{2n+1}\frac{(n+m)!}{(n-m)!}\delta_{nl}. \tag{2.1.58}$$

This property is used for forming an orthonormal basis on a sphere with *spherical harmonics*, which will be considered below.

2.1.2.2 *Spherical Harmonics*

The solutions for the angular variables are usually combined into the spherical harmonics:

$$Y_n^m(\theta,\varphi) = (-1)^m\sqrt{\frac{2n+1}{4\pi}\frac{(n-|m|)!}{(n+|m|)!}}P_n^{|m|}(\cos\theta)e^{im\varphi}, \tag{2.1.59}$$

$$n = 0,1,2,\ldots, \qquad m = -n,\ldots,n.$$

The spherical harmonics are sometimes also called *surface harmonics of the first kind*, *tesseral* for $m < n$ and *sectoral* for $m = n$. Harmonics corresponding to $m = 0$ are also called *zonal* harmonics. The spherical harmonics are complex-valued functions of θ and φ. Functions of negative order are complex conjugates of the function of positive order:

$$Y_n^{-m}(\theta, \varphi) = \overline{Y_n^m(\theta, \varphi)}. \tag{2.1.60}$$

The zonal harmonics do not depend on the angle φ and are real. Also, from the definition (2.1.59) it follows that the real and imaginary parts are related due to the property of trigonometric functions as

$$\text{Re}\{Y_n^m(\theta, \varphi)\} = \text{Im}\left\{Y_n^m\left(\theta, \frac{\pi}{2m} - \varphi\right)\right\}, \qquad m = 1, 2, \ldots, n, \tag{2.1.61}$$

and so the imaginary part can be obtained by rotation about the z-axis of the Cartesian coordinates. To get an idea of how the functions $Y_n^m(\theta, \varphi)$ appear it is sufficient to consider the real part of $Y_n^m(\theta, \varphi)$. Figure 2.4 provides an illustration of the spherical harmonics for several n and m. Here the real part of $Y_n^m(\theta, \varphi)$ is plotted on the surface of a sphere, where the value of $Y_n^m(\theta, \varphi)$ is shown using a color index. It is seen that the zonal harmonics are axisymmetric functions with respect to the z-axis. The real part of the sectorial harmonics for fixed φ preserves the sign, since $P_n^n(\mu)$ retains its sign for any $|\mu| < 1$ (see expression (2.1.50)). The real part of the tesseral harmonics changes its sign with both θ and φ. The zero degree harmonic is both a zonal and sectorial harmonic and is a constant

$$Y_0^0(\theta, \varphi) = \text{const} = \sqrt{\frac{1}{4\pi}}. \tag{2.1.62}$$

The spherical harmonics form a complete orthonormal system in the functional space $L^2(S_u)$, where S_u is the surface of the unit sphere $x^2 + y^2 + z^2 = 1$.

From the definition of the norm of an arbitrary complex-valued function $F(\theta, \varphi)$ in $L^2(S_u)$ we have

$$\|F\|^2 = (F, F) = \int_{S_u} F\bar{F}\, dS = \int_{S_u} |F|^2 dS$$
$$= \int_0^{\pi} \sin\theta\, d\theta \int_0^{2\pi} |F(\theta, \varphi)|^2 d\varphi. \tag{2.1.63}$$

where $\bar{F}$ denotes the complex conjugate of F. Therefore, the scalar product

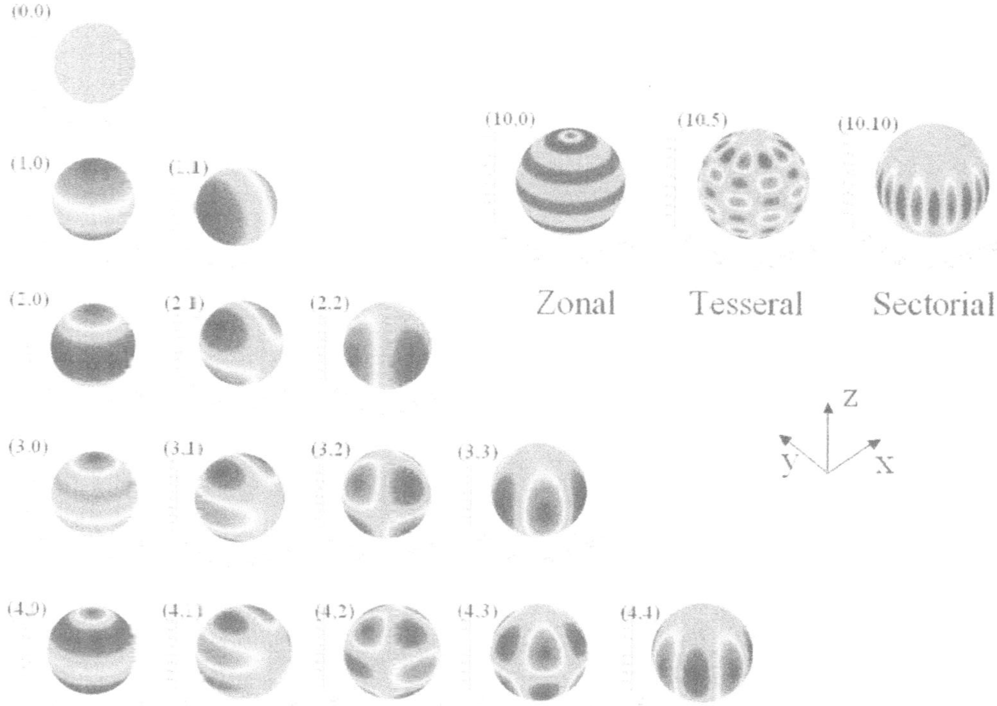

FIGURE 2.4 Spherical harmonics. For color version see Plate A, after Chapter 10. The illustration is made by plotting the real part of the spherical harmonics $Y_n^m(\theta, \varphi)$ on a sphere. The color corresponds to the value of $\mathrm{Re}\{Y_n^m(\theta, \varphi)\}$. The numbers near the spheres show respectively the degree (n) and the order (m) of the harmonics.

of two spherical harmonics in $L^2(S_u)$ is

$$
\begin{aligned}
(Y_n^m, Y_{n'}^{m'}) &= \int_0^{\pi} \sin\theta\, d\theta \int_0^{2\pi} Y_n^m(\theta,\varphi)\bar{Y}_{n'}^{m'}(\theta,\varphi) d\varphi \\
&= \frac{1}{4\pi}\sqrt{(2n+1)\frac{(n-|m|)!}{(n+|m|)!}(2n'+1)\frac{(n'-|m'|)!}{(n'+|m'|)!}} \\
&\quad \times \int_0^{\pi} P_n^{|m|}(\cos\theta)P_{n'}^{|m'|}(\cos\theta)\sin\theta\, d\theta \int_0^{2\pi} e^{i(m-m')\varphi} d\varphi \qquad (2.1.64) \\
&= \delta_{mm'}\frac{1}{2}\sqrt{(2n+1)\frac{(n-|m|)!}{(n+|m|)!}(2n'+1)\frac{(n'-|m'|)!}{(n'+|m'|)!}} \\
&\quad \times \int_{-1}^{1} P_n^{|m'|}(\mu)P_{n'}^{|m'|}(\mu) d\mu = \delta_{mm'}\delta_{nn'}.
\end{aligned}
$$

This proves orthonormality of the system. A proof of the completeness of the system of the spherical harmonics can be found elsewhere.

Consider now expansion of an arbitrary function $F(\theta, \varphi)$ using the spherical harmonics basis

$$
F(\theta,\varphi) = \sum_{n=0}^{\infty}\sum_{m=-n}^{n} F_n^m Y_n^m(\theta,\varphi). \qquad (2.1.65)
$$

The scalar product of F and the spherical harmonic $Y_{n'}^{m'}(\theta, \varphi)$ is

$$
(F, Y_{n'}^{m'}) = \int_0^{\pi} \sin\theta\, d\theta \int_0^{2\pi} F(\theta,\varphi) Y_{n'}^{-m'}(\theta,\varphi) d\varphi. \qquad (2.1.66)
$$

At the same time using Eqs. (2.1.63) and (2.1.64) we have

$$
(F, Y_{n'}^{m'}) = \sum_{n=0}^{\infty}\sum_{m=-n}^{n} F_n^m (Y_n^m, Y_{n'}^{m'}) = \sum_{n=0}^{\infty}\sum_{m=-n}^{n} F_n^m \delta_{mm'}\delta_{nn'} = F_{n'}^{m'}. \qquad (2.1.67)
$$

Therefore, we have

$$
F_{n'}^{m'} = \int_0^{\pi} \sin\theta\, d\theta \int_0^{2\pi} F(\theta,\varphi) Y_{n'}^{-m'}(\theta,\varphi) d\varphi. \qquad (2.1.68)
$$

This determines the expansion coefficients of series (2.1.65). The expansion coefficients also satisfy the Parseval equality:

$$
\|F\|^2 = (F, F) = \sum_{n=0}^{\infty}\sum_{m=-n}^{n} |F_n^m|^2, \qquad (2.1.69)
$$

which can be proven by inserting the series (2.1.65) into the definition of the scalar product and using the orthonormality of the spherical harmonics.

Addition Theorem

One of the most important relations between the spherical harmonics and the Legendre polynomials is provided by the addition theorem. The proof of this theorem can be found elsewhere, and here we just state it as:

$$P_n(\cos\theta) = \frac{4\pi}{2n+1} \sum_{m=-n}^{n} Y_n^{-m}(\theta_1, \varphi_1) Y_n^m(\theta_2, \varphi_2), \qquad (2.1.70)$$

where θ is the angle between points with spherical coordinates (θ_1, φ_1) and (θ_2, φ_2) on the unit sphere (see Fig. 2.5). The addition theorem can also be rewritten in vector form. Indeed, if we denote as $\mathbf{s}$ a unit vector, $|\mathbf{s}| = 1$, and accept the following notation:

$$Y_n^m(\mathbf{s}) = Y_n^m(\theta, \varphi), \qquad \mathbf{s} = (\sin\theta\cos\varphi, \sin\theta\sin\varphi, \cos\theta), \qquad (2.1.71)$$

then relation (2.1.70) can be rewritten in the form

$$P_n(\mathbf{s}_1 \cdot \mathbf{s}_2) = \frac{4\pi}{2n+1} \sum_{m=-n}^{n} Y_n^{-m}(\mathbf{s}_1) Y_n^m(\mathbf{s}_2)$$

$$= \frac{4\pi}{2n+1} \sum_{m=-n}^{n} Y_n^m(\mathbf{s}_1) Y_n^{-m}(\mathbf{s}_2). \qquad (2.1.72)$$

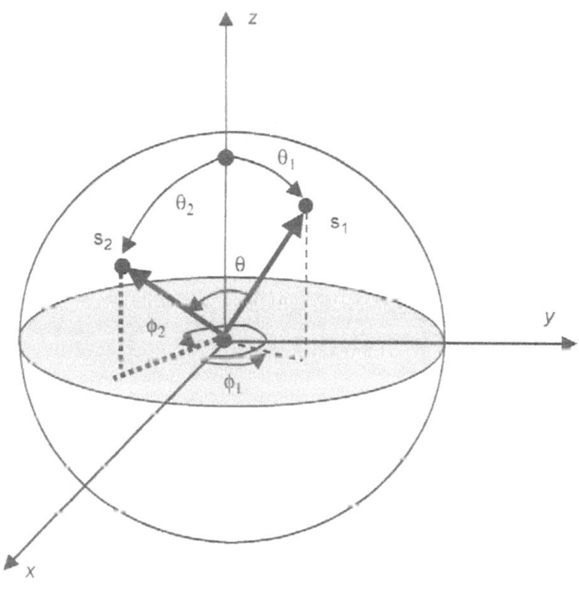

FIGURE 2.5 Illustration to the addition theorem for spherical harmonics.

Indeed, the cosine of the angle between two unit vectors $\mathbf{s}_1$ and $\mathbf{s}_2$ is nothing but their scalar product.

Funk–Hecke Theorem

The addition theorem and orthogonality of the Legendre polynomials and the spherical harmonics in the respective functional spaces yields the following theorem, which can be very helpful for representation of functions in different bases.

THEOREM 3 *Let* $f(\mu) \in C^0[-1,1]$ *and* $\mathbf{s}_1$ *and* $\mathbf{s}_2$ *be two unit vectors. Then*

$$\int_{S_u} f(\mathbf{s}_1\cdot\mathbf{s}_2)Y_n^m(\mathbf{s}_1)dS(\mathbf{s}_1) = \lambda_n Y_n^m(\mathbf{s}_2), \quad \lambda_n = 2\pi\int_{-1}^{1} f(\mu)P_n(\mu)d\mu, \tag{2.1.73}$$

where the first integral is over the surface of the unit sphere traced out by the varying $\mathbf{s}_1$.

PROOF. Because $\{P_n(\mu)\}$ forms a complete orthogonal basis on $L^2[-1,1]$, any continuous function $f(\mu) \in C^0[-1,1] \subset L^2[-1,1]$ can be expanded in a series (see relations (2.1.39) and (2.1.40)):

$$f(\mu) = \sum_{n=0}^{\infty} c_n P_n(\mu), \quad c_n = \frac{2n+1}{2}\int_{-1}^{1} f(\mu)P_n(\mu)d\mu = \frac{2n+1}{4\pi}\lambda_n. \tag{2.1.74}$$

Then, using the addition theorem (2.1.72) and orthonormality of the spherical harmonics, we have

$$\begin{aligned}
\int_{S_u} f(\mathbf{s}_1\cdot\mathbf{s}_2)Y_n^m(\mathbf{s}_1)dS_1 &= \int_{S_u} Y_n^m(\mathbf{s}_1)\sum_{n'=0}^{\infty} c_{n'}P_{n'}(\mathbf{s}_1\cdot\mathbf{s}_2)dS_1 \\
&= \int_{S_u} Y_n^m(\mathbf{s}_1)\sum_{n'=0}^{\infty} c_{n'}\frac{4\pi}{2n'+1}\sum_{m'=-n'}^{n'} Y_{n'}^{-m'}(\mathbf{s}_1)Y_{n'}^{m'}(\mathbf{s}_2)dS_1 \\
&= \sum_{n'=0}^{\infty}\sum_{m'=-n'}^{n'} c_{n'}\frac{4\pi}{2n'+1}Y_l^{m'}(\mathbf{s}_2)\int_{S_u} Y_n^m(\mathbf{s}_1)Y_{n'}^{-m'}(\mathbf{s}_1)dS_1 \\
&= \sum_{n'=0}^{\infty}\sum_{m'=-n'}^{n'} c_{n'}\frac{4\pi}{2n'+1}Y_{n'}^{m'}(\mathbf{s}_2)\delta_{mm'}\delta_{nn'} \\
&= c_n\frac{4\pi}{2n+1}Y_n^m(\mathbf{s}_2) = \lambda_n Y_n^m(\mathbf{s}_2).
\end{aligned}$$

The continuity of the function $f(\mu)$ on $[-1,1]$ provides uniform convergence for the above series and allows interchange of the integration and summation signs. □

2.1.2.3 *Spherical Bessel and Hankel functions*

As mentioned in connection with Rayleigh's formulas (2.1.29) and (2.1.30) the spherical Bessel functions can be expressed via elementary transcendental functions. Due to the symmetry relation between the functions of the first and the second kind,

$$y_n(\rho) = (-1)^{n+1} j_{-n-1}(\rho), \qquad n = 0, \pm 1, \pm 2, \ldots \tag{2.1.75}$$

functions of negative order n are linearly dependent on functions of non-negative order n. So it is sufficient to consider the latter case. Explicit expressions here are

$$j_n(\rho) = \rho^{-1}\left[P(n+\tfrac{1}{2},\rho)\sin(\rho - \tfrac{1}{2}\pi n) + Q(n+\tfrac{1}{2},\rho)\cos(\rho - \tfrac{1}{2}\pi n)\right], \tag{2.1.76}$$

$$\begin{aligned} y_n(\rho) = (-1)^{n+1}\rho^{-1}\Big[&P(n+\tfrac{1}{2},\rho)\cos(\rho + \tfrac{1}{2}\pi n) \\ &- Q(n+\tfrac{1}{2},\rho)\sin(\rho + \tfrac{1}{2}\pi n)\Big], \end{aligned} \tag{2.1.77}$$

where $P(n+\frac{1}{2},\rho)$ and $Q(n+\frac{1}{2},\rho)$ are polynomials in ρ^{-1}:

$$P(n+\tfrac{1}{2},\rho) = \sum_{l=0}^{[n/2]} (-1)^l \frac{(n+2l)!}{l!(n-2l)!}(2\rho)^{-2l}, \tag{2.1.78}$$

$$Q(n+\tfrac{1}{2},\rho) = \sum_{l=0}^{[(n-1)/2]} (-1)^l \frac{(n+2l+1)!}{l!(n-2l-1)!}(2\rho)^{-2l-1}. \tag{2.1.79}$$

In particular, we have the following expressions for the first three functions:

$$\begin{aligned} j_0(\rho) &= \frac{\sin\rho}{\rho}, \qquad j_1(\rho) = \frac{\sin\rho}{\rho^2} - \frac{\cos\rho}{\rho}, \\ j_2(\rho) &= \left(\frac{3}{\rho^3} - \frac{1}{\rho}\right)\sin\rho - \frac{3}{\rho^2}\cos\rho, \end{aligned} \tag{2.1.80}$$

$$y_0(\rho) = -\frac{\cos\rho}{\rho}, \qquad y_1(\rho) = -\frac{\cos\rho}{\rho^2} - \frac{\sin\rho}{\rho},$$
$$y_2(\rho) = \left(-\frac{3}{\rho^3} + \frac{1}{\rho}\right)\cos\rho - \frac{3}{\rho^2}\sin\rho. \tag{2.1.81}$$

These functions are plotted for real positive values of ρ in Fig. 2.6. It is seen that both the spherical Bessel functions oscillate and decay as $\rho \to \infty$. Functions of the first kind, $j_n(\rho)$, are bounded for any $\rho \geq 0$, while functions of the second kind, $y_n(\rho)$, are singular at $\rho = 0$.

The above equations provide explicit expressions for the spherical Hankel functions introduced by relations (2.1.27) and (2.1.28) via elementary functions

$$h_n^{(1)}(\rho) = \mathrm{i}^{-n-1}\frac{e^{\mathrm{i}\rho}}{\rho}\sum_{l=0}^{n}\frac{(n+l)!}{l!(n-l)!}\left(-\frac{1}{2\mathrm{i}\rho}\right)^l, \tag{2.1.82}$$

$$h_n^{(2)}(\rho) = \mathrm{i}^{n+1}\frac{e^{-\mathrm{i}\rho}}{\rho}\sum_{l=0}^{n}\frac{(n+l)!}{l!(n-l)!}\left(\frac{1}{2\mathrm{i}\rho}\right)^l. \tag{2.1.83}$$

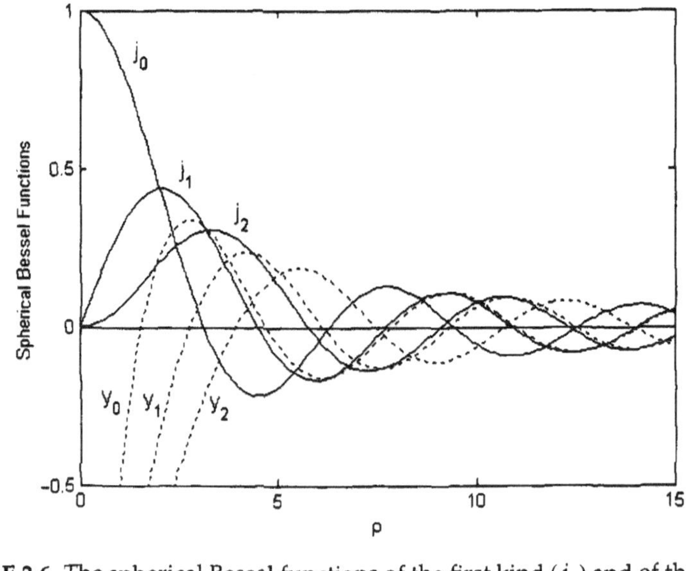

FIGURE 2.6 The spherical Bessel functions of the first kind (j_n) and of the second kind (y_n) of real argument ρ.

From these formulas the first three Hankel functions are

$$h_0^{(1)}(\rho) = \frac{e^{i\rho}}{i\rho}, \qquad h_1^{(1)}(\rho) = -\frac{e^{i\rho}}{\rho}\left(1 - \frac{1}{i\rho}\right),$$
$$h_2^{(1)}(\rho) = \frac{ie^{i\rho}}{\rho}\left(1 - \frac{3}{i\rho} + \frac{3}{\rho}\right), \tag{2.1.84}$$

$$h_0^{(2)}(\rho) = -\frac{e^{-i\rho}}{i\rho}, \qquad h_1^{(2)}(\rho) = -\frac{e^{-i\rho}}{\rho}\left(1 + \frac{1}{i\rho}\right),$$
$$h_2^{(2)}(\rho) = -\frac{ie^{-i\rho}}{\rho}\left(1 + \frac{3}{i\rho} + \frac{3}{\rho}\right). \tag{2.1.85}$$

The spherical Bessel and Hankel functions satisfy many useful relations, some of which will be introduced and discussed later as needed. Here we only mention the following recurrence relations, which are common to all the spherical Bessel and Hankel functions. Accordingly in the following formulas we denote the function by f_n, which is one of $j_n(\rho)$, $y_n(\rho)$, $h_n^{(1)}(\rho)$, or $h_n^{(2)}(\rho)$:

$$f_{n-1} + f_{n+1} = \frac{2n+1}{\rho} f_n, \tag{2.1.86}$$

$$nf_{n-1} - (n+1)f_{n+1} = (2n+1)f'_n, \tag{2.1.87}$$

$$\frac{n+1}{\rho} f_n + f'_n = f_{n-1}, \tag{2.1.88}$$

$$\frac{n}{\rho} f_n - f'_n = f_{n+1}. \tag{2.1.89}$$

Here the prime denotes derivative with respect to the argument.

2.1.3 Spherical basis functions

The above expressions for the spherical Bessel and Hankel functions are valid for arbitrary complex $\rho = kr$, which is the case of arbitrary complex wavenumber k in the Helmholtz equation. Note that the Helmholtz equation is for parameter k^2. This means that if solution is obtained for some k, the same function, where k is replaced by $-k$ also provides a solution of the Helmholtz equation. However, due to symmetry of the spherical Bessel functions with respect to change in the sign of the independent variable

$$j_n(-\rho) = (-1)^n j_n(\rho),$$
$$y_n(-\rho) = (-1)^{n+1} y_n(\rho), \tag{2.1.90}$$

functions of negative ρ can always be expressed via functions of positive ρ. Therefore, we can always consider the case, when the argument of the Bessel or Hankel functions belongs to the half space of the complex plane

$$\mathrm{Re}\{k\} \geq 0, \qquad |k| > 0. \tag{2.1.91}$$

Depending on the value k one should select an appropriate pair of linearly independent solutions of the spherical Bessel equation to satisfy certain conditions as $r \to 0$ and $r \to \infty$. The only solution regular at $\rho = 0$ is provided by the functions of the first kind,

$$\lim_{\rho \to 0} j_n(\rho) < \infty, \qquad n = 0, 1, 2, \ldots \tag{2.1.92}$$

Indeed any solution of order n can be represented as a sum $j_n(\rho)$ and $y_n(\rho)$. Since the latter function is singular as $\rho \to 0$, a superposition of these functions, $c_1 j_n(\rho) + c_2 y_n(\rho)$ will also be singular, unless c_2 is zero. Selection of the second solution is related to a problem considered. In any case the second solution will be singular at $\rho = 0$. Below we consider cases which depend on the argument of the complex wavenumber k.

2.1.3.1 *The case Im{k}=0*

This is the case of the Helmholtz equation obtained by the Fourier transform of the classical wave equation and wave scattering problems are usually the problems under consideration.

For such problems in infinite domains the scattered fields satisfy the Sommerfeld radiation condition (2.2.9). In terms of the radial function in the factored solution (2.1.7) this can be written as

$$\lim_{r \to \infty} r \left[\frac{d\Pi(r)}{dr} - ik\Pi(r) \right] = 0. \tag{2.1.93}$$

The only spherical function satisfying this condition is the spherical Hankel function of the first kind:

$$\lim_{|\rho| \to \infty} \rho \left[\frac{dh_n^{(1)}(\rho)}{d\rho} - ih_n^{(1)}(\rho) \right] = 0, \qquad n = 0, 1, 2, \ldots \tag{2.1.94}$$

Indeed, any linearly independent solution of the spherical Bessel equation is a linear combination of $h_n^{(1)}(\rho)$ and $h_n^{(2)}(\rho)$ and one can check that $h_n^{(2)}(\rho)$ does not satisfy the radiation condition.

We note that despite the fact that condition (2.1.93) holds for negative and positive k so ρ defined by relation (2.1.23) can be, respectively, negative or positive, we can treat k as a positive quantity, assuming that the frequencies $\omega > 0$ and that the phase velocity is positive $\omega/k > 0$. This is consistent with definition (2.1.91). Obviously, this definition contradicts the physics for $\omega < 0$, since that specifies incoming waves instead of outgoing waves. However, this situation can be handled easily, by convention that in all solutions with $k > 0$ containing $h_n^{(1)}$ these functions should be replaced with $h_n^{(2)}$. So if by some reasons we want to use the Fourier integral for transforms with negative frequencies, while using positive arguments for spherical Hankel functions, we need to replace $h_n^{(1)}$ with $h_n^{(2)}$. Another way of tackling this problem is to always use $h_n^{(1)}(kr)$ independent of the sign of k with the understanding that negative k correspond to negative frequencies. In the following chapters we only focus on the case $k > 0$.

2.1.3.2 *The case* $Re\{k\} = 0$

This is the case of the Helmholtz equation describing the Yukawa potential fields, which appear, for example, in molecular dynamics simulations. For infinite domains the solutions should decay as $r \to \infty$. Representing k as

$$k = \mathrm{i}|k|, \tag{2.1.95}$$

(so that the second root is $-\mathrm{i}|k|$) we can see from (2.1.82) and (2.1.83) that for large r

$$h_n^{(1)}(kr) \sim -\mathrm{i}^{-n}\frac{\mathrm{e}^{-|k|r}}{|k|r}, \qquad h_n^{(2)}(kr) \sim \mathrm{i}^{n}\frac{\mathrm{e}^{|k|r}}{|k|r}. \tag{2.1.96}$$

The first function here decays exponentially, while the second one grows exponentially. So, again $h_n^{(1)}(kr)$ should be selected as the basis, assuming that $\mathrm{Im}\{k\} > 0$.

Note that the spherical Bessel functions of purely imaginary argument can be expressed via the *modified spherical Bessel functions* of real argument:

$$j_n(\mathrm{i}|k|r) = \mathrm{i}\sqrt{\frac{\pi}{2|k|r}}I_{n+1/2}(|k|r), \qquad n = 0, 1, 2, \ldots, \tag{2.1.97}$$

$$h_n^{(1)}(\mathrm{i}|k|r) = \mathrm{i}^{-1}\sqrt{\frac{2}{\pi|k|r}}K_{n+1/2}(|k|r), \qquad n = 0, 1, 2, \ldots, \tag{2.1.98}$$

where $I_{n+1/2}$ and $K_{n+1/2}$ are *modified Bessel functions of the first and the second kind*. For real arguments these functions do not oscillate while exponentially growing ($I_{n+1/2}$) or decaying ($K_{n+1/2}$) as $r \to \infty$. The modified Bessel functions of semi-integer order can also be expressed via elementary transcendental functions and satisfy certain recurrence relations. Such representations follow from the relations for the spherical Bessel and Hankel functions and we will not reproduce them, redirecting the interested reader to handbooks on special functions. We also recall that as soon as we are working with complex spherical harmonics and complex-valued expansions the case of complex k can be treated in this general framework.

2.1.3.3 *The case Im{k} > 0, Re{k} > 0*

This case can appear as a result of the Fourier transform of the heat conduction equation or in wave propagation in dissipative media. In this case k can be represented as

$$k = k_r + \mathrm{i}k_i, \qquad k_i > 0, \qquad k_r > 0. \tag{2.1.99}$$

As before we note that this corresponds to positive frequencies in the transform from the time to the frequency domain. The real part of k is related to the velocity of wave propagation, while the imaginary part is responsible for wave attenuation. Note that by convention $k_r > 0$ (outgoing wave) and so the second root, $-k$, has both negative real and imaginary parts. The asymptotics of the Hankel functions for large $|kr|$ are then

$$h_n^{(1)}(kr) \sim \frac{\mathrm{i}^{-n-1}}{kr}\mathrm{e}^{\mathrm{i}k_r r}\,\mathrm{e}^{-k_i r}, \qquad h_n^{(2)}(kr) \sim \frac{\mathrm{i}^{n+1}}{kr}\mathrm{e}^{-\mathrm{i}k_r r}\,\mathrm{e}^{k_i r}. \tag{2.1.100}$$

The requirement that the solution decays at infinity leaves $h_n^{(1)}(kr)$ as the only candidate as basis function.

2.1.3.4 *The case Im{k} < 0, Re{k} > 0*

While this case also can be considered from a formal mathematical viewpoint, as we mentioned in the introduction, physically this means that the outgoing wave ($\mathrm{Re}\{k\} > 0$) should exponentially grow while propagating through the media. Such situations are rather rare, but not impossible (e.g. some specially designed media can release energy and add the energy to the wave, which is opposite to the usual energy consumption or dissipation). In this case an appropriate condition at infinity is that the solution should *grow*, not to decay (!). Therefore, from asymptotics (2.1.100) which are valid for arbitrary complex k we should again select $h_n^{(1)}(kr)$ as the proper functions, keeping in mind that we are working in this case with exponentially growing solutions.

In this context and in the case when the Helmholtz equation is obtained as a Fourier transform of a time dependent equation the root $\mathrm{Re}\{k\} \geq 0$ can be treated as the root corresponding to the process evolving in the positive direction of time, while the other root as that corresponding to the *time reversal*.

2.1.3.5 Basis functions

The above consideration shows that the following functions, which we call the *spherical basis functions*, can be introduced as a basis to represent solutions of the Helmholtz equation

$$R_n^m(\mathbf{r}) = j_n(kr)Y_n^m(\theta, \varphi), \qquad n = 0, 1, 2, \ldots, \qquad m = -n, \ldots, n, \qquad (2.1.101)$$

$$S_n^m(\mathbf{r}) = h_n^{(1)}(kr)Y_n^m(\theta, \varphi), \qquad n = 0, 1, 2, \ldots, \qquad m = -n, \ldots, n. \qquad (2.1.102)$$

We selected the notation R and S to keep in mind that R stands for "regular" solution and S for "singular" solution at $r = 0$. In analogy with the spherical harmonics we will call the subscript n, "degree" and the superscript m, "order" of the basis function. Also we can call the functions with $m = 0$ as "zonal", functions with $n = |m|$ as "sectorial", and intermediate functions for $0 < |m| < n$ as "tesseral" functions. Since mostly the functions $h_n^{(1)}(kr)$ will be considered below, we will drop the superscript (1) for notational simplicity. In cases where both $h_n^{(1)}(kr)$ and $h_n^{(2)}(kr)$ are involved we will return to the original indexing.

Because the functions $h_n(kr)$ and $j_n(kr)$ have similar recurrence properties, and when an expression applies to both types of basis functions, we will use the notation

$$F_n^m(\mathbf{r}) = f_n(kr)Y_n^m(\theta, \varphi), \qquad f = h, j, \qquad F = S, R. \qquad (2.1.103)$$

to denote any of $S_n^m(\mathbf{r})$ or $R_n^m(\mathbf{r})$ and $h_n(kr)$ or $j_n(kr)$.

Figures 7–9 illustrate some features of the spherical basis functions. Figures 7 and 8 plot isosurfaces of

$$\mathrm{Re}\{R_n^m(\mathbf{r})\} = \mathrm{const}, \qquad \mathrm{Re}\{S_n^m(\mathbf{r})\} = \mathrm{const} \qquad (2.1.104)$$

for real k. Of course the shape of these isosurfaces depends on the value chosen for "const". However, the purpose of plotting here is to just give a general feeling about the spatial behavior of these functions and so we selected to plot for each function two surfaces corresponding to two values: a positive constant (red surfaces) and a negative constant of the same absolute value (blue surfaces).

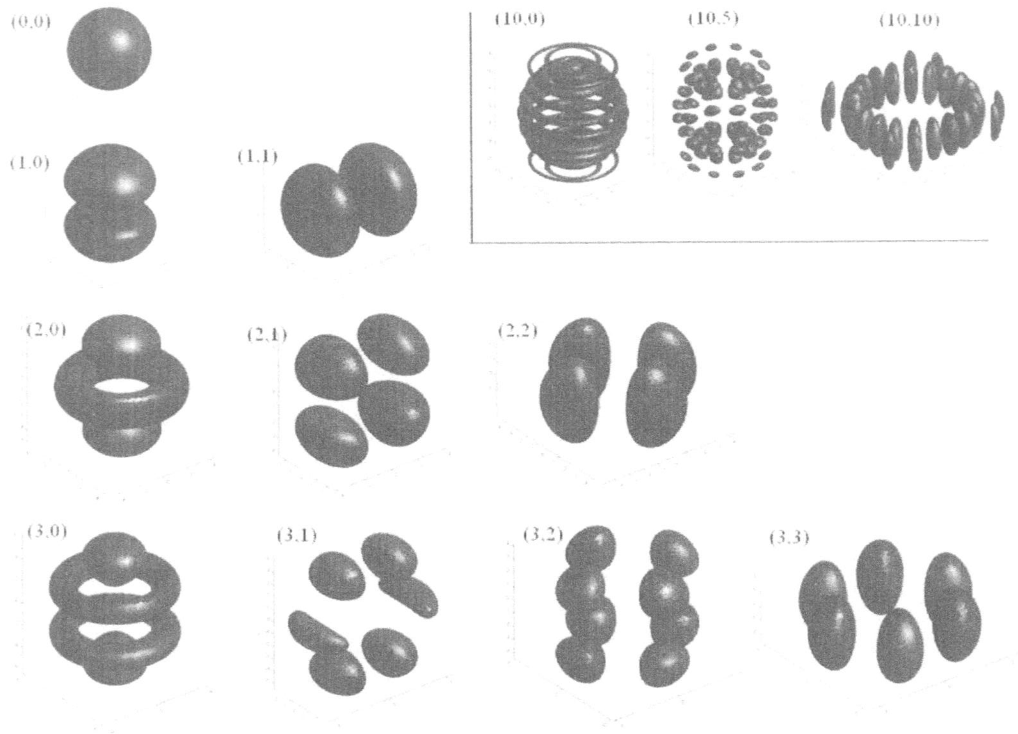

FIGURE 2.7 Isosurfaces of the spherical basis functions $\mathrm{Re}\{R_n^m(\mathbf{r})\} = \text{const}$ for various degrees n and orders m indicated in parenthesis as (n, m) near each mode. For color version, see Plate B after Chapter 10.

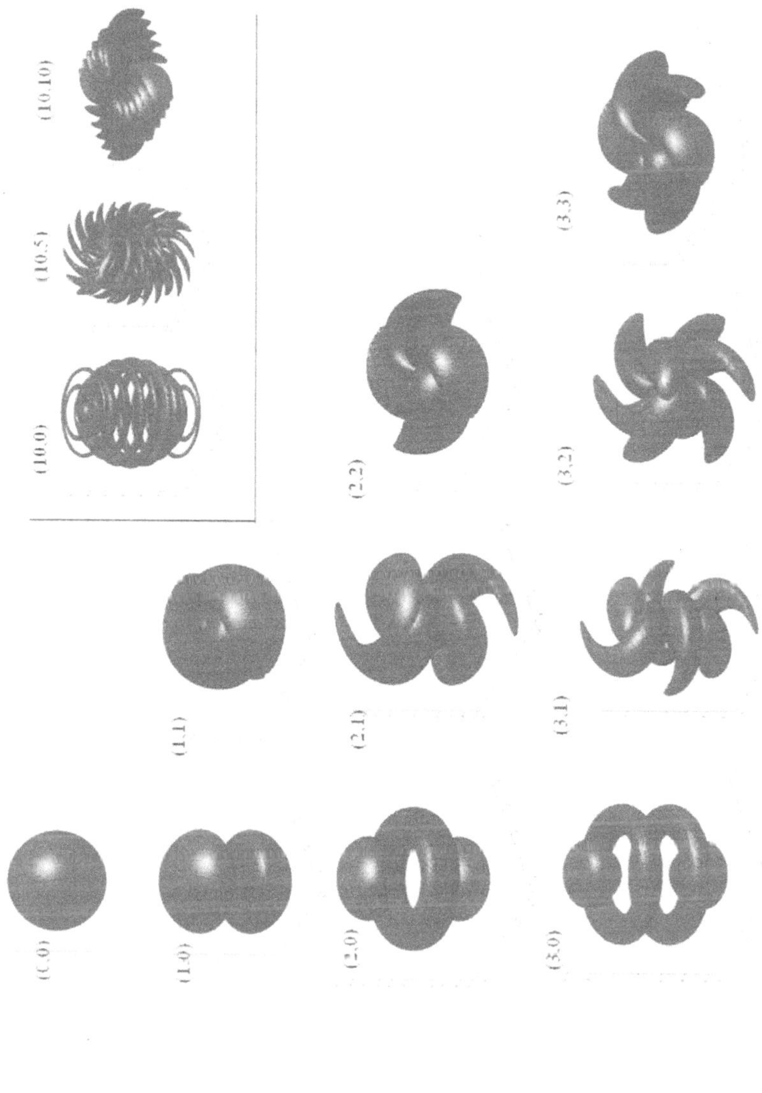

FIGURE 2.8 Isosurfaces of spherical basis functions $\mathrm{Re}\{S_n^m(\mathbf{r})\} = \mathrm{const}$ for various degrees n and orders m indicated in parenthesis as (n, m) near each mode. For color version, see Plate C after Chapter 10.

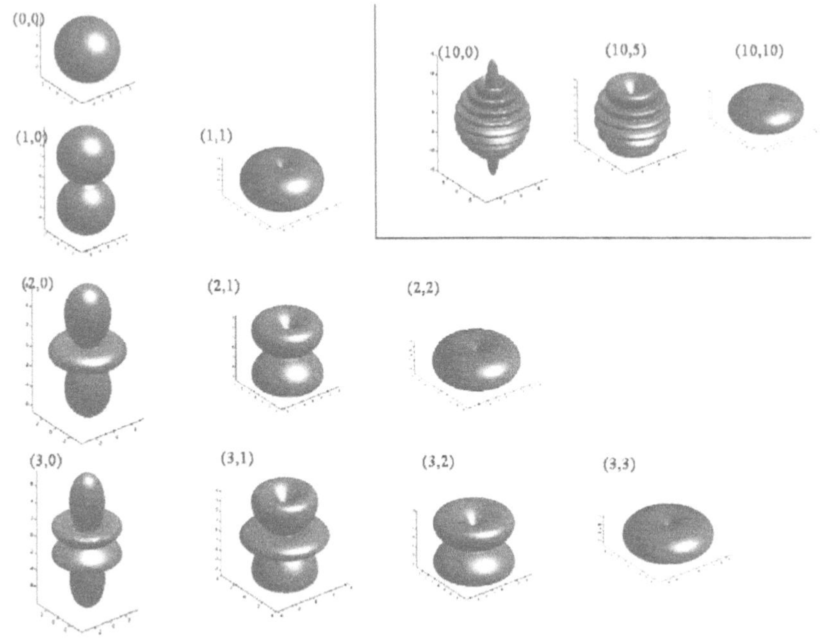

FIGURE 2.9 Isosurfaces of the spherical basis functions $|S_n^m(\mathbf{r})| = \text{const}$ for various degrees n and orders m indicated in parenthesis as (n, m) near each mode.

The zonal functions $R_n^0(\mathbf{r})$ and $S_n^0(\mathbf{r})$ do not depend on the polar angle φ. That is why the isosurfaces for these functions are axisymmetric (bodies of revolution). Moreover, $R_0^0(\mathbf{r})$ and $S_0^0(\mathbf{r})$ do not depend on θ and represent spherically symmetrical functions, so that the isosurfaces for these functions are spheres. Also, the spherical harmonics $Y_n^0(\theta, \varphi)$ are real, and that is the reason why $\text{Re}\{R_n^0(\mathbf{r})\} = \text{Re}\{S_n^0(\mathbf{r})\}$ and the isosurfaces of real parts of regular and singular zonal functions are the same. It is seen that the isosurfaces for zonal functions are multiconnected domains consisting of toroidal and spheroidal objects. The number of such objects increases with n as the spatial frequency of the spherical harmonics increases.

The isosurfaces of the real parts of the tesseral and sectorial regular and singular functions are qualitatively different. If, for the regular basis functions they look like surfaces of multiconnected kidney-type bodies, then for singular basis functions the isosurfaces are sets of vortex-type 3D bodies. Again it is clear that the spatial frequency of these structures is determined by the degree n. The vortex-type isosurfaces for $\text{Re}\{S_n^m(\mathbf{r})\}$ are caused by interaction of the complex-valued Hankel functions and complex exponents of angle φ.

The modulus $|S_n^m(\mathbf{r})|$ does not depend on φ and so this is an axisymmetric function, which isosurfaces, $|S_n^m(\mathbf{r})| = \text{const}$ are some bodies of revolution (see Fig. 2.9).

The spherical basis functions of zero degree obey the spherical symmetry (independence on the angular variables θ and φ)

$$R_0^0(\mathbf{r}) = j_0(kr)Y_0^0(\theta,\varphi) = \sqrt{\frac{1}{4\pi}}\frac{\sin kr}{kr}, \tag{2.1.105}$$

$$S_0^0(\mathbf{r}) = h_0(kr)Y_0^0(\theta,\varphi) = \sqrt{\frac{1}{4\pi}}\frac{e^{ikr}}{ikr}. \tag{2.1.106}$$

The latter basis function is related to the fundamental solution of the Helmholtz equation, or Green's function $G(\mathbf{r})$:

$$G(\mathbf{r}) = \frac{e^{ik|\mathbf{r}|}}{4\pi|\mathbf{r}|} = ik\sqrt{\frac{1}{4\pi}}S_0^0(\mathbf{r}), \tag{2.1.107}$$

Note that the imaginary part of the Green function is a regular one:

$$G_i(\mathbf{r}) = \mathrm{Im}\{G(\mathbf{r})\} = \frac{\sin k|\mathbf{r}|}{4\pi|\mathbf{r}|} = k\sqrt{\frac{1}{4\pi}}R_0^0(\mathbf{r}). \tag{2.1.108}$$

2.2 DIFFERENTIATION OF ELEMENTARY SOLUTIONS

A theory of translation and rotation operators for the Helmholtz equation presented in this book is based on the differential properties of elementary solutions. These properties serve as basis for recurrence relations, fast transforms, sparse matrix decomposition, and other computationally efficient methods, which can be employed for solution of the Helmholtz equation. Here we provide several differentiation theorems for the elementary solutions in spherical coordinates, which presumably were first reported by Chew [Che92] and derived independently in the paper of the authors [GD01], who at the time of their work were not aware of Chew's results. In this section we also introduce the concepts of multipoles and plane waves and show a relation between the spherical basis functions and that solutions.

2.2.1 Differentiation theorems

Let us define the following differential operators that can be expressed in spherical coordinates (2.1.1) as:

$$\partial_z \equiv \frac{1}{k}\frac{\partial}{\partial z} = \frac{1}{k}\left(\mu\frac{\partial}{\partial r} + \frac{1-\mu^2}{r}\frac{\partial}{\partial \mu}\right), \qquad \mu = \cos\theta. \tag{2.2.1}$$

$$\begin{aligned}\partial_{x+iy} &\equiv \frac{1}{k}\left(\frac{\partial}{\partial x} + \mathrm{i}\frac{\partial}{\partial y}\right)\\ &= \frac{e^{\mathrm{i}\varphi}}{kr\sqrt{1-\mu^2}}\left[(1-\mu^2)\left(r\frac{\partial}{\partial r} - \mu\frac{\partial}{\partial \mu}\right) + \mathrm{i}\frac{\partial}{\partial \varphi}\right],\end{aligned} \tag{2.2.2}$$

$$\begin{aligned}\partial_{x-iy} &\equiv \frac{1}{k}\left(\frac{\partial}{\partial x} - \mathrm{i}\frac{\partial}{\partial y}\right)\\ &= \frac{e^{-\mathrm{i}\varphi}}{kr\sqrt{1-\mu^2}}\left[(1-\mu^2)\left(r\frac{\partial}{\partial r} - \mu\frac{\partial}{\partial \mu}\right) - \mathrm{i}\frac{\partial}{\partial \varphi}\right],\end{aligned} \tag{2.2.3}$$

$$\begin{aligned}\frac{1}{k}\nabla &\equiv \frac{1}{k}\left(\mathbf{i}_x\frac{\partial}{\partial x} + \mathbf{i}_y\frac{\partial}{\partial y} + \mathbf{i}_z\frac{\partial}{\partial z}\right)\\ &= \mathbf{i}_x\frac{1}{2}(\partial_{x+iy} + \partial_{x-iy}) - \mathbf{i}_y\frac{\mathrm{i}}{2}(\partial_{x+iy} - \partial_{x-iy}) + \mathbf{i}_z\partial_z\\ &= \frac{1}{2}(\mathbf{i}_x - \mathrm{i}\mathbf{i}_y)\partial_{x+iy} + \frac{1}{2}(\mathbf{i}_x + \mathrm{i}\mathbf{i}_y)\partial_{x-iy} + \mathbf{i}_z\partial_z.\end{aligned} \tag{2.2.4}$$

We also introduce the following notation for the normalization factor of spherical harmonics:

$$\begin{aligned}N_n^m &= N_n^{|m|} = (-1)^m\sqrt{\frac{2n+1}{4\pi}\frac{(n-|m|)!}{(n+|m|)!}},\\ n &= 0, 1, \ldots, \qquad m = -n, \ldots, n.\end{aligned} \tag{2.2.5}$$

Relations (2.1.59) and (2.1.103) then yield

$$F_n^m(\mathbf{r}) = N_n^m f_n(kr) P_n^{|m|}(\mu) e^{\mathrm{i}m\varphi}, \qquad F = S, R, \qquad f = j, h. \tag{2.2.6}$$

Applying the differential operators to the basis functions $S_n^m(\mathbf{r})$ or $R_n^m(\mathbf{r})$ written in this form we obtain the differentiation theorems that are listed below.

THEOREM 4 *For $k \neq 0$ and integer n and m*

$$\partial_z F_n^m(\mathbf{r}) = a_{n-1}^m F_{n-1}^m(\mathbf{r}) - a_n^m F_{n+1}^m(\mathbf{r}), \qquad F = S, R. \tag{2.2.7}$$

where

$$a_n^m = a_n^{-m} = \sqrt{\frac{(n+1+m)(n+1-m)}{(2n+1)(2n+3)}}, \quad \text{for } n \geq |m|.$$
$$a_n^m = 0, \quad \text{for } n < |m|. \tag{2.2.8}$$

PROOF. Using the recurrence relations (2.1.52) and (2.1.53) for the associated Legendre functions and relations (2.1.88) and (2.1.89) for the spherical Bessel or Hankel functions we obtain from definitions (2.2.1) and (2.2.6):

$$\partial_z F_n^m(\mathbf{r}) = N_n^m e^{im\varphi}\left[\mu P_n^{|m|}(\mu) f_n'(kr) + \frac{f_n(kr)}{kr}(1-\mu^2)\frac{dP_n^{|m|}(\mu)}{d\mu}\right]$$

$$= \frac{N_n^m e^{im\varphi}}{2n+1}\left\{(n+|m|)P_{n-1}^{|m|}(\mu)\left[f_n'(kr) + \frac{n+1}{kr}f_n(kr)\right]\right.$$

$$\left. - (n-|m|+1)P_{n+1}^{|m|}(\mu)\left[\frac{n}{kr}f_n(kr) - f_n'(kr)\right]\right\}$$

$$= \frac{N_n^m e^{im\varphi}}{2n+1}\{(n+|m|)f_{n-1}(kr)P_{n-1}^{|m|}(\mu) - (n-|m|+1)P_{n+1}^{|m|}(\mu)f_{n+1}(kr)\}.$$

$$= \begin{cases} -\dfrac{N_0^0}{N_1^0} S_1^0(\mathbf{r}), & n = 0 \\[2ex] \dfrac{N_n^m}{2n+1}\left[\dfrac{n+|m|}{N_{n-1}^m}F_{n-1}^m(\mathbf{r}) - \dfrac{n-|m|+1}{N_{n+1}^m}F_{n+1}^m(\mathbf{r})\right], & n \geq 1 \end{cases}$$

The coefficients a_n^m can be defined as stated, because of (2.1.45) for $|m| > n$

and since for $n \geq 1$ we have according to definitions (2.2.5) and (2.2.8):

$$\frac{N_n^m}{N_{n-1}^m}\frac{n+|m|}{2n+1} = \sqrt{\frac{(n+|m|)(n-|m|)}{(2n-1)(2n+1)}} = a_{n-1}^m,$$

$$\frac{N_n^m}{N_{n+1}^m}\frac{n-|m|+1}{2n+1} = \sqrt{\frac{(n+|m|+1)(n-|m|+1)}{(2n+1)(2n+3)}} = a_n^m.$$

□

THEOREM 5 *For $k \neq 0$ and integer m and n:*

$$\partial_{x+iy}F_n^m(\mathbf{r}) = b_{n+1}^{-m-1}F_{n+1}^{m+1}(\mathbf{r}) - b_n^m F_{n-1}^{m+1}(\mathbf{r}), \qquad F = S, R. \tag{2.2.9}$$

where

$$b_n^m = \sqrt{\frac{(n-m-1)(n-m)}{(2n-1)(2n+1)}} \quad \text{for } 0 \leq m \leq n,$$

$$b_n^m = -\sqrt{\frac{(n-m-1)(n-m)}{(2n-1)(2n+1)}} \quad \text{for } -n \leq m < 0, \tag{2.2.10}$$

$$b_n^m = 0 \quad \text{for } |m| > n,$$

PROOF. Applying ∂_{x+iy} to $F_n^m(\mathbf{r})$, we obtain

$$\partial_{x+iy}F_n^m(\mathbf{r}) = e^{i(m+1)\varphi}N_n^{|m|}1-\mu^2$$
$$\times\left[P_n^{|m|}(\mu)f'_n(kr) - \left(\mu\frac{dP_n^{|m|}(\mu)}{d\mu} + \frac{mP_n^{|m|}(\mu)}{1-\mu^2}\right)\frac{f_n(kr)}{kr}\right].$$

For $m \geq 0$, using recurrence relations (2.1.54), (2.1.55), (2.1.88), and (2.1.89) we have

$$\partial_{x+iy}F_n^m(\mathbf{r}) = \frac{N_n^m e^{i(m+1)\varphi}}{2n+1}\left\{\left[P_{n-1}^{m+1}(\mu) - P_{n+1}^{m+1}(\mu)\right]f'_n(kr)\right.$$
$$\left. + \left[nP_{n+1}^{m+1}(\mu) + (n+1)P_{n-1}^{m+1}(\mu)\right]\frac{f_n(kr)}{kr}\right\}$$
$$= \frac{N_n^m e^{i(m+1)\varphi}}{2n+1}\left[f_{n-1}(kr)P_{n-1}^{m+1}(\mu) + f_{n+1}(kr)P_{n+1}^{m+1}(\mu)\right].$$

In case $n=m$ the first term in the last square brackets is zero. Thus, using relation (2.2.6) we obtain the statement of the theorem for all $m \geq 0$, since for $n \geq 1$ according to definitions (2.2.5) and (2.2.10) we have

$$\frac{1}{2n+1}\frac{N_n^m}{N_{n-1}^{m+1}} = \sqrt{\frac{(n-m-1)(n-m)}{(2n-1)(2n+1)}} = b_n^m,$$
$$\frac{1}{2n+1}\frac{N_n^m}{N_{n+1}^{m+1}} = \sqrt{\frac{(n+m+1)(n+m+2)}{(2n+1)(2n+3)}} = -b_{n+1}^{-m-1},$$

and for $n=0$ we have $b_n^m = 0$.

For $m \leq -1$, $(m = -|m|)$ using recurrence relations (2.1.56), (2.1.57), (2.1.88), and (2.1.89) we have

$$\begin{aligned}
\partial_{x+iy}F_n^m(\mathbf{r}) &= \frac{N_n^{|m|}\, e^{i(m+1)\varphi}}{2n+1} \\
&\times\Bigg\{\Big[(n-|m|+1)(n-|m|+2)P_{n+1}^{|m|-1}(\mu) \\
&- (n+|m|-1)(n+|m|)P_{n-1}^{|m|-1}(\mu)\Big]f_n'(kr) \\
&- \Big[(n+1)(n+|m|-1)(n+|m|)P_{n-1}^{|m|-1}(\mu) \\
&+ n(n-|m|+1)(n-|m|+2)P_{n+1}^{|m|-1}(\mu)\Big]\frac{f_n(kr)}{kr}\Bigg\} \\
&= -\frac{N_n^m\, e^{i(m+1)\varphi}}{2n+1}\Big[(n+|m|-1)(n+|m|)f_{n-1}(kr)P_{n-1}^{|m|-1}(\mu) \\
&+ (n-|m|+1)(n-|m|+2)f_{n+1}(kr)P_{n+1}^{|m|-1}(\mu)\Big].
\end{aligned}$$

Again using the definitions of the coefficients N_n^m and b_n^m we can see that the statement of the theorem holds in this case as well. □

THEOREM 6 *For $k \neq 0$ and integer n and m*

$$\partial_{x-iy}F_n^m(\mathbf{r}) = b_{n+1}^{m-1}F_{n+1}^{m-1}(\mathbf{r}) - b_n^{-m}F_{n-1}^{m-1}(\mathbf{r}), \qquad F = S, R. \tag{2.2.11}$$

where the coefficients b_n^m are as defined by relation (2.2.10).

PROOF. Note that $j_n(kr)$ is a real function for real k and due to the symmetry property of spherical harmonics (2.1.60) we have $\overline{R_n^m(\mathbf{r})} = R_n^{-m}(\mathbf{r})$. Thus, we have from the previous theorem

$$\partial_{x-\mathrm{i}y} R_n^m(\mathbf{r}) = \overline{\partial_{x+\mathrm{i}y} R_n^{-m}(\mathbf{r})} = b_n^{-m}\overline{R_{n-1}^{-m+1}(\mathbf{r})} - b_{n+1}^{m-1}\overline{R_{n+1}^{-m+1}(\mathbf{r})}$$

$$= b_n^{-m} R_{n-1}^{m-1}(\mathbf{r}) - b_{n+1}^{m-1} R_{n+1}^{m-1}(\mathbf{r}),$$

which proves the theorem for $R_n^m(\mathbf{r})$ and real k. For $S_n^m(\mathbf{r})$ and real k the same relation holds since $h_n(kr)$ satisfies the same recurrence relations as $j_n(kr)$ In case of complex k the theorem holds again due to the Hankel functions $h_n(kr)$ are analytical and can be analytically continued. □

THEOREM 7 *For $k \neq 0$ and integer n and m*

$$\frac{1}{k}\nabla F_n^m(\mathbf{r}) = \frac{1}{2}(\mathbf{i}_x - \mathrm{i}\mathbf{i}_y)[b_{n+1}^{-m-1} F_{n+1}^{m+1}(\mathbf{r}) - b_n^m F_{n-1}^{m+1}(\mathbf{r})]$$
$$+ \frac{1}{2}(\mathbf{i}_x + \mathrm{i}\mathbf{i}_y)[b_{n+1}^{m-1} F_{n+1}^{m-1}(\mathbf{r}) - b_n^{-m} F_{n-1}^{m-1}(\mathbf{r})] \tag{2.2.12}$$
$$+ \mathbf{i}_z[a_{n-1}^m F_{n-1}^m(\mathbf{r}) - a_n^m F_{n+1}^m(\mathbf{r})], \qquad F = S, R.$$

where the coefficients a_n^m and b_n^m are defined by (2.2.8) and (2.2.10).

PROOF. Follows from the previous three differentiation theorems. □

THEOREM 8 *For any integer n and m the following relations between the coefficients a_n^m and b_n^m defined by (2.2.8) and (2.2.10) holds:*

$$a_{n-1}^m a_n^m + b_n^{-m-1} b_{n+1}^m = 0, \tag{2.2.13}$$

$$(a_{n-1}^m)^2 + (a_n^m)^2 + (b_{n+1}^{-m-1})^2 + (b_n^m)^2 = 1. \tag{2.2.14}$$

PROOF. While these relations can be checked directly using the definitions (2.2.8) and (2.2.10), we note that they also follow from the fact that any elementary solution, $F_n^m(\mathbf{r})$, satisfies the Helmholtz equation

$$(\partial_z^2 + \partial_{x+\mathrm{i}y}\partial_{x-\mathrm{i}y} + 1)F_n^m(\mathbf{r}) = 0. \tag{2.2.15}$$

Applying the differentiation theorems we can find that this relation is valid only if the relations to be proved hold. □

2.2.2 Multipole solutions

The above differentiation theorems show that derivatives of the elementary solutions in spherical coordinates can be simply expressed via elementary solutions of different degree and order. This provides an idea for use of other basis functions, known in physics as "multipoles".[1]

We define multipoles as derivatives of the Green function (2.1.107):

$$M_{n_1 n_2 n_3}(\mathbf{r}) = \frac{\partial^{n_1+n_2+n_3}}{\partial x^{n_1}\,\partial y^{n_2}\,\partial z^{n_3}} G(\mathbf{r}), \qquad n_1 \geq 0,\ n_2 \geq 0,\ n_3 \geq 0. \quad (2.2.16)$$

In the domain $r > 0$, the function $G(\mathbf{r})$ satisfies the Helmholtz equation, and so the multipoles satisfy the same equations, since the operators $\partial/\partial x$, $\partial/\partial y$, $\partial/\partial z$ commute with the Laplacian. Furthermore, since

$$\frac{\partial^2}{\partial x^2} M_{n_1 n_2 n_3}(\mathbf{r}) + \frac{\partial^2}{\partial y^2} M_{n_1 n_2 n_3}(\mathbf{r}) + \frac{\partial^2}{\partial z^2} M_{n_1 n_2 n_3}(\mathbf{r}) + k^2 M_{n_1 n_2 n_3}(\mathbf{r}) = 0, \quad (2.2.17)$$

we have

$$M_{n_1+2, n_2 n_3}(\mathbf{r}) + M_{n_1 n_2+2, n_3}(\mathbf{r}) + M_{n_1 n_2 n_3+2}(\mathbf{r}) + k^2 M_{n_1 n_2 n_3}(\mathbf{r}) = 0. \quad (2.2.18)$$

This relation shows that the functions $M_{n_1 n_2 n_3}(\mathbf{r})$ are linearly dependent, and so, strictly speaking do not form a basis. The basis we seek should represent any function that is a solution of the Helmholtz equation i.e., it can be uniquely expanded over the basis. In the case of expansion over the set of multipoles, such an expansion is not unique, since any multipole can be represented as a sum of three other multipoles.

Using the differentiation theorems for spherical elementary solutions and expression (2.1.107) we can relate the multipoles and functions $S_n^m(\mathbf{r})$. The lower order multipoles are also referred to in the literature as "monopoles", "dipoles", "quadrupoles", "octupoles", and so on. We have from definitions the following expressions:

Monopole:

$$M_{000}(\mathbf{r}) = G(\mathbf{r}) = ik\sqrt{\frac{1}{4\pi}}\, S_0^0(\mathbf{r}). \quad (2.2.19)$$

[1] In some literature the singular spherical basis functions of the Helmholtz equation are called "multipoles".

Dipoles:

$$M_{100}(\mathbf{r}) = \frac{\partial}{\partial x}G(\mathbf{r}) = \frac{ik^2}{2\sqrt{4\pi}}(\partial_{x+iy} + \partial_{x-iy})S_0^0(\mathbf{r})$$

$$= \frac{ik^2 b_1^{-1}}{2\sqrt{4\pi}}\left[S_1^1(\mathbf{r}) + S_1^{-1}(\mathbf{r})\right], \tag{2.2.20}$$

$$M_{010}(\mathbf{r}) = \frac{\partial}{\partial y}G(\mathbf{r}) = \frac{k^2}{2\sqrt{4\pi}}(\partial_{x+iy} - \partial_{x-iy})S_0^0(\mathbf{r})$$

$$= \frac{k^2 b_1^{-1}}{2\sqrt{4\pi}}\left[S_1^1(\mathbf{r}) - S_1^{-1}(\mathbf{r})\right], \tag{2.2.21}$$

$$M_{001}(\mathbf{r}) = \frac{\partial}{\partial z}G(\mathbf{r}) = \frac{ik^2}{\sqrt{4\pi}}\partial_z S_0^0(\mathbf{r}) = -\frac{k^2 a_0^0}{\sqrt{4\pi}}S_1^0(\mathbf{r}). \tag{2.2.22}$$

Furthermore, we can see that multipoles of degree $l = n_1 + n_2 + n_3$ can be expressed as finite sums of elementary singular solutions in spherical coordinates:

$$M_{n_1 n_2 n_3}(\mathbf{r}) = ik^{l+1}\sqrt{\frac{1}{4\pi}}\sum_{n=0}^{l}\sum_{m=-n}^{n}(M|S)_{n(n_1 n_2 n_3)}^{m}S_n^m(\mathbf{r}), \tag{2.2.23}$$

$$l = n_1 + n_2 + n_3.$$

where coefficients $(M|S)_{n(n_1 n_2 n_3)}^{m}$ can be referred as multipole expansion coefficients over the elementary singular basis. This expression shows that to represent solutions of the Helmholtz equation one can use multipoles instead of the elementary singular basis and the functions $\mathrm{Im}\{M_{n_1 n_2 n_3}(\mathbf{r})\}$ regular at $r = 0$ instead of the elementary regular basis.

In the literature there exist some alternative definitions of multipoles of degree l. It is based on invariant differentiation operation based on a set of vectors $\mathbf{p}_1, \ldots, \mathbf{p}_l$, which sometimes are called "multipole moments". The definition is

$$M_l(\mathbf{r}) = (-1)^l(\mathbf{p}_l\cdot\nabla)(\mathbf{p}_{l-1}\cdot\nabla)\cdots(\mathbf{p}_1\cdot\nabla)G(\mathbf{r}), \qquad M_0(\mathbf{r}) = G(\mathbf{r}). \tag{2.2.24}$$

The operator $\mathbf{p}_j\cdot\nabla$ is nothing but the differentiation operator in the direction determined by vector $\mathbf{p}_j = (p_{jx}, p_{jy}, p_{jz})$:

$$\mathbf{p}_j\cdot\nabla = p_{jx}\frac{\partial}{\partial x} + p_{jy}\frac{\partial}{\partial y} + p_{jz}\frac{\partial}{\partial z}, \qquad j = 1, \ldots, l, \tag{2.2.25}$$

and so a multipole of degree l in this definition can be expressed via the standardized multipoles $M_{n_1 n_2 n_3}(\mathbf{r})$

$$M_l(\mathbf{r}) = \sum_{n_1+n_2+n_3=l} Q^{(l)}_{n_1 n_2 n_3} M_{n_1 n_2 n_3}(\mathbf{r}), \tag{2.2.26}$$

where $Q^{(l)}_{n_1 n_2 n_3}$ are called the "components of the multipole moment" and can be expressed via the multipole moment vectors $\mathbf{p}_1, \ldots, \mathbf{p}_l$.

The use of multipoles has some advantages and disadvantages. The advantages are related to easier differentiation of multipoles, while the disadvantages are related to their larger number and the linear dependence of the functions used for representation of the solution and the non-orthogonality of the set of multipoles on the sphere $r = \text{const}$.

2.3 SUMS OF ELEMENTARY SOLUTIONS

2.3.1 Plane waves

The Helmholtz equation has a very important class of solutions called *plane waves*. Such solutions can be simply expressed in the form

$$\psi(\mathbf{r}) = e^{i\mathbf{k}\cdot\mathbf{r}}, \qquad \mathbf{k} = (k_x, k_y, k_z) \tag{2.3.1}$$

Here the 3D vector $\mathbf{k}$, which can have complex-valued components k_x, k_y, and k_z, is called the "wave vector". To satisfy the Helmholtz equation with the wavenumber k the components of the wave vector should satisfy the following condition

$$k_x^2 + k_y^2 + k_z^2 = k^2. \tag{2.3.2}$$

Note that a non-trivial plane wave solution exists for any k, including for real, complex, and even zero k (in this case the Helmholtz equation degenerates to the Laplace equation). Indeed, any two components of the wave vector can be selected arbitrarily, and the third component can be determined from Eq. (2.3.2). We also notice that if $\mathbf{k} = (k_x, k_y, k_z)$ is the wave vector, any other vector $\mathbf{k}^{\pm} = (\pm k_x, \pm k_y, \pm k_z)$ where the combinations of signs " + " and " − " is arbitrary is also a wave vector, since the plane wave corresponding to $\mathbf{k}^{\pm}$ satisfies the same Helmholtz equation.

Next, let us find a relation between the plane wave solution and elementary solutions in spherical coordinates. For this purpose we use the following well-known *Gegenbauer's expansion* of exponent over the basis of

Legendre polynomials

$$e^{i\rho \cos\theta} = \sum_{n=0}^{\infty} (2n+1) i^n j_n(\rho) P_n(\cos\theta) \tag{2.3.3}$$

where ρ is an arbitrary complex number. If $k \neq 0$ there exists the following non-trivial wave vector

$$\mathbf{k} = k\mathbf{s}, \qquad \mathbf{s} = (\sin\theta_k \cos\varphi_k, \sin\theta_k \sin\varphi_k, \cos\theta_k), \tag{2.3.4}$$

where $\mathbf{s}$ is a unit vector, which determines the direction of the plane wave propagation, and θ_k and φ_k are spherical angles of $\mathbf{s}$. The components of the vector $\mathbf{s}$ are real, while k can be an arbitrary non-zero complex wavenumber. The plane wave solution can be represented then as

$$e^{i\mathbf{k}\cdot\mathbf{r}} = e^{ikr\mathbf{s}\cdot\mathbf{r}/r} = \sum_{n=0}^{\infty} (2n+1) i^n j_n(kr) P_n\left(\mathbf{s}\cdot\frac{\mathbf{r}}{r}\right). \tag{2.3.5}$$

Now using the addition theorem for spherical harmonics (2.1.72) and the definition of the regular spherical basis functions (2.1.101) we obtain

$$\begin{aligned} e^{i\mathbf{k}\cdot\mathbf{r}} &= 4\pi \sum_{n=0}^{\infty} i^n j_n(kr) \sum_{m=-n}^{n} Y_n^{-m}(\mathbf{s}) Y_n^m\left(\frac{\mathbf{r}}{r}\right) \\ &= 4\pi \sum_{n=0}^{\infty} \sum_{m=-n}^{n} i^n Y_n^{-m}(\theta_k, \varphi_k) R_n^m(\mathbf{r}). \end{aligned} \tag{2.3.6}$$

Note that this expansion also yields an integral representation of the regular spherical basis function via the plane waves. Indeed, the above expansion can be interpreted as an expansion of the surface function $e^{i\mathbf{k}\cdot\mathbf{r}}$ over the basis of spherical harmonics $\{Y_n^m(\theta_k, \varphi_k)\}$. Then $4\pi i^n R_n^{-m}(\mathbf{r})$ will be the coefficients of this expansion that can be expressed according to the relation (2.1.68) as

$$R_n^m(\mathbf{r}) = \frac{i^{-n}}{4\pi} \int_{S_u} e^{ik\mathbf{s}\cdot\mathbf{r}} Y_n^m(\mathbf{s}) dS(\mathbf{s}). \tag{2.3.7}$$

2.3.2 Representation of solutions as series

Due to the linearity of the Helmholtz equation any sum of its solutions provides a new solution to the same equation. The above example with plane waves shows that it can be represented as an infinite sum of the regular spherical basis functions. More generally, we can seek solutions of the Helmholtz equation, which can be represented in the following forms:

$$\psi(\mathbf{r}) = \sum_{n=0}^{\infty} \sum_{m=-n}^{n} A_n^m F_n^m(\mathbf{r}) = \sum_{m=-\infty}^{\infty} \sum_{n=|m|}^{\infty} A_n^m F_n^m(\mathbf{r}), \tag{2.3.8}$$

$$F = S, R, \quad A_n^m \in \mathbb{C},$$

Here we assumed that we can change the order of summation and further we will frequently use this operation in various decompositions. However, this procedure is legal only if the series converge *absolutely and uniformly* in the spatial domain of interest $\mathbf{r} \in \Omega$. The domain can be finite or infinite, and the condition of absolute and uniform convergence can be written as

$$\forall \epsilon > 0, \ \exists p(\epsilon), \ |\psi(\mathbf{r}) - \psi^{(p)}(\mathbf{r})| < \epsilon, \qquad \forall \mathbf{r} \in \Omega, \tag{2.3.9}$$

and

$$\forall \epsilon > 0, \ \exists p(\epsilon), \ \sum_{n=p}^{\infty} \sum_{m=-n}^{n} |A_n^m F_n^m(\mathbf{r})| < \epsilon, \qquad \forall \mathbf{r} \in \Omega, \tag{2.3.10}$$

where $\psi^{(p)}(\mathbf{r})$ is a *p-truncated sum*:

$$\psi^{(p)}(\mathbf{r}) = \sum_{n=0}^{p-1} \sum_{m=-n}^{n} A_n^m F_n^m(\mathbf{r}) = \sum_{m=-p+1}^{p-1} \sum_{n=|m|}^{p-1} A_n^m F_n^m(\mathbf{r}), \qquad F = S, R. \tag{2.3.11}$$

Obviously, if the sums (2.3.8) consist of a finite number of terms, the conditions (2.3.9) and (2.3.10) are always satisfied.

The *p-truncated* or *band-limited* functions, such as $\psi^{(p)}(\mathbf{r})$ play an important role in the theory of functional approximation. While in many cases we will be interested with functions of this type, first we will try to answer the question, whether *any* solution of the Helmholtz equation can be represented by absolutely and uniformly convergent series or at least approximated with sums of the spherical basis functions. To answer this question, we consider first *far field expansions* of solutions of the Helmholtz equation.

2.3.3 Far field expansions

2.3.3.1 Asymptotic expansion

Let us assume that k is real (and positive) and the Helmholtz equation appears from the Fourier transform of the wave equation to the frequency domain. The far field expansions are appropriate for the scattered fields or fields generated by any finite sets of monopoles, dipoles, or multipole sources. As soon as the primary or the secondary waves are generated by some sources and scatterers located in a finite domain they propagate outside and go to infinity. To describe the field far from this domain, we can consider a sphere, S, of arbitrarily large radius, which completely surrounds the scatterer or a finite set of scatterers and multipoles. To describe the field we then can introduce a reference frame with the origin at the center of this sphere and spherical coordinates (r, θ, φ). We will seek a solution of the Helmholtz equation in the form

$$\psi(\mathbf{r}) = h_0(kr)\Psi(\mathbf{r}), \tag{2.3.12}$$

where $h_0(kr)$ is the spherical Hankel function of zero order, and $\Psi(\mathbf{r})$ is a function, which we will try to build.

We are looking for solutions that satisfy the radiation conditions. Substituting the above decomposition into the radiation condition (2.2.9), we obtain

$$\begin{aligned} 0 &= \lim_{r\to\infty} r\left(\frac{\partial\psi}{\partial r} - ik\psi\right) = \lim_{r\to\infty} r\left[\frac{\partial(h_0\Psi)}{\partial r} - ikh_0\Psi\right] \\ &= \lim_{r\to\infty}\left[\Psi r\left(\frac{dh_0}{dr} - ikh_0\right) + h_0 r\frac{\partial\Psi}{\partial r}\right] \\ &= \lim_{r\to\infty} rh_0(kr)\left[\frac{\partial\Psi}{\partial r} - k\frac{h_1(kr) + ih_0(kr)}{h_0(kr)}\Psi\right] \\ &= \lim_{r\to\infty}\frac{e^{ikr}}{ik}\left(\frac{\partial\Psi}{\partial r} - \frac{1}{r}\Psi\right). \end{aligned} \tag{2.3.13}$$

Here we used the fact that $h_0'(kr) = -h_1(kr)$ (see the recurrence relation (2.1.89)) and explicit expressions for the spherical Hankel functions (2.1.84). Since for real k, the complex exponent is a bounded function, the radiation condition leads to the following condition for function Ψ:

$$\lim_{r\to\infty}\left(\frac{\partial\Psi}{\partial r} - \frac{1}{r}\Psi\right) = 0. \tag{2.3.14}$$

This shows that the derivatives of Ψ as $r \to \infty$ should behave as

$$\frac{\partial\Psi}{\partial r} = O\left(\frac{1}{r}\Psi\right), \quad \frac{\partial^2\Psi}{\partial r^2} = O\left(\frac{1}{r^2}\Psi\right), \ldots, \frac{\partial^l\Psi}{\partial r^l} = O\left(\frac{1}{r^l}\Psi\right), \ldots \tag{2.3.15}$$

The first asymptotic result here follows from condition (2.3.14), while others can be obtained by subsequent differentiation with respect to r.

Now substituting decomposition (2.3.12) into the Helmholtz equation (2.1.6) and using the fact that $h_0(kr)$ also satisfies the Helmholtz equation in the considered domain, we have

$$\begin{aligned}0 &= \nabla^2\psi + k^2\psi = \nabla^2(h_0\Psi) + k^2 h_0\Psi \\ &= h_0\nabla^2\Psi + 2\nabla h_0\cdot\nabla\Psi + (\nabla^2 h_0 + k^2 h_0)\Psi = h_0\nabla^2\Psi + 2\nabla h_0\cdot\nabla\Psi \\ &= h_0\nabla^2\Psi + 2kh_0'(kr)\frac{\partial\Psi}{\partial r} = h_0\left[\nabla^2\Psi - 2k\frac{h_1(kr)}{h_0(kr)}\frac{\partial\Psi}{\partial r}\right].\end{aligned} \tag{2.3.16}$$

Using explicit expressions for the Hankel functions, we can rewrite this equation in spherical coordinates (2.1.5) as

$$\frac{\partial^2\Psi}{\partial r^2} + 2ik\frac{\partial\Psi}{\partial r} + \frac{1}{r^2}\nabla^2_{(\theta,\varphi)}\Psi = 0, \tag{2.3.17}$$

where

$$\nabla^2_{(\theta,\varphi)} = \frac{1}{\sin\theta}\frac{\partial}{\partial\theta}\left(\sin\theta\frac{\partial}{\partial\theta}\right) + \frac{1}{\sin^2\theta}\frac{\partial^2}{\partial\varphi^2}, \tag{2.3.18}$$

is called the *Beltrami operator* which is the angular part of the Laplace operator. Since this operator does not depend on r it does not change the asymptotic behavior of Ψ, as soon as $\Psi(r, \theta, \varphi)$ is a regular function of the spherical polar angles. So as $r \to \infty$ we have

$$\nabla^2_{(\theta,\varphi)}\Psi = O(\Psi). \tag{2.3.19}$$

Consider now the asymptotic behavior of solutions of equation (2.3.17) as $r \to \infty$. The derivatives of Ψ behave as provided by asymptotics (2.3.15) and (2.3.19). It is easy to see that the principal term Ψ_0 of the far field asymptotic expansion of Ψ should satisfy the following equation:

$$\frac{\partial\Psi_0}{\partial r} = 0. \tag{2.3.20}$$

This means that this principal term is a function of the angular variables (θ, φ) only, or

$$\Psi_0 = \Psi_0(\theta, \varphi). \tag{2.3.21}$$

Returning to Eq. (2.3.12) we find that the principal term of the far field expansion of an arbitrary scattered field is

$$\psi(\mathbf{r}) \sim h_0(kr)\Psi_0(\theta, \varphi). \tag{2.3.22}$$

The function $\Psi_0(\theta, \varphi)$ plays a special role in the theory and we will return to this function later to consider it in detail. For now we can continue our asymptotic procedure and try to obtain more terms based on Eq. (2.3.17) and the conditions on the derivatives (2.3.15) and (2.3.19). These equations contain only functions r^{-l} and this is a hint that the asymptotic expansion as $r \to \infty$ can be sought in the form

$$\Psi(r, \theta, \varphi) = \Psi_0(\theta, \varphi) + \frac{1}{r}\Psi_1(\theta, \varphi) + \cdots + \frac{1}{r^l}\Psi_l(\theta, \varphi) + \cdots$$

$$= \sum_{n=0}^{\infty} \frac{1}{r^l}\Psi_l(\theta, \varphi). \tag{2.3.23}$$

Concerning the asymptotic expansions and series we should always keep in mind that the asymptotic series can be convergent or divergent. Normally we consider only a few terms (p) of the expansion and the pth term provides evaluation of the error. So the infinite sum (2.3.23), in fact, means

$$\Psi(r, \theta, \varphi) = \sum_{l=0}^{p-1} \frac{1}{r^l}\Psi_l(\theta, \varphi) + O\left(\frac{1}{r^p}\right). \tag{2.3.24}$$

Substituting this series into Eq. (2.3.19), and grouping terms near the same powers of r we obtain

$$\sum_{n=0}^{\infty}\left[l(l+1)\Psi_l(\theta,\varphi) - 2ik(l+1)\Psi_{l+1}(\theta,\varphi) + \nabla^2_{(\theta,\varphi)}\Psi_l(\theta,\varphi)\right]\frac{1}{r^{l+2}} = 0. \tag{2.3.25}$$

Now we note that this is an asymptotic series, and if we consider the limit $r \to \infty$ subsequently for $l = 0, 1, 2, \ldots$ the term in the square brackets should turn to zero for any l. So we have

$$2ik(l+1)\Psi_{l+1}(\theta,\varphi) = \left[l(l+1) + \nabla^2_{(\theta,\varphi)}\right]\Psi_l(\theta,\varphi), \quad l = 0, 1, 2, \ldots \tag{2.3.26}$$

This, first, proves that the chosen form of asymptotic expansion (2.3.23) is correct, since we generate only terms of type $r^{-l}\Psi_l(\theta,\varphi)$ and do not generate other type of terms. Second, this provides a recursion formula for computation of all terms in the far field expansion, from the known function $\Psi_0(\theta,\varphi)$. This function, therefore, determines not only the principal term of the expansion, but the *entire* asymptotic series(!).

Substituting Eq. (2.3.23) into Eq. (2.3.12) we obtain then the following far field expansion

$$\psi(\mathbf{r}) = h_0(kr)\sum_{l=0}^{\infty} \frac{1}{r^l}\Psi_l(\mathbf{s}), \qquad \mathbf{r} = r\mathbf{s}, \qquad |\mathbf{s}| = 1, \qquad r > a. \tag{2.3.27}$$

This result, presumably, should be referred as the essence of the *Atkinson theorem* [Atk49], who not only derived such type of formula, but also showed that these series converge *absolutely and uniformly* in all space exterior of a sphere of radius a completely, surrounding the scatterer. The radius a is sometimes called as the *characteristic dimension* of the scatterer. The recursive procedure for obtaining the subsequent terms in expansion (2.3.26) was obtained by Wilcox [Wl156].

While the asymptotic expansion (2.3.27) is derived for real k we note that it also holds for solutions decaying at infinity with complex k and $\mathrm{Im}\{k\} > 0$. Indeed, $\Psi_l(\mathbf{s})$ are angular functions while $h_0(kr)$ and r^{-l} are decaying functions in this case. The recurrence relation (2.3.26) holds as it follows from Eq. (2.3.27) and the Helmholtz equation for arbitrary k. The case of complex k, $\mathrm{Im}\{k\} \geq 0$, was also included in the formulation of the expansion theorem by Wilcox [Wl156], and the theorem is formulated with weaker assumptions than in the Atkinson paper [Atk49].

2.3.3.2 *Relation to expansion over singular spherical basis functions*

Let us now relate these results with expansions over singular spherical basis function. We assume that the surface functions $\psi_l(\mathbf{s})$ belong to class $L^2(S_u)$ and can be expanded in convergent series over the orthonormal basis of spherical harmonics:

$$\Psi_l(\mathbf{s}) = \sum_{n=0}^{\infty} \sum_{m=-n}^{n} B_n^{(l)m} Y_n^m(\mathbf{s}), \qquad l = 0, 1, 2, \ldots \tag{2.3.28}$$

where $B_n^{(l)m}$ are the expansion coefficients. To relate the representation of the Helmholtz equation solution given by series (2.3.27) and (2.3.28) to the expansion over the spherical basis functions, we assume (in fact, this can be proved) that the following series converges absolutely and uniformly:

$$\psi(\mathbf{r}) = h_0(kr) \sum_{l=0}^{\infty} \sum_{n=0}^{\infty} \sum_{m=-n}^{n} \frac{1}{r^l} B_n^{(l)m} Y_n^m(\mathbf{s}). \tag{2.3.29}$$

Particularly this means that the order of summation can be changed, e.g.

$$\psi(\mathbf{r}) = h_0(kr) \sum_{n=0}^{\infty} \sum_{m=-n}^{n} \sum_{l=0}^{\infty} \frac{1}{r^l} B_n^{(l)m} Y_n^m(\mathbf{s}). \tag{2.3.30}$$

As soon as all surface functions $\Psi_l(\mathbf{s})$ are related, all coefficients $B_n^{(l)m}$ can be generated from coefficients $B_n^{(0)m}$. To find these relations, we substitute these series into the recurrence relation (2.3.26) to obtain

$$\sum_{n=0}^{\infty}\sum_{m=-n}^{n} 2ik(l+1)B_n^{(l+1)m}Y_n^m(\mathbf{s}) = \sum_{n=0}^{\infty}\sum_{m=-n}^{n} B_n^{(l)m}\left[l(l+1)+\nabla^2_{(\theta,\varphi)}\right]Y_n^m(\mathbf{s}), \tag{2.3.31}$$

$l = 0, 1, 2, \ldots$

We note now that the spherical harmonics are the eigenfunctions of the Beltrami operator:

$$\nabla^2_{(\theta,\varphi)}Y_n^m(\mathbf{s}) = -n(n+1)Y_n^m(\mathbf{s}), \tag{2.3.32}$$

and Eq. (2.3.31) can be rewritten as

$$\sum_{n=0}^{\infty}\sum_{m=-n}^{n}\left\{2ik(l+1)B_n^{(l+1)m} - \left[l(l+1)-n(n+1)\right]B_n^{(l)m}\right\}Y_n^m(\mathbf{s}) = 0, \tag{2.3.33}$$

$l = 0, 1, 2, \ldots$

The orthonormality of the basis $\{Y_n^m\}$ yields then

$$2ik(l+1)B_n^{(l+1)m} = \left[l(l+1)-n(n+1)\right]B_n^{(l)m}, \qquad l, n = 0, 1, 2, \ldots, \qquad m = -n, \ldots, n. \tag{2.3.34}$$

For given $B_n^{(0)m}$ we can compute the rest of the coefficients. The interesting thing about this recurrence is that for given n we can compute coefficients $B_n^{(l+1)m}$ for $l = 0, 1, \ldots, n-1$. But as soon as we reach the point $l = n$, the term in the square brackets in the right-hand side of Eq. (2.3.34) turns to zero, which yields $B_n^{(n+1)m} = 0$. All subsequent $B_n^{(l+1)m}$ for $l = n+1, n+2, \ldots$ will also be zeros according to the recurrence (2.3.34). So we have

$$B_n^{(l)m} = 0, \qquad n = 0, 1, 2, \ldots, \qquad m = -n, \ldots, n, \qquad l > n \tag{2.3.35}$$

The non-zero coefficients can be computed from the recurrence relation

(2.3.34) as

$$\begin{aligned} B_n^{(l)m} &= \frac{(n-l+1)(n+l)}{l(-2ik)} B_n^{(l-1)m} \\ &= \frac{(n-l+1)(n-l+2)(n+l)(n+l-1)}{(l+1)l(-2ik)(-2ik)} B_n^{(l-2)m} = \cdots \\ &= \frac{(n-l+1)\cdots n(n+l)\cdots(n+1)}{l!(-2ik)^l} B_n^{(0)m} \\ &= \frac{(n+l)!}{l!(n-l)!(-2ik)^l} B_n^{(0)m}. \end{aligned} \quad (2.3.36)$$

Now we can rewrite sum (2.3.30) in the form

$$\begin{aligned} \psi(\mathbf{r}) &= \sum_{n=0}^{\infty} \sum_{m=-n}^{n} Y_n^m(\mathbf{s}) h_0(kr) \sum_{l=0}^{n} \frac{1}{r^l} B_n^{(l)m} \\ &= \sum_{n=0}^{\infty} \sum_{m=-n}^{n} B_n^{(0)m} Y_n^m(\mathbf{s}) h_0(kr) \sum_{l=0}^{n} \frac{(n+l)!}{l!(n-l)!(-2ikr)^l} \\ &= \sum_{n=0}^{\infty} \sum_{m=-n}^{n} i^n B_n^{(0)m} h_n(kr) Y_n^m(\mathbf{s}) = \sum_{n=0}^{\infty} \sum_{m=-n}^{n} i^n B_n^{(0)m} S_n^m(\mathbf{r}). \end{aligned} \quad (2.3.37)$$

Here we used the explicit expression for $h_n(kr)$ (Eq. (2.1.82)). So we can see that series (2.3.29) are, in fact, series (2.3.8) for singular basis functions, where we should set

$$A_n^m = i^n B_n^{(0)m}, \qquad n = 0, 1, 2, \ldots, \qquad m = -n, \ldots, n. \quad (2.3.38)$$

This proves that any radiating solution of the Helmholtz equation generated by distribution of sources and scatters in a finite domain Ω, such that sum (2.3.29) converges absolutely and uniformly, can be represented by series (2.3.8) over singular basis functions in a domain exterior of a sphere completely surrounding Ω. The principal term of the far field asymptotics is described then by Eq. (2.3.22), where

$$\Psi_0(\mathbf{s}) = \sum_{n=0}^{\infty} \sum_{m=-n}^{n} i^{-n} A_n^m Y_n^m(\mathbf{s}) \quad (2.3.39)$$

is a surface function and the sum converges. This function can be built immediately, if the expansion coefficients over the spherical basis functions, A_n^m, of function $\psi(\mathbf{r})$ are known. Conversely, if $\Psi_0(\mathbf{s})$ is known, then A_n^m can be found (see Eqs. (2.1.65)–(2.1.68)) and $\psi(\mathbf{r})$ determined. Due to this correspondence between $\psi(\mathbf{r})$ and $\Psi_0(\mathbf{s})$, the latter function sometimes is called the *far field signature function of* $\psi(\mathbf{r})$, or simply, *signature function*.

2.3.4 Local expansions

2.3.4.1 Asymptotic expansion

Consider now the asymptotic expansion of an arbitrary solution $\psi(\mathbf{r})$ of the Helmholtz equation about some point $\mathbf{r} = \mathbf{r}_*$. Without loss of generality we can accept that $\mathbf{r}_* = \mathbf{0}$, otherwise we can consider the radius vector related to the center of expansion, $\mathbf{r}' = \mathbf{r} - \mathbf{r}_*$, where $\mathbf{r}' = \mathbf{0}$ corresponds to the center. Assume that within a sphere of radius $r < a$ there are no singularities or boundaries of the domain, and so $\psi(\mathbf{r})$ is a regular function in the interior of this sphere. We can build the asymptotic expansion in several ways which lead to the same result, e.g. we can consider the three-dimensional Taylor series in the vicinity of $\mathbf{r} = \mathbf{0}$. Another way is to consider the asymptotic series with respect to the small parameter $r \to 0$ in the following form:

$$\psi(\mathbf{r}) = \Phi_0 + r\Phi_1(\mathbf{s}) + \cdots + r^l\Phi_l(\mathbf{s}) + \cdots = \sum_{l=0}^{\infty} r^l\Phi_l(\mathbf{s}), \qquad (2.3.40)$$

$$\mathbf{r} = r\mathbf{s}, \qquad |\mathbf{s}| = 1.$$

Here the functions $\Phi_l(\mathbf{s})$ are functions of the angular variables $\mathbf{s} = (\theta, \varphi)$ only, since the dependence on r is expanded into the asymptotic series. Note that the zero order approximation does not depend on s and so this is a constant, which can be found by setting $r = 0$ in series (2.3.40):

$$\Phi_0 = \psi(\mathbf{0}). \qquad (2.3.41)$$

Further, since the function $\psi(\mathbf{r})$ is regular, or infinitely differentiable, then taking its derivatives with respect to r and setting $r = 0$ we obtain

$$\Phi_l(\mathbf{s}) = \frac{1}{l!}\left.\frac{\partial^l \psi}{\partial r^l}\right|_{r=0}, \qquad l = 0, 1, 2, \ldots \qquad (2.3.42)$$

Substituting expansion (2.3.40) into the Helmholtz equation (2.1.6), we find

$$
\begin{aligned}
0 &= (\nabla^2 + k^2)\sum_{l=0}^{\infty} r^l \Phi_l(\mathbf{s}) \\
&= \sum_{l=0}^{\infty} r^{l-2}\left[l(l+1) + \nabla^2_{(\theta,\varphi)}\right]\Phi_l(\mathbf{s}) + k^2 \sum_{l=0}^{\infty} r^l \Phi_l(\mathbf{s}) \\
&= \frac{1}{r}(2 + \nabla^2_{(\theta,\varphi)})\Phi_1(\mathbf{s}) + \sum_{l=0}^{\infty} r^l \left\{\left[(l+2)(l+3) + \nabla^2_{(\theta,\varphi)}\right]\Phi_{l+2}(\mathbf{s}) + k^2\Phi_l(\mathbf{s})\right\},
\end{aligned}
\tag{2.3.43}
$$

where $\nabla^2_{(\theta,\varphi)}$ is the Beltrami operator (2.3.18). The coefficient of r^{-1} should be zero since there are no singularities as $r \to 0$. This leaves no other opportunities for $\Phi_1(\mathbf{s})$, except to be the eigenfunction of the Beltrami operator with eigenvalue -2. The only eigenfunctions of this operator, which are 2π-periodical with respect to θ and φ are the spherical harmonics of degree 1 (see relation (2.3.32)). So $\Phi_1(\mathbf{s})$ is a linear combination of these harmonics:

$$
\Phi_1(\mathbf{s}) = \sum_{m=-1}^{1} C_1^{(1)m} Y_1^m(\mathbf{s}), \tag{2.3.44}
$$

where the coefficients $C_1^{(1)m}$ can be arbitrary (depend on function Ψ, see Eq. (2.3.42)).

Since the series (2.3.43) are asymptotic in nature, the multipliers of the other powers of r should be zero also. This yields the following recursion formula for $\Phi_l(\mathbf{s})$:

$$
\left[(l+2)(l+3) + \nabla^2_{(\theta,\varphi)}\right]\Phi_{l+2}(\mathbf{s}) = -k^2\Phi_l(\mathbf{s}), \qquad l = 0, 1, 2, \ldots \tag{2.3.45}
$$

2.3.4.2 *Relation to expansion over the regular spherical basis functions*

As surface functions, $\Phi_l(\mathbf{s})$ can be expanded in the following series:

$$
\Phi_l(\mathbf{s}) = \sum_{n=0}^{\infty} \sum_{m=-n}^{n} C_n^{(l)m} Y_n^m(\mathbf{s}). \tag{2.3.46}
$$

Substituting these series into the recursion (2.3.45) and using the orthonormality of the system of spherical harmonics, we obtain

$$\left[(l+2)(l+3) - n(n+1)\right]C_n^{(l+2)m} = -k^2 C_n^{(l)m}, \qquad l, n = 0, 1, 2, \ldots, \quad m = -n, \ldots, n. \tag{2.3.47}$$

For $n = l + 2$ the left-hand side of this equation turns to zero. This means that the right-hand side should also be zero, or

$$C_{l+2}^{(l)m} = 0. \tag{2.3.48}$$

Setting here $l = l + 2$ and substituting into the left-hand side of the recurrence relation (2.3.47) we find that $C_{l+4}^{(l)m} = 0$. Such a recursive substitution can be repeated as many times as we want, so we obtain

$$C_n^{(l)m} = 0, \qquad n = l + 2, l + 4, \ldots \tag{2.3.49}$$

We now note that if $n + l$ is odd, then the coefficient in the square brackets in Eq. (2.3.47) never turns to zero. This means that we can apply this recursion in this case as many times as we want and can express $C_n^{(l+2)m}$ for odd n via $C_n^{(0)m}$ or for even n via $C_n^{(1)m}$. However, these coefficients are zero (see Eq. (2.3.41), where we can set $\Phi_0 = C_0^{(0)0} Y_0^0(\mathbf{s})$, and Eq. (2.3.46)). Thus,

$$C_n^{(l)m} = 0, \qquad n = l \pm 1, l \pm 3, \ldots \tag{2.3.50}$$

Combining this result with relation (2.3.49), we can see that the surface functions (2.3.46) are represented by finite sums:

$$\Phi_l(\mathbf{s}) = \sum_{n=0}^{l} \sum_{m=-n}^{n} C_n^{(l)m} Y_n^m(\mathbf{s}), \tag{2.3.51}$$

and the asymptotic series (2.3.40) can be represented as

$$\psi(\mathbf{r}) = \sum_{l=0}^{\infty} r^l \sum_{n=0}^{l} \sum_{m=-n}^{n} C_n^{(l)m} Y_n^m(\mathbf{s}). \tag{2.3.52}$$

For $l > n$ the non-zero coefficients of degree n can be expressed via $C_n^{(n)m}$ as follows from the recursion (2.3.47):

$$\begin{aligned} C_n^{(l)m} &= -\frac{k^2}{(l-n)(l+n+1)} C_n^{(l-2)m} \\ &= \frac{k^4}{(l-n)(l-2-n)(l+n+1)(l+n-1)} C_n^{(l-4)m} = \cdots \\ &= \frac{(-1)^{(l-n)/2} k^{l-n}}{(l-n)\cdots 2(l+n+1)\cdots(2n+3)} C_n^{(n)m}, \end{aligned} \tag{2.3.53}$$

$$l - n = 2l', \quad l' = 1, 2, \ldots$$

Assuming that the required series converge absolutely and uniformly we can change the order of summation in series (2.3.52) and substitute there expression (2.3.53) to obtain

$$\begin{aligned} \psi(\mathbf{r}) &= \sum_{n=0}^{\infty} \sum_{m=-n}^{n} C_n^{(n)m} Y_n^m(\mathbf{s}) \left[r^n + \sum_{l=n+2}^{\infty} r^l \frac{(-1)^{(l-n)/2} k^{l-n}}{(l-n)\cdots 2(l+n+1)\cdots(2n+3)} \right] \\ &= \sum_{n=0}^{\infty} \sum_{m=-n}^{n} \frac{(2n+1)!!}{k^n} C_n^{(n)m} j_n(kr) Y_n^m(\mathbf{s}) = \sum_{n=0}^{\infty} \sum_{m=-n}^{n} A_n^m R_n^m(\mathbf{r}), \end{aligned} \tag{2.3.54}$$

since [AS64]

$$\begin{aligned} & r^n + \sum_{l=n+2}^{\infty} r^l \frac{(-1)^{(l-n)/2} k^{l-n}}{(l-n)\cdots 2(l+n+1)\cdots(2n+3)} \\ &= r^n \left[1 + \sum_{l'=1}^{\infty} \frac{(-1)^{l'} (kr)^{2l'}}{2\cdot 4\cdots 2l'(2n+3)\cdots(2l'+2n+1)} \right] \\ &= r^n \left\{ 1 - \frac{1}{1!(2n+3)} \left[\frac{(kr)^2}{2} \right] + \frac{1}{2!(2n+3)(2n+5)} \left[\frac{(kr)^2}{2} \right]^2 \right\} \\ &= \frac{1\cdot 3\cdots(2n+1)}{k^n} j_n(kr). \end{aligned} \tag{2.3.55}$$

These series converge absolutely and uniformly for any kr.

We showed that an arbitrary infinitely differentiable solution of the Helmholtz equation $\psi(\mathbf{r})$ in the interior of a sphere can be expanded into a series over the regular basis functions. Coefficients of this expansion A_n^m are related to asymptotic coefficients $C_n^{(n)m}$ according to Eq. (2.3.54) which in turn can be expressed via the derivatives of $\psi(\mathbf{r})$ at the center of expansion.

2.3.5 Uniqueness

It is not difficult to see that the expansion coefficients $\{A_n^m\}$ for infinitely differentiable solutions of the Helmholtz equation, $\psi(\mathbf{r})$ in the domain exterior or interior to the sphere are unique. Indeed, if we assume the contrary, then $\psi(\mathbf{r})$ can be represented by two series

$$\psi(\mathbf{r}) = \sum_{n=0}^{\infty} \sum_{m=-n}^{n} A_n^m F_n^m(\mathbf{r}) = \sum_{n=0}^{\infty} \sum_{m=-n}^{n} B_n^m F_n^m(\mathbf{r}), \qquad \mathbf{r} \in \Omega, \qquad (2.3.56)$$

where Ω is the domain, where $\psi(\mathbf{r})$ is regular. So there exist m and n such that $C_n^m = A_n^m - B_n^m \neq 0$. Subtracting these two series we have

$$\exists C_n^m \neq 0, \; \forall \mathbf{r} \in \Omega, \qquad \phi(\mathbf{r}) = \sum_{n=0}^{\infty} \sum_{m=-n}^{n} C_n^m F_n^m(\mathbf{r}) = 0.$$

Due to orthonormality of the basis of spherical harmonics on a sphere which is located completely inside Ω and can be selected so that $f_n(kr) \neq 0$, we have (see Eq. (2.1.68)):

$$C_n^m = \frac{1}{f_n(kr)} \int_0^{\pi} \sin\theta \, d\theta \int_0^{2\pi} \phi(\mathbf{r}) Y_n^{-m}(\theta, \varphi) d\varphi = 0,$$
$$n = 0, 1, \ldots, \qquad m = -n, \ldots, n.$$

This contradicts the assumption that $\exists C_n^m \neq 0$ and proves the statement.

We note that the spherical Bessel functions have a countable number of zeros. Therefore, in the vicinity of any zero there exists some r, such that $f_n(kr) \neq 0$ for any n, so we can always slightly change the radius of a sphere over which we integrate to provide $f_n(kr) \neq 0$.

2.4 SUMMARY

In this chapter we considered the elementary solutions of the 3D Helmholtz equation in spherical coordinates. We introduced and described some properties of special functions constituting such solutions. Particularly, we realized two basic types of solutions: regular and singular. We proved several differentiation theorems, which enable to represent derivatives of the elementary solutions via the elementary solutions of different degree and order. We also considered two different types of solutions: multipoles and plane waves, which can be expressed through the elementary spherical basis functions for the Helmholtz equation. Finally, we considered how convergent infinite series over the spherical basis functions appear from the far field and local expansions of an arbitrary solution of the Helmholtz equation.

CHAPTER 3

Translations and Rotations of Elementary Solutions

When the reference frame in which the solutions are expressed is changed, the various solutions discussed in Chapter 2 can be re-expressed in terms of elementary solutions that are expressed in the new coordinate system. While this is seemingly a simple operation, it is in fact the basis for many numerical methods for solving boundary value problems for the Helmholtz equation. This chapter provides the basic theory for such expansions and essential theoretical background to the rest of the book.

3.1 EXPANSIONS OVER SPHERICAL BASIS FUNCTIONS

In Chapter 2 we considered expansions of solutions of the Helmholtz equation over sets of elementary solutions in spherical coordinates $\{R_n^m(\mathbf{r})\}$ or $\{S_n^m(\mathbf{r})\}$:

$$\psi(\mathbf{r}) = \sum_{n=0}^{\infty} \sum_{m=-n}^{n} A_n^m R_n^m(\mathbf{r}) \quad \text{or} \quad \psi(\mathbf{r}) = \sum_{n=0}^{\infty} \sum_{m=-n}^{n} A_n^m S_n^m(\mathbf{r}). \qquad (3.1.1)$$

For finite sums we can assume that all $A_n^m = 0$ for $n \geq p$. Of course, if the sum contains an infinite number of terms we require that such a series converges in some spatial domain of our interest, and this imposes some limitations on the coefficients A_n^m, which we call *expansion coefficients*. The coefficients of expansion depend on the selection of the coordinate system, while the *field*, $\psi(\mathbf{r})$, is an invariant object. This raises a question as to how, knowing the expansion coefficients for one reference frame, we can determine the expansion coefficients in some other reference frame.

3.1.1 Translations

The functions $R_n^m(\mathbf{r})$ and $S_n^m(\mathbf{r})$ are specified in spherical coordinates, which have peculiarities at $\mathbf{r} = \mathbf{0}$. E.g. for $\mathbf{r} = \mathbf{0}$ the direction of the zero vector is not defined, and functions $S_n^m(\mathbf{r})$ are singular. Therefore, the reason for using sums of type (3.1.1) is because $\mathbf{r} = \mathbf{0}$ is in some sense a special, or preferred, point or vector. We call such a point the *center of expansion*. Obviously one may look for expansion of a function $\psi(\mathbf{r})$ near some other point, e.g. near $\mathbf{r} = \mathbf{r}_*$. Since functions $R_n^m(\mathbf{r} - \mathbf{r}_*)$ and $S_n^m(\mathbf{r} - \mathbf{r}_*)$ satisfy the Helmholtz equation, they can be selected as a basis for representation of function $\psi(\mathbf{r})$. In this case we have

$$\psi(\mathbf{r}) = \sum_{n=0}^{\infty} \sum_{m=-n}^{n} C_n^m R_n^m(\mathbf{r} - \mathbf{r}_*) \qquad \text{or}$$

$$\psi(\mathbf{r}) = \sum_{n=0}^{\infty} \sum_{m=-n}^{n} C_n^m S_n^m(\mathbf{r} - \mathbf{r}_*), \tag{3.1.2}$$

where C_n^m are expansion coefficients of the function $\psi(\mathbf{r})$ near the center $\mathbf{r} = \mathbf{r}_*$. If $\psi(\mathbf{r})$ in Eqs. (3.1.1) and (3.1.2) is the same, then the expansion coefficients C_n^m depend on $\mathbf{r}_*$. This fact can be denoted as $C_n^m(\mathbf{r}_*)$. Moreover, since the expansion (3.1.1) is a particular case of Eq. (3.1.2) at $\mathbf{r}_* = \mathbf{0}$, we have $A_n^m = C_n^m(\mathbf{0})$.

Assume now that we have the expansion of the same function near the center $\mathbf{r} = \mathbf{r}_{*1}$ and near the center $\mathbf{r} = \mathbf{r}_{*2}$, e.g.

$$\psi(\mathbf{r}) = \sum_{n=0}^{\infty} \sum_{m=-n}^{n} C_n^m(\mathbf{r}_{*1}) R_n^m(\mathbf{r} - \mathbf{r}_{*1}) = \sum_{n=0}^{\infty} \sum_{m=-n}^{n} C_n^m(\mathbf{r}_{*2}) R_n^m(\mathbf{r} - \mathbf{r}_{*2}). \tag{3.1.3}$$

and we try to relate the coefficients $C_n^m(\mathbf{r}_{*2})$ and $C_n^m(\mathbf{r}_{*1})$. This can be achieved by *reexpansion* of the basis the functions. Indeed, assume that we can reexpand function $R_n^m(\mathbf{r} - \mathbf{r}_{*1})$ over the functions $R_n^m(\mathbf{r} - \mathbf{r}_{*2})$ as follows

$$R_n^m(\mathbf{r} - \mathbf{r}_{*1}) = \sum_{n'=0}^{\infty} \sum_{m'=-n'}^{n'} B_{n'n}^{m'm} R_{n'}^{m'}(\mathbf{r} - \mathbf{r}_{*2}), \tag{3.1.4}$$

where $B_{n'n}^{m'm}$ are the *reexpansion coefficients* which we return to later. Then we can rewrite expansion of $\psi(\mathbf{r})$ near center $\mathbf{r}_{*1}$ as:

$$\begin{aligned}\psi(\mathbf{r}) &= \sum_{n=0}^{\infty}\sum_{m=-n}^{n} C_n^m(\mathbf{r}_{*1}) R_n^m(\mathbf{r}-\mathbf{r}_{*1}) \\ &= \sum_{n=0}^{\infty}\sum_{m=-n}^{n} C_n^m(\mathbf{r}_{*1}) \sum_{n'=0}^{\infty}\sum_{m'=-n'}^{n'} B_{n'n}^{m'm} R_{n'}^{m'}(\mathbf{r}-\mathbf{r}_{*2}) \\ &= \sum_{n'=0}^{\infty}\sum_{m'=-n'}^{n'} \left[\sum_{n=0}^{\infty}\sum_{m=-n}^{n} B_{n'n}^{m'm} C_n^m(\mathbf{r}_{*1})\right] R_{n'}^{m'}(\mathbf{r}-\mathbf{r}_{*2}) \\ &= \sum_{n'=0}^{\infty}\sum_{m'=-n'}^{n'} C_{n'}^{m'}(\mathbf{r}_{*2}) R_{n'}^{m'}(\mathbf{r}-\mathbf{r}_{*2}). \end{aligned} \tag{3.1.5}$$

Here we changed the order of summation, which is legal in the case when the series converge *absolutely* and *uniformly* in the domain of consideration. As shown in Chapter 2, if the domain includes a spherical shell concentric with the expansion center, then the expansion of $\psi(\mathbf{r})$ is *unique*. Therefore, comparing Eqs. (3.1.3) and (3.1.5) we have

$$C_n^m(\mathbf{r}_{*2}) = \sum_{n'=0}^{\infty}\sum_{m'=-n'}^{n'} B_{nn'}^{mm'} C_{n'}^{m'}(\mathbf{r}_{*1}). \tag{3.1.6}$$

So if we know the reexpansion coefficients then we can express expansion coefficients near the center $\mathbf{r}_{*2}$ via expansion coefficients near the center $\mathbf{r}_{*1}$. We also call this procedure *translation* and we can say that we *translate* the expansion coefficients from $\mathbf{r}_{*1}$ to $\mathbf{r}_{*2}$. We also use special terminology for the reexpansion coefficients $B_{nn'}^{mm'}$, and call them *translation coefficients*, as soon as we are concerned with reexpansions related to translations.

Looking at relation (3.1.4) we can say that the translation coefficients should, generally speaking, depend on the centers of expansion $\mathbf{r}_{*1}$ and $\mathbf{r}_{*2}$. However, it is not difficult to see that they should not depend on particular values of $\mathbf{r}_{*1}$ and $\mathbf{r}_{*2}$, but depend on their difference only:

$$\mathbf{t} = \mathbf{r}_{*2} - \mathbf{r}_{*1}. \tag{3.1.7}$$

This is a vector directed from the expansion center $\mathbf{r}_{*1}$ to the expansion center $\mathbf{r}_{*2}$ and we call it the *translation vector*. The fact that $B_{n'n}^{m'm} = B_{n'n}^{m'm}(\mathbf{t})$ can be checked straightforward. Indeed, if we denote

$$\mathbf{r}' = \mathbf{r} - \mathbf{r}_{*2}, \qquad \mathbf{r} - \mathbf{r}_{*1} = (\mathbf{r} - \mathbf{r}_{*2}) + (\mathbf{r}_{*2} - \mathbf{r}_{*1}) = \mathbf{r}' + \mathbf{t}, \tag{3.1.8}$$

then Eq. (3.1.4) turns into

$$R_n^m(\mathbf{r}' + \mathbf{t}) = \sum_{n'=0}^{\infty} \sum_{m'=-n'}^{n'} B_{n'n}^{m'm} R_{n'}^{m'}(\mathbf{r}') \tag{3.1.9}$$

and there is no other chance for $B_{n'n}^{m'm}$, but to be a function of $\mathbf{t}$ only.

3.1.2 Rotations

Returning to sums (3.1.1) providing solution of the Helmholtz equation, we note that the spherical coordinates in Eq. (2.1.1) are not symmetric with respect to the Cartesian coordinates (x, y, z). For example, the z-coordinate can be expressed only in terms of the distance r from the center and the spherical polar angle θ, while x and y depend also on the spherical polar angle φ. If we rename coordinates as $x \to y$, $y \to z$, $z \to x$, we should use other values of angles θ and φ to specify the *same* point in the space. Moreover, we can make an arbitrary rotation of the reference frame by our convenience, for example, to direct the axis z towards a particular point in space. Such a selection of the reference frame does not affect the physical quantities, such as fields, and the Helmholtz equation, written in Cartesian coordinates, will be the Helmholtz equation in Cartesian coordinates, independent of how the basis vectors of the frame of reference are oriented in space.

Let some spatial point M have coordinates $\mathbf{r} = (x, y, z)$ and $\hat{\mathbf{r}} = (\hat{x}, \hat{y}, \hat{z})$ in two Cartesian reference frames with the same origin O and basis vectors $(\mathbf{i}_x, \mathbf{i}_y, \mathbf{i}_z)$ and $(\mathbf{i}_{\hat{x}}, \mathbf{i}_{\hat{y}}, \mathbf{i}_{\hat{z}})$, respectively. Then we can write sums (3.1.1) as

$$\psi(\mathbf{r}) = \sum_{n=0}^{\infty} \sum_{m=-n}^{n} A_n^m R_n^m(\mathbf{r}) = \sum_{n=0}^{\infty} \sum_{m=-n}^{n} C_n^m R_n^m(\hat{\mathbf{r}}), \qquad \text{or}$$
$$\psi(\mathbf{r}) = \sum_{n=0}^{\infty} \sum_{m=-n}^{n} A_n^m S_n^m(\mathbf{r}) = \sum_{n=0}^{\infty} \sum_{m=-n}^{n} C_n^m S_n^m(\hat{\mathbf{r}}). \tag{3.1.10}$$

The expansion coefficients C_n^m depend on the *rotation matrix* Q, whose entries are the scalar products of the basis vectors of the two reference frames:

$$Q = \begin{pmatrix} \mathbf{i}_{\hat{x}} \cdot \mathbf{i}_x & \mathbf{i}_{\hat{x}} \cdot \mathbf{i}_y & \mathbf{i}_{\hat{x}} \cdot \mathbf{i}_z \\ \mathbf{i}_{\hat{y}} \cdot \mathbf{i}_x & \mathbf{i}_{\hat{y}} \cdot \mathbf{i}_y & \mathbf{i}_{\hat{y}} \cdot \mathbf{i}_z \\ \mathbf{i}_{\hat{z}} \cdot \mathbf{i}_x & \mathbf{i}_{\hat{z}} \cdot \mathbf{i}_y & \mathbf{i}_{\hat{z}} \cdot \mathbf{i}_z \end{pmatrix}. \tag{3.1.11}$$

We can denote this fact as $C_n^m = C_n^m(Q)$. Obviously, if there is no rotation, or in other words Q is the identity matrix, I, the expansion coefficients C_n^m turn into A_n^m, and we have $A_n^m = C_n^m(I)$.

In general, we can consider the problem of relating the coefficients C_n^m and A_n^m. This problem, as in the case of translation, can be solved with the aid of reexpansion of the basis functions. For example, if we have the following reexpansion of a regular function

$$R_n^m(\mathbf{r}) = \sum_{n'=0}^{\infty} \sum_{m'=-n'}^{n'} B_{n'n}^{m'm} R_{n'}^{m'}(\hat{\mathbf{r}}), \tag{3.1.12}$$

then repeating procedure (3.1.5) we have

$$\begin{aligned}\sum_{n=0}^{\infty} \sum_{m=-n}^{n} A_n^m R_n^m(\mathbf{r}) &= \sum_{n'=0}^{\infty} \sum_{m'=-n'}^{n'} \left[\sum_{n=0}^{\infty} \sum_{m=-n}^{n} B_{n'n}^{m'm} A_n^m \right] R_{n'}^{m'}(\hat{\mathbf{r}}) \\ &= \sum_{n'=0}^{\infty} \sum_{m'=-n'}^{n'} C_{n'}^{m'} R_{n'}^{m'}(\hat{\mathbf{r}}),\end{aligned} \tag{3.1.13}$$

So

$$C_n^m(Q) = \sum_{n'=0}^{\infty} \sum_{m'=-n'}^{n'} B_{nn'}^{mm'}(Q) A_{n'}^m. \tag{3.1.14}$$

It will be shown in this chapter that the summation over n' disappears here due to $B_{nn'}^{mm'}(Q) = 0$ for $n \neq n'$, so

$$C_n^m(Q) = \sum_{m'=-n'}^{n'} T_n^{mm'}(Q) A_n^m, \qquad B_{nn'}^{mm'}(Q) = \delta_{nn'} T_n^{mm'}(Q), \tag{3.1.15}$$

where $\delta_{nn'}$ is the Kronecker delta. We call the reexpansion coefficients $B_{nn'}^{mm'}$ or $T_n^{mm'}$ the *rotation coefficients* when as we are concerned with rotations.

3.2 TRANSLATIONS OF SPHERICAL BASIS FUNCTIONS

We first note that the regular elementary solutions $R_n^m(\mathbf{r})$ introduced by expression (2.1.101) are regular for any $\mathbf{r}$. Therefore, without any other limitation we can seek expansions of the shifted functions $R_n^m(\mathbf{r}+\mathbf{t})$ over the same basis, and assume that such expansions are valid for any $\mathbf{r}$.

Therefore, for some given degree n and order m, we can write

$$R_n^m(\mathbf{r}+\mathbf{t}) = \sum_{n'=0}^{\infty} \sum_{m'=-n'}^{n'} (R|R)_{n'n}^{m'm}(\mathbf{t}) R_{n'}^{m'}(\mathbf{r}), \quad n = 0, 1, 2, \ldots, \quad m = -n, \ldots, n. \tag{3.2.1}$$

Here we introduced notation $(R|R)_{n'n}^{m'm}(\mathbf{t})$ for translation coefficients. The symbol $(R|R)$ stays here for "regular-to-regular". Translation coefficients depend on the translation vector $\mathbf{t}$ and the degrees and orders of the basis functions and are characterized by *two* degrees and *two* orders.

If we try to write a similar expansion for the singular elementary solutions $S_n^m(\mathbf{r})$, the problem we encounter immediately is that the shifted functions $S_n^m(\mathbf{r}+\mathbf{t})$ are singular at $\mathbf{r} = -\mathbf{t}$, while the basis functions $S_n^m(\mathbf{r})$ are singular at $\mathbf{r} = \mathbf{0}$. While such expansions are possible, they definitely should diverge at the singularity (and possibly somewhere else). To figure out what is going on, let us look closer on the following classical result (e.g. [MF53], also known as *addition theorem*) for the expansion of monopole or Green's function over a series of elementary solutions. Using the definitions of the basis functions (2.1.101), (2.1.102), and (2.1.107), we can write the addition theorem as

$$\begin{aligned} ik\sqrt{\frac{1}{4\pi}} S_0^0(\mathbf{r}_1+\mathbf{r}_2) = G(\mathbf{r}_1+\mathbf{r}_2) &= ik \sum_{n'=0}^{\infty} \sum_{m'=-n'}^{n'} S_{n'}^{-m'}(-\mathbf{r}_1) R_{n'}^{m'}(\mathbf{r}_2) \\ &= ik \sum_{n'=0}^{\infty} \sum_{m'=-n'}^{n'} R_{n'}^{-m'}(-\mathbf{r}_2) S_{n'}^{m'}(\mathbf{r}_1), \quad |\mathbf{r}_2| < |\mathbf{r}_1|. \end{aligned} \tag{3.2.2}$$

Here $\mathbf{r}_1$ and $\mathbf{r}_2$ are two arbitrary vectors, and so we can assign them values $\mathbf{r}$ and $\mathbf{t}$ also arbitrarily. Then we obtain the following expressions for translations of $S_0^0(\mathbf{r}+\mathbf{t})$:

$$S_0^0(\mathbf{r}+\mathbf{t}) = \sqrt{4\pi} \sum_{n'=0}^{\infty} \sum_{m'=-n'}^{n'} \begin{cases} S_{n'}^{-m'}(-\mathbf{t}) R_{n'}^{m'}(\mathbf{r}), & |\mathbf{r}| < |\mathbf{t}| \\ R_{n'}^{-m'}(-\mathbf{t}) S_{n'}^{m'}(\mathbf{r}), & |\mathbf{r}| > |\mathbf{t}|. \end{cases} \tag{3.2.3}$$

This shows that there are two spatial domains, which should be treated differently, since in one domain the function $S_0^0(\mathbf{r}+\mathbf{t})$ should be expanded over the basis of regular functions, and in the other domain it should be expanded over the basis of singular functions. Obviously, the first domain is internal to a sphere, $|\mathbf{r}| = |\mathbf{t}|$, and the second is external to the same sphere (see the figure below).

We note now that series (3.2.2) converge absolutely for $|\mathbf{r}_2| < |\mathbf{r}_1|$ (see Chapter 9 for error bounds) and can be differentiated as many times as we want, and this will not change the radius of their convergence. Due to the differentiation theorems of Chapter 2, we see that any function $S_n^m(\mathbf{r}+\mathbf{t})$ can be obtained by differentiations of $S_0^0(\mathbf{r}+\mathbf{t})$. This shows that for the singular solution of arbitrary order and degree, we have

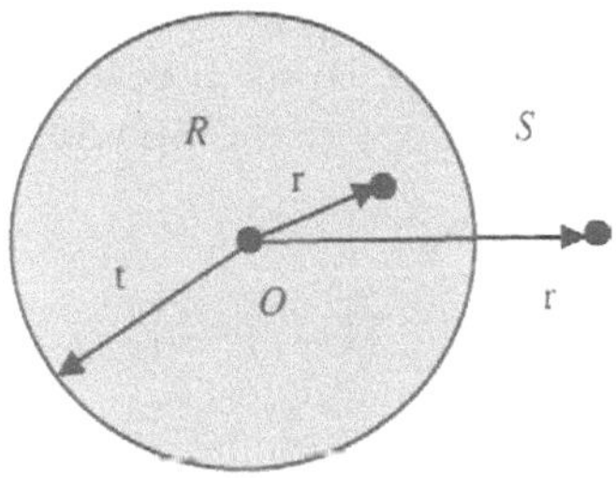

domains of convergence similar to $S_0^0(\mathbf{r}+\mathbf{t})$. In other words translated singular elementary solutions can be expanded over bases of regular or singular spherical basis functions according to the following relations:

$$S_n^m(\mathbf{r}+\mathbf{t}) = \sum_{n'=0}^{\infty}\sum_{m'=-n'}^{n'} \begin{cases} (S|R)_{n'n}^{m'm}(\mathbf{t})R_{n'}^{m'}(\mathbf{r}), & |\mathbf{r}| < |\mathbf{t}|, \\ (S|S)_{n'n}^{m'm}(\mathbf{t})S_{n'}^{m'}(\mathbf{r}), & |\mathbf{r}| > |\mathbf{t}|, \end{cases} \qquad (3.2.4)$$

$$n = 0, 1, 2, \ldots, \quad m = -n, \ldots, n.$$

The meaning of symbols $(S|R)$ and $(S|S)$ in this formula is similar to the meaning of $(R|R)$—they stay for "singular-to-regular" and "singular-to-singular" expansions. As $(R|R)_{n'n}^{m'm}$ translation coefficients $(S|R)_{n'n}^{m'm}$ and $(S|S)_{n'n}^{m'm}$ are functions of the translation vector only.

Some initial idea what the functions $(S|R)_{n'n}^{m'm}(\mathbf{t})$ and $(S|S)_{n'n}^{m'm}(\mathbf{t})$ look like can be obtained comparing expressions (3.2.3) and (3.2.4). This comparison shows that

$$(S|R)_{n'0}^{m'0}(\mathbf{t}) = \sqrt{4\pi}(-1)^{n'} S_{n'}^{-m'}(\mathbf{t}), \quad n' = 0, 1, 2, \ldots, \quad m' = -n', \ldots, n'. \qquad (3.2.5)$$

$$(S|S)_{n'0}^{m'0}(\mathbf{t}) = \sqrt{4\pi}(-1)^{n'} R_{n'}^{-m'}(\mathbf{t}), \qquad n' = 0, 1, 2, \ldots, \quad m' = -n', \ldots, n'. \qquad (3.2.6)$$

So these functions are of the same type as singular and regular basis functions, and coefficients $(S|R)_{n'0}^{m'0}(\mathbf{t})$ are singular at $\mathbf{t} = \mathbf{0}$, while coefficients $(S|S)_{n'0}^{m'0}(\mathbf{t})$ are regular at $\mathbf{t} = \mathbf{0}$. To change the sign of $\mathbf{t}$, we used the definitions of the basis functions (2.1.101) and (2.1.102), and the properties of spherical harmonics, which show that

$$S_n^m(-\mathbf{t}) = (-1)^n S_n^m(\mathbf{t}), \qquad R_n^m(-\mathbf{t}) = (-1)^n R_n^m(\mathbf{t}), \tag{3.2.7}$$
$$n = 0, 1, 2, \ldots, \quad m = -n, \ldots, n.$$

We can also find a similar expression for the $(R|R)$-coefficients. The addition theorem for regular functions is the following (an interested reader can derive this formula from relation (3.2.3); hint: consider conjugate and use the fact that the Bessel functions are the real parts of the Hankel functions):

$$R_0^0(\mathbf{r} + \mathbf{t}) = \sqrt{4\pi} \sum_{n'=0}^{\infty} \sum_{m'=-n'}^{n'} R_{n'}^{-m'}(-\mathbf{t}) R_{n'}^{m'}(\mathbf{r}). \tag{3.2.8}$$

Then comparing this result with expression (3.2.1), we obtain

$$(R|R)_{n'0}^{m'0}(\mathbf{t}) = \sqrt{4\pi}(-1)^{n'} R_{n'}^{-m'}(\mathbf{t}), \qquad n' = 0, 1, 2, \ldots, \quad m' = -n', \ldots, n'. \tag{3.2.9}$$

Comparing this with relation (3.2.6), we see that

$$(R|R)_{n'0}^{m'0}(\mathbf{t}) = (S|S)_{n'0}^{m'0}(\mathbf{t}). \tag{3.2.10}$$

Before considering the efficient evaluation of the translation coefficients $(S|R)_{n'n}^{m'm}(\mathbf{t})$, etc., we mention their integral representations, which immediately follow from definitions (3.2.3) and (3.2.4), and orthonormality of spherical harmonics (2.1.68):

$$(R|R)_{n'n}^{m'm}(\mathbf{t}) = \frac{1}{j_{n'}(k|\mathbf{r}|)} \int_{S_u} j_n(k|\mathbf{r} + \mathbf{t}|) Y_n^m(\mathbf{s}) Y_{n'}^{-m'}(\hat{\mathbf{s}}) dS(\hat{\mathbf{s}}), \tag{3.2.11}$$
$$\mathbf{s} = \frac{\mathbf{r} + \mathbf{t}}{|\mathbf{r} + \mathbf{t}|}, \quad \hat{\mathbf{s}} = \frac{\mathbf{r}}{|\mathbf{r}|},$$

$$(S|R)_{n'n}^{m'm}(\mathbf{t}) = \frac{1}{j_{n'}(k|\mathbf{r}|)} \int_{S_u} h_n(k|\mathbf{r} + \mathbf{t}|) Y_n^m(\mathbf{s}) Y_{n'}^{-m'}(\hat{\mathbf{s}}) dS(\hat{\mathbf{s}}), \tag{3.2.12}$$
$$|\mathbf{r}| < |\mathbf{t}|,$$

$$(S|S)_{n'n}^{m'm}(\mathbf{t}) = \frac{1}{h_{n'}(k|\mathbf{r}|)} \int_{S_u} h_n(k|\mathbf{r} + \mathbf{t}|) Y_n^m(\mathbf{s}) Y_{n'}^{-m'}(\hat{\mathbf{s}}) dS(\hat{\mathbf{s}}), \tag{3.2.13}$$
$$|\mathbf{r}| > |\mathbf{t}|.$$

3.2.1 Structure of translation coefficients

3.2.1.1 *Relation to spherical basis functions*

While the addition theorem for monopoles provides us with an explicit formula for the translation coefficients of zero order and degree, the question about what type of functions represent higher order coefficients remains unclear at this point. To clarify this issue, let us consider first the $(S|R)$-reexpansion (3.2.4). The function $S_n^m(\mathbf{r}+\mathbf{t})$ is regular inside the sphere $|\mathbf{r}| = |\mathbf{t}|$ and satisfies the Helmholtz equation:

$$(\nabla^2 + k^2)S_n^m(\mathbf{r}+\mathbf{t}) = 0. \tag{3.2.14}$$

The Laplace operator here can be considered to be acting either at fixed $\mathbf{t}$ or at fixed $\mathbf{r}$. To be certain we denote this by the respective subscript. In the former case the $(S|R)$-expansion (3.2.4) yields

$$(\nabla_{\mathbf{r}}^2 + k^2)R_{n'}^{m'}(\mathbf{r}) = 0, \tag{3.2.15}$$

which is an obvious fact, since functions $R_{n'}^{m'}(\mathbf{r})$ are introduced as solutions of the Helmholtz equation. In the latter case we have

$$(\nabla_{\mathbf{t}}^2 + k^2)(S|R)_{n'n}^{m'm}(\mathbf{t}) = 0. \tag{3.2.16}$$

So first thing we learn is that the translation coefficients satisfy the Helmholtz equation (this statement is true, of course not only for the $(S|R)$ coefficients, but for $(S|S)$ and $(R|R)$ coefficients as well). We note now that since the functions $(S|R)_{n'n}^{m'm}(\mathbf{t})$ appear as functions representing $S_n^m(\mathbf{r}+\mathbf{t})$, they should have the same type of asymptotic behavior as $S_n^m(\mathbf{r}+\mathbf{t})$ at $|\mathbf{t}| \to \infty$. For example, if k is real and, so $S_n^m(\mathbf{r}+\mathbf{t})$ are radiating functions, $(S|R)_{n'n}^{m'm}(\mathbf{t})$ should also be radiating, or if k is purely imaginary, $(S|R)_{n'n}^{m'm}(\mathbf{t})$ should decay as $|\mathbf{t}| \to \infty$.

This means that as solutions of the Helmholtz equation these coefficients can be sought in the form of expansion over functions $S_n^m(\mathbf{t})$:

$$(S|R)_{n'n}^{m'm}(\mathbf{t}) = \sum_{n''=0}^{\infty} \sum_{m''=-n''}^{n''} (s|r)_{n''n'n}^{m''m'm} S_{n''}^{m''}(\mathbf{t}). \tag{3.2.17}$$

Similar considerations for the other reexpansion pairs yields

$$\begin{aligned}
(S|S)_{n'n}^{m'm}(\mathbf{t}) &= \sum_{n''=0}^{\infty} \sum_{m''=-n''}^{n''} (s|s)_{n''n'n}^{m''m'm} R_{n''}^{m''}(\mathbf{t}), \\
(R|R)_{n'n}^{m'm}(\mathbf{t}) &= \sum_{n''=0}^{\infty} \sum_{m''=-n''}^{n'} (r|r)_{n''n'n}^{m''m'm} R_{n''}^{m''}(\mathbf{t}).
\end{aligned} \tag{3.2.18}$$

The coefficients $(s|r)$, $(s|s)$, and $(r|r)$ can be called *structural reexpansion coefficients* or *structural translation coefficients.* As follows from further consideration they are all the same.

3.2.1.2 *Addition theorems for spherical basis functions*

Substituting expression (3.2.17) into relation (3.2.4) we obtain the following expansion:

$$S_n^m(\mathbf{r}+\mathbf{t}) = \sum_{n'=0}^{\infty}\sum_{m'=-n'}^{n'}\sum_{n''=0}^{\infty}\sum_{m''=-n''}^{n''} (s|r)_{n''n'n}^{m''m'm} S_{n''}^{m''}(\mathbf{t})R_{n'}^{m'}(\mathbf{r}), \qquad |\mathbf{r}| < |\mathbf{t}|, \tag{3.2.19}$$

which is a form of the addition theorem for multipole solutions of the Helmholtz equation. Similarly, we have two more addition theorems by substituting expansions (3.2.18) into relations (3.2.1) and (3.2.4):

$$R_n^m(\mathbf{r}+\mathbf{t}) = \sum_{n'=0}^{\infty}\sum_{m'=-n'}^{n'}\sum_{n''=0}^{\infty}\sum_{m''=-n''}^{n''} (r|r)_{n''n'n}^{m''m'm} R_{n''}^{m''}(\mathbf{t})R_{n'}^{m'}(\mathbf{r}), \tag{3.2.20}$$

$$S_n^m(\mathbf{r}+\mathbf{t}) = \sum_{n'=0}^{\infty}\sum_{m'=-n'}^{n'}\sum_{n''=0}^{\infty}\sum_{m''=-n''}^{n''} (s|s)_{n''n'n}^{m''m'm} R_{n''}^{m''}(\mathbf{t})S_{n'}^{m'}(\mathbf{r}), \qquad |\mathbf{r}| > |\mathbf{t}|. \tag{3.2.21}$$

The latter expression reveals the relation between the coefficients $(s|r)$ and $(s|s)$. Indeed, comparing expressions (3.2.19) and (3.2.21), we see that these are indeed the same expansions of $S_n^m(\mathbf{r}+\mathbf{t})$, where symbols $\mathbf{r}$ and $\mathbf{t}$, and indices (n', m') and (n'', m'') are simply exchanged. Thus,

$$(s|s)_{n''n'n}^{m''m'm} = (s|r)_{n'n''n}^{m'm''m}. \tag{3.2.22}$$

The coefficients $(r|r)$ can be also expressed via $(s|r)$. As it follows from our further analysis,

$$(S|S)_{n'n}^{m'm}(\mathbf{t}) = (R|R)_{n'n}^{m'm}(\mathbf{t}), \tag{3.2.23}$$

in which case we should have from relation (3.2.18):

$$(r|r)_{n''n'n}^{m''m'm} = (s|s)_{n''n'n}^{m''m'm}. \tag{3.2.24}$$

The above addition theorems include four summations. As it will be shown in Section 3.2.1.3 one sum always disappears and the other sum turns from infinite to finite as some $(s|r)$ coefficients are zero.

3.2.1.3 *Relation to Clebsch–Gordan coefficients*

The structural coefficients $(s|r)_{n''n'n}^{m''m'm}$ are purely numerical. If so, one can be interested in their values (by the way, the analytical or exact expression for these numbers are usually not very useful in practice, since they are rather unwieldy, and some faster methods, such as recursions are used for their computations). Without going into details we just state here that they can be related to the vector-addition, or *Clebsch–Gordan coefficients* (see e.g. [Ste61]), or to the *Wigner* 3-j *symbols* (e.g. [EW31]), which are a more symmetrical form of the vector-addition coefficients. Below we provide just several useful relations and definitions.

The Clebsch–Gordan coefficients are denoted by eight indices as $(j_1j_2m_1m_2|j_1j_2jm)$ and can be defined explicitly as [AS64]:

$$(j_1j_2m_1m_2|j_1j_2jm)$$
$$=\delta_{m,m_1+m_2}\sqrt{\frac{(2j+1)(j_1+j_2-j)!(j+j_1-j_2)!(j+j_2-j_1)!}{(j+j_1+j_2+1)!}}\sum_l \alpha_l,$$
$$\alpha_l=\frac{(-1)^l\sqrt{(j_1+m_1)!(j_1-m_1)!(j_2+m_2)!(j_2-m_2)!(j+m)!(j-m)!}}{l!(j_1+j_2-j-l)!(j_1-m_1-l)!(j_2+m_2-l)!(j-j_2+m_1+l)!(j-j_1-m_2+l)!}. \quad (3.2.25)$$

Here δ is the Kronecker delta and summation should be taken over all possible l, with the notice that $1/(-n)!=0$, for $n=1,2,\ldots$.

These coefficients are related to the Wigner 3-j symbols denoted by the 2×3 matrix

$$(j_1j_2m_1m_2|j_1j_2jm)=(-1)^{-j_1+j_2-m}(2j+1)^{1/2}\begin{pmatrix} j_1 & j_2 & j \\ m_1 & m_2 & -m \end{pmatrix}. \quad (3.2.26)$$

The Wigner symbols can be related then to the so-called *E-symbols*:

$$E\begin{pmatrix} m_1 & m_2 & m \\ j_1 & j_2 & j \end{pmatrix}=4\pi\varepsilon_{m_1}\varepsilon_{m_2}\varepsilon_m\begin{pmatrix} j_1 & j_2 & j \\ 0 & 0 & 0 \end{pmatrix}\begin{pmatrix} j_1 & j_2 & j \\ m_1 & m_2 & m \end{pmatrix},$$
$$\varepsilon_m=\begin{cases}(-1)^m, & m\geq 0\\ 1, & m\leq 0\end{cases}, \quad (3.2.27)$$

where the symbol E is defined as

$$E\begin{pmatrix} m_1 & m_2 & m \\ j_1 & j_2 & j \end{pmatrix}=4\pi\left[\frac{4\pi}{(2j+1)(2j_1+1)(2j_2+1)}\right]^{1/2}\int_0^{2\pi}\mathrm{d}\varphi \quad (3.2.28)$$
$$\times\int_0^{\pi} Y_j^m(\theta,\varphi)Y_{j_1}^{m_1}(\theta,\varphi)Y_{j_2}^{m_2}(\theta,\varphi)\sin\theta\,\mathrm{d}\theta.$$

Here $Y_j^m(\theta,\varphi)$ is an orthonormal spherical harmonic defined by Eq. (2.1.59).

With these definitions of symbols and coefficients, Epton and Dembart [ED95] provide the following expression (rewritten in the present notation) for the reexpansion coefficients:

$$(S|R)_{n'n}^{m'm}(\mathbf{t}) = \sum_{n''=0}^{\infty} \sum_{m''=-n''}^{n''} \left[\frac{(2n+1)(2n'+1)(2n''+1)}{4\pi} \right]^{1/2} \mathrm{i}^{n'+n''-n} \times E\begin{pmatrix} m & -m' & -m'' \\ n & n' & n'' \end{pmatrix} S_{n''}^{m''}(\mathbf{t}). \tag{3.2.29}$$

Comparing expressions (3.2.17) and (3.2.29), we have the following relation between our numerical coefficients and the E-symbols:

$$(s|r)_{n''n'n}^{m''m'm} = \left[\frac{(2n+1)(2n'+1)(2n''+1)}{4\pi} \right]^{1/2} \mathrm{i}^{n'+n''-n} E\begin{pmatrix} m & -m' & -m'' \\ n & n' & n'' \end{pmatrix}. \tag{3.2.30}$$

This also can be expressed directly via integrals (3.2.28) as

$$(s|r)_{n''n'n}^{m''m'm} = 4\pi \mathrm{i}^{n'+n''-n} \int_{S_u} Y_n^m(\mathbf{s}) Y_{n'}^{-m'}(\mathbf{s}) Y_{n''}^{-m''}(\mathbf{s}) dS. \tag{3.2.31}$$

Equations (3.2.26), (3.2.27), and (3.2.30) provide a direct relation between the structural coefficients $(s|r)_{n''n'n}^{m''m'm}$ and the Clebsch–Gordan coefficients:

$$(s|r)_{n''n'n}^{m''m'm} = \varepsilon_m \varepsilon_{-m'} \varepsilon_{m''} \mathrm{i}^{n'+n''-n} \left[\frac{4\pi(2n+1)(2n'+1)}{(2n''+1)} \right]^{1/2} (nn'00|nn'n''0) \times (nn'm - m'|nn'n''m''). \tag{3.2.32}$$

What is noteworthy here is that the Clebsch–Gordan coefficients have a multiplier in the form of the Kronecker delta, which means that E-symbols has a multiplier $\delta_{m'',m-m'}$, and results in

$$(s|r)_{n''n'n}^{m''m'm} = 0, \qquad \text{for } m'' \neq m - m'. \tag{3.2.33}$$

This simplifies expression (3.2.17), since one summation then disappears

$$(S|R)_{n'n}^{m'm}(\mathbf{t}) = \sum_{n''=0}^{\infty} (s|r)_{n''n'n}^{m-m',m'm} S_{n''}^{m-m'}(\mathbf{t}). \tag{3.2.34}$$

Also the sum will consist of the finite number of terms, since non-zero Clebsch–Gordan coefficients are achieved only when all three of the quantities below are non-negative simultaneously:

$$j_1 + j_2 - j \geq 0, \qquad j_1 - j_2 + j \geq 0, \qquad -j_1 + j_2 + j \geq 0. \tag{3.2.35}$$

So

$$(S|R)_{n'n}^{m'm}(\mathbf{t}) = \sum_{n''=|n-n'|}^{n+n'} (s|r)_{n''n'n}^{m-m',m'm} S_{n''}^{m-m'}(\mathbf{t}). \tag{3.2.36}$$

Note that $(j_1 j_2 00 | j_1 j_2 j 0) = 0$ if $j_1 + j_2 + j$ is an odd number. This results in

$$(s|r)_{n''n'n}^{m''m'm} = 0, \qquad \text{for } n + n' + n'' = 2l + 1, \; l = 0, 1, 2, \ldots \tag{3.2.37}$$

and so neglecting the zero terms in the sum (3.2.36) we can rewrite it as

$$(S|R)_{n'n}^{m'm}(\mathbf{t}) = \sum_{l=0}^{(n+n'-|n-n'|)/2} (s|r)_{|n-n'|+2l,n'n}^{m-m',m'm} S_{|n-n'|+2l}^{m-m'}(\mathbf{t}). \tag{3.2.38}$$

It is good to have some particular values of the structural coefficients. This can be easily done when one of the degrees of these coefficients is zero. Indeed, from the definition (3.2.28) and orthonormality of the spherical harmonics we have

$$E\begin{pmatrix} 0 & m & m' \\ 0 & n & n' \end{pmatrix} = E\begin{pmatrix} m & 0 & m' \\ n & 0 & n' \end{pmatrix} = E\begin{pmatrix} m & m' & 0 \\ n & n' & 0 \end{pmatrix}$$

$$= 4\pi \left[\frac{1}{(2n+1)(2n'+1)} \right]^{1/2} \delta_{m,-m'} \delta_{nn'}, \tag{3.2.39}$$

leading to

$$(s|r)_{n''n'0}^{m''m'0} = \sqrt{4\pi}(-1)^{n'} \delta_{m'',-m'} \delta_{n''n'}, \qquad (s|r)_{n''0n}^{m''0m} = \sqrt{4\pi}\,\delta_{m''m} \delta_{n''n},$$
$$(s|r)_{0n'n}^{0m'm} = \sqrt{4\pi}\,\delta_{m'm} \delta_{n'n}. \tag{3.2.40}$$

3.2.1.4 *Symmetries of translation coefficients*

The translation coefficients obey a number of symmetries, which are particularly important for their computations, since one can substantially reduce the amount of computational work. These symmetries can be derived and proved using the recurrence relations considered below and properties of coefficients a_n^m and b_n^m specified by expressions (3.2.8) and (3.2.10). They also can be derived from symmetries of the Clebsch–Gordan coefficients or related symbols.

Let us start consideration of symmetries from the fact that according to the definition (3.2.28) the E-symbols are symmetrical with respect to arbitrary permutation of columns of these symbols. Three columns can be permuted by $3! = 6$ ways, which can be denoted as $(1,2,3)$, $(1,3,2)$, $(2,1,3)$, $(2,3,1)$, $(3,1,2)$, and $(3,2,1)$, where $(1,2,3)$ denotes original order of columns (identity or no permutations), and other triples show the order of permuted columns. Using Eq. (3.2.30) we have the following symmetries for $(s|r)_{n''n'n}^{m''m'm}$ for the five non-identity permutations:

$$(1,3,2) \Rightarrow (s|r)_{n''n'n}^{m''m'm} = (s|r)_{n'n''n}^{m'm''m}, \tag{3.2.41}$$

$$(2,1,3) \Rightarrow (s|r)_{n''n'n}^{m''m'm} = (-1)^{n+n'}(s|r)_{n'',n,n'}^{m'',-m,-m'}, \tag{3.2.42}$$

$$(2,3,1) \Rightarrow (s|r)_{n''n'n}^{m''m'm} = (-1)^{n+n'}(s|r)_{n,n''n'}^{-m,m'',-m'}, \tag{3.2.43}$$

$$(3,1,2) \Rightarrow (s|r)_{n''n'n}^{m''m'm} = (-1)^{n+n''}(s|r)_{n',n,n''}^{m',-m,-m''}, \tag{3.2.44}$$

$$(3,2,1) \Rightarrow (s|r)_{n''n'n}^{m''m'm} = (-1)^{n+n''}(s|r)_{n,n'n''}^{-m,m',-m''}. \tag{3.2.45}$$

Comparing the symmetry (3.2.41) with relations of the $(s|s)$ and $(r|r)$ coefficients to the $(s|r)$ coefficients (3.2.22) and (3.2.24), we establish that all these coefficients are the same:

$$(r|r)_{n''n'n}^{m''m'm} = (s|s)_{n''n'n}^{m''m'm} = (s|r)_{n''n'n}^{m''m'm}. \tag{3.2.46}$$

Particularly, this immediately results in

$$(R|R)_{n'n}^{m'm}(\mathbf{t}) = \sum_{n''=|n-n'|}^{n+n'} (s|r)_{n''n'n}^{m-m',m'm} R_{n''}^{m-m'}(\mathbf{t}), \tag{3.2.47}$$

or

$$(R|R)_{n'n}^{m'm}(\mathbf{t}) = \sum_{l=0}^{(n+n'-|n-n'|)/2} (s|r)_{|n-n'|+2l,n'n}^{m-m',m'm} R_{|n-n'|+2l}^{m-m'}(\mathbf{t}), \tag{3.2.48}$$

which follows from relations (3.2.17), (3.2.18), (3.2.36), and (3.2.38).

The symmetry (3.2.42) shows that

$$\begin{aligned}(S|R)_{n'n}^{m'm}(\mathbf{t}) &= \sum_{n''=0}^{\infty} \sum_{m''=-n''}^{n''} (s|r)_{n''n'n}^{m''m'm} S_{n''}^{m''}(\mathbf{t}) \\ &= (-1)^{n+n'} \sum_{n''=0}^{\infty} \sum_{m''=-n''}^{n''} (s|r)_{n''n,n'}^{m'',-m,-m'} S_{n''}^{m''}(\mathbf{t}) \\ &= (-1)^{n+n'} (S|R)_{nn'}^{-m,-m'}(\mathbf{t}). \qquad (3.2.49)\end{aligned}$$

Because the coefficients $(r|r)$, $(s|s)$, and $(s|r)$ are the same, we have from definition (3.2.18) similar relations for the $(S|R)$ and $(S|R)$ translation coefficients:

$$\begin{aligned}(R|R)_{n'n}^{m'm}(\mathbf{t}) &= (-1)^{n+n'} (R|R)_{nn'}^{-m,-m'}(\mathbf{t}), \\ (S|S)_{n'n}^{m'm}(\mathbf{t}) &= (-1)^{n+n'} (S|S)_{nn'}^{-m,-m'}(\mathbf{t}).\end{aligned} \qquad (3.2.50)$$

These relations show that particularly

$$(S|R)_{0n}^{0m}(\mathbf{t}) = (-1)^{n} (S|R)_{n0}^{-m0}(\mathbf{t}) = \sqrt{4\pi} S_{n}^{m}(\mathbf{t}), \qquad (3.2.51)$$

$$(S|S)_{0n}^{0m}(\mathbf{t}) = (R|R)_{0n}^{0m}(\mathbf{t}) = (-1)^{n} (R|R)_{n0}^{-m0}(\mathbf{t}) = \sqrt{4\pi} R_{n}^{m}(\mathbf{t}), \qquad (3.2.52)$$

where we used relations (3.2.5), (3.2.6), and (3.2.10).

It can be observed that transposition of indices is connected to the sign of translation. Indeed, due to property (3.2.7) of spherical basis functions, we have for $(S|R)$-reexpansion

$$\begin{aligned}S_{n}^{m}(\mathbf{r}+\mathbf{t}) &= (-1)^{n} S_{n}^{m}(-\mathbf{r}-\mathbf{t}) = (-1)^{n} \sum_{n'=0}^{\infty} \sum_{m'=-n'}^{n'} (S|R)_{n'n}^{m'm}(-\mathbf{t}) R_{n'}^{m'}(-\mathbf{r}) \\ &= \sum_{n'=0}^{\infty} \sum_{m'=-n'}^{n'} (-1)^{n+n'} (S|R)_{n'n}^{m'm}(-\mathbf{t}) R_{n'}^{m'}(\mathbf{r}). \qquad (3.2.53)\end{aligned}$$

Comparing this with definition (3.2.4), we obtain

$$(S|R)_{n'n}^{m'm}(-\mathbf{t}) = (-1)^{n+n'} (S|R)_{n'n}^{m'm}(\mathbf{t}). \qquad (3.2.54)$$

Now using symmetry (3.2.49), we see that

$$(S|R)_{n'n}^{m'm}(-\mathbf{t}) = (S|R)_{nn'}^{-m,-m'}(\mathbf{t}). \qquad (3.2.55)$$

Obviously, the same symmetries hold for the $(S|S)$ and $(R|R)$ translation coefficients.

The symmetry relation (3.2.55) reveals more symmetries for the coefficients $(s|r)_{n''n'n}^{m''m'm}$. Indeed we have from the definitions and (3.2.55):

$$(S|R)_{n'n}^{m'm}(-\mathbf{t}) = \sum_{n''=0}^{\infty} \sum_{m''=-n''}^{n''} (s|r)_{n''n'n}^{m''m'm} S_{n''}^{m''}(-\mathbf{t})$$

$$= \sum_{n''=0}^{\infty} \sum_{m''=-n''}^{n''} (-1)^{n''} (s|r)_{n''n'n}^{m''m'm} S_{n''}^{m''}(\mathbf{t}). \quad (3.2.56)$$

$$(S|R)_{n'n}^{m'm}(-\mathbf{t}) = (S|R)_{nn'}^{-m,-m'}(\mathbf{t}) = \sum_{n''=0}^{\infty} \sum_{m''=-n''}^{n''} (s|r)_{n''n,n'}^{m''-m,-m'} S_{n''}^{m''}(\mathbf{t}). \quad (3.2.57)$$

Since these are expansions of the same function over the same basis, we obtain

$$(s|r)_{n''n'n}^{m''m'm} = (-1)^{n''} (s|r)_{n''n,n'}^{m''-m,-m'}. \quad (3.2.58)$$

This relation can be compared with the symmetry (3.2.42). The comparison shows that these relations are consistent for arbitrary $(s|r)$ when $n + n' + n''$ is some even number, otherwise the coefficients $(s|r)$ should be zero. So we have

$$(s|r)_{n''n'n}^{m''m'm} = 0, \qquad n + n' + n'' = 2l + 1, \;\; l = 0, 1, 2, \ldots \quad (3.2.59)$$

Note also the symmetry relations for complex conjugates, which take place for real k and regular translation coefficients. In this case we have

$$\overline{R_n^m(\mathbf{r}+\mathbf{t})} = \sum_{n'=0}^{\infty} \sum_{m'=-n'}^{n'} \overline{(R|R)_{n'n}^{m'm}(\mathbf{t}) R_{n'}^{m'}(\mathbf{r})}$$

$$= \sum_{n'=0}^{\infty} \sum_{m'=-n'}^{n'} \overline{(R|R)_{n'n}^{m'm}(\mathbf{t})} R_{n'}^{-m'}(\mathbf{r}), \quad (3.2.60)$$

$$\overline{R_n^m(\mathbf{r}+\mathbf{t})} = R_n^{-m}(\mathbf{r}+\mathbf{t}) = \sum_{n'=0}^{\infty} \sum_{m'=-n'}^{n'} (R|R)_{n'n}^{m'-m}(\mathbf{t}) R_{n'}^{m'}(\mathbf{r})$$

$$= \sum_{n'=0}^{\infty} \sum_{m'=-n'}^{n'} (R|R)_{n'n}^{-m'-m}(\mathbf{t}) R_{n'}^{-m'}(\mathbf{r}). \quad (3.2.61)$$

This yields

$$\overline{(R|R)_{n'n}^{m'm}(\mathbf{t})} = (R|R)_{n'n}^{-m'-m}(\mathbf{t}) = (R|R)_{nn'}^{mm'}(-\mathbf{t}). \quad (3.2.62)$$

Here the last equality follows from properties (3.2.55), which is valid for all types of the reexpansion coefficients. As coefficients $(R|R)$ and $(S|S)$ are the same, the properties of the $(R|R)$-coefficients can be used for $(S|S)$-translation.

3.2.2 Recurrence relations for translation coefficients

Recurrence relations among the fundamental solutions of the Helmholtz equation produce recurrence relations for the reexpansion coefficients due to invariance of the differential operators ∂_z, $\partial_{x\pm iy}$ with respect to translations. Since S_n^m and R_n^m satisfy the same recurrence relations, the reexpansion coefficients $(S|R)_{n'n}^{m'm}$, $(S|S)_{n'n}^{m'm}$ and $(R|R)_{n'n}^{m'm}$ also satisfy the same recurrence relations. To avoid repeating theorems and recurrence relations for every combination of regular and singular functions, we denote the generic translation coefficient as $(E|F)_{ln}^{sm}(\mathbf{t})$ for any of the reexpansion coefficients ($(E|F) = (S|R)$, $(S|S)$ or $(R|R)$), i.e. E and F can be any of the functions S and R. Thus the following reexpansion holds:

$$E_n^m(\mathbf{r}+\mathbf{t}) = \sum_{n'=0}^{\infty} \sum_{m'=-n'}^{n'} (E|F)_{n'n}^{m'm}(\mathbf{t}) F_{n'}^{m'}(\mathbf{r}), \qquad E, F = S, R. \tag{3.2.63}$$

Denoting by ∂_α any of the operators ∂_z or $\partial_{x\pm iy}$ at fixed translation vector $\mathbf{t}$, we have:

$$\partial_\alpha E_n^m(\mathbf{r}+\mathbf{t}) = \sum_{n'=0}^{\infty} \sum_{m'=-n'}^{n'} (E|F)_{n'n}^{m'm}(\mathbf{t}) \partial_\alpha F_{n'}^{m'}(\mathbf{r}), \qquad E, F = S, R. \tag{3.2.64}$$

The following theorems establish general recurrence relations. Their proofs are based on the theorems for differentiation of elementary solutions in spherical coordinates. We provide the proof only for the first theorem. The proof of other theorems are very similar and the interested reader can do this as an exercise or may be obtained from our report [GD01]. Note that Chew [Che92] was presumably the first who published the recurrence relations stated by these theorems.

THEOREM 9 *For $k \neq 0$ the following recurrence relation holds for $(E|F)_{ln}^{sm}(\mathbf{t})$:*

$$\begin{aligned} &a_{n-1}^m (E|F)_{n',n-1}^{m'm}(\mathbf{t}) - a_n^m (E|F)_{n',n+1}^{m'm}(\mathbf{t}) \\ &= a_{n'}^{m'} (E|F)_{n'+1,n}^{m'm}(\mathbf{t}) - a_{n'-1}^{m'} (E|F)_{n'-1,n}^{m'm}(\mathbf{t}), \end{aligned} \tag{3.2.65}$$

$$E, F = S, R, \quad n', n = 0, 1, \ldots, \quad m' = -n', \ldots, n', \quad m = -n, \ldots, n,$$

where coefficients a_n^m are given by expression (2.2.8).

PROOF. Using differentiation formula (2.2.7) and definition of translation coefficients (3.2.63), we find

$$\partial_z E_n^m(\mathbf{r}+\mathbf{t}) = a_{n-1}^m E_{n-1}^m(\mathbf{r}+\mathbf{t}) - a_n^m E_{n+1}^m(\mathbf{r}+\mathbf{t})$$

$$= \sum_{n'=0}^{\infty} \sum_{m'=-n'}^{n'} \left[a_{n-1}^m (E|F)_{n',n-1}^{m'm}(\mathbf{t}) - a_n^m (E|F)_{n',n+1}^{m'm}(\mathbf{t}) \right] F_{n'}^{m'}(\mathbf{r}).$$

On the other hand using the same differentiation formula (2.2.7) and relation (3.2.64), we have

$$\begin{aligned}
\partial_z E_n^m(\mathbf{r}+\mathbf{t}) &= \sum_{n'=0}^{\infty} \sum_{m'=-n'}^{n'} (E|F)_{n'n}^{m'm}(\mathbf{t}) \partial_z F_{n'}^{m'}(\mathbf{r}) \\
&= \sum_{n'=0}^{\infty} \sum_{m'=-n'}^{n'} (E|F)_{n'n}^{m'm}(\mathbf{t}) \left[a_{n'-1}^{m'} F_{n'-1}^{m'}(\mathbf{r}) - a_{n'}^{m'} F_{n'+1}^{m'}(\mathbf{r}) \right] \\
&= \sum_{n'=-1}^{\infty} \sum_{m'=-n'-1}^{n'+1} a_{n'}^{m'} (E|F)_{n'+1,n}^{m'm}(\mathbf{t}) F_{n'}^{m'}(\mathbf{r}) \\
&\quad - \sum_{n'=1}^{\infty} \sum_{m'=-n'+1}^{n'-1} a_{n'-1}^{m'} (E|F)_{n'-1,n}^{m'm}(\mathbf{t}) F_{n'}^{m'}(\mathbf{r}) \\
&= \sum_{n'=0}^{\infty} \sum_{m'=-n'}^{n'} \left[a_{n'}^{m'} (E|F)_{n'+1,n}^{m'm}(\mathbf{t}) - a_{n'-1}^{m'} (E|F)_{n'-1,n}^{m'm}(\mathbf{t}) \right] F_{n'}^{m'}(\mathbf{r}).
\end{aligned}$$

The last equality holds according to definition (2.2.8)

$$a_{-1}^{m'} = a_{n'}^{n'+1} = a_{n'}^{-n'-1} = a_{n'-1}^{n'} = a_{n'-1}^{-n'} = 0.$$

Comparing these two expressions and using the orthogonality and completeness of the surface harmonics we obtain the statement of the theorem. □

COROLLARY 3 *For* $n = |m|$:

$$a_{|m|}^m (E|F)_{n',|m|+1}^{m'm}(\mathbf{t}) = a_{n'-1}^{m'} (E|F)_{n'-1,|m|}^{m'm}(\mathbf{t}) - a_{n'}^{m'} (E|F)_{n'+1,|m|}^{m'm}(\mathbf{t}),$$
$$E, F = S, R, \quad n' = 0, 1, \ldots, \quad m' = -n', \ldots, n', \quad m = 0, \pm 1, \pm 2, \ldots \tag{3.2.66}$$

For $n' = |m'|$:

$$a_{|m'|}^{m'} (E|F)_{|m'|+1,n}^{m'm}(\mathbf{t}) = a_{n-1}^{m} (E|F)_{|m'|,n-1}^{m'm}(\mathbf{t}) - a_{n}^{m} (E|F)_{|m'|,n+1}^{m'm}(\mathbf{t}),$$
$$E, F = S, R, \quad n = 0, 1, \ldots, \quad m = -n, \ldots, n, \quad m' = 0, \pm 1, \pm 2, \ldots \tag{3.2.67}$$

For $n = |m|$ and $n' = |m'|$:

$$a_{|m'|}^{m'}(E|F)_{|m'|+1,|m|}^{m'm}(\mathbf{t}) = -a_{|m|}^{m}(E|F)_{|m'|,|m|+1}^{m'm}(\mathbf{t}),$$
$$E,F = S,R, \quad m = 0, \pm 1, \pm 2, \ldots, \quad m' = 0, \pm 1, \pm 2, \ldots \tag{3.2.68}$$

THEOREM 10 *For $k \neq 0$ the following recurrence relation holds for $(E|F)_{n'n}^{m'm}(\mathbf{t})$:*

$$b_n^m (E|F)_{n',n-1}^{m',m+1}(\mathbf{t}) - b_{n+1}^{-m-1}(E|F)_{n',n+1}^{m',m+1}(\mathbf{t})$$
$$= b_{n'+1}^{m'-1}(E|F)_{n'+1,n}^{m'-1,m}(\mathbf{t}) - b_{n'}^{-m'}(E|F)_{n'-1,n}^{m'-1,m}(\mathbf{t}), \tag{3.2.69}$$

$$E,F = S,R, \quad n',n = 0,1,\ldots, \quad m' = -n',\ldots,n', \quad m = -n,\ldots,n,$$

where coefficients b_n^m are given by expression (2.2.10).

COROLLARY 4 *For $n = m$*

$$b_{m+1}^{-m-1}(E|F)_{n',m+1}^{m',m+1}(\mathbf{t}) = b_{n'}^{-m'}(E|F)_{n'-1,m}^{m'-1,m}(\mathbf{t}) - b_{n'+1}^{m'-1}(E|F)_{n'+1,m}^{m'-1,m}(\mathbf{t}),$$
$$E,F = S,R, \quad n' = 0.1,\ldots, \quad m' = -n',\ldots,n', \quad m = 0,1,2,\ldots \tag{3.2.70}$$

For $n' = |m'|$, $m' \leq 0$

$$b_{|m'|+1}^{-|m'|-1}(E|F)_{|m'|+1,n}^{-|m'|-1,m}(\mathbf{t}) = b_n^m (E|F)_{|m'|,n-1}^{-|m'|,m+1}(\mathbf{t}) - b_{n+1}^{-m-1}(E|F)_{|m'|,n+1}^{-|m'|,m+1}(\mathbf{t}),$$
$$E,F = S,R, \quad n = 0,1,\ldots, \quad m = -n,\ldots,n, \quad m' = 0,-1,-2,\ldots \tag{3.2.71}$$

For $n = m$ and $n' = |m'|$, $m' \leq 0$

$$b_{m+1}^{-m-1}(E|F)_{|m'|,m+1}^{-|m'|,m+1}(\mathbf{t}) = -b_{|m'|+1}^{-|m'|-1}(E|F)_{|m'|+1,m}^{-|m'|-1,m}(\mathbf{t}),$$
$$E,F = S,R, \quad m = 0,1,2,\ldots, \quad m' = 0,-1,-2,\ldots \tag{3.2.72}$$

THEOREM 11 *For $k \neq 0$ the following recurrence relation holds for $(E|F)_{n'n}^{m'm}(\mathbf{t})$:*

$$b_{n+1}^{m-1}(E|F)_{n',n+1}^{m',m-1}(\mathbf{t}) - b_n^{-m}(E|F)_{n',n-1}^{m',m-1}(\mathbf{t})$$
$$= b_{n'}^{m'}(E|F)_{n'-1,n}^{m'+1,m}(\mathbf{t}) - b_{n'+1}^{-m'-1}(E|F)_{n'+1,n}^{m'+1,m}(\mathbf{t}), \tag{3.2.73}$$

$$E,F = S,R, \quad n',n = 0,1,\ldots, \quad m' = -n',\ldots,n', \quad m = -n,\ldots,n,$$

where coefficients b_n^m are given by expression (2.2.10).

COROLLARY 5 *For* $n = |m|$, $m \leq 0$

$$b_{|m|+1}^{-|m|-1}(E|F)_{n',|m|+1}^{m',-|m|-1}(\mathbf{t})$$
$$= b_{n'}^{m'}(E|F)_{n'-1,|m|}^{m'+1,-|m|}(\mathbf{t}) - b_{n'+1}^{-m'-1}(E|F)_{n'+1,|m|}^{m'+1,-|m|}(\mathbf{t}), \quad (3.2.74)$$

$$E,F = S,R, \; n' = 0,1,\ldots, \; m' = -n',\ldots,n', \; m = 0,-1,-2,\ldots$$

For $n' = m'$

$$b_{m'+1}^{-m'-1}(E|F)_{m'+1,n}^{m'+1,m}(\mathbf{t}) = b_n^{-m}(E|F)_{m',n-1}^{m',m-1}(\mathbf{t}) - b_{n+1}^{m-1}(E|F)_{m',n+1}^{m',m-1}(\mathbf{t}), \quad (3.2.75)$$

$$E,F = S,R, \; n = 0,1,\ldots, \; m = -n,\ldots,n, \; m' = 0,1,2,\ldots$$

For $n = |m|$, $m \leq 0$ *and* $n' = m'$

$$b_{|m|+1}^{-|m|-1}(E|F)_{m',|m|+1}^{m',-|m|-1}(\mathbf{t}) = -b_{m'+1}^{-m'-1}(E|F)_{m'+1,|m|}^{m'+1,-|m|}(\mathbf{t}), \quad (3.2.76)$$

$$E,F = S,R, \; m = 0,-1,-2,\ldots, \; m' = 0,1,2,\ldots$$

3.2.2.1 Sectorial coefficients

In analogy with the surface spherical harmonics we will call translation coefficients of the type $(E|F)_{n'|m|}^{m'm}$ and $(E|F)_{|m'|n}^{m'm}$ as "sectorial" translation coefficients, since they involve reexpansion of sectorial harmonics or represent coefficients near sectorial harmonics in reexpansions (so coefficients of general type $(E|F)_{n'n}^{m'm}$ we can call the "tesseral" translation coefficients). For such coefficients we will use simplified notation

$$(E|F)_{n',}^{m'm} = (E|F)_{n'|m|}^{m'm}, \qquad (E|F)_{,n}^{m'm} = (E|F)_{|m'|n}^{m'm}, \qquad E,F = S,R. \quad (3.2.77)$$

These coefficients are described with only three indices, opposed to the four indices in the general case and their computation and properties are simpler compared to general translation coefficients. Note that only coefficients $(E|F)_{n,}^{m'm}$ can be computed while $(E|F)_{,n}^{m'm}$ can be determined using symmetries (3.2.49) and (3.2.50). As follows from general recurrence relationships for translation coefficients, the sectorial coefficients satisfy the following recursions:

$$b_{m+1}^{-m-1}(E|F)_{n',}^{m',m+1} = b_{n'}^{-m'}(E|F)_{n'-1,}^{m'-1,m} - b_{n'+1}^{m'-1}(E|F)_{n'+1,}^{m'-1,m}, \quad (3.2.78)$$

$$E,F = S,R,$$

$$b_{m+1}^{-m-1}(E|F)_{n',}^{m',-m-1} = b_{n'}^{m'}(E|F)_{n'-1,}^{m'+1,-m} - b_{n'+1}^{-m'-1}(E|F)_{n'+1,}^{m'+1,-m},$$
$$n' = 0, 1, \ldots, \quad m' = -n', \ldots, n', \quad m = 0, 1, 2, \ldots \tag{3.2.79}$$

Particular kind of sectorial coefficients (3.2.77), or two-index coefficients of the type $(E|F)_{|m'||m|}^{m'm}$ can be referred as "double sectorial" coefficients, and they represent relations between the sectorial-to-sectorial basis function transforms. The following simplified notation can be used for the double sectorial coefficients:

$$(E|F)^{m'm} = (E|F)_{|m'||m|}^{m'm}, \qquad E, F = S, R. \tag{3.2.80}$$

These coefficients satisfy a very simple recursion following from relations (3.2.78) and (3.2.79)

$$b_{m+1}^{-m-1}(E|F)^{-m',m+1} = -b_{m'+1}^{-m'-1}(E|F)^{-m'-1,m},$$
$$E, F = S, R, \quad m, m' = 0, 1, 2, \ldots \tag{3.2.81}$$

Recursive application of this formula and knowledge of the values of coefficients b_m^{-m} from Eq. (2.2.10) enable explicit expressions for the $(E|F)^{-m'm}$ coefficients $(m, m' = 0, 1, 2, \ldots)$ via $(E|F)^{-m'-m,0}$, whose values are known from Eqs. (3.2.5) and (3.2.6):

$$(S|R)^{-m',m}(\mathbf{t}) = (-1)^{m'}\frac{(m'+m)!}{m'!m!}\sqrt{\frac{4\pi(2m+1)!(2m'+1)!}{(2m'+2m+1)!}}S_{m'+m}^{m'+m}(\mathbf{t}), \tag{3.2.82}$$

$$(R|R)^{-m',m}(\mathbf{t}) = (S|S)^{-m',m}(\mathbf{t})$$
$$= (-1)^{m'}\frac{(m'+m)!}{m'!m!}\sqrt{\frac{4\pi(2m+1)!(2m'+1)!}{(2m'+2m+1)!}}R_{m'+m}^{m'+m}(\mathbf{t}), \tag{3.2.83}$$
$$m = 0, 1, 2, \ldots, \quad m' = 0, 1, 2, \ldots$$

Using this and symmetry relations (3.2.49) and (3.2.50) we can also determine $(S|R)^{m',-m}(\mathbf{t})$ and $(R|R)^{m',-m}(\mathbf{t})$.

3.2.2.2 *Computation of translation coefficients*

Translation of solutions of the Helmholtz equation in three dimensions can be performed efficiently without direct computation of general translation coefficients (we will consider operator decomposition methods in Chapters 7 and 8). However, some practical problems in exact formulations can be reduced to computation of matrices consisting of translation coefficients (see Chapter 4). Moreover, for low truncation numbers, p, translation methods based on direct application of the truncated translation matrix can be faster than other methods. So researchers may encounter a task of how to compute the translation coefficients explicitly. While dealing with such a task we normally are interested in computation of a set of coefficients $(E|F)_{n'n}^{m'm}(\mathbf{t})$ in a finite specified range of indices n, n', m, and m', say for

$$-(p-1) \leq m, m' \leq p-1, \qquad 0 \leq n, n' \leq p-1, \tag{3.2.84}$$

where p is the maximum number of harmonics sufficient for function representation within some error bounds. Further we would like to perform these computations rapidly and accurately.

A straightforward idea, which was used successfully in several studies dealing with translation of spherical basis functions is to compute the translation coefficients using the addition theorems (3.2.19)–(3.2.21) in which structural coefficients of the type $(s|r)$ can be precomputed using their connection with the Clebsh–Gordan or relative coefficients. Equation (3.2.36) shows that such computations can be performed exactly (within computer precision), since only a finite number of terms are needed to compute each translation coefficients. Even though according to relation (3.2.59) half of the terms in Eq. (3.2.36) are zero, computation of each sum for functions of order $\sim p$ requires $O(p)$ operations. Since the total number of coefficients $(E|F)_{n'n}^{m'm}(\mathbf{t})$ to compute for range (3.2.84) is $O(p^4)$ the total complexity for computation of the set of these coefficients will be $O(p^5)$. This imposes an obvious computational efficiency problem as soon as one is concerned with problems of even moderate complexity (say that need $p = 10$) and there is need for computation of $(E|F)_{n'n}^{m'm}(\mathbf{t})$ for a variety of translations $\mathbf{t}$. On the other hand even if only one operation is spent in computing one translation coefficient $(E|F)_{n'n}^{m'm}$, the complexity of the computation for the whole set will be $O(p^4)$, which is the theoretical limit for complexity of computation of all $(E|F)_{n'n}^{m'm}$ in the specified range of indices.

Algorithms based on the recurrence relations that were derived above achieve this complexity, $O(p^4)$. Moreover, since these relations are exact there is no loss of accuracy and errors are determined by the precision of computer arithmetic. Of course, as for any recurrent computation (say recurrent computations of the Bessel functions or spherical harmonics) the question about the stability of recurrence formula is valid. However, we will not go into the details of such stability analysis for the following practical reason. If an instability exists, it will manifest itself at larger p. Since the computational algorithm is anyway $O(p^4)$ such high values of p are precluded, and other methods for solution of the Helmholtz equation (or other problem related to this equation) should be adopted for efficiency.

There are several ways to organize the recurrent computations of the translations. Below we represent one of the possible algorithms, which we have tested and found to be fast and accurate for multiple scattering problems [GD02].

First, we note that the range of non-zero coefficients $(E|F)_{n'n}^{m'm}$ is bounded by $|m| \leq n$ and $|m'| \leq n'$. The process starts with specification of initial values $(E|F)_{n'0}^{m'0}$ using Eq. (3.2.5) or (3.2.6) for $n' = 0, \ldots, 2p-2$, and $m' = -n', \ldots, n'$. Relation (3.2.78) shows that these data provide computation of the sectorial coefficients $(E|F)_{n',}^{m'm}$ at $m = 1$ for $n' = 0, \ldots, 2p-3$ and $m' = -n', \ldots, n'$ and further till $m = p-1$, so the range $n' = 0, \ldots, p-1$, $m' = -n', \ldots, n'$ is covered. Similarly, relation (3.2.79) enables computation of the sectorial coefficients $(E|F)_{n',}^{m'm}$ for $m = -1, \ldots, -(p-1)$ and the same range of n' and m'. Symmetries (3.2.49) and (3.2.50) are used to find sectorial coefficients $(E|F)_{,n}^{m'm}$ for $m' = -(p-1), \ldots, p-1$ and $n = 0, \ldots, 2p-2-|m'|$, $m = -n, \ldots, n$. A scheme for computation of the sectorial translation coefficients is shown in Fig. 3.1.

Consider now the computation of the other (tesseral) coefficients. For this purpose we use Eq. (3.2.65) at the layer $m = \text{const}$, $m' = \text{const}$. Assume that $|m'| \leq |m|$. For such a layer we have coefficients known at $n = |m|$ and $n' = |m'|, \ldots, 2p-2-|m|$. At $n = |m|+1$ Eq. (3.2.65) yields values of the translation coefficients for $n' = |m'|, \ldots, 2p-3-|m|$ and further, until $n = p-1$ and $n' = |m'|, \ldots, p-1$. This fills some trapezoidal

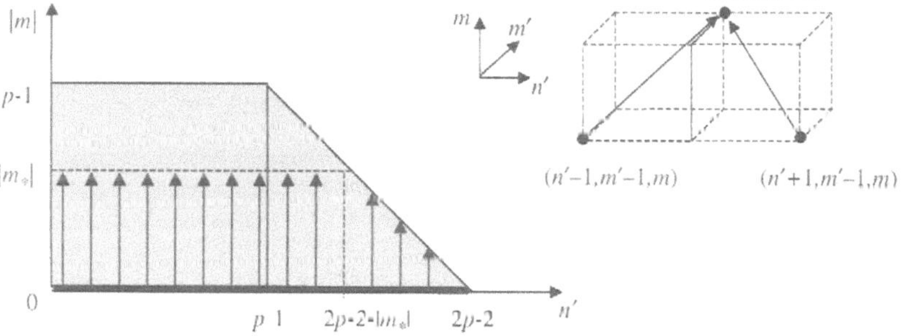

FIGURE 3.1 Schematic of recursive computation of the three-index sectorial translation coefficients. Recurrence relation (3.2.78) enables propagation in (n', m', m)-space for increasing m (shown on the right). Filling a block of coefficients in this space starts from initial values specified at $m = 0$ and propagates to $|m| > 0$ (shown on the left in projection to the (n', m)-plane).

domain in the (n, n')-plane from the left to the right (see Fig. 3.2). The rest of the domain can be filled by applying the same recurrence relation, but filling from the bottom to the top by propagation with respect to n'. So we use known values of the sectorial coefficients at $n' = |m'|$ and $n = |m|, \ldots, 2p - 2 - |m'|$ and use the recurrence equation (3.2.65) resolved with respect to coefficients for $n' + 1$. A similar procedure holds for $|m| \leq |m'|$ (see Fig. 3.2). Thus this algorithm enables computation of all translation coefficients inside the specified domain.

Note that the presentation of the explicit algorithm for computation of all translation coefficients from specified initial values $(E|F)_{n'0}^{m'0}$ immediately proves the statement (3.2.23) that the $(R|R)$ and $(S|S)$ coefficients for the 3D Helmholtz equation are the same. Indeed their initial values are the same (see Eq. (3.2.10)) and they satisfy the same recurrence relations sufficient for computation of $(E|F)_{n'n}^{m'm}$. This fact can also be derived from the symmetries of the structural coefficients as we showed above.

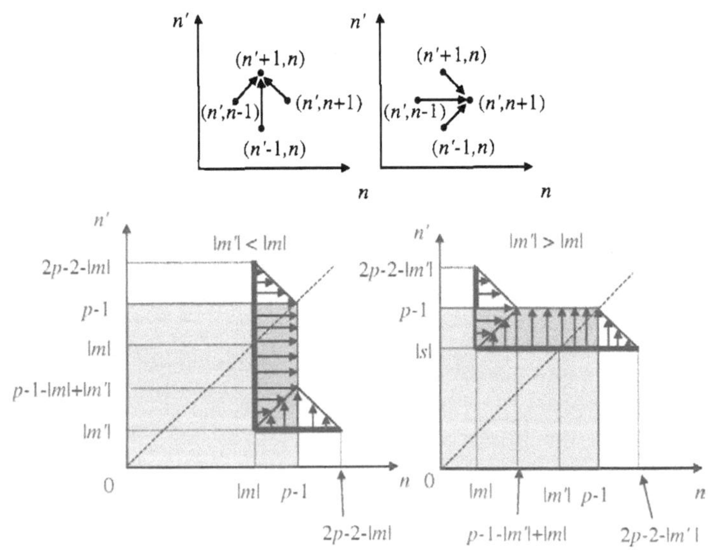

FIGURE 3.2 A scheme for recursive computation of the translation coefficients at layer $m = \text{const}$, $m' = \text{const}$. Propagation according Eq. (3.2.65) can be performed with respect to n' or to n as shown in the top graphs. Using precomputed sectorial coefficients at $n' = |m'|$ and $n = |m|$ and these propagation schemes one can compute all translation coefficients in the specified range, $|m'| \leq n' \leq p - 1$, $|m| \leq n \leq p - 1$. To reduce the amount of unnecessary computations, two cases, $|m'| \leq |m|$ and $|m| < |m'|$, can be considered as shown in figures on the bottom right and left.

3.2.3 Coaxial translation coefficients

In one important case the translation of the spherical basis functions can be performed with lesser number of coefficients, and that is why this case deserves special consideration. Assume that the translation vector is oriented along the z-axis:

$$\mathbf{t} = t\mathbf{i}_z, \qquad t = |\mathbf{t}|. \tag{3.2.85}$$

where t is the translation distance. We will call such a translation "coaxial", since it is along the coordinate *axis*. We can observe that the basis functions $E_n^m(\mathbf{r} + t\mathbf{i}_z)$ and $E_n^m(\mathbf{r})$, $E = S, R$ have the same dependence on the spherical polar angle φ, $E_n^m(\mathbf{r} + t\mathbf{i}_z) \sim e^{im\varphi}$, $E_n^m(\mathbf{r}) \sim e^{im\varphi}$. Due to the orthogonality of the functions $e^{im\varphi}$, $m = 0, \pm 1, \pm 2, \ldots$ we can see that the basis functions of order m can be expanded over basis function of order m only. Moreover, the translation coefficients in this case will depend on the translation distance only. So we have

$$S_n^m(\mathbf{r} + t\mathbf{i}_z) = \sum_{n'=|m|}^{\infty} (S|R)_{n'n}^m(t) R_{n'}^m(\mathbf{r}), \qquad |\mathbf{r}| < t, \tag{3.2.86}$$

$$S_n^m(\mathbf{r} + t\mathbf{i}_z) = \sum_{n'=|m|}^{\infty} (S|S)_{n'n}^m(t) S_{n'}^m(\mathbf{r}), \qquad |\mathbf{r}| > t, \tag{3.2.87}$$

$$R_n^m(\mathbf{r} + t\mathbf{i}_z) = \sum_{n'=|m|}^{\infty} (R|R)_{n'n}^m(t) R_{n'}^m(\mathbf{r}). \tag{3.2.88}$$

Here the lower bound of summation is $n' = |m|$, is due to $Y_{n'}^m = 0$ for $n' < |m|$.

The coaxial coefficients $(E|F)_{n'n}^m(t)$, $E, F = S, R$ introduced in this way, obviously, can be related to the general translation coefficients. Indeed, the coaxial translation along the z-axis means that the spherical polar angle of the translation vector $\mathbf{t}$ is zero, $\theta_t = 0$. In other words,

$$\begin{aligned} &(E|F)_{n'n}^m(t) = (E|F)_{n'n}^{mm}(\mathbf{t})|_{\theta_t=0}, \\ &E, F = S, R; \quad n = 0, 1, \ldots, \quad m = -n, \ldots, n, \quad n' = |m|, |m|+1, \ldots \end{aligned} \tag{3.2.89}$$

3.2.3.1 *Recurrences*

The latter expression shows that recurrence relations and symmetry properties for $(E|F)_{n'n}^{mm}(\mathbf{t})$ can be used for computation of three-index coefficients $(E|F)_{n'n}^m(t)$. The recurrence formula (3.2.65) does not act on the orders of the translation coefficients, so setting $m' = m$ there yields

$$a_{n-1}^m (E|F)_{n',n-1}^m - a_n^m (E|F)_{n',n+1}^m = a_{n'}^m (E|F)_{n'+1,n}^m - a_{n'-1}^m (E|F)_{n'-1,n}^m, \quad E, F = S, R. \tag{3.2.90}$$

In relation (3.2.69) we set $m' = m+1$ to obtain

$$b_n^m (E|F)_{n',n-1}^{m+1} - b_{n+1}^{-m-1} (E|F)_{n',n+1}^{m+1} = b_{n'+1}^m (E|F)_{n'+1,n}^m - b_{n'}^{-m-1} (E|F)_{n'-1,n}^m, \tag{3.2.91}$$

where E, $F = S$, R. If coefficients $(E|F)_{n'n}^0$ are known one can compute $(E|F)_{ln}^m$ for $m = 1,2,\ldots$ to obtain all coaxial translation coefficients for positive m.

3.2.3.2 Symmetries

Now we note that the coaxial coefficients can be computed only for positive m due to the additional symmetry

$$(E|F)_{n'n}^m = (E|F)_{n'n}^{-m} = (E|F)_{n'n}^{|m|}, \qquad E, F = S, R. \tag{3.2.92}$$

This symmetry can be proved by several ways.

For example, if we consider expression (3.2.36) in the case $m = m'$, we obtain

$$(S|R)_{n'n}^{mm}(\mathbf{t}) = \sum_{n''=|n-n'|}^{n+n'} (s|r)_{n''n'n}^{0mm} S_{n''}^0(\mathbf{t}). \tag{3.2.93}$$

Now we can show that

$$(s|r)_{n''n'n}^{0mm} = (s|r)_{n''n'n}^{0,-m,-m}. \tag{3.2.94}$$

This follows from relation (3.2.30) and the definition of the E-symbols (3.2.28):

$$\begin{aligned}
E\begin{pmatrix} m & -m & 0 \\ n & n' & n'' \end{pmatrix} &= 4\pi \left[\frac{4\pi}{(2n+1)(2n'+1)(2n''+1)} \right]^{1/2} \int_{-\pi}^{\pi} d\varphi \\
&\quad \times \int_0^{\pi} Y_{n''}^0(\theta,\varphi) Y_{n'}^{-m}(\theta,\varphi) Y_n^m(\theta,\varphi) \sin\theta \, d\theta \\
&= 4\pi \left[\frac{4\pi}{(2n+1)(2n'+1)(2n''+1)} \right]^{1/2} \int_{-\pi}^{\pi} d\varphi \\
&\quad \times \int_0^{\pi} Y_{n''}^0(\theta,\varphi) Y_{n'}^{m}(\theta,\varphi) Y_n^{-m}(\theta,\varphi) \sin\theta \, d\theta \\
&= E\begin{pmatrix} -m & m & 0 \\ n & n' & n'' \end{pmatrix},
\end{aligned}$$

where according to definition of spherical harmonics (2.1.59) we have

$$Y_{n'}^{-m}(\theta,\varphi)Y_n^m(\theta,\varphi) = e^{-im\varphi}f_{n'}^{|m|}(\theta)e^{im\varphi}f_n^{|m|}(\theta) = e^{im\varphi}f_{n'}^{|m|}(\theta)e^{-im\varphi}f_n^{|m|}(\theta)$$
$$= Y_{n'}^m(\theta,\varphi)Y_n^{-m}(\theta,\varphi), \tag{3.2.95}$$

and $f_n^{|m|}(\theta)$ represents dependence of spherical harmonics on angle θ, which does not depend on the sign of m.

Symmetry (3.2.92) combined with symmetry (3.2.49) or (3.2.50) yields

$$(E|F)_{n'n}^m = (-1)^{n+n'}(E|F)_{nn'}^m, \qquad E, F = S, R. \tag{3.2.96}$$

Note also that coefficients of type $(E|F)_{n'n}^{mm}(\mathbf{t})$ for an arbitrary translation vector $\mathbf{t}$ depend only on the translation angle θ_t and translation distance t, and do not depend on the translation angle φ_t. This follows from relation (3.2.93), since basis functions of zero order $S_{n''}^0(\mathbf{t})$ depend on θ_t and t only and similar formula holds for $(R|R)$-coefficients. So we have

$$(S|R)_{n'n}^{mm}(\mathbf{t}) = \sum_{n''=|n-n'|}^{n+n'} (s|r)_{n''n'n}^{0mm} S_{n''}^0(\mathbf{t})$$
$$= \sum_{n''=|n-n'|}^{n+n'} (s|r)_{n''n'n}^{0mm} h_{n''}(kt) Y_{n''}^0(\theta_t,\varphi_t)$$
$$= \sum_{n''=|n-n'|}^{n+n'} \sqrt{\frac{2n''+1}{4\pi}} (s|r)_{n''n'n}^{0mm} h_{n''}(kt) P_{n''}(\cos\theta_t), \tag{3.2.97}$$

and

$$(R|R)_{n'n}^{mm}(\mathbf{t}) = \sum_{n''=|n-n'|}^{n+n'} \sqrt{\frac{2n''+1}{4\pi}} (s|r)_{n''n'n}^{0mm} j_{n''}(kt) P_{n''}(\cos\theta_t). \tag{3.2.98}$$

At $\theta_t = 0$ these coefficients turn into coaxial coefficients (3.2.89), so we use $P_{n''}(1) = 1$:

$$(S|R)_{n'n}^m(t) = \sum_{n''=|n-n'|}^{n+n'} \sqrt{\frac{2n''+1}{4\pi}} (s|r)_{n''n'n}^{0mm} h_{n''}(kt), \tag{3.2.99}$$

$$(R|R)_{n'n}^m(t) = \sum_{n''=|n-n'|}^{n+n'} \sqrt{\frac{2n''+1}{4\pi}} (s|r)_{n''n'n}^{0mm} j_{n''}(kt). \tag{3.2.100}$$

3.2.3.3 Computations

As in general case of translation coefficients the most efficient way to compute coaxial translation coefficients is via recurrence relations. The total number of coaxial translation coefficients in computational domain

$$m = -(p-1), \ldots, p-1, \qquad n, n' = |m|, |m|+1, \ldots, p-1 \tag{3.2.101}$$

is $O(p^3)$, and the recurrent method of computation is exact and requires $O(p^3)$, operations. Due to the symmetry relations $(E|F)^m_{ln}$ can be computed only for

$$p - 1 \geq n' \geq n \geq m \geq 0. \tag{3.2.102}$$

The process of recurrent computation of the coefficients $\{(E|F)^m_{n'n}\}$ can be performed by computing the entries corresponding to the degrees n' and n followed by advancement with respect to the order m. We need to initialize the procedure by providing values for $m = 0$ and employing the recurrence relation (3.2.90). According to Eqs. (3.2.5) and (3.2.6) we have

$$\begin{aligned} (S|R)^0_{n'0}(t) &= \sqrt{4\pi}(-1)^{n'} S^0_{n'}(\mathbf{t})|_{\theta_t=0} = (-1)^{n'}\sqrt{2n'+1}h_{n'}(kt), \\ (R|R)^0_{n'0}(t) &= (-1)^{n'}\sqrt{2n'+1}j_{n'}(kt). \end{aligned} \tag{3.2.103}$$

For advancement with respect to m it is convenient to use relation (3.2.91) for $n = m$:

$$\begin{aligned} &b^{-m-1}_{m+1}(E|F)^{m+1}_{n',m+1} = b^{-m-1}_{n'}(E|F)^m_{n'-1,m} - b^m_{n'+1}(E|F)^m_{n'+1,m}, \\ &n' = m+1, m+2, \ldots, \end{aligned} \tag{3.2.104}$$

where $E, F = S, R$, and obtain other $(E|F)^{m+1}_{n'n}$ using Eq. (3.2.90) in the same way as $(E|F)^0_{n'n}$ are computed.

Formulae (3.2.103) and (3.2.104) employ sectorial coefficients of the type $(E|F)^m_{lm}$, which can conveniently be denoted as

$$(E|F)^m_{n'} = (E|F)^m_{n'm}, \qquad l = m, m+1, \ldots \tag{3.2.105}$$

and called *coaxial sectorial coefficients*, and satisfy the following relations:

$$\begin{aligned} &(S|R)^0_{n'}(t) = (-1)^{n'}\sqrt{2n'+1}h_{n'}(kt), \\ &(R|R)^0_{n'}(t) = (-1)^{n'}\sqrt{2n'+1}j_{n'}(kt), \\ &b^{-m-1}_{m+1}(E|F)^{m+1}_{n'} = b^{-m-1}_{n'}(E|F)^m_{n'-1} - b^m_{n'+1}(E|F)^m_{n'+1}, \\ &E, F = S, R, \quad n' = m+1, m+2, \ldots \end{aligned} \tag{3.2.106}$$

Note that for computation of the translation coefficients inside an (n', n, m) cube of size (p, p, p) the coefficients $(E|F)^{0}_{n'0}$ must be computed for $n' = 0, \ldots, 2p-2$. This is because the recurrence relations for increase of n and for increase of n' Eqs. (3.2.90) and (3.2.106) require knowledge of $(E|F)^{m}_{n'+1,n}$ to compute $(E|F)^{m}_{n',n+1}$ and $(E|F)^{m+1}_{n'n}$.

3.3 ROTATIONS OF ELEMENTARY SOLUTIONS

3.3.1 Angles of rotation

It is often convenient (and physically meaningful) to link the entries the of the matrix Q given by expression (3.1.11) to the physical angles between the axes. This is usually done via one of many representations such as the *Euler angles*. We also may wish to rotate a set of axes so that the old z-axis is rotated to a specified $\hat{z}$ direction, in view of the simplicity of the coaxial translation. A simple expression for Q in terms of the direction cosines of the $\hat{z}$ direction that achieves this objective can be derived from elementary geometric considerations. We recall from Euler's theorem that any rotation of a rigid body can be uniquely specified by providing an axis of rotation and the angle of rotation through that axis.

Referring to the figure below, the origin and the two z-axes form a given plane $(Oz\hat{z})$. In this case the vector that is normal to this plane, and passes through the origin is obviously the axis of rotation. Let the direction cosines of the new $\hat{z}$-axis (or $\mathbf{i}_{\hat{z}}$) be e_x, e_y, e_z. Then the angle β through which we must rotate the original system about the rotation axis is specified by

$$\cos\beta = e_z. \tag{3.3.1}$$

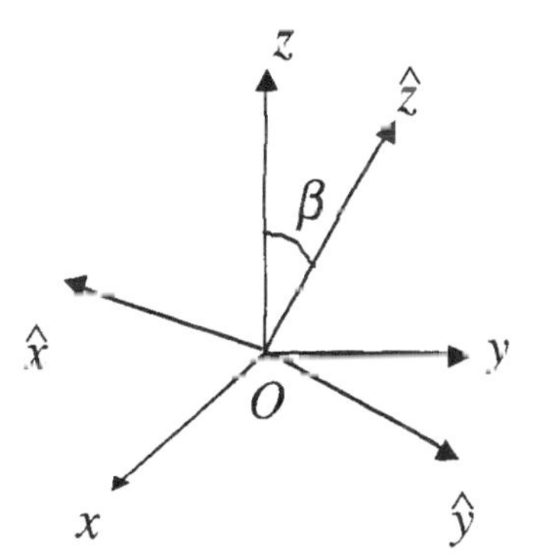

The direction of the axis of rotation can be specified as

$$\mathbf{n} = \mathbf{i}_z \times \mathbf{i}_{\hat{z}} = \begin{vmatrix} \mathbf{i}_x & \mathbf{i}_y & \mathbf{i}_z \\ 0 & 0 & 1 \\ e_x & e_y & e_z \end{vmatrix} = -e_y \mathbf{i}_x + e_x \mathbf{i}_y. \tag{3.3.2}$$

Let us make a choice that the new $\mathbf{i}_{\hat{x}}$ direction is along $\mathbf{n}$. The unit vector along this direction is

$$\mathbf{i}_{\hat{x}} = \frac{-e_y \mathbf{i}_x + e_x \mathbf{i}_y}{\sqrt{e_x^2 + e_y^2}}. \tag{3.3.3}$$

We then have the remaining axis chosen by the cyclic order of coordinate vectors as

$$\begin{aligned} \mathbf{i}_{\hat{y}} = \mathbf{i}_{\hat{z}} \times \mathbf{i}_{\hat{x}} &= \frac{1}{\sqrt{e_x^2 + e_y^2}} \begin{vmatrix} \mathbf{i}_x & \mathbf{i}_y & \mathbf{i}_z \\ e_x & e_y & e_z \\ -e_y & e_x & 0 \end{vmatrix} \\ &= \frac{-e_z e_x \mathbf{i}_x - e_z e_y \mathbf{i}_y}{\sqrt{e_x^2 + e_y^2}} + \sqrt{e_x^2 + e_y^2}\,\mathbf{i}_z. \end{aligned} \tag{3.3.4}$$

We can now evaluate the matrix Q using expression (3.1.11) as

$$Q = \begin{pmatrix} \mathbf{i}_{\hat{x}}\cdot\mathbf{i}_x & \mathbf{i}_{\hat{x}}\cdot\mathbf{i}_y & \mathbf{i}_{\hat{x}}\cdot\mathbf{i}_z \\ \mathbf{i}_{\hat{y}}\cdot\mathbf{i}_x & \mathbf{i}_{\hat{y}}\cdot\mathbf{i}_y & \mathbf{i}_{\hat{y}}\cdot\mathbf{i}_z \\ \mathbf{i}_{\hat{z}}\cdot\mathbf{i}_x & \mathbf{i}_{\hat{z}}\cdot\mathbf{i}_y & \mathbf{i}_{\hat{z}}\cdot\mathbf{i}_z \end{pmatrix} = \begin{pmatrix} -\dfrac{e_y}{\sqrt{e_x^2 + e_y^2}} & \dfrac{e_x}{\sqrt{e_x^2 + e_y^2}} & 0 \\ -\dfrac{e_z e_x}{\sqrt{e_x^2 + e_y^2}} & -\dfrac{e_z e_y}{\sqrt{e_x^2 + e_y^2}} & \sqrt{e_x^2 + e_y^2} \\ e_x & e_y & e_z \end{pmatrix}. \tag{3.3.5}$$

Of course, here the choice of the $\hat{x}$ and the $\hat{y}$ axes was arbitrary. If we have a specification for the orientation of these axes (thereby fixing the 0° meridian in the rotated coordinate system), we can compute the Q matrix as a composition of two rotations as

$$Q = \begin{pmatrix} \cos\phi & -\sin\phi & 0 \\ \sin\phi & \cos\phi & 0 \\ 0 & 0 & 1 \end{pmatrix} \begin{pmatrix} -\dfrac{e_y}{\sqrt{e_x^2+e_y^2}} & \dfrac{e_x}{\sqrt{e_x^2+e_y^2}} & 0 \\ -\dfrac{e_z e_x}{\sqrt{e_x^2+e_y^2}} & -\dfrac{e_z e_y}{\sqrt{e_x^2+e_y^2}} & \sqrt{e_x^2+e_y^2} \\ e_x & e_y & e_z \end{pmatrix}. \tag{3.3.6}$$

where ϕ is the rotation angle about the $\mathbf{i}_{\hat{z}}$-axis.

For computation of rotations of spherical harmonics it is convenient to represent the rotation matrix using spherical polar angles in both coordinate systems $(\mathbf{i}_x, \mathbf{i}_y, \mathbf{i}_z)$ and $(\mathbf{i}_{\hat{x}}, \mathbf{i}_{\hat{y}}, \mathbf{i}_{\hat{z}})$. Let β and γ be the spherical polar angles of the axis $\mathbf{i}_z$ in the reference frame $(\mathbf{i}_{\hat{x}}, \mathbf{i}_{\hat{y}}, \mathbf{i}_{\hat{z}})$, and let β and α be the spherical polar angles of the axis $\mathbf{i}_{\hat{z}}$ in the reference frame $(\mathbf{i}_x, \mathbf{i}_y, \mathbf{i}_z)$ (see Fig. 3.3). As $\mathbf{i}_z \cdot \mathbf{i}_{\hat{z}} = \mathbf{i}_{\hat{z}} \cdot \mathbf{i}_z = \cos\beta = e_z$ we do not need to introduce four angles for description, and the three independent angles β, γ, and α uniquely specify arbitrary rotation.

The relation between the components of the rotation matrix (3.1.11) and the angles β and γ are provided by the following expressions:

$$\begin{aligned} \cos\beta &= \mathbf{i}_z \cdot \mathbf{i}_{\hat{z}} = Q_{33} = e_z, \\ \cos\gamma \sin\beta &= \mathbf{i}_z \cdot \mathbf{i}_{\hat{x}} = Q_{13} = -\sqrt{e_x^2+e_y^2}\sin\phi, \\ \sin\gamma \sin\beta &= \mathbf{i}_z \cdot \mathbf{i}_{\hat{y}} = Q_{23} = \sqrt{e_x^2+e_y^2}\cos\phi. \end{aligned} \tag{3.3.7}$$

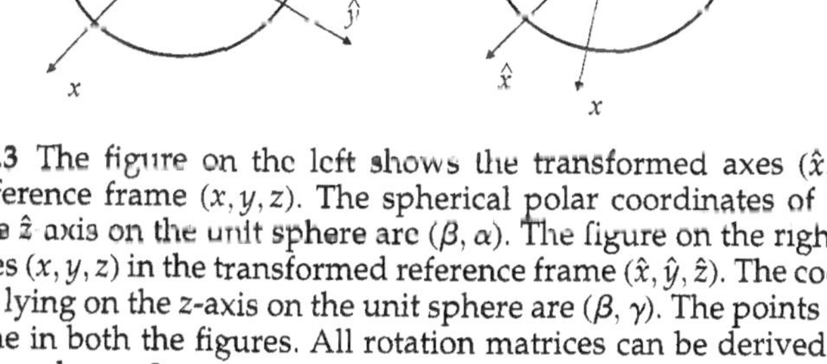

FIGURE 3.3 The figure on the left shows the transformed axes $(\hat{x}, \hat{y}, \hat{z})$ in the original reference frame (x, y, z). The spherical polar coordinates of the point $\hat{A}$ lying on the $\hat{z}$ axis on the unit sphere are (β, α). The figure on the right shows the original axes (x, y, z) in the transformed reference frame $(\hat{x}, \hat{y}, \hat{z})$. The coordinates of the point A lying on the z-axis on the unit sphere are (β, γ). The points O, A, and $\hat{A}$ are the same in both the figures. All rotation matrices can be derived in terms of these three angles α, β, γ.

Thus, the rotation angle ϕ and the polar angle γ are related as

$$\phi = \gamma - \frac{\pi}{2}. \tag{3.3.8}$$

At the same time we have

$$\begin{aligned} e_x &= \mathbf{i}_{\hat{z}} \cdot \mathbf{i}_x = \sin\beta\cos\alpha, \\ e_y &= \mathbf{i}_{\hat{z}} \cdot \mathbf{i}_y = \sin\beta\sin\alpha. \end{aligned} \tag{3.3.9}$$

The matrix Q representing the rotation between the axes can be represented in terms of these angles in the form

$$Q(\alpha,\beta,\gamma) = \begin{pmatrix} \sin\gamma & \cos\gamma & 0 \\ -\cos\gamma & \sin\gamma & 0 \\ 0 & 0 & 1 \end{pmatrix} \begin{pmatrix} -\sin\alpha & \cos\alpha & 0 \\ -\cos\beta\cos\alpha & -\cos\beta\sin\alpha & \sin\beta \\ \sin\beta\cos\alpha & \sin\beta\sin\alpha & \cos\beta \end{pmatrix}$$

$$= \begin{pmatrix} -\sin\gamma\sin\alpha - \cos\beta\cos\gamma\cos\alpha & \sin\gamma\cos\alpha - \cos\beta\cos\gamma\sin\alpha & \sin\beta\cos\gamma \\ \cos\gamma\sin\alpha - \cos\beta\sin\gamma\cos\alpha & -(\cos\gamma\cos\alpha + \cos\beta\sin\gamma\sin\alpha) & \sin\beta\sin\gamma \\ \sin\beta\cos\alpha & \sin\beta\sin\alpha & \cos\beta \end{pmatrix}. \tag{3.310}$$

Since Q is an orthogonal rotation matrix, it satisfies:

$$[Q(\alpha,\beta,\gamma)]^{-1} = [Q(\alpha,\beta,\gamma)]^{\mathrm{T}} = Q(\gamma,\beta,\alpha). \tag{3.3.11}$$

The last equality holds because we can exchange the symbols α and γ if we exchange symbols $(\mathbf{i}_x, \mathbf{i}_y, \mathbf{i}_z)$ and $(\mathbf{i}_{\hat{x}}, \mathbf{i}_{\hat{y}}, \mathbf{i}_{\hat{z}})$. Note also that Q can be represented as a composition of three rotations:

$$Q(\alpha,\beta,\gamma) = \begin{pmatrix} \sin\gamma & \cos\gamma & 0 \\ -\cos\gamma & \sin\gamma & 0 \\ 0 & 0 & 1 \end{pmatrix} \begin{pmatrix} -1 & 0 & 0 \\ 0 & -\cos\beta & \sin\beta \\ 0 & \sin\beta & \cos\beta \end{pmatrix}$$

$$\times \begin{pmatrix} \sin\alpha & -\cos\alpha & 0 \\ \cos\alpha & \sin\alpha & 0 \\ 0 & 0 & 1 \end{pmatrix}$$

$$= A(\gamma)B(\beta)A^{\mathrm{T}}(\alpha). \tag{3.3.12}$$

Here

$$B(\beta) = [B(\beta)]^{-1} = [B(\beta)]^{T} \tag{3.3.13}$$

and

$$[Q(\alpha,\beta,\gamma)]^{-1} = [A(\gamma)B(\beta)A^{T}(\alpha)]^{-1} = [A^{T}(\alpha)]^{-1}[B(\beta)]^{-1}[A(\gamma)]^{-1}$$
$$= A(\alpha)B(\beta)A^{T}(\gamma) = Q(\gamma,\beta,\alpha). \tag{3.3.14}$$

Such a symmetrical representation using the angles α, β, and γ has advantages in that it leads to compact expressions for transforming the spherical coordinates of vectors in the rotated reference frames.

The angles (α, β, γ) are simply related to the Euler rotation angles. There are several different definitions of these angles in the literature. We define them as angles for right-hand rotations near positive coordinate axes. First, we rotate the reference frame around $\mathbf{i}_z$ by angle α_E. This transforms the basis vectors as $(\mathbf{i}_x, \mathbf{i}_y, \mathbf{i}_z) \to (\tilde{\mathbf{i}}_x, \tilde{\mathbf{i}}_y, \mathbf{i}_z)$. Second, we rotate the axes around $\tilde{\mathbf{i}}_y$ by angle β_E. This transforms the basis vectors as $(\tilde{\mathbf{i}}_x, \tilde{\mathbf{i}}_y, \mathbf{i}_z) \to (\tilde{\mathbf{i}}'_x, \tilde{\mathbf{i}}_y, \mathbf{i}_{\hat{z}})$. Finally, we rotate the reference frame near $\mathbf{i}_{\hat{z}}$ by angle γ_E. This transforms the basis vectors as $(\tilde{\mathbf{i}}'_x, \tilde{\mathbf{i}}_y, \mathbf{i}_{\hat{z}}) \to (\mathbf{i}_{\hat{x}}, \mathbf{i}_{\hat{y}}, \mathbf{i}_{\hat{z}})$. The Euler rotation angles are shown in Fig. 3.4. The relation of these angles with the angles in spherical polar coordinates is

$$\alpha_E = \pi - \alpha, \qquad \beta_E = \beta, \qquad \gamma_E = \gamma. \tag{3.3.15}$$

This shows that β and γ indeed are Euler angles β_E and γ_E, while α is the angle supplementary to the Euler angle α_E.

3.3.2 Rotation coefficients

A feature of the rotation transform is that the point $\mathbf{r} = \mathbf{0}$ does not change, and so the value of the function $\psi(\mathbf{r})$ does not change at this point after rotation. As far as elementary solutions are concerned, we can see then that due to rotations functions regular at $\mathbf{r} = \mathbf{0}$ remain regular functions at $\hat{\mathbf{r}} = \mathbf{0}$, and functions singular at $\mathbf{r} = \mathbf{0}$ remain singular at the center of rotation. Another important feature of the rotation transform is that

$$r = |\mathbf{r}| = |\hat{\mathbf{r}}| = \hat{r}, \tag{3.3.16}$$

where $\mathbf{r} = (x, y, z)$ and $\hat{\mathbf{r}} = (\hat{x}, \hat{y}, \hat{z})$ are coordinates of the spatial point M in the original and rotated frames. Indeed, the distance from the origin O of the frame of reference to M does not depend on the selection of the system of coordinates, and since this distance is r for the original and $\hat{r}$ for the rotated frame we have the latter relation.

Consider now the rotation transform of an arbitrary spherical basis function of the Helmholtz equation, denoted by $F_n^m(\mathbf{r})$, where under F as earlier we can mean either singular (S) or regular (R) basis functions. As stated above under the transform $F_n^m(\mathbf{r})$ can be expressed via functions of the same type, $F_n^m(\hat{\mathbf{r}})$ only, so we can write in general

$$F_n^m(\mathbf{r}) = \sum_{n'=0}^{\infty} \sum_{m'=-n'}^{n'} T_{n'n}^{m'm}(Q) F_{n'}^{m'}(\hat{\mathbf{r}}), \tag{3.3.17}$$
$$n = 0, 1, 2, \ldots, \quad m = -n, \ldots, n, \quad F = S, R.$$

Now we can use the following factorization of $F_n^m(\mathbf{r})$:

$$F_n^m(\mathbf{r}) = f_n(kr) Y_n^m(\mathbf{s}), \qquad \mathbf{s} = \frac{\mathbf{r}}{|\mathbf{r}|}, \quad (F, f) = (S, h), (R, j), \tag{3.3.18}$$

where $\mathbf{s}$ is the unit vector directed along $\mathbf{r}$, and f denotes the spherical Hankel and Bessel functions for singular or regular functions F, respectively. With this factorization, expression (3.3.17) can be rewritten as

$$F_n^m(\mathbf{r}) = f_n(kr) Y_n^m(\mathbf{s}) = \sum_{n'=0}^{\infty} f_{n'}(kr) \sum_{m'=-n'}^{n'} T_{n'n}^{m'm}(Q) Y_{n'}^{m'}(\hat{\mathbf{s}}), \quad \hat{\mathbf{s}} = \frac{\hat{\mathbf{r}}}{|\hat{\mathbf{r}}|}. \tag{3.3.19}$$

We note then that the spherical Hankel or Bessel functions are linearly independent. This means that no functions $f_n(kr)$ can be expressed as a linear combination of basis functions $f_{n'}(kr)$ if $n' \neq n$, and the only opportunity to satisfy relation (3.3.19) is

$$\sum_{m'=-n'}^{n'} T_{n'n}^{m'm}(Q) Y_{n'}^{m'}(\hat{\mathbf{s}}) = \delta_{nn'} Y_n^m(\hat{\mathbf{s}}), \quad n = 0, 1, 2, \ldots, \ m = -n, \ldots, n, \tag{3.3.20}$$

where $\delta_{nn'}$ is the Kronecker delta. For $n \neq n'$ the right-hand side of this equation turns to zero. Equation (3.3.20) can be viewed in this case as expansion of zero over the basis of spherical harmonics $Y_{n'}^{m'}(\hat{\mathbf{s}})$. Since this basis is complete, the only opportunity is that all expansion coefficients in this case should be zero, or

$$T_{n'n}^{m'm}(Q) = 0, \quad n' \neq n. \tag{3.3.21}$$

In case $n = n'$, expression (3.3.20) turns to

$$Y_n^m(\mathbf{s}) = \sum_{m'=-n}^{n} T_n^{m'm}(Q) Y_n^{m'}(\hat{\mathbf{s}}), \quad n = 0, 1, 2, \ldots, \ m = -n, \ldots, n, \tag{3.3.22}$$

where we dropped the subscript n', since as follows from the discussion above, Eq. (3.3.20) is non-trivial only for $n = n'$. The three indices $T_n^{m'm}(Q)$ then are the *rotation coefficients*. Note that the expansion (3.3.22) is unique, due to the orthonormality of the basis of spherical harmonics on a unit sphere. Using this notation and (3.3.21), we find from expansion (3.3.17) the following formula for spherical basis function rotation transform:

$$F_n^m(\mathbf{r}) = \sum_{m'=-n}^{n} T_n^{m'm}(Q) F_n^{m'}(\hat{\mathbf{r}}),$$
$$n=0,1,2,\ldots, m=-n,\ldots,n, F=S,R. \tag{3.3.23}$$

This reveals several interesting facts. First, that the rotation coefficients are common for both regular or singular functions and depend only on the rotation matrix Q. Second, the sum is finite and the degree of the basis function n does not change under rotation. Third, that the rotation coefficients are actually rotation coefficients for the spherical harmonics, and related to two-dimensional transforms on the unit sphere. Finally, we note that while we derived the result from ideas of function expansions over some bases, the fact that the degree of spherical harmonics is conserved with rotations is a classical result in quantum mechanics and usually is proven within the framework of the *group theory* [EW31].

The rotation matrix depends only on the three angles of rotation α, β, and γ. So Eq. (3.3.22) can also be written in the following form:

$$Y_n^m(\mathbf{s}) = \sum_{m'=-n}^{n} T_n^{m'm}(\alpha,\beta,\gamma) Y_n^{m'}(\hat{\mathbf{s}}),$$
$$n = 0,1,2,\ldots, \quad m = -n,\ldots,n, \tag{3.3.24}$$

or

$$Y_n^m(\theta,\varphi) = \sum_{m'=-n}^{n} T_n^{m'm}(\alpha,\beta,\gamma) Y_n^{m'}(\hat{\theta},\hat{\varphi}),$$
$$n = 0,1,2,\ldots, \quad m = -n,\ldots,n. \tag{3.3.25}$$

Consider a particular case of this expansion for $m = 0$. We have

$$\sqrt{\frac{2n+1}{4\pi}} P_n(\cos\theta) = Y_n^0(\theta,\varphi) = Y_n^0(\mathbf{s}) = \sum_{m'=-n}^{n} T_n^{m'0}(\alpha,\beta,\gamma) Y_n^{m'}(\hat{\mathbf{s}})$$
$$= \sum_{m'=-n}^{n} T_n^{m'0}(\alpha,\beta,\gamma) Y_n^{m'}(\hat{\theta},\hat{\varphi}), \qquad n = 0,1,2,\ldots \tag{3.3.26}$$

We note that by definition the vectors $\mathbf{s}$ and $\hat{\mathbf{s}}$ are coordinates of the same spatial point M on a unit sphere in the original and the rotated reference frame (see Fig. 3.5). So we can apply the addition theorem for spherical harmonics in the form (2.1.70) or (2.1.72). In this theorem, we set the vector $\mathbf{s}_1$ as the basis vector $\mathbf{i}_z$ of the original frame of reference, and $\mathbf{s}_2$ as the radius-vector of the point M (compare Figs. 2.5 and 3.5). This unit radius-vector has coordinates (θ, φ) in the original frame, and $(\hat{\theta}, \hat{\varphi})$ in the rotated frame. At the same time the vector $\mathbf{i}_z$ has coordinates (β, γ) in the rotated system of coordinates (see Fig. 3.1). Therefore, the relation (2.1.72) yields

$$P_n(\cos\theta) = \frac{4\pi}{2n+1} \sum_{m'=-n}^{n} Y_n^{-m'}(\beta, \gamma) Y_n^{m'}(\hat{\theta}, \hat{\varphi}), \qquad n = 0, 1, 2, \ldots \quad (3.3.27)$$

Comparing Eqs. (3.3.26) and (3.3.27), we obtain the following expression for the rotation coefficients $T_n^{m'0}(\alpha, \beta, \gamma)$:

$$T_n^{m'0}(\alpha, \beta, \gamma) = \sqrt{\frac{4\pi}{2n+1}} Y_n^{-m'}(\beta, \gamma), \qquad (3.3.28)$$

$$n = 0, 1, 2, \ldots, \quad m' = -n, \ldots, n.$$

This shows that the coefficients $T_n^{m'0}(\alpha, \beta, \gamma)$ do not depend on the rotation angle α.

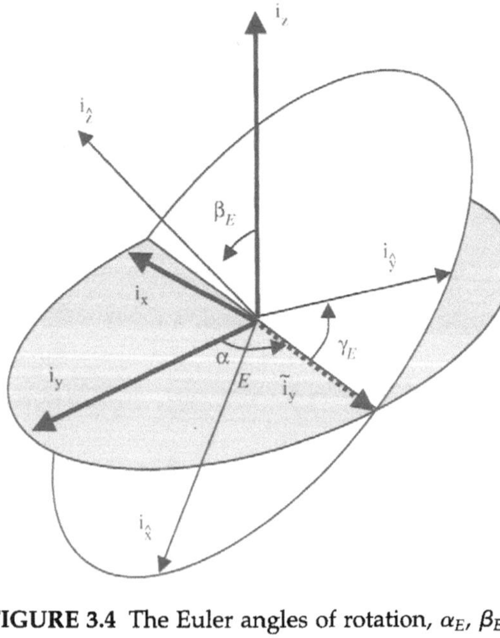

FIGURE 3.4 The Euler angles of rotation, α_E, β_E, γ_E.

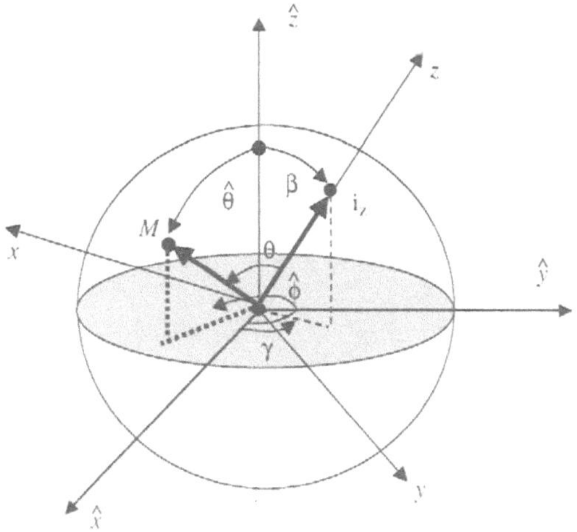

FIGURE 3.5 Illustration of the addition theorem for spherical harmonics applied to rotation of the reference frame (compare with Fig. 2.5).

It is worth noting how vectors were treated in the derivation of Eq. (3.3.28). On the one hand the vectors can be considered as invariant objects, which do not depend on the choice of the reference frame (as the radius-vector of point M or basis vector $\mathbf{i}_z$). On the other hand, we can associate the vector with their coordinates in some reference frame (as vectors $\mathbf{s}$ and $\hat{\mathbf{s}}$). In this case the vector changes with coordinate transforms. In fact both approaches are valid, since we provide a *one-to-one mapping* of the "physical" space of points and vectors to the "mathematical" space of coordinates. If we plot vectors as $\mathbf{s} = (\theta, \varphi)$ and $\hat{\mathbf{s}} = (\hat{\theta}, \hat{\varphi})$ in the *same* "standard" coordinate system we will have a picture as shown in Fig. 3.6. These two different vectors in coordinate space correspond to the same point M in the "physical" space, and vector $\hat{\mathbf{s}}$ is obtained from vector $\mathbf{s}$ by a rotation of the space as a rigid body. Note also that there is a *duality* of the "physical" and "mathematical" spaces. These space can change the roles and the "physical" space will be the "mathematical" once the "mathematical" space becomes the "physical". Indeed, invariant objects (vectors) in the space of coordinates change in the "physical" space, if we rotate the basis vectors. This change again can be described by a rotation matrix. So even the labeling of spaces as "mathematical" or "physical" is relative (and it may make sense to just denote them as "space 1" and "space 2", or "space" and "adjoint space"). Written in vector form, the relations are valid for both spaces, and can be interpreted at our convenience (that is why we use vector notation and

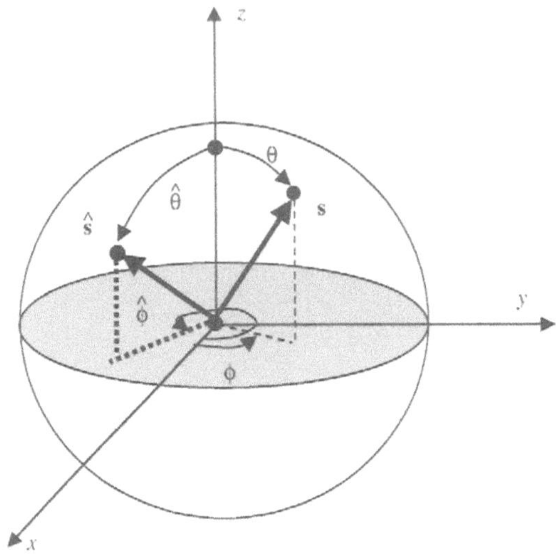

FIGURE 3.6 Two vectors **s** and **ŝ** obtained by a rotation transform.

vector algebra). When we come to computations, we should in any case work with some numerical values, like angles, distances, etc. and work with the coordinates of these vectors (that is why we are interested in the values of coordinates).

Note that formula (3.3.25) provides expansion of some surface function $Y_n^m(\theta, \varphi)$ over orthonormal system of spherical harmonics $Y_n^{m'}(\hat{\theta}, \hat{\varphi})$. The coefficients of this expansion can be found by integration of the product of this surface function with respective conjugate harmonics as given by relation (2.1.68). This provides the following integral representation of the rotation coefficients:

$$T_n^{m'm}(\alpha, \beta, \gamma) = \int_0^{2\pi} \mathrm{d}\hat{\varphi} \int_0^{\pi} Y_n^m(\theta, \varphi) Y_n^{-m'}(\hat{\theta}, \hat{\varphi}) \sin \hat{\theta} \, \mathrm{d}\hat{\theta}. \qquad (3.3.29)$$

3.3.3 Structure of rotation coefficients

The structure of the rotation coefficients can be found from the properties of the rotation matrix. Since the rotation matrix can be decomposed into three rotations about the axes with angles α, β, and γ we can see how the spherical harmonics change with such a transformation. First, we note that the rotation of the coordinates with the angles α and γ do not change quantities that depend on β. For example, rotation with angle α conserves the original z-axis (see formula (3.3.15) and Fig. 3.4) and denoting the angle φ in the original coordinates as φ_1 in the rotated coordinates, we have

$$\varphi = \varphi_1 + \alpha. \tag{3.3.30}$$

The spherical harmonics then change as

$$Y_n^m(\theta, \varphi) = Y_n^m(\theta, \varphi_1 + \alpha) = e^{im\alpha} Y_n^m(\theta, \varphi_1). \tag{3.3.31}$$

The transform of the spherical harmonics from the α-rotated coordinates to the final coordinates can be described by

$$Y_n^m(\theta, \varphi_1) = \sum_{m'=-n}^{n} T_n^{(1)m'm}(\beta, \gamma) Y_n^{m'}(\hat{\theta}, \hat{\varphi}). \tag{3.3.32}$$

When we perform the last rotation of coordinates to arrive at the angle $\phi = \gamma - (\pi/2)$, we have the same situation, and so

$$\hat{\varphi} = \varphi_2 + \phi = \varphi_2 + \gamma - \frac{\pi}{2}, \tag{3.3.33}$$

where φ_2 is the angle corresponding to the given point in coordinates prior to the last rotation being performed. So

$$Y_n^{m'}(\hat{\theta}, \hat{\varphi}) = Y_n^{m'}\left(\hat{\theta}, \varphi_2 + \gamma - \frac{\pi}{2}\right) = (-i)^{m'} e^{im'\gamma} Y_n^{m'}(\hat{\theta}, \varphi_2). \tag{3.3.34}$$

The transform from $Y_n^m(\theta, \varphi_1)$ to $Y_n^m(\hat{\theta}, \varphi_2)$ occurs only because of the rotation related to angle β, and so

$$Y_n^m(\theta, \varphi_1) = \sum_{m'=-n}^{n} T_n^{(12)m'm}(\beta) Y_n^{m'}(\hat{\theta}, \varphi_2). \tag{3.3.35}$$

Combining these results we find that

$$Y_n^m(\theta, \varphi) = e^{im\alpha} \sum_{m'=-n}^{n} i^{m'} e^{-im'\gamma} T_n^{(12)m'm}(\beta) Y_n^{m'}(\hat{\theta}, \hat{\varphi}). \tag{3.3.36}$$

This shows that

$$T_n^{m'm}(\alpha, \beta, \gamma) = e^{im\alpha} e^{-im'\gamma} H_n^{m'm}(\beta), \tag{3.3.37}$$

where

$$H_n^{m'm}(\beta) = i^{m'} T_n^{(12)m'm}(\beta). \tag{3.3.38}$$

An explicit expression for $H_n^{m'm}(\beta)$ can be found, e.g. in the paper of Stein [Ste61], which in the notation of the present paper[1] can be represented in the form

$$H_n^{m'm}(\beta) = \varepsilon_m \varepsilon_{m'} [(n+m')!(n-m')!(n+m)!(n-m)!]^{1/2} \times$$
$$\sum_{\sigma=\max(0,-(m+m'))}^{\min(n-m,n-m')} \frac{(-1)^{n-\sigma} \cos^{2\sigma+m'+m} \frac{1}{2}\beta \sin^{2n-2\sigma-m'-m} \frac{1}{2}\beta}{\sigma!(n-m-\sigma)!(n-m'-\sigma)!(m+m'+\sigma)!}, \quad (3.3.39)$$

where the symbol ε_m is defined in Eq. (3.2.27). This shows that the functions $H_n^{m'm}(\beta)$ are real (which is important for computations). Equations (3.3.37) and (3.3.28) yield the following particular values of $H_n^{m'm}(\beta)$ at $m = 0$:

$$H_n^{m'0}(\beta) = (-1)^{m'} \sqrt{\frac{(n-|m'|)!}{(n+|m'|)!}} P_n^{|m'|}(\cos\beta), \quad (3.3.40)$$
$$n = 0, 1, \ldots, \quad m' = -n, \ldots, n.$$

While the relations (3.3.37) and (3.3.39) provide explicit expressions for calculation of the rotation coefficients $T_n^{m'm}(\alpha, \beta, \gamma)$, the computations based on these relations are rather inefficient, since to compute each coefficient $H_n^{m'm}(\beta)$ we need compute $O(n)$ terms in the sum. Therefore, if $n = O(p)$ and the total number of rotation coefficients $H_n^{m'm}(\beta)$ is $O(p^3)$ computational complexity for obtaining the set of these coefficients is $O(p^4)$. As will be shown below, a procedure based on recurrence relations has computational complexity $O(p^3)$ to obtain this set, which obviously is a theoretical minimum for the complexity (assuming we need to spend at least one operation for the computation of each coefficient).

3.3.3.1 *Symmetries of rotation coefficients*

The rotation coefficients obey a number of symmetries. Below we mention the following:

THEOREM 12 *The following symmetry holds*

$$T_n^{-m',-m}(\alpha, \beta, \gamma) = \overline{T_n^{m'm}(\alpha, \beta, \gamma)}, \quad n = 0, 1, \ldots, \ m', m = -n, \ldots, n. \quad (3.3.41)$$

---

[1] Note that in Ref. [Ste61] the Euler angles are defined with sign opposite to our definitions of α_E, β_E and γ_E.

PROOF. Since $Y_n^{-m'} = \overline{Y_n^{m'}}$ for any $n = 0, 1, \ldots, m' = -n, \ldots, n$, we have from Eq. (3.3.24)

$$Y_n^m(\theta,\varphi) = \overline{Y_n^{-m}(\theta,\varphi)} = \sum_{m'=-n}^{n} \overline{T_n^{m',-m} Y_n^{m'}(\hat{\theta},\hat{\varphi})}$$

$$= \sum_{m'=-n}^{n} \overline{T_n^{m',-m}} Y_n^{-m'}(\hat{\theta},\hat{\varphi}) = \sum_{m'=-n}^{n} \overline{T_n^{-m',-m}} Y_n^{m'}(\hat{\theta},\hat{\varphi}). \quad (3.3.42)$$

Comparing this result with the expansion of $Y_n^m(\theta,\varphi)$ given by Eq. (3.3.24) we obtain the statement of the theorem, since $Y_n^{m'}(\hat{\theta},\hat{\varphi})$ is orthonormal and the representation (3.3.24) is unique. □

This theorem generates the following relation for the coefficients $H_n^{m'm}(\beta)$:

$$H_n^{m'm}(\beta) = H_n^{-m',-m}(\beta), \qquad n = 0, 1, \ldots, \quad m', m = -n, \ldots, n. \quad (3.3.43)$$

This follows from relation (3.3.41), where we substitute expression (3.3.37) and use the fact that the coefficients $H_n^{m'm}(\beta)$ are real.

Another symmetry relation follows from the explicit formula (3.3.39):

$$H_n^{m'm}(\beta) = H_n^{mm'}(\beta), \qquad n = 0, 1, \ldots, \quad m', m = -n, \ldots, n, \quad (3.3.44)$$

Here we can exchange m and m' and obtain the same expression. If in Eq. (3.3.39) we change the sign of β, then one more symmetry property appears:

$$H_n^{m'm}(-\beta) = (-1)^{m+m'} H_n^{m'm}(\beta), \qquad n = 0, 1, \ldots, \quad m', m = -n, \ldots, n. \quad (3.3.45)$$

3.3.3.2 *Relation to Clebsch–Gordan coefficients*

To obtain relations between the rotation coefficients and the Clebsch–Gordan or similar coefficients (3j-Wigner or E-symbols, whose relation to the Clebsch–Gordan coefficients was established earlier in the chapter), we consider, first, expansion of a product of two spherical harmonics into a series of spherical harmonics of the same argument:

$$Y_{j_1}^{m_1}(\theta,\varphi) Y_{j_2}^{m_2}(\theta,\varphi) = \sum_{j=0}^{\infty} \sum_{m=-j}^{j} t_{jj_1j_2}^{mm_1m_2} Y_j^m(\theta,\varphi). \quad (3.3.46)$$

Since $Y_{j_1}^{m_1}(\theta,\varphi)Y_{j_2}^{m_2}(\theta,\varphi)$ is some surface function and $\{Y_j^m(\theta,\varphi)\}$ is an orthonormal basis on a sphere, such an expansion is meaningful, and the coefficients of this expansion can be found using Eq. (2.1.68):

$$t_{jj_1j_2}^{mm_1m_2} = \int_0^{\pi} \sin\theta \int_0^{2\pi} Y_{j_1}^{m_1}(\theta,\varphi)Y_{j_2}^{m_2}(\theta,\varphi)Y_j^{-m}(\theta,\varphi)d\varphi. \qquad (3.3.47)$$

Comparing this result with the definition of the E-symbols (Eq. (3.2.28)), we obtain

$$t_{jj_1j_2}^{mm_1m_2} = \frac{1}{4\pi}\left[\frac{(2j+1)(2j_1+1)(2j_2+1)}{4\pi}\right]^{1/2} E\begin{pmatrix} m_1 & m_2 & -m \\ j_1 & j_2 & j \end{pmatrix}. \qquad (3.3.48)$$

Let us now express spherical harmonics of argument (θ,φ) via spherical harmonics of argument $(\hat{\theta},\hat{\varphi})$ according to series (3.3.25):

$$\sum_{m_1'=-j_1}^{j_1}\sum_{m_2'=-j_2}^{j_2} T_{j_1}^{m_1'm_1}(\alpha,\beta,\gamma)T_{j_2}^{m_2'm_2}(\alpha,\beta,\gamma)Y_{j_1}^{m_1'}(\hat{\theta},\hat{\varphi})Y_{j_2}^{m_2'}(\hat{\theta},\hat{\varphi})$$

$$=\sum_{j=0}^{\infty}\sum_{m=-j}^{j}\sum_{m'=-j}^{j} t_{jj_1j_2}^{mm_1m_2}T_j^{m'm}(\alpha,\beta,\gamma)Y_j^{m'}(\hat{\theta},\hat{\varphi}).$$

Here we can substitute expansion (3.3.46) for the product of spherical harmonics and find

$$\sum_{j'=0}^{\infty}\sum_{m''=-j'}^{j'}\sum_{m_1'=-j_1}^{j_1}\sum_{m_2'=-j_2}^{j_2} T_{j_1}^{m_1'm_1}(\alpha,\beta,\gamma)T_{j_2}^{m_2'm_2}(\alpha,\beta,\gamma)t_{j'j_1j_2}^{m''m_1'm_2'}Y_{j''}^{m''}(\hat{\theta},\hat{\varphi})$$

$$=\sum_{j=0}^{\infty}\sum_{m'=-j}^{j}\sum_{m=-j}^{j} t_{jj_1j_2}^{mm_1m_2}T_j^{m'm}(\alpha,\beta,\gamma)Y_j^{m'}(\hat{\theta},\hat{\varphi}).$$

We can see now that this is two expansions of some function over the same orthonormal basis of spherical harmonics. So the coefficients of these expansions should be the same. Changing the symbol j' to j and m'' to m' in the left-hand side we then have

$$\sum_{m_1'=-j_1}^{j_1}\sum_{m_2'=-j_2}^{j_2} t_{jj_1j_2}^{m'm_1'm_2'}T_{j_1}^{m_1'm_1}(\alpha,\beta,\gamma)T_{j_2}^{m_2'm_2}(\alpha,\beta,\gamma)$$

$$=\sum_{m=-j}^{j} t_{jj_1j_2}^{mm_1m_2}T_j^{m'm}(\alpha,\beta,\gamma). \qquad (3.3.49)$$

Note that $t_{jj_1j_2}^{m'm_1'm_2'}$ and $t_{jj_1j_2}^{mm_1m_2}$ are non-zero, only if

$$m = m_1 + m_2, \qquad m_1' + m_2' = m'. \tag{3.3.50}$$

So the summation in the right-hand side disappears, and we need to keep only one summation in the left-hand side of relation (3.3.49):

$$\sum_{m_1'+m_2'=m'} t_{jj_1j_2}^{m'm_1'm_2'} T_{j_1}^{m_1'm_1}(\alpha,\beta,\gamma) T_{j_2}^{m_2'm_2}(\alpha,\beta,\gamma)$$

$$= t_{jj_1j_2}^{(m_1+m_2)m_1m_2} T_j^{m'(m_1+m_2)}(\alpha,\beta,\gamma). \tag{3.3.51}$$

Now, let us substitute here expression (3.3.37) to find a relation for $H_n^{m'm}(\beta)$. Due to Eq. (3.3.50) functions of α and γ cancel out, and we have

$$\sum_{m_1'+m_2'=m'} t_{jj_1j_2}^{m'm_1'm_2'} H_{j_1}^{m_1'm_1}(\beta) H_{j_2}^{m_2'm_2}(\beta) = t_{jj_1j_2}^{(m_1+m_2)m_1m_2} H_j^{m'(m_1+m_2)}(\beta). \tag{3.3.52}$$

Using the relations (3.2.26)–(3.2.28) and (3.3.48) between the coefficients $t_{jj_1j_2}^{m'm_1'm_2'}$ and the Clebsch–Gordan coefficients, we can rewrite this equation in the form

$$t_{jj_1j_2}^{mm_1m_2} H_j^{m'm}(\beta) = \sum_{m_1'+m_2'=m'} (-1)^{m+m'} \frac{\varepsilon_{m_1'}\varepsilon_{m_2'}\varepsilon_{-m'}}{\varepsilon_{m_1}\varepsilon_{m_2}\varepsilon_{-m}} \frac{(j_1j_2m_1'm_2'|j_1j_2jm')}{(j_1j_2m_1m_2|j_1j_2jm)}$$

$$\times H_{j_1}^{m_1'm_1}(\beta) H_{j_2}^{m_2'm_2}(\beta). \tag{3.3.53}$$

Formulas of this type can be used in some recursive procedures, since they express higher order coefficients ($m = m_1 + m_2$) via products of lower order coefficients (m_1 and m_2). While there exist more relations between functions $H_n^{m'm}(\beta)$ which involve the Clebsch–Gordan coefficients we will not proceed with this approach further and refer the interested reader to a special literature [VMK88].

3.3.4 Recurrence relations for rotation coefficients

The rotation coefficients satisfy many recurrence relations, and there are many strategies for their computations based on recurrences. Many recurrences can be obtained from the differentiation theorems for elementary solutions in spherical coordinates. Below we provide several theorems and results, which show how these recurrences appear.

THEOREM 13 *The following recurrence relations holds for* $T_n^{m'm}(Q)$*:*

$$\frac{1}{2}(\mathbf{i}_x - i\mathbf{i}_y)b_{n+1}^{-m-1}T_{n+1}^{m',m+1} + \frac{1}{2}(\mathbf{i}_x + i\mathbf{i}_y)b_{n+1}^{m-1}T_{n+1}^{m',m-1} - \mathbf{i}_z a_n^m T_{n+1}^{m'm}$$

$$= \frac{1}{2}(\mathbf{i}_{\hat{x}} - i\mathbf{i}_{\hat{y}})b_{n+1}^{-m'}T_n^{m'-1,m} + \frac{1}{2}(\mathbf{i}_{\hat{x}} + i\mathbf{i}_{\hat{y}})b_{n+1}^{m'}T_n^{m'+1,m} - \mathbf{i}_{\hat{z}} a_n^{m'} T_n^{m'm}, \tag{3.3.54}$$

where $n = 0, 1, 2, \ldots,\ m = -n, \ldots, n,\ m' = -n-1, \ldots, n+1$*, and*

$$\frac{1}{2}(\mathbf{i}_x - i\mathbf{i}_y)b_n^m T_{n-1}^{m',m+1} + \frac{1}{2}(\mathbf{i}_x + i\mathbf{i}_y)b_n^{-m}T_{n-1}^{m',m-1} - \mathbf{i}_z a_{n-1}^m T_{n-1}^{m'm}$$

$$= \frac{1}{2}(\mathbf{i}_{\hat{x}} - i\mathbf{i}_{\hat{y}})b_n^{m'-1}T_n^{m'-1,m} + \frac{1}{2}(\mathbf{i}_{\hat{x}} + i\mathbf{i}_{\hat{y}})b_n^{-m'-1}T_n^{m'+1,m}$$

$$- \mathbf{i}_{\hat{z}} a_{n-1}^{m'} T_n^{m'm}. \tag{3.3.55}$$

where $n = 0, 1, 2, \ldots,\ m = -n, \ldots, n,\ m' = -n+1, \ldots, n-1$*, and coefficients* a_n^m *and* b_n^m *are defined by expressions (2.2.8) and (2.2.10).*

PROOF. Applying the operator $k^{-1}\nabla$ to any of the relations (3.3.23), we obtain

$$k^{-1}\nabla F_n^m(\mathbf{r}) = \sum_{m'=-n}^{n} T_n^{m'm}(Q)k^{-1}\nabla F_n^{m'}(\hat{\mathbf{r}}), \qquad |\hat{\mathbf{r}}| = |\mathbf{r}|,\ F = S, R. \tag{3.3.56}$$

The operator ∇ is independent of the reference frame it is represented in. This applies to rotations as well, and so we can use Eq. (2.2.12), to represent the left and right-hand sides of Eq. (3.3.56). Grouping terms corresponding to the same basis functions $F_n^{m'}(\hat{\mathbf{r}})$ we obtain the statement of the theorem.[2] □

THEOREM 14 *The following recurrence relations hold for* $T_n^{m'm}(Q)$*:*

$$b_{n+1}^{-m-1}T_{n+1}^{m',m+1} + b_{n+1}^{m-1}T_{n+1}^{m',m-1}$$

$$= W_{11}b_{n+1}^{-m'}T_n^{m'-1,m} + W_{12}b_{n+1}^{m'}T_n^{m'+1,m} + W_{13}a_n^{m'}T_n^{m'm}, \tag{3.3.57}$$

$$b_{n+1}^{-m-1}T_{n+1}^{m',m+1} - b_{n+1}^{m-1}T_{n+1}^{m',m-1}$$

$$= W_{21}b_{n+1}^{-m'}T_n^{m'-1,m} + W_{22}b_{n+1}^{m'}T_n^{m'+1,m} + W_{23}a_n^{m'}T_n^{m'm}, \tag{3.3.58}$$

[2] A detailed proof is provided in our technical report, Gumerov and Duraiswami [GD01].

$$a_n^m T_{n+1}^{m'm} = W_{31} b_{n+1}^{-m'} T_n^{m'-1,m} + W_{32} b_{n+1}^{m'} T_n^{m'+1,m} + W_{33} a_n^{m'} T_n^{m'm}, \qquad (3.3.59)$$

where $n = 0, 1, 2, \ldots$, $m = -n, \ldots, n$, $m' = -n-1, \ldots, n+1$, *and* $W_{\alpha\beta}$ *are the elements of the following complex rotation matrix:*

$$\mathrm{W} = \begin{pmatrix} \mathbf{i}_{x'}\cdot(\mathbf{i}_{\hat{x}} - i\mathbf{i}_{\hat{y}}) & \mathbf{i}_{x'}\cdot(\mathbf{i}_{\hat{x}} + i\mathbf{i}_{\hat{y}}) & -2\mathbf{i}_{x'}\cdot\mathbf{i}_{\hat{z}} \\ i\mathbf{i}_{y'}\cdot(\mathbf{i}_{\hat{x}} - i\mathbf{i}_{\hat{y}}) & i\mathbf{i}_{y'}\cdot(\mathbf{i}_{\hat{x}} + i\mathbf{i}_{\hat{y}}) & -2i\mathbf{i}_{y'}\cdot\mathbf{i}_{\hat{z}} \\ -\frac{1}{2}\mathbf{i}_{z'}\cdot(\mathbf{i}_{\hat{x}} - i\mathbf{i}_{\hat{y}}) & -\frac{1}{2}\mathbf{i}_{z'}\cdot(\mathbf{i}_{\hat{x}} + i\mathbf{i}_{\hat{y}}) & \mathbf{i}_{z'}\cdot\mathbf{i}_{\hat{z}} \end{pmatrix}, \qquad (3.3.60)$$

and coefficients a_n^m *and* b_n^m *are defined by expressions (2.2.8) and (2.2.10).*

PROOF. Taking scalar product with $\mathbf{i}_x$, $i\mathbf{i}_y$, and $\mathbf{i}_z$ of both sides of Eq. (3.3.54), we obtain the statement of the theorem. □

COROLLARY 6 *Summing and subtracting relations (3.3.57) and (3.3.58) we have*

$$2b_{n+1}^{-m-1} T_{n+1}^{m',m+1} = (W_{11} + W_{21}) b_{n+1}^{-m'} T_n^{m'-1,m} + (W_{12} + W_{22}) b_{n+1}^{m'} T_n^{m'+1,m} + (W_{13} + W_{23}) a_n^{m'} T_n^{m'm}, \qquad (3.3.61)$$

$$2b_{n+1}^{m-1} T_{n+1}^{m',m-1} = (W_{11} - W_{21}) b_{n+1}^{-m'} T_n^{m'-1,m} + (W_{12} - W_{22}) b_{n+1}^{m'} T_n^{m'+1,m} + (W_{13} - W_{23}) a_n^{m'} T_n^{m'm}. \qquad (3.3.62)$$

For $m' = n + 1$ *we have*

$$2b_{n+1}^{-m-1} T_{n+1}^{n+1,m+1} = (W_{11} + W_{21}) b_{n+1}^{-n-1} T_n^{nm}, \qquad (3.3.63)$$

$$2b_{n+1}^{m-1} T_{n+1}^{n+1,m-1} = (W_{11} - W_{21}) b_{n+1}^{-n-1} T_n^{nm}, \qquad (3.3.64)$$

$$a_n^m T_{n+1}^{n+1,m} = W_{31} b_{n+1}^{-n-1} T_n^{nm}, \qquad (3.3.65)$$

For $m' = -n - 1$ *we have*

$$2b_{n+1}^{-m-1} T_{n+1}^{-n-1,m+1} = (W_{12} + W_{22}) b_{n+1}^{-n-1} T_n^{-n,m}, \qquad (3.3.66)$$

$$2b_{n+1}^{m-1} T_{n+1}^{-n-1,m-1} = (W_{12} - W_{22}) b_{n+1}^{-n-1} T_n^{-n,m}, \qquad (3.3.67)$$

$$a_n^m T_{n+1}^{-n-1,m} = W_{32} b_{n+1}^{-n-1} T_n^{-n,m}. \qquad (3.3.68)$$

THEOREM 15 *The following recurrence relations holds for* $T_n^{m'm}(Q)$:

$$b_n^m T_{n-1}^{m',m+1} + b_n^{-m} T_{n-1}^{m',m-1}$$

$$= W_{11} b_n^{m'-1} T_n^{m'-1,m} + W_{12} b_n^{-m'-1} T_n^{m'+1,m} + W_{13} a_{n-1}^{m'} T_n^{m'm}, \qquad (3.3.69)$$

$$b_n^m T_{n-1}^{m',m+1} - b_n^{-m} T_{n-1}^{m',m-1}$$

$$= W_{21} b_n^{m'-1} T_n^{m'-1,m} + W_{22} b_n^{-m'-1} T_n^{m'+1,m} + W_{23} a_{n-1}^{m'} T_n^{m'm}. \qquad (3.3.70)$$

$$a_{n-1}^m T_{n-1}^{m'm} = W_{31} b_n^{m'-1} T_n^{m'-1,m} + W_{32} b_n^{-m'-1} T_n^{m'+1,m} + W_{33} a_{n-1}^{m'} T_n^{m'm}, \qquad (3.3.71)$$

where $n=0,1,2,\ldots$, $m=-n,\ldots,n$, $m'=-n+1,\ldots,n-1$, a_n^m *and* b_n^m *are defined by expressions (2.2.8) and (2.2.10), and* $W_{\alpha\beta}$ *are components of complex rotation matrix (3.3.60).*

PROOF. Taking the scalar product of both sides of Eq. (3.3.55) with $\mathbf{i}_x$, $\mathbf{ii}_y$, and $\mathbf{i}_z$, we obtain the statement of the theorem. □

COROLLARY 7 *Summing and subtracting relations (3.3.69) and (3.3.70) we have*

$$2b_n^m T_{n-1}^{m',m+1} = (W_{11} + W_{21}) b_n^{m'-1} T_n^{m'-1,m} + (W_{12} + W_{22}) b_n^{-m'-1} T_n^{m'+1,m}$$

$$+ (W_{13} + W_{23}) a_{n-1}^{m'} T_n^{m'm}, \qquad (3.3.72)$$

$$2b_n^{-m} T_{n-1}^{m',m-1} = (W_{11} - W_{21}) b_n^{m'-1} T_n^{m'-1,m} + (W_{12} - W_{22}) \times b_n^{-m'-1} T_n^{m'+1,m} + (W_{13} - W_{23}) a_{n-1}^{m'} T_n^{m'm}, \qquad (3.3.73)$$

For $m=n$ *we have*

$$(W_{11} + W_{21}) b_n^{m'-1} T_n^{m'-1,n} + (W_{12} + W_{22}) b_n^{-m'-1} T_n^{m'+1,n}$$

$$+ (W_{13} + W_{23}) a_{n-1}^{m'} T_n^{m'n} = 0, \qquad (3.3.74)$$

$$W_{31} b_n^{m'-1} T_n^{m'-1,n} + W_{32} b_n^{-m'-1} T_n^{m'+1,n} + W_{33} a_{n-1}^{m'} T_n^{m'n} = 0. \qquad (3.3.75)$$

For $m=-n$ *we have*

$$(W_{11} - W_{21}) b_n^{m'-1} T_n^{m'-1,-n} + (W_{12}$$

$$- W_{22}) b_n^{-m'-1} T_n^{m'+1,-n} + (W_{13} - W_{23}) a_{n-1}^{m'} T_n^{m',-n} = 0, \qquad (3.3.76)$$

$$W_{31} b_n^{m'-1} T_n^{m'-1,-n} + W_{32} b_n^{-m'-1} T_n^{m'+1,-n} + W_{33} a_{n-1}^{m'} T_n^{m',-n} = 0. \qquad (3.3.77)$$

THEOREM 16 *The following recurrence relation holds for* $T_n^{m'm}(\alpha, \beta, \gamma)$:

$$
\begin{aligned}
T_{n-1}^{m',m+1} &= \frac{e^{i\alpha}}{b_n^m} \\
&\times \left\{ \frac{1}{2} \left[b_n^{-m'-1}\, e^{i\gamma}(1-\cos\beta) T_n^{m'+1,m} - b_n^{m'-1}(1+\cos\beta) e^{-i\gamma} T_n^{m'-1,m} \right] \right. \\
&\quad \left. - a_{n-1}^{m'} \sin\beta\, T_n^{m'm} \right\}, \\
&\quad n = 2, 3, \ldots, \quad m' = -n+1, \ldots, n-1, \; m = 0, \ldots, n-2,
\end{aligned}
\tag{3.3.78}
$$

where α, β *and* γ *are the rotation angles, and* a_n^m *and* b_n^m *are defined by expressions (2.2.8) and (2.2.10).*

PROOF. Elements of complex and real rotation matrices W (Eq. (3.3.60)) and Q (Eq. (3.1.11)) are related as

$$
\begin{aligned}
W_{11} + W_{21} &= Q_{11} + Q_{22} + i(Q_{12} - Q_{21}), \\
W_{12} + W_{22} &= Q_{11} - Q_{22} + i(Q_{12} + Q_{21}), \\
W_{13} + W_{23} &= -2(Q_{31} + iQ_{32}),
\end{aligned}
\tag{3.3.79}
$$

The elements of Q can be expressed via rotation angles α, β, and γ using Eq. (3.3.10). Relation (3.3.72) then provides the statement of the theorem. □

COROLLARY 8 *The following recurrence relation holds for* $H_n^{m'm}(\beta)$:

$$
\begin{aligned}
H_{n-1}^{m',m+1} &= \frac{1}{b_n^m} \\
&\times \left\{ \frac{1}{2} [b_n^{-m'-1}(1-\cos\beta) H_n^{m'+1,m} - b_n^{m'-1}(1+\cos\beta) H_n^{m'-1,m}] \right. \\
&\quad \left. - a_{n-1}^{m'} \sin\beta\, H_n^{m'm} \right\}, \\
&\quad n - 2, 3, \ldots, \quad m' = -n+1, \ldots, n-1, \quad m = 0, \ldots, n-2.
\end{aligned}
\tag{3.3.80}
$$

PROOF. Follows from relations (3.3.37) and (3.3.78). □

3.3.4.1 Computational procedure

We describe briefly a computational procedure to obtain the coefficients $T_n^{m'm}(Q)$ via recurrence relations. It is noteworthy to remark that these

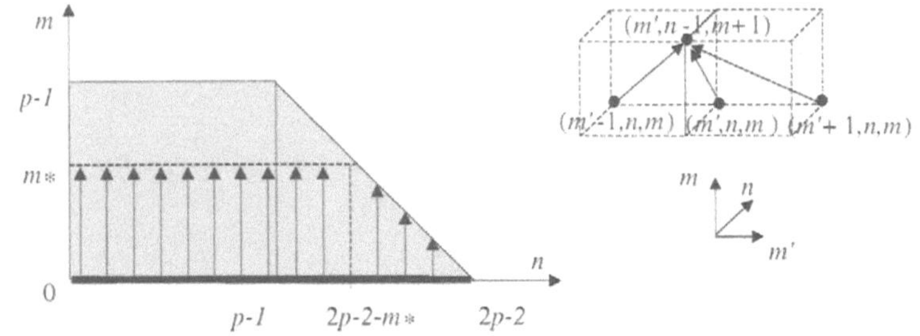

FIGURE 3.7 A scheme for computation of the rotation coefficients according to Eq. (3.3.80) (the chart on the right). Once the coefficients on layer m are known, they can be propagated to layer $m+1$. The process starts with known initial values at $m = 0$. Coefficients for negative m can be found then from symmetry relations. Propagation in the (n, m)-plane is shown on the chart on the left.

coefficients are not a property of the Helmholtz equation but purely a property of the spherical harmonics. Due to their dependence only on the angular part, these results should be the same for the Laplace and other equations, which can be expressed in terms of series involving spherical harmonics. The spherical harmonics have been studied in greater depth, than have the translation coefficients for the Helmholtz equation (see e.g. Ref. [Ste61] for addition theorems, and explicit relations to Wigner's symbols). However, this classical problem is still of interest, and research is still underway into the stable and rapid computation of the rotation coefficients based on the recurrence relations for real spherical harmonics [IR96] and general complex case [CIGR99]. Our derivation of the recurrence relations differs from these cited papers and has comparable or superior complexity.[3]

While we can use different recursion relationships for computations, the most convenient from the one listed above is given by Eq. (3.3.78), since it is expressed via the rotation angles. One can use it for non-negative m and then apply symmetry (3.3.41) to compute the rotation coefficients for negative m. However, instead of immediate computation of complex rotation coefficients, we recommend to compute the real-valued coefficients $H_n^{m'm}(\beta)$ using recursion (3.3.80), and then obtain the complex coefficients simply multiplying those coefficients by complex exponents (see Eq. (3.3.37)). Such a procedure is more efficient both in terms of speed and memory, since one only needs to store and multiply real numbers, instead of complex. Here we also have an additional

[3] Care should be taken when comparing our results with those of these papers, as we use different definitions of the spherical harmonics.

symmetry relation (3.3.44). A scheme for computation of the rotation coefficients according to the recursion (3.3.80) is shown in Fig. 3.7. Once the coefficients on a layer m are known, they can be propagated to layer $m + 1$. The process starts with known initial values at $m = 0$. Coefficients for negative m can then be found from symmetry relations. This scheme shows that if we need to compute coefficients for p degrees ($n = 0, \ldots, p - 1$) then we need to initialize the process for $n = 0, \ldots, 2p - 2$, since the recurrence scheme for each subsequent layer of m reduces the degree by 1.

3.4 SUMMARY

In this chapter we considered translation and rotation transforms, which enable expansion of solutions of the Helmholtz equation in various domains, and express solutions in various reference frames. We considered structures of translation and rotation coefficients, some of their properties, and methods for their computation. We proved several theorems and derived recurrence relations for the translation and rotation coefficients. These relations provide faster computation of these coefficients than methods based on expansions involving Clebsch–Gordan or similar coefficients.

CHAPTER 4

Multipole Methods

In this chapter we will consider several example problems to illustrate how multipole reexpansion/translation theory can be applied to obtain their solution.

4.1 ROOM ACOUSTICS: FAST SUMMATION OF SOURCES

The room impulse response is a quantity that characterizes the reverberant structure of a room. A method to approximate this quantity was proposed by Allen and Berkley [AB79], by using the image method. The method has been used by many authors to compute the impulse response. Using multipole methods this computation can be performed much faster, as was first shown in Ref. [DGZD01]. This simple example also serves as an illustration of the factorization idea that is central to fast multipole methods.

4.1.1 Formulation

Consider a room of length L, width W, and height H (see Fig. 4.1). A Cartesian reference frame is connected with the center of this room and the axes are parallel to the walls, so the six walls have coordinates $x = -L/2, L/2$; $y = -W/2, W/2$; $z = -H/2, H/2$. Inside the room we have a monopole sound source of intensity Q_0 placed at $x = x_0$, $y = y_0$, $z = z_0$ which in the absence of the walls generates the acoustic field:

$$\psi_0 = \frac{Q_0}{4\pi|\mathbf{r} - \mathbf{r}_0|} e^{ik|\mathbf{r}-\mathbf{r}_0|}. \qquad (4.1.1)$$

This field satisfies the Helmholtz equation with the Sommerfeld radiation conditions.

The presence of walls can be modeled by placing image sources which mirror the actual source and all other images with respect to the walls. The location of these image sources is denoted as $\mathbf{r}_q$, $q = 1, 2, \ldots$

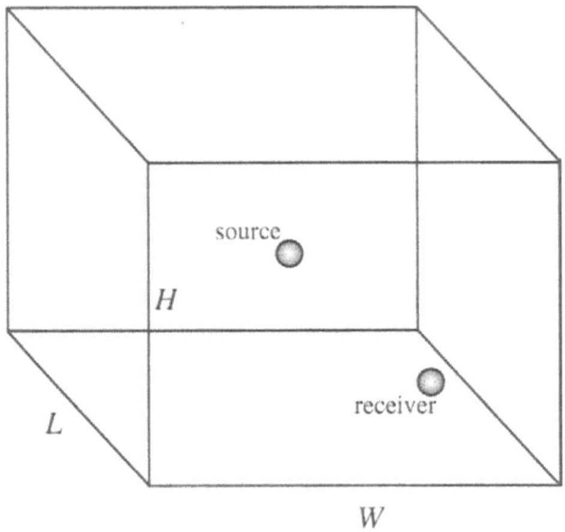

FIGURE 4.1 Problem formulation.

Generally, there should be an infinite number of sources to model reflection of signals of arbitrary frequency and duration. In practice the duration of acoustic signals is limited by the finite time over which the sound reflections need be considered. We denote t_{max} the maximum time considered. Since acoustic waves propagate with finite speed, only image sources with $|\mathbf{r}_q - \mathbf{r}| \leq ct_{max}$ influence the sound field inside the room, where $\mathbf{r}$ is the radius vector of an arbitrary point inside the room.

Denote $a = \frac{1}{2}\sqrt{L^2 + W^2 + H^2}$ which is the distance from the center of the room to any corner (vertex of the rectangle). Denote by Ω_a the sphere of radius a centered at the origin of the reference frame. This sphere includes the entire room. Consider also a sphere Ω_b of radius $b = ct_{max} + D$ concentric to Ω_a (see Fig. 4.2). Assume that we have N_b image sources located inside Ω_b. The image sources located outside Ω_b do not influence the sound field in the room for time interval $[0, t_{max}]$. Therefore, the field inside the room within specified time limits can be represented as

$$\psi(\mathbf{r}) = \sum_{r_q < b} \psi_q(\mathbf{r}), \quad \psi_q(\mathbf{r}) = \frac{Q_q}{4\pi|\mathbf{r} - \mathbf{r}_q|} e^{ik|\mathbf{r}-\mathbf{r}_q|}, \quad r_q = |\mathbf{r}_q|,\ r = |\mathbf{r}|, \tag{4.1.2}$$

where Q_q is the strength of the qth source. For sound-hard walls the intensity of the image sources is the same as the intensity of the real source, $Q_q = Q_0$. Some heuristic techniques for changing the strength of the image sources to account for losses due to reflections are considered in Ref. [AB79], but will be ignored here, since they are not essential to the demonstration of the acceleration of the summation.

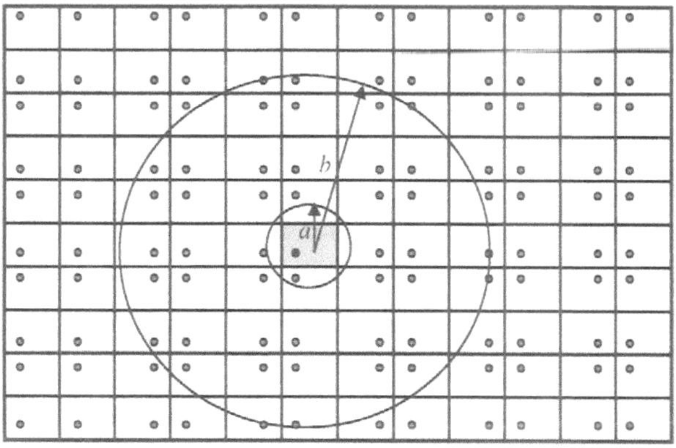

FIGURE 4.2 Schematic for computation of room acoustics using the method of images.

The practical shortcoming of this solution is that the number of image sources within Ω_b can be large (usually several thousand) and evaluation of the sum over N sources in Ω_b for any given location in the room can be an expensive computational procedure. Further, we may need the solution to be evaluated at many points, say M, in the room (for example, in an auralization calculation). In general, this requires computation of $O(NM)$ distances and subsequent multiplications and additions.

The problem can be solved much faster using a multipole reexpansion technique. This provides a fast algorithm requiring $O(N+M)$ operations for evaluation of the sound field and this solution is illustrated below.

4.1.2 Solution

Let us break the sum (4.1.2) into two terms:

$$\psi(\mathbf{r}) = \psi_s(\mathbf{r}) + \psi_r(\mathbf{r}), \tag{4.1.3}$$

where

$$\psi_s(\mathbf{r}) = \sum_{r_q<a} \psi_q(\mathbf{r}), \qquad \psi_r(\mathbf{r}) = \sum_{a<r_q<b} \psi_q(\mathbf{r}), \tag{4.1.4}$$

and $\psi_s(\mathbf{r})$ and $\psi_r(\mathbf{r})$ are the singular and regular parts of the field inside the sphere Ω_a. The singular part is generated by the real source, plus image sources $\psi_q(\mathbf{r})$ that are located inside the sphere Ω_a. The regular part is generated by the image sources located in the domain bounded by the spheres Ω_a and Ω_b. Below we consider the evaluation of the regular part.

A monopole source can be expanded into a series of spherical harmonics (3.2.2):

$$\psi_q(\mathbf{r}) = ik \sum_{n=0}^{\infty} \sum_{m=-n}^{n} S_n^{-m}(\mathbf{r}_q) R_n^m(\mathbf{r}), \qquad r \leq r_q. \tag{4.1.5}$$

Thus, the regular part of the field can be written in the form:

$$\psi_r(\mathbf{r}) = ik \sum_{n=0}^{\infty} \sum_{m=-n}^{n} C_n^m R_n^m(\mathbf{r}), \qquad C_n^m = \sum_{D<r_q<R_{\max}} Q_q S_n^{-m}(\mathbf{r}_q), \tag{4.1.6}$$

where the coefficients C_n^m can be precomputed for given image source locations. The series can be truncated at $n = p - 1$, and the evaluation of $\psi_r(\mathbf{r})$ at any given point $\mathbf{r}$ will require summation of p^2 compared to summation over N_s terms representing N_s sources with $D < r_q < R_{\max}$ if Eq. (4.1.4) is used directly. Note that p^2 can be orders of magnitude smaller than N_s reducing the number of operations significantly. This gain is more sensible for large number of field points where $\psi_r(\mathbf{r})$ should be evaluated.

4.1.3 Computations and discussion

To check the accuracy and performance of the proposed algorithm, we implemented it using Matlab and compared results with those obtained using the straightforward summation algorithm. Figure 4.3 shows an example comparison between results of computations. In this example the room size was $2m \times 2m \times 2m$ and evaluation points were distributed over the entire room. The radius b was about 12 m, which corresponds to $t_{\max} = 30$ ms, and includes the monopole source and its 910 images. The wavenumber k was $3\ \mathrm{m}^{-1}$ (corresponds to 164 Hz).

In computations we selected $a = 1.73$ m and varied the truncation number p. Results shown in Fig. 4.3 were obtained for $p = 8$. Even for low values of p the accuracy of computations is remarkable. Use of large values of p does not make sense, since it does not substantially improve accuracy, requires larger computation times and may cause computational problems due to the exponential growth of coefficients of the multipole expansion C_n^m.

The computational time for the case discussed for $M = 1331$ evaluation points was about 2.7 times smaller when using the proposed algorithm compared to the straightforward algorithm. The gain in computational time increases with the increase of M and N. For example, for $b = 19$ m, which corresponds to $t_{\max} = 50$ ms, and includes the source

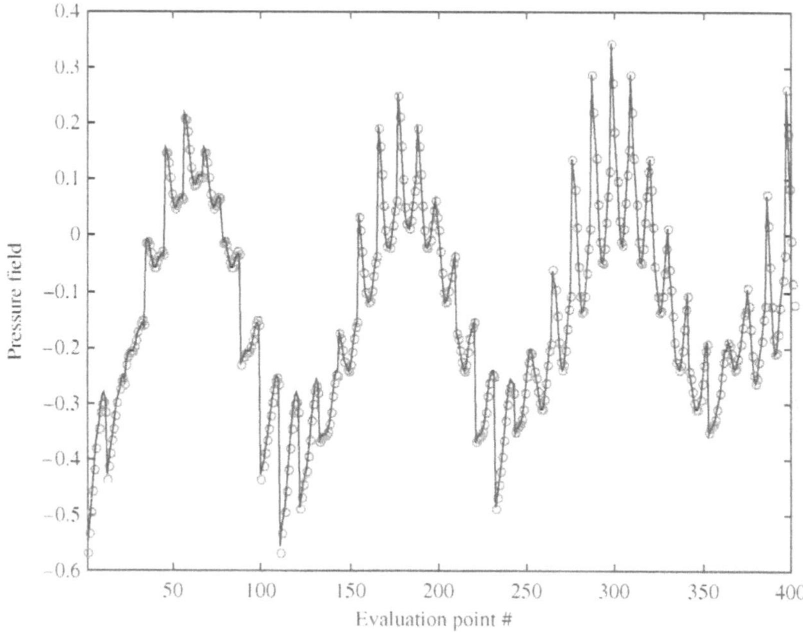

FIGURE 4.3 Comparison of the pressure field in a $2m \times 2m \times 2m$ reverberant room using the algorithm of Ref. [AB79] (solid curve) and the proposed multipole algorithm (circles) for 400 evaluation points ($k = 3\ \mathrm{m}^{-1}$, $t_{\max} = 30$ ms). No difference can be observed.

and its 3534 images, the multipole algorithm performed the same evaluations 6.5 times faster than the straightforward algorithm.

4.2 SCATTERING FROM A SINGLE SPHERE

An understanding of this basic problem is a key to scattering theory.

4.2.1 Formulation

Consider a sphere of radius a centered at $\mathbf{r} = \mathbf{r}_0$ which is exposed to an incident field $\psi_{\text{in}}(\mathbf{r})$ (see Fig. 4.4) that satisfies the Helmholtz equation in the vicinity of the sphere. The total potential that satisfies the Helmholtz equation in the domain outside the sphere can be decomposed as

$$\psi(\mathbf{r}) = \psi_{\text{in}}(\mathbf{r}) + \psi_{\text{scat}}(\mathbf{r}), \tag{4.2.1}$$

where $\psi_{\text{scat}}(\mathbf{r})$ is the potential of the scattered field, which is a radiating

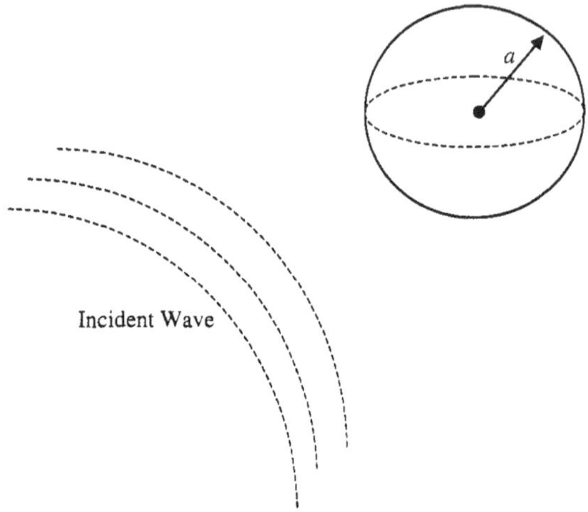

FIGURE 4.4 The geometry of the problem of wave scattering by a single sphere.

function and has no singularities at $|\mathbf{r} - \mathbf{r}_0| > a$. The problem is to find $\psi_{\text{scat}}(\mathbf{r})$ or $\psi(\mathbf{r})$ for a given $\psi_{\text{in}}(\mathbf{r})$, assuming that the total field satisfies the following boundary condition on the surface of the sphere:

$$|\mathbf{r} - \mathbf{r}_0| = a: \quad \frac{\partial \psi(\mathbf{r})}{\partial n} + \mathrm{i}\sigma\psi(\mathbf{r}) = 0, \tag{4.2.2}$$

where σ is the complex admittance and $\partial/\partial n$ is the derivative in the direction of the outer normal to the surface. For $0 < |\sigma| < \infty$ we have the Robin problem, which reduces to the Neumann and Dirichlet problems for $\sigma = 0$ and $\sigma \to \infty$, respectively.

4.2.2 Solution

4.2.2.1 Determination of expansion coefficients

Since $\psi_{\text{in}}(\mathbf{r})$ is a regular function inside a sphere of radius $b > a$ concentric with the scatterer, this function can be expanded in the following series about $\mathbf{r} = \mathbf{r}_0$:

$$\psi_{\text{in}}(\mathbf{r}) = \sum_{n=0}^{\infty} \sum_{m=-n}^{n} E_n^m R_n^m(\mathbf{r} - \mathbf{r}_0), \tag{4.2.3}$$

where the coefficients E_n^m depend on the incident field. For example, for a plane wave of complex intensity $\psi_{\text{in}}(\mathbf{r}_0) = Q$, according to Eq. (2.3.6), we have

$$\psi_{\text{in}}(\mathbf{r}) = Q\,\mathrm{e}^{\mathrm{i}\mathbf{k}\cdot(\mathbf{r}-\mathbf{r}_0)} = 4\pi Q \sum_{n=0}^{\infty} \sum_{m=-n}^{n} \mathrm{i}^n Y_n^{-m}(\theta_k, \varphi_k) R_n^m(\mathbf{r} - \mathbf{r}_0),$$
$$E_n^m = 4\pi Q \mathrm{i}^n Y_n^{-m}(\theta_k, \varphi_k), \qquad n = 0, 1, \ldots, \quad m = -n, \ldots, n, \tag{4.2.4}$$

where the spherical polar angles θ_k and φ_k characterize the direction of the plane wave (see relation (2.3.4)). In the case, when the incident wave is generated by a source of intensity Q located at $\mathbf{r} = \mathbf{r}_s$, $|\mathbf{r}_s - \mathbf{r}_0| \geq b$, we have from expansion (3.2.2):

$$\psi_{\text{in}}(\mathbf{r}) = QG(\mathbf{r} - \mathbf{r}_s) = Qik \sum_{n=0}^{\infty} \sum_{m=-n}^{n} S_n^{-m}(\mathbf{r}_s - \mathbf{r}_0) R_n^m(\mathbf{r} - \mathbf{r}_0),$$
$$E_n^m = QikS_n^{-m}(\mathbf{r}_s - \mathbf{r}_0), \qquad n = 0, 1, \ldots, \quad m = -n, \ldots, n. \tag{4.2.5}$$

It is not difficult to build the local expansion of the incident field generated by an arbitrary multipole source or a sum of multipole sources and plane waves coming from different directions.

Since the scattered field is a radiating function, we construct it in the form of a sum of singular spherical basis functions:

$$\psi_{\text{scat}}(\mathbf{r}) = \sum_{n=0}^{\infty} \sum_{m=-n}^{n} A_n^m S_n^m(\mathbf{r} - \mathbf{r}_0), \tag{4.2.6}$$

where the expansion coefficients A_n^m should be determined from the boundary conditions.

Using definitions of the spherical basis functions (2.1.101) and (2.1.102) the total potential can be represented as

$$\psi(\mathbf{r}) = \sum_{n=0}^{\infty} \sum_{m=-n}^{n} [E_n^m j_n(k|\mathbf{r} - \mathbf{r}_0|) + A_n^m h_n(k|\mathbf{r} - \mathbf{r}_0|)] Y_n^m(\mathbf{s}),$$
$$\mathbf{s} = \frac{\mathbf{r} - \mathbf{r}_0}{|\mathbf{r} - \mathbf{r}_0|}. \tag{4.2.7}$$

To satisfy boundary condition (4.2.2) at $|\mathbf{r} - \mathbf{r}_0| = a$, we need to have

$$\sum_{n=0}^{\infty} \sum_{m=-n}^{n} \{k[E_n^m j_n'(ka) + A_n^m h_n'(ka)] + \sigma[E_n^m j_n(ka)$$
$$+ A_n^m h_n(ka)]\} Y_n^m(\mathbf{s}) = 0, \tag{4.2.8}$$

where $\mathbf{s} = (\mathbf{r} - \mathbf{r}_0)/a$. Due to orthogonality and completeness of the spherical harmonics, each term in this sum should be zero.

This determines A_n^m as

$$A_n^m = -\frac{j_n'(ka)+(\sigma/k)j_n(ka)}{h_n'(ka)+(\sigma/k)h_n(ka)}E_n^m, \quad n=0,1,\ldots, \; m=-n,\ldots,n. \tag{4.2.9}$$

In the particular cases of sound-hard and sound-soft surfaces we have

$$\sigma=0: \; A_n^m = -\frac{j_n'(ka)}{h_n'(ka)}E_n^m, \quad \sigma=\infty: \; A_n^m = -\frac{j_n(ka)}{h_n(ka)}E_n^m, \\ n=0,1,\ldots, \; m=-n,\ldots,n. \tag{4.2.10}$$

We can also see that these limiting cases are realized for $\sigma \ll k$ and $\sigma \gg k$, respectively. This introduces the scale and defines "small" and "large" σ.

In any case, the above equations show that coefficients of the scattered field are proportional to coefficients of the incident field, and the proportionality coefficient depends only on the degree of the spherical functions n. In a more general problem of scattering from a body of an arbitrary shape, the solution for the scattered field in the region outside a sphere surrounding the scatterer can be represented in the form (4.2.6) (see also Eq. (2.3.37)). In the general case the expansion coefficients are related to the coefficients of the incident field linearly, due to the linearity of the Helmholtz equation and boundary conditions. Therefore, we can write

$$A_n^m = \sum_{n'=0}^{\infty} \sum_{m'=-n'}^{n'} T_{nn'}^{mm'} E_{n'}^{m'}, \qquad n = 0, 1, \ldots, \; m = -n, \ldots n, \tag{4.2.11}$$

where $T_{nn'}^{mm'}$ are elements of the matrix $\mathbb{T}$, which is sometimes called the *T-matrix*. This matrix depends on the size and the shape of the object, wavenumber, k, and admittance (or impedance) of the scatterer. As we can see from expressions (4.2.9) and (4.2.11) in the case of the sphere, this matrix is diagonal:

$$T_{nn'}^{mm'} = -\frac{j_n'(ka) + (\sigma/k)j_n(ka)}{h_n'(ka) + (\sigma/k)h_n(ka)}\delta_{nn'}\delta_{mm'}, \\ n, n' = 0, 1, \ldots, \; m = -n, \ldots, n, \; m' = -n', \ldots, n'. \tag{4.2.12}$$

4.2.2.2 *Surface function*

In many practical problems the potential and its derivative (say, the complex acoustic pressure and the normal velocity) need to be determined on the surface of the body. The expressions for these functions can be simplified using the *Wronskian* for spherical Bessel functions:

$$W\{j_n(ka), h_n(ka)\} = j_n(ka)h_n'(ka) - j_n'(ka)h_n(ka) = \mathrm{i}(ka)^{-2}. \qquad (4.2.13)$$

Indeed, substituting Eq. (4.2.9) into Eq. (4.2.7) at $|\mathbf{r} - \mathbf{r}_0| = a$ we obtain

$$\psi|_S = \sum_{n=0}^{\infty}\sum_{m=-n}^{n}\left[j_n(ka) - \frac{j_n'(ka) + (\sigma/k)j_n(ka)}{h_n'(ka) + (\sigma/k)h_n(ka)}h_n(ka)\right]E_n^m Y_n^m(\mathbf{s})$$
$$= \frac{\mathrm{i}}{(ka)^2}\sum_{n=0}^{\infty}\frac{1}{h_n'(ka) + (\sigma/k)h_n(ka)}\sum_{m=-n}^{n} E_n^m Y_n^m(\mathbf{s}). \qquad (4.2.14)$$

The normal derivative can be then determined using boundary condition (4.2.2):

$$\left.\frac{\partial\psi}{\partial n}\right|_S = -\mathrm{i}\sigma\psi|_S = \frac{\sigma}{(ka)^2}\sum_{n=0}^{\infty}\frac{1}{h_n'(ka)+(\sigma/k)h_n(ka)}\sum_{m=-n}^{n} E_n^m Y_n^m(\mathbf{s}). \qquad (4.2.15)$$

In limiting cases corresponding to the Neumann and Dirichlet boundary conditions, we obtain

$$\sigma=0:\ \psi|_S = \frac{\mathrm{i}}{(ka)^2}\sum_{n=0}^{\infty}\frac{1}{h_n'(ka)}\sum_{m=-n}^{n} E_n^m Y_n^m(\mathbf{s}), \qquad \left.\frac{\partial\psi}{\partial n}\right|_S = 0, \qquad (4.2.16)$$

$$\sigma=\infty:\ \psi|_S = 0, \qquad \left.\frac{\partial\psi}{\partial n}\right|_S = \frac{1}{ka^2}\sum_{n=0}^{\infty}\frac{1}{h_n(ka)}\sum_{m=-n}^{n} E_n^m Y_n^m(\mathbf{s}). \qquad (4.2.17)$$

Consider now some typical incident fields. In the case of a plane wave, when coefficients E_n^m are given by expression (4.2.4), using the addition theorem for spherical harmonics (2.1.70) or (2.1.72), we have

$$\sum_{m=-n}^{n} E_n^m Y_n^m(\mathbf{s}) = Q(2n+1)\mathrm{i}^n P_n(\cos\theta), \qquad \cos\theta = \frac{\mathbf{k}\cdot\mathbf{s}}{k}. \qquad (4.2.18)$$

Therefore, in this case we have

$$\psi|_S = \frac{\mathrm{i}Q}{(ka)^2}\sum_{n=0}^{\infty}\frac{(2n+1)\mathrm{i}^n P_n(\cos\theta)}{h_n'(ka)+(\sigma/k)h_n(ka)},$$
$$\left.\frac{\partial\psi}{\partial n}\right|_S = \frac{\sigma Q}{(ka)^2}\sum_{n=0}^{\infty}\frac{(2n+1)\mathrm{i}^n P_n(\cos\theta)}{h_n'(ka)+(\sigma/k)h_n(ka)}. \qquad (4.2.19)$$

In the case when the incident field is generated by a monopole and E_n^m are specified by expression (4.2.5), we can again use the addition theorem to obtain

$$\sum_{m=-n}^{n} E_n^m Y_n^m(\mathbf{s}) = \frac{ikQ}{4\pi} h_n(kd)(2n+1)P_n(\cos\theta),$$
$$\cos\theta = \frac{(\mathbf{r}_s - \mathbf{r}_0)\cdot\mathbf{s}}{d}, \qquad d = |\mathbf{r}_s - \mathbf{r}_0|. \tag{4.2.20}$$

So the surface potential and its normal derivative in this case are

$$\psi|_S = -\frac{Q}{4\pi ka^2}\sum_{n=0}^{\infty}\frac{(2n+1)h_n(kd)P_n(\cos\theta)}{h_n'(ka)+(\sigma/k)h_n(ka)},$$
$$\left.\frac{\partial\psi}{\partial n}\right|_S = \frac{i\sigma Q}{4\pi ka^2}\sum_{n=0}^{\infty}\frac{(2n+1)h_n(kd)P_n(\cos\theta)}{h_n'(ka)+(\sigma/k)h_n(ka)}. \tag{4.2.21}$$

Note that in both the cases the surface potential and its normal derivative depend only on the angle θ, which is also called as the *incidence angle* and is an angle between the radius vector of the surface point, where the potential is measured, and the direction of the incident wave. Obviously, that for the plane wave and a monopole source the problem is axisymmetric, so the above results could be obtained by consideration of axisymmetric problem, when the z-axis is selected to be colinear to the direction of the incident wave.

Equations (4.2.19) and (4.2.21) can also be simplified for the sound-hard and sound-soft bodies. In the limiting cases ($\sigma = 0, \infty$) these solutions are classical and can be found elsewhere [Han81]. Note that the solution of the problem of plane wave scattering by a rigid sphere was published by Lord Rayleigh a century ago [Ray04].

4.2.3 Computations and discussion

Figure 4.5 illustrates some computational results for scattering by a sound-hard sphere placed in the field of a monopole source. For illustrations we took the real part of the complex potential and plotted this function in the plane which passes through the center of the scatterer and the source. In this plane the field of the monopole source is represented by concentric circles centered at the source location (see the upper row in Fig. 4.5). As we can see the presence of the scatterer (the white disc) modifies the field and $\psi(\mathbf{r}) = \psi_{\text{in}}(\mathbf{r}) + \psi_{\text{scat}}(\mathbf{r})$ is different from $\psi_{\text{in}}(\mathbf{r})$. Particularly, for a sound-hard sphere the wave fronts are orthogonal to the sphere surface, since in this case $\partial\psi/\partial n|_S = 0$. For large ka a "shadow" of the sphere is seen.

Figures 4.6 and 4.7 show computational results for the amplitude of the surface potential $|\psi|_S|$ for various ka and fixed $d/a = 5$. It is seen that the problem is axisymmetric and only the dependence on the incidence

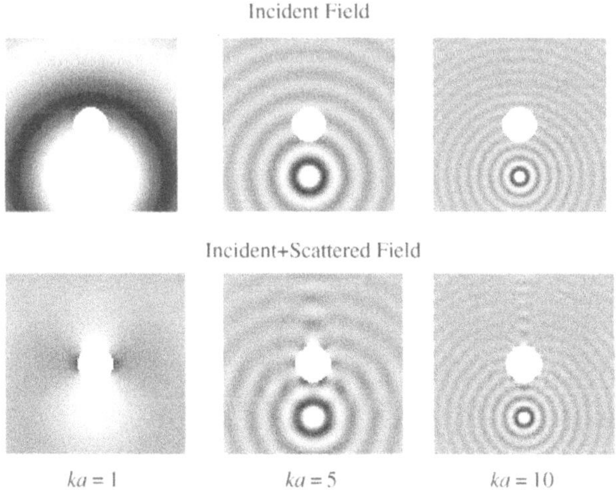

FIGURE 4.5 The incident and scattered fields for a sphere, when the incident field is generated by a monopole source. Computations are made with different non-dimensional wavenumbers shown near the pictures.

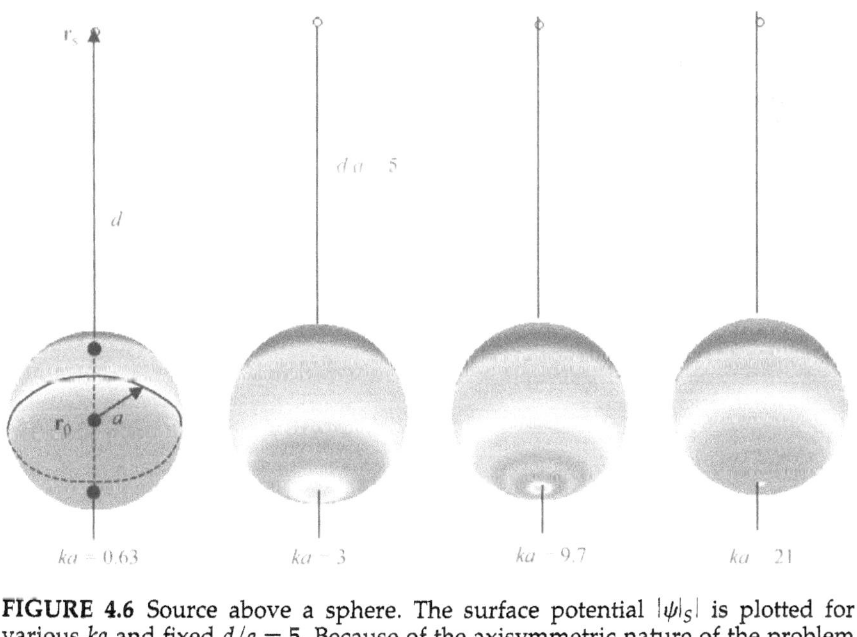

FIGURE 4.6 Source above a sphere. The surface potential $|\psi|_S|$ is plotted for various *ka* and fixed $d/a = 5$. Because of the axisymmetric nature of the problem only a dependence on the incidence angle θ is observed. For color version, see Plate D after Chapter 10.

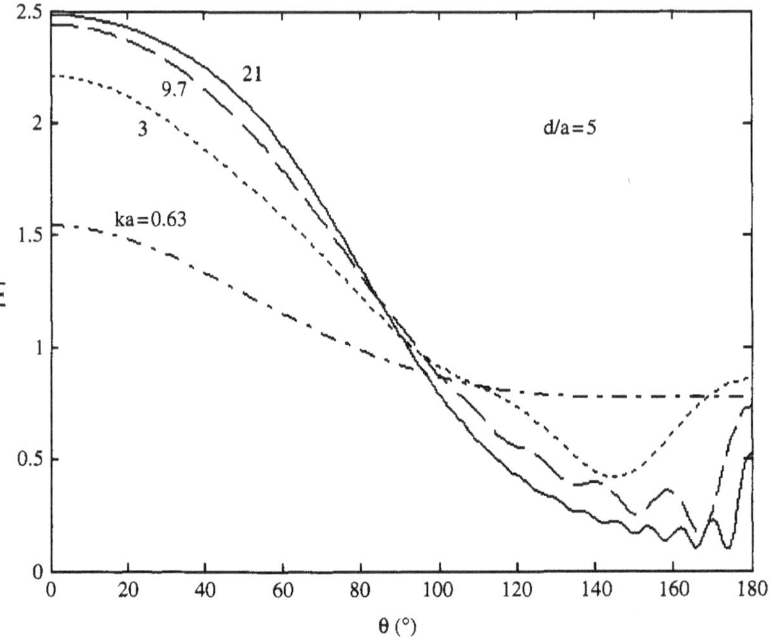

FIGURE 4.7 Convergence of the computations with truncation number for a source above a sphere. The truncation number depends on ka and we selected $p = \min(3, eka)$. A smooth behavior of the field is observed.

angle θ is observed. In computations we truncated expansion (4.2.21) to the first p modes. The truncation number depends on ka and while we selected $p = \min(3, eka)$ we found experimentally that there is no visible difference in figures produced if $p \geq \frac{1}{2}eka$.

The truncation number required to provide computations with a prescribed accuracy can be, in fact, evaluated rigorously. Several lemmas for bounds of the spherical Hankel functions in Chapter 9. Here we just note that for $n > ka + 1$, $d > a$ the ratios $|h_n(kd)|/|h_n(ka)|$ and $|h_n(kd)|/|h_n'(ka)|$ decay exponentially with n, so for large ka we need to have $p > ka$ and selection of $p = cka$, with some $c > 1$ depending on the prescribed error provides accurate results.

4.3 SCATTERING FROM TWO SPHERES

4.3.1 Formulation

The problem of wave scattering by two spheres is interesting from a mathematical point of view and has several practical applications.

For example, if a spherical scatterer is placed near a rigid wall, the wall can be replaced by an image sphere. In this case the problem is equivalent to scattering from a doublet of spheres placed in the incident field, which is a sum of the original field and its mirror image (see Ref. [GD01b]). Another problem is related to modeling of some scatterers with two spheres (e.g. head and torso model for spatial audio [ADD02, GDT02]).

The geometry of the problem is shown in Fig. 4.8. Here we have two spheres of radii a_1 and a_2 whose centers, $\mathbf{r}_1'$ and $\mathbf{r}_2'$ are located at the distance $|\mathbf{r}_1' - \mathbf{r}_2'| = t$, and which are placed in the incident field, generated, for example, by a monopole source. As discussed above, the nature of the incident field, $\psi_{\text{in}}(\mathbf{r})$, is not important (sums of plane waves, multipole sources, etc.), and we just assume that this field can be expanded in series of type (4.2.3) inside spheres of radii $b_1 > a_1$ and $b_2 > a_2$ concentric with the scatterers. We also introduce two reference frames $(O_1x_1y_1z_1)$ and $(O_2x_2y_2z_2)$ with a common z-axis and origins at the centers of the spheres. These frames can be made to coincide by translation along the z-axis.

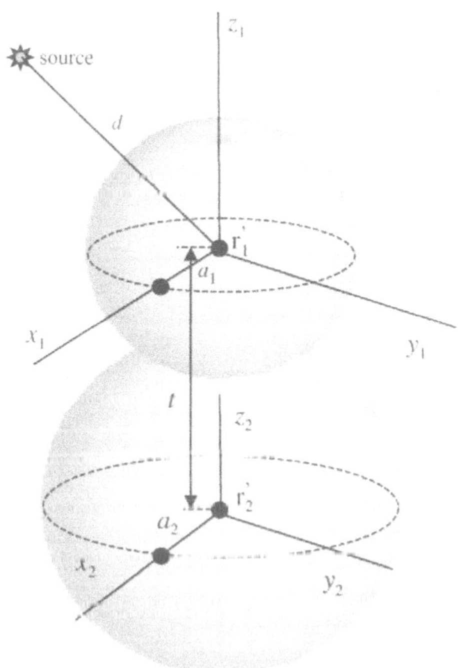

FIGURE 4.8 Formulation of the two-sphere ("snowman") problem.

As in the case of the single sphere, the total field that satisfies the Helmholtz equation in the domain external to the scatterers and singularities of the incident field can be represented as a superposition of the incident and scattered fields (4.2.1). The scattered field satisfies the Sommerfeld radiation condition far from a large sphere which includes the scatterers. We assume that the spheres has some admittances σ_1 and σ_2, respectively, and the total potential satisfies the impedance boundary conditions on the surface of each sphere:

$$\begin{aligned} |\mathbf{r}-\mathbf{r}'_1| = a_1: \quad & \frac{\partial \psi(\mathbf{r})}{\partial n_1} + \mathrm{i}\sigma_1 \psi(\mathbf{r}) = 0, \\ |\mathbf{r}-\mathbf{r}'_2| = a_2: \quad & \frac{\partial \psi(\mathbf{r})}{\partial n_2} + \mathrm{i}\sigma_2 \psi(\mathbf{r}) = 0. \end{aligned} \tag{4.3.1}$$

Here the normal derivatives are taken in the direction of the outer normals to the surfaces $\mathbf{n}_1$ and $\mathbf{n}_2$.

4.3.2 Solution

4.3.2.1 Determination of expansion coefficients

Let us denote

$$\mathbf{r}_1 = \mathbf{r} - \mathbf{r}'_1, \quad \mathbf{r}_2 = \mathbf{r} - \mathbf{r}'_2, \quad \mathbf{r}'_{12} = \mathbf{r}'_2 - \mathbf{r}'_1 = \mathbf{r}_1 - \mathbf{r}_2 = -\mathbf{r}'_{21}. \tag{4.3.2}$$

Therefore, $\mathbf{r}_1$ and $\mathbf{r}_2$ represent coordinates of the same field point in the reference frames $(O_1x_1y_1z_1)$ and $(O_2x_2y_2z_2)$, respectively. We decompose the total field potential into the following sum:

$$\psi(\mathbf{r}) = \psi_{\mathrm{in}}(\mathbf{r}) + \psi_1(\mathbf{r}) + \psi_2(\mathbf{r}), \tag{4.3.3}$$

where $\psi_1(\mathbf{r}) + \psi_2(\mathbf{r})$ is the total scattered field and we associate $\psi_1(\mathbf{r})$ with sphere 1 and $\psi_2(\mathbf{r})$ with sphere 2. Note that while $\psi_{\mathrm{in}}(\mathbf{r})$ is a prescribed function, $\psi_1(\mathbf{r})$ and $\psi_2(\mathbf{r})$ depend on one other, and on $\psi_{\mathrm{in}}(\mathbf{r})$ due to boundary conditions (4.3.1). So $\psi_1(\mathbf{r})$ can be taken as the potential of the field scattered by the sphere 1 in the presence of sphere 2, and $\psi_2(\mathbf{r})$ as the potential of the field scattered by the sphere 2 in the presence of sphere 1.

Since the scattered field is radiating, we can represent the potentials of the scattered field in corresponding reference frames as

$$\psi_1(\mathbf{r}) = \sum_{n=0}^{\infty} \sum_{m=-n}^{n} A_n^{(1)m} S_n^m(\mathbf{r}_1), \qquad \psi_2(\mathbf{r}) = \sum_{n=0}^{\infty} \sum_{m=-n}^{n} A_n^{(2)m} S_n^m(\mathbf{r}_2). \tag{4.3.4}$$

The problem now is to determine the expansion coefficients, which satisfy boundary conditions (4.3.1).

First, consider the sphere 1. We can expand $\psi_{\text{in}}(\mathbf{r})$ and reexpand $\psi_2(\mathbf{r})$ about the center of this sphere as follows:

$$\psi_{\text{in}}(\mathbf{r}) = \sum_{n=0}^{\infty} \sum_{m=-n}^{n} E_n^{(1)m} R_n^m(\mathbf{r}_1), \qquad \psi_2(\mathbf{r}) = \sum_{n=0}^{\infty} \sum_{m=-n}^{n} B_n^{(1)m} R_n^m(\mathbf{r}_1), \quad (4.3.5)$$

where

$$B_n^{(1)m} = \sum_{n'=0}^{\infty} \sum_{m'=-n'}^{n'} (S|R)_{nn'}^{mm'}(\mathbf{r}'_{21}) A_{n'}^{(2)m'},$$

$$n = 0, 1, \ldots, \quad m = -n, \ldots, n, \qquad (4.3.6)$$

since at $t = |\mathbf{r}'_{21}| > |\mathbf{r}_1|$ we have according to expression (3.2.4) the following reexpansion:

$$S_n^m(\mathbf{r}_2) = S_n^m(\mathbf{r}_1 + \mathbf{r}'_{21}) = \sum_{n'=0}^{\infty} \sum_{m'=-n'}^{n'} (S|R)_{n'n}^{m'm}(\mathbf{r}'_{21}) R_{n'}^{m'}(\mathbf{r}_1) \qquad (4.3.7)$$

and relation (4.3.6) follows from the regrouping of the terms in representation of $\psi_2(\mathbf{r})$ (substitute Eq. (4.3.7) into Eq. (4.3.4)). This also can be interpreted as translation of the expansion coefficients.

Due to expression (4.3.5), the sum of functions $\psi_{\text{in}}(\mathbf{r})$ and $\psi_2(\mathbf{r})$ can be represented in the form:

$$\psi_{\text{in}}^{(\text{eff})(1)}(\mathbf{r}) = \psi_{\text{in}}(\mathbf{r}) + \psi_2(\mathbf{r}) = \sum_{n=0}^{\infty} \sum_{m=-n}^{n} E_n^{(\text{eff})(1)m} R_n^m(\mathbf{r}_1),$$

$$E_n^{(\text{eff})(1)m} = E_n^{(1)m} + B_n^{(1)m}, \qquad n = 0, 1, \ldots, \quad m = -n, \ldots, n. \qquad (4.3.8)$$

This form shows that the problem can be reformulated for the first scatterer as the problem for a single sphere which is placed in an "effective" incident field $\psi_{\text{in}}^{(\text{eff})(1)}(\mathbf{r})$, given by expansion coefficients $E_n^{(\text{eff})(1)m}$. Therefore, we can use Eq. (4.2.9) to relate $A_n^{(1)m}$ and $E_n^{(\text{eff})(1)m}$:

$$A_n^{(1)m} = -\frac{j'_n(ka_1) + (\sigma_1/k) j_n(ka_1)}{h'_n(ka_1) + (\sigma_1/k) h_n(ka_1)} E_n^{(\text{eff})(1)m},$$

$$n = 0, 1, \ldots, \quad m = -n, \ldots, n. \qquad (4.3.9)$$

The same procedure can be done for sphere 2. One can also notice that the problem is symmetrical with respect to an exchange of subscripts 1 and 2, or indices of the sphere. So we have

$$A_n^{(2)m} = -\frac{j'_n(ka_2) + (\sigma_2/k)j_n(ka_2)}{h'_n(ka_2) + (\sigma_2/k)h_n(ka_2)} E_n^{(\mathrm{eff})(2)m},$$

$$E_n^{(\mathrm{eff})(2)m} = E_n^{(2)m} + B_n^{(2)m}, B_n^{(2)m} = \sum_{n'=0}^{\infty} \sum_{m'=-n'}^{n'} (S|R)_{nn'}^{mm'}(\mathbf{r}'_{12}) A_{n'}^{(1)m'}, \tag{4.3.10}$$

$$n = 0, 1, \ldots, \quad m = -n, \ldots, n.$$

Equations (4.3.6) and (4.3.8)–(4.3.10) form a closed linear system which allows us to determine coefficients $A_n^{(1)m}$ and $A_n^{(2)m}$ and so solve the problem. The issue here is that this linear system is infinite and in practice we replace it with a truncated system for the first p modes, $n = 0, 1, \ldots, p - 1$, with a hope that solution of this system converges at $p \to \infty$ to the solution of the problem.

We note that, in the case of two spheres, the solution and matrices can be substantially simplified by selection of the reference frame as described above. Indeed, in this case we have

$$\mathbf{r}'_{21} = t\mathbf{i}_z, \qquad \mathbf{r}'_{12} = -t\mathbf{i}_z. \tag{4.3.11}$$

This means that the translations are coaxial. In this case translations for each order m can be performed independently of other orders m (see expansion (3.2.86)). This shows that

$$B_n^{(1)m} = \sum_{n'=|m|}^{\infty} (S|R)_{nn'}^{m}(t) A_{n'}^{(2)m}, \qquad m = 0, \pm 1, \ldots, \quad n = |m|, |m| + 1, \ldots,$$

$$B_n^{(2)m} = \sum_{n'=|m|}^{\infty} (-1)^{n+n'} (S|R)_{nn'}^{m}(t) A_{n'}^{(1)m} = \sum_{n'=|m|}^{\infty} (S|R)_{n'n}^{m}(t) A_{n'}^{(1)m},$$

$$m = 0, \pm 1, \ldots, \quad n = |m|, |m| + 1, \ldots. \tag{4.3.12}$$

Here we used the symmetry property of the translation coefficients (3.2.54) and (3.2.96), which show that the coaxial translation matrix for the opposite direction is equal to the transpose of the translation matrix in the direction of $\mathbf{i}_z$.

Let us rewrite the obtained equations in a p-truncated matrix–vector form. Denoting

$$(\mathbf{S}|\mathbf{R})^m(t) = \begin{pmatrix} (S|R)^m_{|m|,|m|} & (S|R)^m_{|m|,|m|+1} & \cdots & (S|R)^m_{|m|,p-1} \\ (S|R)^m_{|m|+1,|m|} & (S|R)^m_{|m|+1,|m|+1} & \cdots & (S|R)^m_{|m|+1,p-1} \\ \cdots & \cdots & \cdots & \cdots \\ (S|R)^m_{p-1,|m|} & (S|R)^m_{p-1,|m|+1} & \cdots & (S|R)^m_{p-1,p-1} \end{pmatrix},$$

$$\Lambda^{(q)m}(t) = \begin{pmatrix} \Lambda^{(q)}_{|m|} & 0 & \cdots & 0 \\ 0 & \Lambda^{(q)}_{|m|+1} & \cdots & 0 \\ \cdots & \cdots & \cdots & \cdots \\ 0 & 0 & \cdots & \Lambda^{(q)}_{p-1} \end{pmatrix}, \tag{4.3.13}$$

$$\mathbf{B}^{(q)m} = (B^{(q)m}_{|m|}, \ldots, B^{(q)m}_{p-1})^{\mathrm{T}}, \qquad \mathbf{A}^{(q)m} = (A^{(q)m}_{|m|}, \ldots, A^{(q)m}_{p-1})^{\mathrm{T}},$$

$$\mathbf{E}^{(q)m} = (E^{(q)m}_{|m|}, \ldots, E^{(q)m}_{p-1})^{\mathrm{T}}, \qquad \Lambda^{(q)}_n = -\frac{j'_n(ka_q) + (\sigma_q/k)j_n(ka_q)}{h'_n(ka_q) + (\sigma_q/k)h_n(ka_q)},$$

$$m = 0, \pm 1, \ldots, \pm(p-1), \quad n = |m|, \ldots, p-1, \quad q = 1, 2,$$

where the superscript T denotes the transpose. We can see that the system can be written in the form:

$$\mathbf{B}^{(1)m} = (\mathbf{S}|\mathbf{R})^m(t)\mathbf{A}^{(2)m}, \qquad \mathbf{B}^{(2)m} = [(\mathbf{S}|\mathbf{R})^m(t)]^{\mathrm{T}}\mathbf{A}^{(1)m},$$

$$\mathbf{A}^{(1)m} = \Lambda^{(1)m}(\mathbf{E}^{(1)m} + \mathbf{B}^{(1)m}), \qquad \mathbf{A}^{(2)m} = \Lambda^{(2)m}(\mathbf{E}^{(2)m} + \mathbf{B}^{(2)m}), \tag{4.3.14}$$

$$m = 0, \pm 1, \ldots, \pm(p-1)$$

We can reduce the size of the system by excluding the unknowns $\mathbf{B}^{(1)m}$ and $\mathbf{B}^{(2)m}$:

$$\begin{aligned} \mathbf{A}^{(1)m} &= \Lambda^{(1)m}\{\mathbf{E}^{(1)m} + (\mathbf{S}|\mathbf{R})^m(t)\mathbf{A}^{(2)m}\}, \\ \mathbf{A}^{(2)m} &= \Lambda^{(2)m}\{\mathbf{E}^{(2)m} + [(\mathbf{S}|\mathbf{R})^m(t)]^{\mathrm{T}}\mathbf{A}^{(1)m}\}. \end{aligned} \tag{4.3.15}$$

We can also represent this system as a system with a single matrix:

$$\begin{pmatrix} [\Lambda^{(1)m}]^{-1} & -(\mathbf{S}|\mathbf{R})^m(t) \\ -[(\mathbf{S}|\mathbf{R})^m]^{\mathrm{T}}(t) & [\Lambda^{(2)m}]^{-1} \end{pmatrix} \begin{pmatrix} \mathbf{A}^{(1)m} \\ \mathbf{A}^{(2)m} \end{pmatrix} = \begin{pmatrix} \mathbf{E}^{(1)m} \\ \mathbf{E}^{(2)m} \end{pmatrix}, \tag{4.3.16}$$
$$m = 0, \pm 1, \ldots, \pm(p-1).$$

Note that for order m the size of vectors $\mathbf{A}^{(1)m}$ and $\mathbf{A}^{(2)m}$ is $p - |m|$ and therefore we have $2(p - |m|)$ unknowns. The size of the matrices $(\mathbf{S}|\mathbf{R})^m(t)$, $\Lambda^{(1)m}$ and $\Lambda^{(2)m}$ are $(p - |m|) \times (p - |m|)$, which produces $2(p - |m|) \times 2(p - |m|)$ matrices for system of order m. Standard numerical methods based on Gauss elimination or *LU*-decomposition require of $C(p - |m|)^3$ operations, where C is some constant. Therefore, the total computational complexity for solution of the above linear system is

$$\text{Cost} = C \sum_{m=-p+1}^{p-1} (p - |m|)^3 = C\frac{1}{2}p^2(p^2 + 1) = O(p^4). \tag{4.3.17}$$

Note that computation of elements of matrices $(\mathbf{S}|\mathbf{R})^m(t)$ for all $|m| < p$ with the recurrence procedures described in the previous chapter also requires $O(p^4)$ operations. Therefore, the total computational cost for solution of the problem truncated to the first p modes is $O(p^4)$.

4.3.2.2 *Surface function*

Once the expansion coefficients $\mathbf{A}^{(1)m}$ and $\mathbf{A}^{(2)m}$ are computed, the potential and its derivative can be determined in any point of space outer to the boundary or immediately on the boundary. In the latter case some simplifications are possible based on Wronskians of the spherical Bessel functions, as considered in the case of a single sphere. Expressions (4.2.14) and (4.2.15) are written in terms of the coefficients of the incident field expanded about the sphere center. In the case of two spheres, we can use the same expressions with the parameters for spheres 1 and 2, respectively, while we need to replace the actual incident field with the "effective" incident field. Since the expansion coefficients of the effective field can be simply expressed via coefficients $\mathbf{A}^{(1)m}$ and $\mathbf{A}^{(2)m}$ (see expressions (4.3.9) and (4.3.10)), we can replace relations (4.2.14) and (4.2.15) with the following:

$$\psi|_{S_q} = -\frac{\mathrm{i}}{(ka_q)^2} \sum_{n=0}^{\infty} \frac{1}{j_n'(ka_q) + (\sigma_q/k)j_n(ka_q)} \sum_{m=-n}^{n} A_n^{(q)m} Y_n^m(\mathbf{s}), \qquad q = 1, 2,$$
$$\left.\frac{\partial \psi}{\partial n_q}\right|_{S_q} = -\mathrm{i}\sigma_q \psi|_{S_q} = -\frac{\sigma_q}{(ka_q)^2} \sum_{n=0}^{\infty} \frac{1}{j_n'(ka_q) + (\sigma_q/k)j_n(ka_q)} \sum_{m=-n}^{n} A_n^{(q)m} Y_n^m(\mathbf{s}),$$
$$q = 1, 2. \tag{4.3.18}$$

4.3.3 Computations and discussion

To test the validity of the solution obtained using multipole reexpansions, we first performed some computations and comparisons with a commercial Boundary Element Method code[1] designed for solution of 3D acoustic problems. In the validation example we considered the case $t = a_1 + a_2$ with the ratio of sphere radii 1.3253. The incident field was generated by a monopole source located at $\mathbf{r}_1 = (0, d, 0)$. Computations were performed with $ka_1 \approx 3$, $\sigma_1 = \sigma_2 = 0$. The mesh for computations using the BEM contained 5400 triangular elements (2700 elements for each sphere). A picture of this two sphere configuration with computational mesh and distribution of the acoustic pressure is shown in Fig. 4.9.

Figure 4.10 shows a comparison between the BEM and the MultisphereHelmholtz (which is our code implementing the multipole reexpansion technique) computational results for function H calculated for sphere 1 according to

$$H = 20 \lg \left| \frac{\psi|_{S_1}}{\psi_{\text{in}}(\mathbf{r}'_1)} \right|. \tag{4.3.19}$$

Here angle θ_1 is the incidence angle (4.2.20) defined with respect to sphere 1. This angle can also be interpreted as a spherical polar angle in the reference frame where the z-axis is directed towards the source point. We directed the y-axis of this frame as the z_1-axis in $(O_1x_1y_1z_1)$ (see Fig. 4.8) and introduced spherical polar angle φ_1, correspondingly (see Fig. 4.1). The selection of a new reference frame is dictated by a desire to see the influence of the second sphere on the solution, since in the absence of the sphere 2, the solution is axisymmetric and depends on θ_1 only (see the solution for a single sphere). Each curve in Fig. 4.10 corresponds to a fixed value of the spherical polar angle θ_1 and demonstrates dependence on the angle φ_1. The comparison shows a good agreement between the results obtained by different methods. Some small dispersion of the points obtained using BEM is due to the mesh discretization of the sphere surface, which normally can be avoided by additional smoothing/interpolation procedures (we did not apply such smoothing in the results plotted). Our code far outperformed the BEM computationally, both in achieving much faster computations (seconds as opposed to tens of minutes on the same computer) and memory usage.

For computation of this case we used truncation number $p = 10$, which results in 100 A_n^m coefficients for each sphere. Proper selection of

[1] COMET 4.0 by Collins and Aikman Automotive Interior Systems, Plymoth, MI.

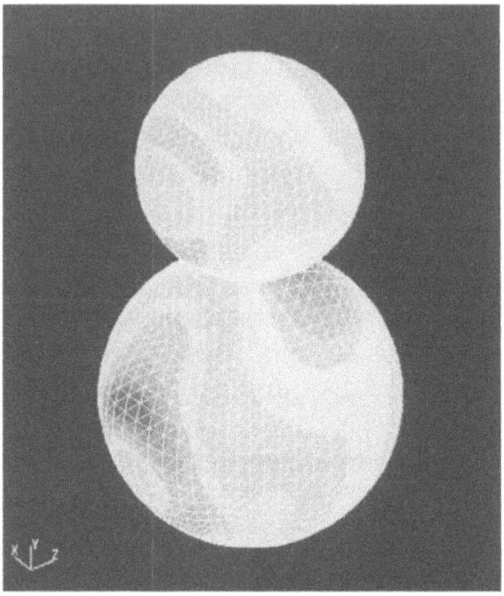

FIGURE 4.9 An example of BEM (COMET 4.0) computations of potential distribution over the surface of two spheres generated by a monopole source. Each sphere surface is discretized to 2700 triangular elements. The ratio of sphere radii is 1.3253 and they touch in one point.

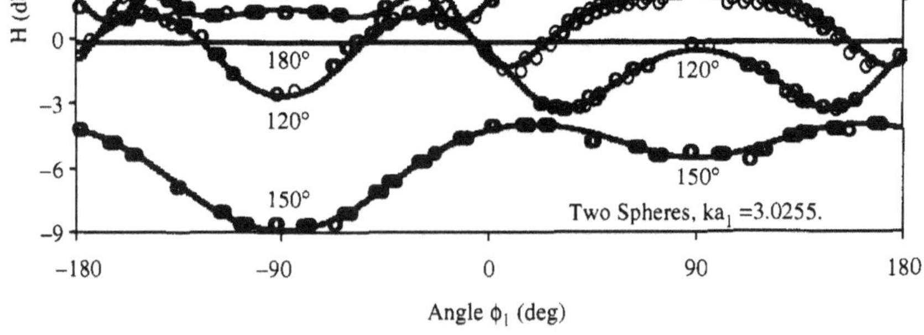

FIGURE 4.10 Comparison of the angular dependence of function H, Eq. (4.3.19), over the surface of a smaller sphere computed for the two-sphere geometry shown in Fig. 4.9 using the BEM and the multipole reexpansion technique. Both spheres have zero admittance.

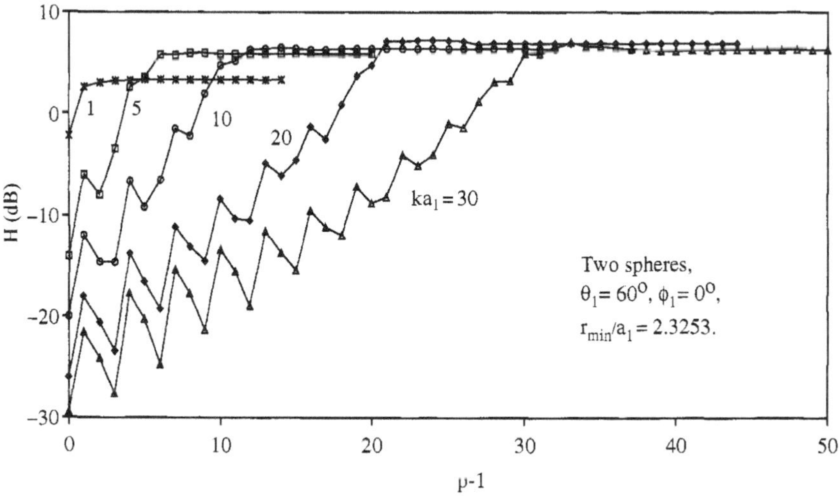

FIGURE 4.11 Dependence of the computations of H, Eq. (4.3.19), for the two-sphere geometry (see Fig. 4.9) on the truncation number, p, for different non-dimensional wavenumbers.

the truncation number is an important issue for applications of multipole translation techniques. Figure 4.11 shows convergence of the computations with increasing truncation numbers for H at a specified point on the surface ($\theta_1 = 60^\circ$ and $\varphi_1 = 0^\circ$ in the case shown in the figure). Computations with low truncation numbers may provide poor accuracy. At some particular truncation number (depending on the non-dimensional wavenumber ka) the computational results stabilize (note that the solution H depends on the wavenumber and, therefore, for each ka the solution asymptotes to the corresponding value). Further increase in the truncation number increases both the accuracy of the results and computational time/memory, since the computational complexity grows in proportion to p^4 for coaxial spheres.

Figure 4.11 shows that stabilization starts for sphere 1 at values $p \sim ka_1$. As we found from the error analysis (see relations (9.1.31) and (9.1.32)) this corresponds to the values when the spherical Bessel functions $j_n(ka)$ change their dependence on n from oscillatory to exponentially decaying. More accurate selection of p at large ka depends on the prescribed error, while we should always have $p > k \max(a_1, a_2)$ to provide an exponential absolute and uniform convergence of the expansion series.

Figure 4.12 illustrates the influence of the sphere 2 on sphere 1 for spheres of the same size and impedances. The pictures were

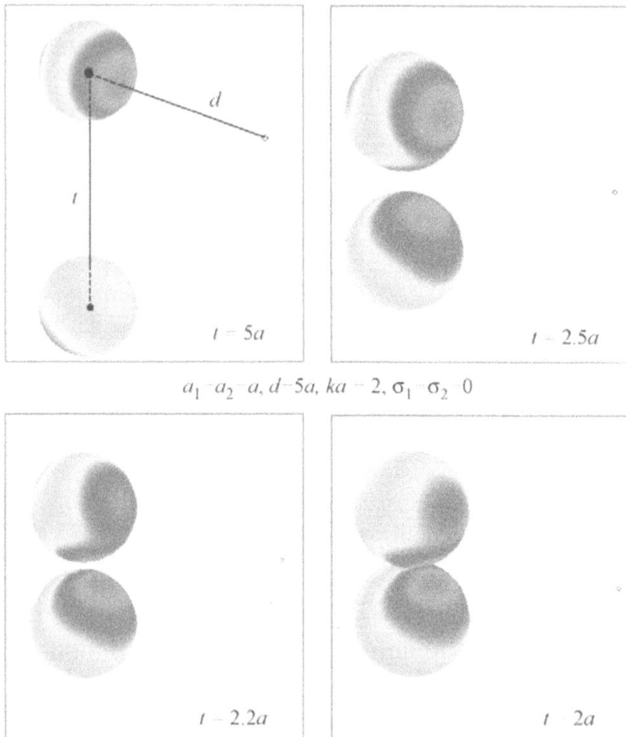

FIGURE 4.12 Surface distributions of the modulus of the potential of the Helmholtz equation for two spheres of equal radius exposed to the field of monopole source as the spheres approach each other. For color version, see Plate E after Chapter 10.

obtained using the solution described above. If the distance between the spheres is large enough, this influence is small and the surface potential for the sphere 1 has an axisymmetric distribution pattern close to that of the single sphere. When sphere 2 approaches sphere 1, the distribution is not axisymmetric anymore and interaction between the spheres becomes important (physically this can be interpreted as multiple reflections of the waves between the spheres). The colors show the modulus of the surface potential with a rainbow color map (red denotes the highest value and blue the lowest value). It is seen that the proximity of the second sphere changes colors of some pieces from red to blue, which can be explained by the wave superposition and cancellation phenomenon (due to the difference in phases).

4.4 SCATTERING FROM N SPHERES

4.4.1 Formulation

The case of sound and electromagnetic wave scattering by many arbitrarily located spheres has numerous practical and theoretical applications, including acoustics and hydrodynamics of multiphase flows, sound propagation in composite materials, electromagnetic waves in clouds and systems of micro- and nanoparticles, and inverse problems, such as the detection of buried objects, medical tomography, etc.

The solution obtained above for scattering off two spheres can be generalized for the case of arbitrary number, N, of spheres of radii $a_1, \ldots, a_N$ and complex admittances $\sigma_1, \ldots, \sigma_N$ whose centers are located at $\mathbf{r}'_1, \ldots, \mathbf{r}'_N$. We are seeking the field potential $\psi(\mathbf{r})$ for a given incident field $\psi_{\rm in}(\mathbf{r})$, which satisfies the Helmholtz equation in the domain outer to the spheres and possible singularities of $\psi_{\rm in}(\mathbf{r})$ (if the incident field is generated by a set of sources). The total potential $\psi(\mathbf{r})$ can be decomposed according to Eq. (4.2.1) to the incident and scattered parts, where the latter part is a radiating function. On the surface of each sphere we impose the impedance boundary condition:

$$|\mathbf{r} - \mathbf{r}'_q| = a_q : \quad \frac{\partial \psi(\mathbf{r})}{\partial n_q} + {\rm i}\sigma_q \psi(\mathbf{r}) = 0, \qquad q = 1, \ldots, N, \tag{4.4.1}$$

where the normal derivative is taken in the direction of the outer normal to the surface S_q of the qth sphere.

4.4.2 Solution

Let us introduce N local reference frames centered at $\mathbf{r}'_1, \ldots, \mathbf{r}'_N$ and use notation consistent with notations (4.3.2):

$$\mathbf{r}_q = \mathbf{r} - \mathbf{r}'_q, \quad \mathbf{r}'_{qq'} = \mathbf{r}'_{q'} - \mathbf{r}'_q = \mathbf{r}_q - \mathbf{r}_{q'} = -\mathbf{r}'_{q'q}, \quad q, q' = 1, \ldots, N. \tag{4.4.2}$$

As in the case of two spheres we decompose the total potential into a sum of potentials related to each sphere:

$$\psi(\mathbf{r}) = \psi_{\rm in}(\mathbf{r}) + \psi_1(\mathbf{r}) + \cdots + \psi_N(\mathbf{r}) = \psi_{\rm in}(\mathbf{r}) + \sum_{q=1}^{N} \psi_q(\mathbf{r}). \tag{4.4.3}$$

Here $\psi_q(\mathbf{r})$ is a radiating function, which has a singularity at the center of the qth sphere and is regular everywhere else. Therefore, we can seek these potentials in the form:

$$\psi_q(\mathbf{r}) = \sum_{n=0}^{\infty} \sum_{m=-n}^{n} A_n^{(q)m} S_n^m(\mathbf{r}_q), \quad q = 1, \ldots, N, \tag{4.4.4}$$

and the problem reduces to determination of the expansion coefficients.

For a sphere q we rewrite decomposition (4.4.3) in the form:

$$\psi(\mathbf{r}) = \psi_{\text{in}}^{(\text{eff})(q)}(\mathbf{r}) + \psi_q(\mathbf{r}), \qquad \psi_{\text{in}}^{(\text{eff})(q)}(\mathbf{r}) = \psi_{\text{in}}(\mathbf{r}) + \sum_{q' \neq q} \psi_{q'}(\mathbf{r}), \tag{4.4.5}$$

where the summation is taken over all sphere indices, except q. In the vicinity of the sphere q, including its surface, potential $\psi_{\text{in}}^{(\text{eff})(q)}(\mathbf{r})$ is a regular function and so it can be decomposed in a sum of regular spherical basis functions:

$$\psi_{\text{in}}^{(\text{eff})(q)}(\mathbf{r}) = \sum_{n=0}^{\infty} \sum_{m=-n}^{n} E_n^{(\text{eff})(q)m} R_n^m(\mathbf{r}_q), \qquad q = 1, \ldots, N. \tag{4.4.6}$$

The expansion coefficients here can be related to the coefficients of expansion near $\mathbf{r}'_q$ of the actual incident field, $E_n^{(q)m}$, and coefficients $B_n^{(q)m}$:

$$E_n^{(\text{eff})(q)m} = E_n^{(q)m} + B_n^{(q)m}, \qquad n = 0, 1, \ldots, \quad m = -n, \ldots, n, \quad q = 1, \ldots, N. \tag{4.4.7}$$

Coefficients $B_n^{(q)m}$ can be obtained by translation of coefficients $A_n^{(q')m}$, $q' \neq q$ to the expansion center $\mathbf{r}'_q$:

$$B_n^{(q)m} = \sum_{q' \neq q} \sum_{n'=0}^{\infty} \sum_{m'=-n'}^{n'} (S|R)_{nn'}^{mm'}(\mathbf{r}'_{q'q}) A_{n'}^{(q')m'}, \qquad n = 0, 1, \ldots, \quad m = -n, \ldots, n, \quad q = 1, \ldots, N. \tag{4.4.8}$$

Solution of the problem (4.4.4)–(4.4.6) where $\psi(\mathbf{r})$ satisfies boundary condition (4.4.1) on the sphere surface is provided by relation (4.2.9), which we modify as

$$A_n^{(q)m} = \Lambda_n^{(q)} E_n^{(\text{eff})(q)m}, \qquad \Lambda_n^{(q)} = -\frac{j'_n(ka_q) + (\sigma_q/k) j_n(ka_q)}{h'_n(ka_q) + (\sigma_q/k) h_n(ka_q)}, \qquad n = 0, 1, \ldots, \quad m = -n, \ldots, n, \quad q = 1, \ldots, N. \tag{4.4.9}$$

Since the incident field, and so its expansion coefficients $E_n^{(q)m}$ are given, Eqs. (4.4.7)–(4.4.9) form a closed linear system to determine unknowns $\{E_n^{(\text{eff})(q)m}\}$, $\{B_n^{(q)m}\}$, and $\{A_n^{(q)m}\}$. As in the case of two spheres this infinite system can be solved using truncation of matrices and modes considered with some truncation number p. We also assume that at $p \to \infty$ there is a convergence of the solution to a true solution and selection of an appropriate finite p will provide an acceptable error ϵ.

Also we can represent the system in a matrix–vector form. To avoid dealing with four-dimensional matrices and two-dimensional vectors, such as $\{(S|R)_{nn'}^{mm'}\}$ and $\{A_n^m\}$, we can align these objects into conventional matrices and vectors. Indeed, coefficients $\{A_n^m\}$ can be stacked into one column vector as

$$\mathbf{A} = \{A_n^m\} = (A_0^0, A_1^{-1}, A_1^0, A_1^1, A_2^{-2}, A_2^{-1}, A_2^0, A_2^1, A_2^2, \ldots)^{\mathrm{T}}, \qquad (4.4.10)$$

where the superscript T denotes the transpose. In this representation the elements of the vector $\mathbf{A}$ are related to coefficients A_n^m by

$$A_l = A_n^m, \qquad l = (n+1)^2 - (n-m),$$
$$n = 0, 1, 2, \ldots, \quad m = -n, \ldots, n, \quad l = 1, 2, \ldots. \qquad (4.4.11)$$

The same reduction in dimension can be applied to the reexpansion coefficients $(S|R)_{nn'}^{mm'}$. The coefficients can then be stacked in a two-dimensional matrix as

$$(\mathbf{S}|\mathbf{R}) = \{(S|R)_{nn'}^{mm'}\}$$
$$= \begin{pmatrix} (S|R)_{00}^{00} & (S|R)_{01}^{0-1} & (S|R)_{01}^{00} & (S|R)_{01}^{01} & (S|R)_{02}^{0-2} & \cdots \\ (S|R)_{10}^{-10} & (S|R)_{11}^{-1-1} & (S|R)_{11}^{-10} & (S|R)_{11}^{-11} & (S|R)_{12}^{-1-2} & \cdots \\ (S|R)_{10}^{00} & (S|R)_{11}^{0-1} & (S|R)_{11}^{00} & (S|R)_{11}^{01} & (S|R)_{12}^{0-2} & \cdots \\ (S|R)_{10}^{10} & (S|R)_{11}^{1-1} & (S|R)_{11}^{10} & (S|R)_{11}^{11} & (S|R)_{12}^{1-2} & \cdots \\ (S|R)_{20}^{-20} & (S|R)_{21}^{-2-1} & (S|R)_{21}^{-20} & (S|R)_{21}^{-21} & (S|R)_{22}^{-2-2} & \cdots \\ \cdots & \cdots & \cdots & \cdots & \cdots & \cdots \end{pmatrix}, \qquad (4.4.12)$$

with the following correspondence of the matrix elements and coefficients:

$$(S|R)_{ll'} = (S|R)_{nn'}^{mm'}, \qquad l = (n+1)^2 - (n-m), \quad l' = (n'+1)^2 - (n'-m'),$$
$$n, n' = 0, 1, 2, \ldots, \quad m = -n, \ldots, n, \quad m' = -n', \ldots, n', \quad l, l' = 1, 2, \ldots. \qquad (4.4.13)$$

The same stacking rule can be applied to the elements of diagonal T-matrix (4.2.12), which is specific for each sphere (so we have N different T-matrices if the sizes and admittances of the spheres are different).

The T-matrix for the qth sphere can be defined as

$$T_{nn'}^{(q)mm'} = \Lambda_n^{(q)} \delta_{nn'} \delta_{mm'},$$
$$n, n' = 0, 1, \ldots, \quad m = -n, \ldots, n, \quad m' = -n', \ldots, n', \quad q = 1, \ldots, N, \tag{4.4.14}$$

with $\Lambda_n^{(q)}$ specified in Eq. (4.4.9). Using this representation we introduce the following vectors and matrices:

$$\mathbf{A}^{(q)} = \{A_l^{(q)}\}^{\mathrm{T}}, \qquad \mathbf{B}^{(q)} = \{B_l^{(q)}\}^{\mathrm{T}}, \qquad \mathbf{E}^{(q)} = \{E_l^{(q)}\}^{\mathrm{T}},$$
$$(\mathbf{S}|\mathbf{R})(\mathbf{r}'_{qq'}) = \{(S|R)_{ll'}(\mathbf{r}'_{q'q})\}, \qquad \mathbf{T}^{(q)} = \{\mathbf{T}_{ll'}^{(q)}\}, \qquad q, q' = 1, \ldots, N. \tag{4.4.15}$$

Equations (4.4.7)–(4.4.9) can then be represented in the form:

$$\mathbf{B}^{(q)} = \sum_{q' \neq q} (\mathbf{S}|\mathbf{R})(\mathbf{r}'_{q'q})\mathbf{A}^{(q')}, \qquad \mathbf{A}^{(q)} = \mathbf{T}^{(q)}(\mathbf{B}^{(q)} + \mathbf{E}^{(q)}), \qquad q = 1, \ldots, N. \tag{4.4.16}$$

Coefficients $\mathbf{B}^{(q)}$ can be excluded to reduce the size of the system:

$$(\mathbf{T}^{(q)})^{-1}\mathbf{A}^{(q)} - \sum_{q' \neq q} (\mathbf{S}|\mathbf{R})(\mathbf{r}'_{q'q})\mathbf{A}^{(q')} = \mathbf{E}^{(q)}, \qquad q = 1, \ldots, N. \tag{4.4.17}$$

This system can also be represented as a single equation:

$$\mathbf{LA} = \mathbf{E}, \tag{4.4.18}$$

where the vectors and matrices are stacked as

$$\mathbf{L} = \begin{pmatrix} (\mathbf{T}^{(1)})^{-1} & -(\mathbf{S}|\mathbf{R})(\mathbf{r}'_{21}) & \cdots & -(\mathbf{S}|\mathbf{R})(\mathbf{r}'_{N1}) \\ -(\mathbf{S}|\mathbf{R})(\mathbf{r}'_{12}) & (\mathbf{T}^{(2)})^{-1} & \cdots & -(\mathbf{S}|\mathbf{R})(\mathbf{r}'_{N2}) \\ \cdots & \cdots & \cdots & \cdots \\ -(\mathbf{S}|\mathbf{R})(\mathbf{r}'_{1N}) & -(\mathbf{S}|\mathbf{R})(\mathbf{r}'_{2N}) & \cdots & (\mathbf{T}^{(N)})^{-1} \end{pmatrix},$$
$$\mathbf{A} = \begin{pmatrix} \mathbf{A}^{(1)} \\ \mathbf{A}^{(2)} \\ \cdots \\ \mathbf{A}^{(N)} \end{pmatrix}, \qquad \mathbf{E} = \begin{pmatrix} \mathbf{E}^{(1)} \\ \mathbf{E}^{(2)} \\ \cdots \\ \mathbf{E}^{(N)} \end{pmatrix}. \tag{4.4.19}$$

This linear system can be solved numerically using standard routines, such as the LU decomposition. The block-structure can also be exploited using block-oriented solvers, though we do not pursue this here.

Note also that elements of matrices $(\mathbf{S}|\mathbf{R})(\mathbf{r}'_{qq'})$ do not require additional computations if $(\mathbf{S}|\mathbf{R})(\mathbf{r}'_{q'q})$ is computed due to symmetry relations (3.2.55).

If we proceed with p truncation the size of vector $\mathbf{A}$ will be p^2N and the size of matrix $\mathbf{L}$ will be $p^2N \times p^2N$. Using straightforward methods for solution of the linear system we get an algorithm of complexity $O(p^6N^3)$. If the elements of $p^2 \times p^2$ matrices $(\mathbf{S}|\mathbf{R})(\mathbf{r}'_{q'q})$ are computed recursively, we need of order $O(p^4N^2)$ operations to compute them. Therefore, this cost is not limiting for the straightforward methods. Assuming that the matrix $\mathbf{L}$ computed one can consider the use of iterative methods. The complexity of such methods is for the matrix we have $O(p^4N^2N_{\text{iter}})$ which is better than $O(p^6N^3)$ for the number of iteration steps $N_{\text{iter}} \ll p^2N$. We can see that in this case the computation of all matrix elements using recurrences, which is of the order of the cost of one matrix–vector multiplication, makes a small contribution to the total cost for $N_{\text{iter}} \gg 1$.

Some special symmetric configurations may be computed faster than the general case. For example, if we have a chain of N coaxial spheres translations can be performed with a coaxial translation matrix, which can be computed for every order m independently on other modes. This case is similar to the case of two spheres that always can be computed with coaxial translation matrix at proper selection of the reference frame. In the case of N coaxial spheres, the complexity can be found by the same way as shown in Eq. (4.3.17) and for N spheres it is $O(p^4N^3)$ plus substantial memory reduction. So in this case we can solve the same problem with the same truncation number p^2 times faster than in the general case.

Finally, we note that due to substantial memory requirements and scaling of the method as N^3 some faster methods should be considered for problems with large N. However, our tests, some of which are presented below, show that for moderate N the present method is fast enough to be practical and far outperforms such methods as standard versions of the BEM.

4.4.3 Computations and discussion

As in the case of two spheres we started tests with validation of the above solution against a solution obtained by some other reliable method (BEM). For this computational example we placed an additional sphere (3) to the case described above for two spheres. The distance between the centers of spheres 1 and 3 was the same as the distance between the centers of spheres 1 and 2, the radius of sphere 3 was the same as the radius of sphere 1 and the center of the added sphere was located at the line connecting the source and the center of sphere 1. The mesh for computations using the BEM contained 5184 triangular elements, 1728 elements for each sphere. A picture of this configuration with the computational mesh and distribution of the acoustic pressure is shown in Fig. 4.13. In the computations the three spheres were all taken to be sound hard. Results of comparisons between the BEM and MultisphereHelmholtz computations with $p = 10$ are shown in Fig. 4.14. The comparison is as good as in the case of two spheres. Since Figs. 4.10 and 4.14 represent similar dependences, we can notice that the presence of the third sphere reduced (at some points by 3–4 dB) the amplitude of the

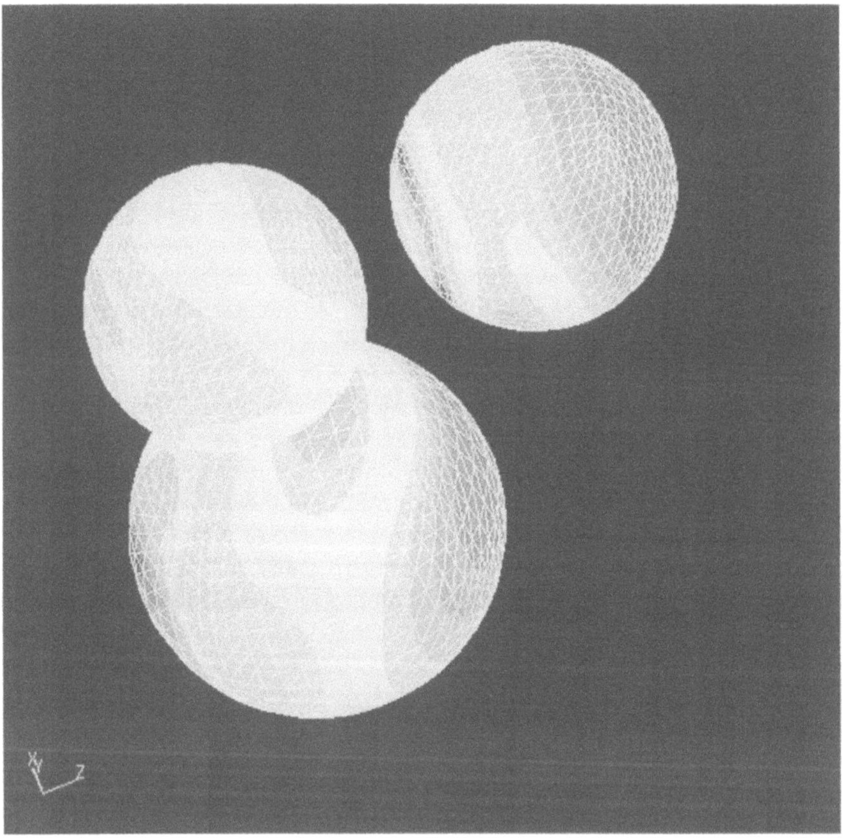

FIGURE 4.13 BEM (COMET 4.0) computations of the potential distribution over the surface of three spheres generated by a monopole source. Each sphere surface is discretized using 1728 triangular elements. Two spheres of non-dimensional radii 1 (sphere 1) and 1.3253 (sphere 2) touch at one point. Sphere 3, with non-dimensional radius 1, is located at a distance 2.3253 from the center of sphere 1, on the line connecting the source and the center of sphere 1.

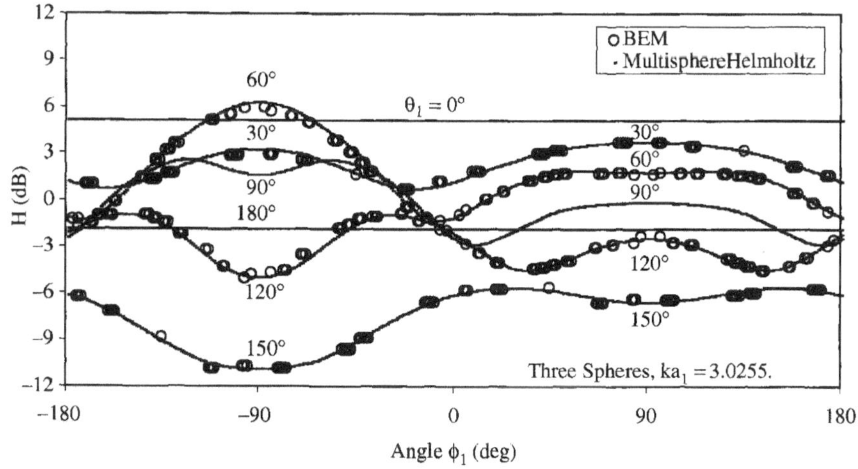

FIGURE 4.14 Comparison of the angular dependence of *H* (Eq. (4.3.19)) over the surface of smaller sphere computed for the three-sphere geometry shown in Fig. 4.13 using the BEM and the multipole reexpansion technique. All three spheres have sound-hard surfaces.

sound field on sphere 1 which is explainable as an effect of acoustic shadow.

Figure 4.15 demonstrates an example computation based on multipole reexpansion technique of scattering of a monochromatic wave from four rigid spheres, whose centers are arranged in a tetrahedron. These spheres are exposed to an incident plane wave that comes from the direction shown by the arrow (the angle of the arrow with the normal to the imaging plane is 45° and it belongs to a symmetry plane). In computations we selected parameter $ka = 15.2$ and used truncation number $p = 42$, which is more than two times larger than ka (in fact, we selected $p = [eka] + 1$) and ensures that results are accurate. The picture on the left-hand side shows distribution of the amplitude of the acoustic field over the sphere surfaces. The patterns here show a complex character of interaction between the spheres. The imaging plane serves only for visualization of the field. By tracking the lines of constant phase, we can animate the wave propagation in time. Some shots of this movie are shown as the three shadow pictures on the right-hand side. In the absence of spheres the plane wave is imaged as moving shadow strips. The presence of spheres results in diffraction patterns (the middle shadow picture). The difference between the total field and the incident field amplitudes yields the patterns of the scattered field.

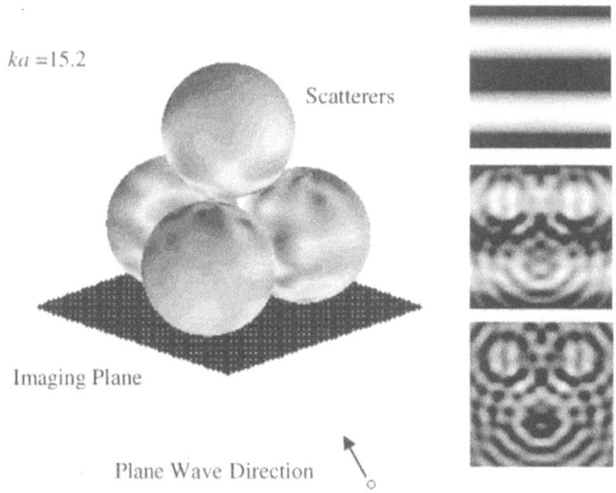

FIGURE 4.15 Scattering of a monochromatic plane wave ($ka = 15.2$) from four spheres of radius a which form a tetrahedron with the size $2a$. The grayscale shadow images on the right-hand side show the incident, total, and scattered fields at some t, $\text{Re}\{e^{-i\omega t}\psi(\mathbf{r})\}$, where $\psi(\mathbf{r})$ is the potential of the 3D Helmholtz equation at the points of the imaging plane, shown by the dots on the left-hand side.

We also studied some cases with larger number of spheres ($N = 10, \ldots, 200$) randomly distributed in space and found that the results converge at increasing p, which normally should be selected $p > ka$. Some results of this study can be found in Ref. [GD02].

4.5 ON MULTIPLE SCATTERING FROM N ARBITRARY OBJECTS

A general case of interaction of multiple scatterers, which are not necessarily spheres, can be also handled by a technique described above, and which is a variant of the so-called *T-matrix method*. Indeed, assume that N objects are separated in a sense that each object can be surrounded by a sphere of some radius a_q, $q = 1, \ldots, N$, and these spheres do not intersect (while they may touch). Then assume that an analytical or numerical solution of scattering problem for an arbitrary incident field for each object is available. By "solution" Under solution here we mean that if an incident field $\psi_{\text{in}}(\mathbf{r})$ is specified in the form (4.2.3), where we should assume $\mathbf{r}_0 = \mathbf{r}_q$ to be a center of the sphere enclosing the

qth scatterer we can determine coefficients $A_n^{(q)m}$ of the scattered wave (4.2.6). This is provided by the T-matrix (4.2.11), which can be computed by different methods. One of them is described below.

As in the case of spheres we seek for solution in form (4.4.3), where each potential can be represented outside the sphere of radius a_q as given by Eq. (4.4.4). This results in the system (4.4.18) and (4.4.19) with the T-matrices for the objects considered. The truncated approximation to the system can be solved and the first p^2 modes for each vector $\mathbf{A}^{(q)}$ can be found.

With this solution available we can determine $\psi(\mathbf{r})$ at any point $\mathbf{r}$ outside the spheres surrounding the scatterers. In case, when we need to obtain $\psi(\mathbf{r})$ for $\mathbf{r}$ which is inside the sphere surrounding the qth object, we determine coefficients of the "effective" incident field:

$$\mathbf{E}^{(\mathrm{eff})(q)} = (\mathbf{T}^{(q)})^{-1}\mathbf{A}^{(q)}, \tag{4.5.1}$$

build corresponding "effective" incident potentials according to (4.4.6) and solve the Helmholtz equation with specified boundary conditions for the qth scatterer, which yields $\psi(\mathbf{r})$ for this $\mathbf{r}$.

4.5.1 A method for computation of the T-matrix

To solve this problem in a practical setting, when only a numerical solution is available for a complex-shaped object, one can formulate it as follows.

Assuming that the incident field is

$$\psi_{\mathrm{in}}(\mathbf{r}) = R_n^m(\mathbf{r}), \tag{4.5.2}$$

determine the radiating function $\psi_{\mathrm{scat}}(\mathbf{r})$ on a surface of sphere a centered at $\mathbf{r} = \mathbf{r}_0 = 0$ and completely surrounding the object with surface S, subject to impedance boundary conditions:

$$\mathbf{r} \in S: \quad \frac{\partial \psi(\mathbf{r})}{\partial n} + \mathrm{i}\sigma\psi(\mathbf{r}) = 0, \qquad \psi_{\mathrm{scat}}(\mathbf{r}) = \psi(\mathbf{r}) - \psi_{\mathrm{in}}(\mathbf{r}). \tag{4.5.3}$$

Once $\psi_{\mathrm{scat}}(\mathbf{r})$ is determined at a set of surface points on the sphere, we can perform a spherical harmonic analysis to determine a necessary number of surface modes for this function. By "analysis" we mean determination of coefficients of the surface function expansion over spherical harmonics (2.1.65), which for example can be determined numerically by computation of integrals (2.1.68). This enables determination of, say, the first p degree expansion coefficients $C_{n'}^{m'}$ in series:

$$\psi_{\mathrm{scat}}(\mathbf{r}) = \sum_{n'=0}^{p-1} \sum_{m'=-n'}^{n'} C_{n'}^{m'} Y_{n'}^{m'}(\mathbf{s}), \qquad \mathbf{r} = a\mathbf{s}. \tag{4.5.4}$$

Now, since the scattered field can be represented in the form (4.2.6) we can determine coefficients of the scattered field as

$$A_{n'}^{m'} = \frac{C_{n'}^{m'}}{h_{n'}(ka)}, \qquad n' = 0, \ldots, p-1, \;\; m' = -n', \ldots, n' \tag{4.5.5}$$

due to the relation between the spherical basis functions and spherical harmonics (2.1.102). This provides a "response" of the body on incident wave (4.5.2) where all $E_{n''}^{m''}$ are zero, except of $n'' = n$, $m'' = m$ for which $E_n^m = 1$, and therefore according to Eq. (4.2.11) we have

$$T_{n'n}^{m'm} = A_{n'}^{m'}, \qquad n' = 0, 1, \ldots, p-1, \;\; m' = -n', \ldots, n'. \tag{4.5.6}$$

Varying n and m for the incident field function in range $n = 0, 1, \ldots, p-1$, $m = -n, \ldots, n$, we can determine all the elements of the truncated T-matrix for the scatterer.

Note that instead of taking $\psi_{\text{in}}(\mathbf{r})$ as a spherical basis function, one can consider other functions, which can be expanded in series of the spherical functions. For example, the problem can be solved for plane waves,

$$\psi_{\text{in}}(\mathbf{r}) = e^{i\mathbf{k}\cdot\mathbf{r}}, \tag{4.5.7}$$

at, say, p^2 different directions of the plane waves. Since these solutions are linearly independent and can be represented as linear combinations of $R_n^m(\mathbf{r})$ the elements of the T-matrix can be found as linear combinations of respective response functions. This method is similar to the methods of tomography, when the information about the object shape can be obtained by exposing the object to the waves coming from different directions and measuring the scattered field.

4.6 SUMMARY

On the base of solution of several physical problems the chapter illustrates how the multipole methods can be applied. Here we considered speeding up of summation of sources with the aid of multipole expansions and solution of boundary value problems for spherical scatterers. We also introduced the T-matrix method which can be used for computation of fields in multiple scattering problems with arbitrary scatterers. Comparisons of the multipole methods for these cases with other methods based on direct summation or boundary discretization show efficiency of the multipole methods. However the use of the "direct" multipole methods for problems where the number of scatterers is large requires development of faster algorithms.

CHAPTER 5

Fast Multipole Methods

Originally developed for the fast summation of multipole solutions of the Laplace equation, fast multipole methods (FMMs) are currently applied for the solution of many problems, which have a mathematical rather than physical nature (matrix–vector multiplication in d dimensions). Despite the fact that the focus of this book is the Helmholtz equation, we will introduce the FMM, more precisely the multilevel FMM (MLFMM) from a more or less general point of view, so that the reader will learn the mathematical tools for the implementation of this method for general problems. This includes the formulation of the problem, introduction to data structures, algorithms, evaluations of the complexity of the method, and optimization strategies. We also consider some particular issues related to coupling of the FMM with integral equation solvers for the Helmholtz equation, and some special issues that arise due to the properties of the elementary solutions of the Helmholtz equation.

5.1 PRELIMINARY IDEAS

Many problems that require solution of the Helmholtz equation in some domain can be reduced to summation of a large number of monopole or multipole sources. For example, problems were formulated in the previous chapters when the boundary of the domain was discretized by a large number of elements or when we had many scatterers. For such problems, usually, the most expensive computational part is related to performing a matrix–vector multiplication or summation of the type:

$$\psi(\mathbf{y}_\beta) = \sum_{\alpha=1}^{N} Q_\alpha \Phi(\mathbf{y}_\beta - \mathbf{x}_\alpha), \qquad \beta = 1, \ldots, M, \tag{5.1.1}$$

where $\Phi(\mathbf{y}-\mathbf{x})$ is some function (e.g. the Green's function for the Helmholtz equation) centered at $\mathbf{x}$, which should be evaluated at $\mathbf{y}$, and Q_α are some coefficients (e.g. the intensities of monopole sources). Straightforward computation of these sums, which can also be treated as multiplication of an $M \times N$ matrix, with elements $\Phi_{\beta\alpha} = \Phi(\mathbf{y}_\beta - \mathbf{x}_\alpha)$, by an N vector with components Q_α to obtain an M vector with components $\psi_\beta = \psi(\mathbf{y}_\beta)$ obviously requires $O(MN)$ operations. The FMM seeks faster methods with computational complexity such as $O(M + N)$ or $O(M \log \times M + N \log N)$ to perform this operation. This is achieved by seeking to perform the computation approximately, but to a specified accuracy. Thus, the FMM is an approximate method, and a feature of its analysis and implementation is the evaluation of the errors (this is a general problem for any approximate numerical method). One should not be misled into thinking that the FMM is not an accurate method. Many implementations and tests showed that even machine precision can be realized with substantial speedup.

5.1.1 Factorization (Middleman method)

The first key idea of the FMM comes from *factorization*. To illustrate it let us consider the case when all *sources* are located in some domain Ω_s and all *evaluation points* are located in some domain Ω_e and that these domains are *well separated*. Examples of two such situations are shown in Fig. 5.1. First, assume that we need to compute sums (5.1.1) when function Φ is Green's function of the Helmholtz equation and all the sources are located outside a sphere of radius a centered at $\mathbf{x}_*$. The potential of each source can then be expanded about the center of domain Ω_e according to Eq. (5.2.2) as

$$G(\mathbf{y} - \mathbf{x}_\alpha) = G(-(\mathbf{x}_\alpha - \mathbf{x}_*) + (\mathbf{y} - \mathbf{x}_*))$$

$$= ik \sum_{n=0}^{\infty} \sum_{m=-n}^{n} S_n^{-m}(\mathbf{x}_\alpha - \mathbf{x}_*) R_n^m(\mathbf{y} - \mathbf{x}_*), \qquad (5.1.2)$$

$$|\mathbf{y} - \mathbf{x}_*| < |\mathbf{x}_\alpha - \mathbf{x}_*|.$$

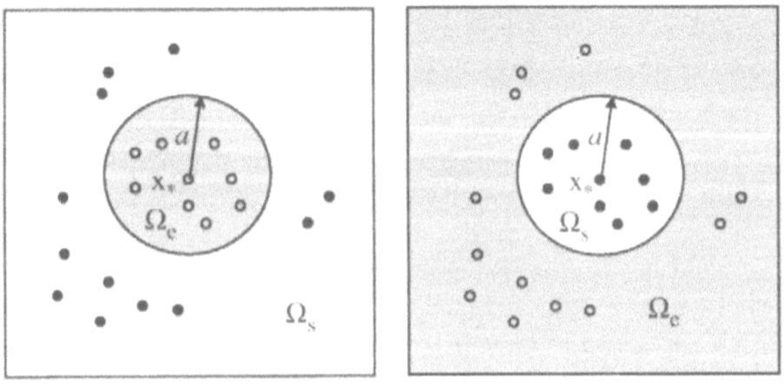

FIGURE 5.1 Examples of well-separated sets of sources (filled circles) and evaluation points (open circles).

If the sum is truncated at p terms, where p is chosen so that the error in the approximation of $G(\mathbf{y} - \mathbf{x}_\alpha)$ with the truncated series is acceptable inside the sphere, and substitute the series into Eq. (5.1.1), then we obtain

$$\psi(\mathbf{y}_\beta) = ik \sum_{\alpha=1}^{N} Q_\alpha \sum_{n=0}^{p-1} \sum_{m=-n}^{n} S_n^{-m}(\mathbf{x}_\alpha - \mathbf{x}_*) R_n^m(\mathbf{y}_\beta - \mathbf{x}_*), \quad \beta = 1, \ldots, M. \tag{5.1.3}$$

Now we can rearrange the summation as follows:

$$\psi(\mathbf{y}_\beta) = \sum_{n=0}^{p-1} \sum_{m=-n}^{n} C_n^m R_n^m(\mathbf{y}_\beta - \mathbf{x}_*), \qquad \beta = 1, \ldots, M, \tag{5.1.4}$$

where

$$C_n^m = ik \sum_{\alpha=1}^{N} Q_\alpha S_n^{-m}(\mathbf{x}_\alpha - \mathbf{x}_*), \quad n = 0, 1, \ldots, p-1, \quad m = -n, \ldots, n. \tag{5.1.5}$$

Computation of the p^2 coefficients C_n^m requires $O(Np^2)$ operations, and computation of $\psi(\mathbf{y}_\beta)$ requires $O(Mp^2)$ operations. So the total number of operations ("complexity") of evaluation of the sum at all evaluation points will take $O(p^2(M+N))$ operations. If $p^2 \ll \max(M, N)$ then this is a much cheaper procedure compared to the $O(MN)$ procedure required for obtaining $\psi(\mathbf{y}_\beta)$ directly from Eq. (5.1.1). Similarly, one can obtain formulae to compute the second case, $|\mathbf{y}_\beta - \mathbf{x}_*| > |\mathbf{x}_\alpha - \mathbf{x}_*|$, shown in Fig. 5.1. Here we need to exchange the S and R functions.

We call the trick used above to reduce the complexity of computation as the "Middleman" method, since it reminds activities of a sales person, who first collected all the inputs from the sources (manufacturers), and then distributed it among the evaluation points (consumers). This is illustrated in Fig. 5.2. The coefficients C_n^m play the role of the "Middleman". When we compute them according to Eq. (5.1.5), the information about all the sources is collected and compressed it into p^2 coefficients. These coefficients are then used to compute the value of function ψ at all locations of the evaluation points. Despite its simplicity, the Middleman method can be extremely efficient and we illustrated its application to computations of the room impulse response in room acoustics in the previous chapter.

5.1.2 Space partitioning (modified Middleman method)

Unfortunately, the Middleman method can only be used when the sets of the source and evaluation points are *well separated*. Of course, if there are

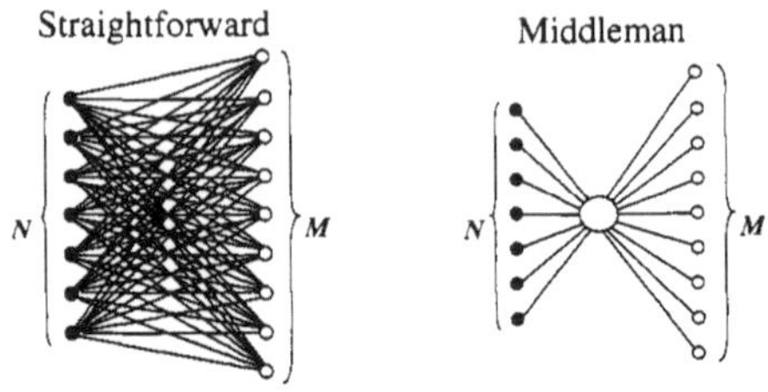

FIGURE 5.2 Illustration of complexity of the straightforward computation and the Middleman algorithm.

just a few "outlier" sources or evaluation points, which are not "well separated", the contribution of such sources or the field at such evaluation points can be computed directly which leaves the complexity of the Middleman algorithm $O(p^2(M+N))$. More precisely, the complexity of the Middleman algorithm will remain $O(p^2(M+N))$ if the number of the "outliers" does not exceed $O(p^2)$. In general case the sources and evaluation points can be sufficiently close and in this case it is impossible to build a single expansion which is acceptable at most locations of the evaluation points. This obstacle to the method can be overcome if we use *space partitioning*.

5.1.2.1 *Space partitioning with respect to evaluation set*

An example of space partitioning (breaking the computational domain into square (2D) or cubic (3D) boxes) is shown in Fig. 5.3. In this case a modification of the Middleman algorithm can be as follows. Let K be the number of boxes which contain evaluation points. We can call such boxes as *target* or *receiver* boxes and number them $R_1, \ldots, R_K$. For each receiver box R_n we subdivide the set of sources $\{\mathbf{x}_\alpha\}$ into two subsets R_n^- and R_n^+. The first subset consists of the sources located outside the *neighborhood* of box n, and the other subset consists of the sources located inside the neighborhood of the given box n. The sum (5.1.1) can then be separated into two parts:

$$\psi(\mathbf{y}_\beta) = \sum_{\mathbf{x}_\alpha \in R_n^+} Q_\alpha \Phi(\mathbf{y}_\beta - \mathbf{x}_\alpha) + \sum_{\mathbf{x}_\alpha \in R_n^-} Q_\alpha \Phi(\mathbf{y}_\beta - \mathbf{x}_\alpha), \quad \mathbf{y}_\beta \in R_n. \qquad (5.1.6)$$

The first sum can be calculated directly and if $N_n^{(s)}$ is the number of sources in the neighborhood of R_n this requires $O(N_n^{(s)})$ operations. The second term can be treated using the Middleman method, since R_n^- is well separated from R_n, which means that there exists a sphere which is enclosed into the neighborhood of the receiver box R_n and encloses this

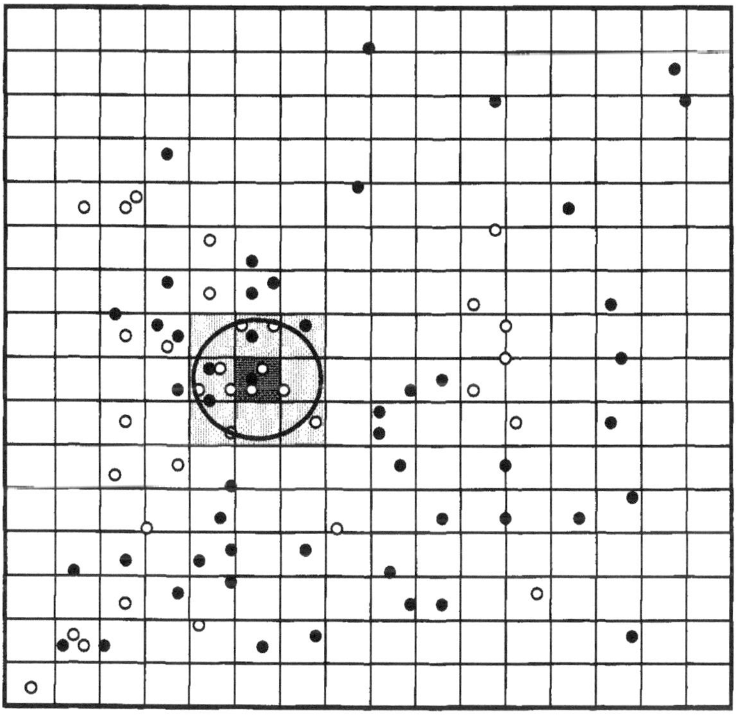

FIGURE 5.3 Example of space partitioning. The filled and open circles show the sources and the evaluation points. A particular box is shown in dark gray and the neighborhood of this box is shown in light gray.

receiver box (see Fig. 5.3). Using Eqs. (5.1.3) and (5.1.4) we can then write

$$\sum_{\mathbf{x}_\alpha \in R_n^-} Q_\alpha \Phi(\mathbf{y}_\beta - \mathbf{x}_\alpha) = \sum_{n'=0}^{p-1} \sum_{m'=-n'}^{n'} C_{n'}^{m'} R_{n'}^{m'}(\mathbf{y}_\beta - \mathbf{x}_{n*}),$$
$$C_{n'}^{m'} = ik \sum_{\mathbf{x}_\alpha \in R_n^-} Q_\alpha S_n^{-m}(\mathbf{x}_\alpha - \mathbf{x}_{n*}), \tag{5.1.7}$$

where $\mathbf{x}_{n*}$ is the center of the box R_n. Computation of $C_{n'}^{m'}$ requires $O((N - N_n^{(\mathrm{s})})p^2)$ operations and evaluation of the sum for all points $\mathbf{y}_\beta \in R_n$ takes $O((N - N_n^{(\mathrm{s})})p^2 + N_n^{(\mathrm{e})}p^2)$ operations. Here $N_n^{(\mathrm{e})}$ is the number of evaluation points in the receiver box R_n. Thus, the total complexity of computation of $\psi(\mathbf{y}_\beta)$ for all evaluation points can be estimated as

$$\text{Complexity} = O(MN^{(\mathrm{s})}) + O(K(N - N^{(\mathrm{s})})p^2 + Mp^2). \tag{5.1.8}$$

Here we used the fact

$$N_1^{(e)} + \cdots + N_K^{(e)} = M, \tag{5.1.9}$$

and introduced a parameter

$$N^{(s)} = \frac{1}{K}\left(N_1^{(s)} + \cdots + N_K^{(s)}\right), \tag{5.1.10}$$

which is the average number of sources in the neighborhood of receiver boxes. This parameter strongly depends on K and the source and evaluation point distributions. It is zero, if there are no sources in the neighborhoods of the receiver boxes (in this case we come to the case of well-separated sets for the given space subdivision). The other limit can be found assuming a *uniform* distribution of sources. In this case the number of boxes in the neighborhood, Pow(d), which we also call the *power* of the neighborhood, plays an important role. This number depends on the space dimensionality d, and the location of the box in the computational domain. For boxes not adjacent to boundary we have Pow(2) = 9 and Pow(3) = 27. Since each box can enter into not more than Pow(d) neighborhoods, we have

$$N_1^{(s)} + \cdots + N_K^{(s)} \leq N \operatorname{Pow}(d) \tag{5.1.11}$$

and so

$$N^{(s)} \leq \frac{N}{K} \operatorname{Pow}(d). \tag{5.1.12}$$

In the limiting case we can estimate complexity for equality in this equation, which shows that

$$\text{Complexity} = O\left(\frac{MN}{K}\operatorname{Pow}(d) + (K - \operatorname{Pow}(d))Np^2 + Mp^2\right). \tag{5.1.13}$$

This estimation shows that the increase of K reduces the first term in the sum and increases the second term. So this number should be selected carefully to provide the minimum of the sum. In fact, this minimum can be obtained easily, since the complexity as a function of K is a sum of hyperbola and linear function (see Fig. 5.4), and has its only minimum at

$$K_{\text{opt}} = \left[\frac{MN \operatorname{Pow}(d)}{Np^2}\right]^{1/2} = \sqrt{\frac{M \operatorname{Pow}(d)}{p^2}}. \tag{5.1.14}$$

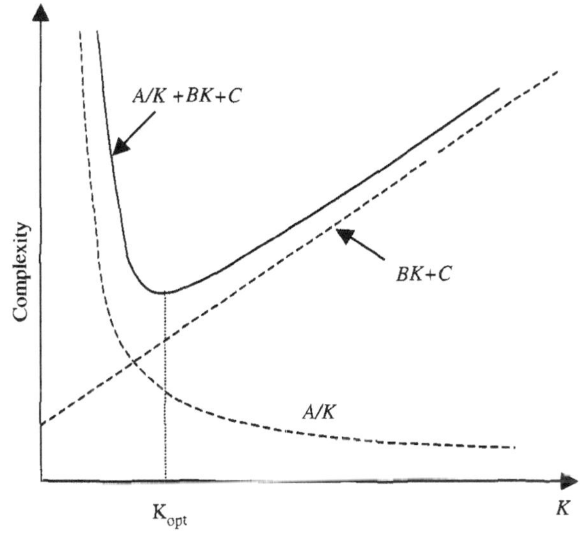

FIGURE 5.4 Dependence of the computational complexity of the method on the number of target boxes for modified Middleman method with space partitioning.

For such optimum K we have

$$\text{Complexity}_{\text{opt}} = O\left(Np\left(2\sqrt{\frac{M}{\text{Pow}(d)}} - \text{Pow}(d)p\right) + Mp^2\right). \quad (5.1.15)$$

For $M \sim N \gg p^2$ this scales as

$$\text{Complexity}_{\text{opt}} = O(pN^{3/2}). \quad (5.1.16)$$

Compared to the $O(N^2)$ complexity required using the straightforward method we can see that the Middleman method with space partitioning results in a faster algorithm. This algorithm, however, scales as $\sim N^{3/2}$ not as N which is the case for the well separated sets of the source and evaluation points. A remark we wish to make is that at the optimum K the cost of the direct evaluations is balanced with the cost of the evaluations using the series. This is also illustrated in Fig. 5.4.

5.1.2.2 Space partitioning with respect to source set

Another strategy for space partitioning is grouping with respect to the source set. In fact this is almost the same method as considered above. Imagine that in Fig. 5.3 the filled circles correspond to the evaluation points and the open circles to the sources. We will call boxes containing sources as *source boxes*. If we now build an S expansion near the center of each source box combining the expansion coefficients of all sources inside this box, it will be valid outside the neighborhood of the box. Thus, the effect of the group of the sources on any evaluation point outside the neighborhood can be computed using the Middleman method. For the evaluation points inside the neighborhood of the source box the field generated by these sources can be evaluated directly. The computational complexity of this algorithm is the same as before and for uniform distributions of sources and evaluation points with $M \sim N \gg p^2$ we obtain Eq. (5.1.16) if the number of boxes selected in an optimal way (when exchanging source and evaluation sets we need to exchange the symbols S and R in the expansion functions and M and N denoting number of sources and evaluation points).

Schematically the modified Middleman method with space partitioning with respect to the evaluation and source sets can be represented as shown in Fig. 5.5. Here for simplicity of the picture we dropped lines connecting source and evaluation points which reside in the same neighborhood of a given receiver or source box. This picture illustrates how computational savings appear in the method with space partitioning. Instead of evaluation of all pairwise interactions of the source and evaluation points, we can evaluate interactions between the sources and the receiver boxes or between the source boxes and evaluation points.

Note that the method, which we call the modified Middleman method, was proposed in a few papers (e.g. [CRW93]). Sometimes it is also referred to as "single level FMM" (SLFMM). In this book we use a slightly different terminology, and by the FMM we mean a method which requires translations, and the SLFMM is described in the following section.

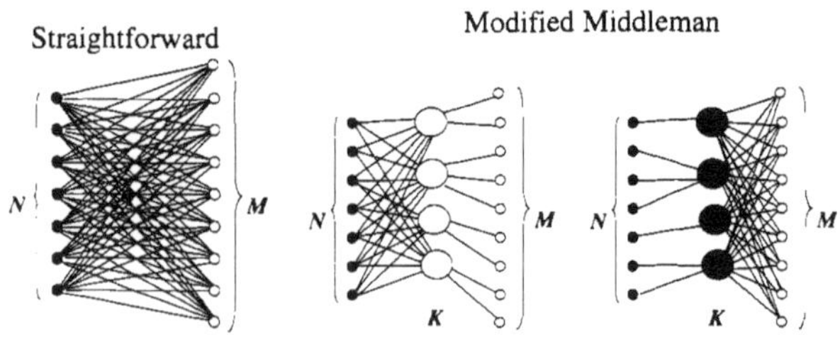

FIGURE 5.5 A scheme for straightforward computation of the field generated by N sources at M evaluation points is shown in the left picture. The picture in the center illustrates the scheme for modified Middleman method with space partitioning with respect to the evaluation points, where K is the number of target boxes. The picture on the right shows the modified Middleman method with the space partitioning with respect to the source points, where K is the number of the source boxes. Direct computations for the evaluation points in the neighborhood of sources are not shown in the chart.

5.1.3 Translations (SLFMM)

The advantage of the modified Middleman method compared to the straightforward computation appears due to grouping of sources or evaluation points. This grouping is, however, partial in a sense that we combine only sources or only evaluation points into groups. A more or less obvious idea is then to combine both sources and evaluation points into groups and consider interaction of groups with a hope to reduce further the complexity of computations. To demonstrate how this works, we subdivide the entire computational domain containing all the sources and evaluation points, enclosed in a large box, into boxes as before. For this division we indicate all the boxes containing the evaluation points as the receiver boxes and all the boxes containing the source points as the source boxes. Of course, some boxes contain both sources and evaluation points, so they are both receiver and source boxes. Some boxes may be empty and can be skipped in the process of computations.

For each source box we build the S expansion near the center of the box, which represents the field due to all sources in the source box. In the case of monopole sources we can determine a field generated by those sources as

$$\begin{aligned} \psi_n(\mathbf{y}) &= \sum_{\mathbf{x}_\alpha \in S_n} Q_\alpha \Phi(\mathbf{y} - \mathbf{x}_\alpha) = \sum_{n'=0}^{p-1} \sum_{m'=-n'}^{n'} C_{n'}^{(n)m'} S_{n'}^{m'}(\mathbf{y} - \mathbf{x}_{n*}^{(s)}), \\ n &= 1, \ldots, K_s, \\ C_{n'}^{(n)m'} &= ik \sum_{\mathbf{x}_\alpha \in S_n} Q_\alpha R_{n'}^{-m'}(\mathbf{x}_\alpha - \mathbf{x}_{n*}^{(s)}), \\ n' &= 0, 1, \ldots, p-1, \quad m' = -n', \ldots, n', \end{aligned} \tag{5.1.17}$$

where K_s is the number of source boxes and $\mathbf{x}_{n*}^{(s)}$ is the center of the nth source box. Function $\psi_n(\mathbf{y})$ can then be reexpanded near the center of the mth receiver box, $\mathbf{y}_{n*}^{(r)}$, which does not belong to the neighborhood of S_n (which we denote as S_n^-) as follows:

$$\psi_n(\mathbf{y}) = \sum_{n'=0}^{p-1} \sum_{m'=-n'}^{n'} C_{n'}^{(nm)m'} R_{n'}^{m'}(\mathbf{y} - \mathbf{y}_{m*}^{(r)}), \qquad n = 1, \ldots, K_s. \tag{5.1.18}$$

Here the expansion coefficients $C_{n'}^{(nm)m'}$ can be found from known expansion coefficients $C_{n'}^{(n)m'}$ using the $S|R$ *translation operator*:

$$\{C_{n'}^{(nm)m'}\} = (S|R)(\mathbf{t}_{nm})\{C_{n'}^{(n)m'}\}, \qquad \mathbf{t}_{nm} = \mathbf{y}_{m*}^{(r)} - \mathbf{x}_{n*}^{(s)}. \tag{5.1.19}$$

This operator depends on the translation vector, which is the vector directed from $\mathbf{x}_{n*}^{(s)}$ to $\mathbf{y}_{m*}^{(r)}$. As shown in Chapter 3 this operator can be thought of as a matrix acting on coefficients $\{C_{n'}^{(nm)m'}\}$ and producing coefficients $\{C_{n'}^{(n)m'}\}$ so

$$C_{n'}^{(nm)m'} = \sum_{n''=0}^{\infty} \sum_{m''=-n''}^{n''} (S|R)_{n'n''}^{m'm''}(\mathbf{t}_{nm})C_{n''}^{(n)m''}, \quad (5.1.20)$$

where the series can be truncated, say, at $n'' = p - 1$, which provides the required accuracy. For the computational method, however, the translation operator can be thought of as a black box subroutine which takes as input $\{C_{n'}^{(n)m'}\}$ and $\mathbf{t}_{nm}$ and returns $\{C_{n'}^{(nm)m'}\}$. This black box is not necessarily the matrix–vector multiplier, and in Chapters 7 and 8 we discuss various methods of translation, which are computationally more efficient than the matrix–vector multiplication based on a straightforward use of Eq. (5.1.20).

With a translation technique available we can represent sums (5.1.1) in the form:

$$\psi(\mathbf{y}_\beta) = \sum_{\mathbf{x}_\alpha \in R_n^+} Q_\alpha \Phi(\mathbf{y}_\beta - \mathbf{x}_\alpha) + \sum_{\mathbf{x}_{n*}^{(s)} \in R_n^-} \psi_n(\mathbf{y}_\beta), \qquad \mathbf{y}_\beta \in R_n. \quad (5.1.21)$$

Computation of all $C_{n'}^{m'}$ requires $O(Np^2)$ operations, translations require $O(K_s K_r \text{ TranslationCost}(p))$ operations, where K_r is the number of the receiver boxes and TranslationCost(p) denotes the number of operation per one translation for truncation number p, evaluation of the latter sum (5.1.21) for all points $\mathbf{y}_\beta$ takes $O(Mp^2)$ operations and the direct computation of interactions in the neighborhood, represented by the first term takes $O(MN^{(s)})$ operations as in Eq. (5.1.8). This results in the total complexity for the method:

$$\text{Complexity} = O(MN^{(s)}) + O(Np^2 + K_s K_r \text{ TranslationCost}(p) + Mp^2). \quad (5.1.22)$$

Again we see that this estimate depends on the source and evaluation point distributions and on how well these sets are separated for the space subdivision used. Assume a "worst" case when the sources and evaluation points are distributed more or less evenly, $K_s \sim K_r = K$, and $N^{(s)}$ can be evaluated with the upper bound of inequality (5.1.12). In this case we have

$$\text{Complexity} = O(F(K; M, N, p, d)),$$
$$F = \frac{MN}{K}\text{Pow}(d) + K^2 \text{ TranslationCost}(p) + (M + N)p^2. \quad (5.1.23)$$

The function F considered as a function of K is a superposition of a hyperbola and parabola (Fig. 5.6). It has one minimum, which can be found from

$$\left.\frac{\partial F}{\partial K}\right|_{K_{\text{opt}}} = 0, \tag{5.1.24}$$

and so

$$K_{\text{opt}} = \left[\frac{MN\,\text{Pow}(d)}{2\,\text{TranslationCost}\,(p)}\right]^{1/3}. \tag{5.1.25}$$

The complexity of the method, which we refer as the SLFMM, with the optimum choice of K is then

$$\text{Complexity}_{\text{opt}} = O\Big([MN]^{2/3}\,\text{TranslationCost}^{1/3}(p) + (M+N)p^2\Big). \tag{5.1.26}$$

For $M \sim N$ this scales as

$$\text{Complexity}_{\text{opt}} = O(N^{4/3}\,\text{TranslationCost}^{1/3}(p) + Np^2). \tag{5.1.27}$$

Therefore, the complexity of the SLFMM is asymptotically smaller than that of the straightforward method with complexity $O(N^2)$ if

$$\text{TranslationCost}(p) \ll N^2, \qquad p \ll N^{1/2}. \tag{5.1.28}$$

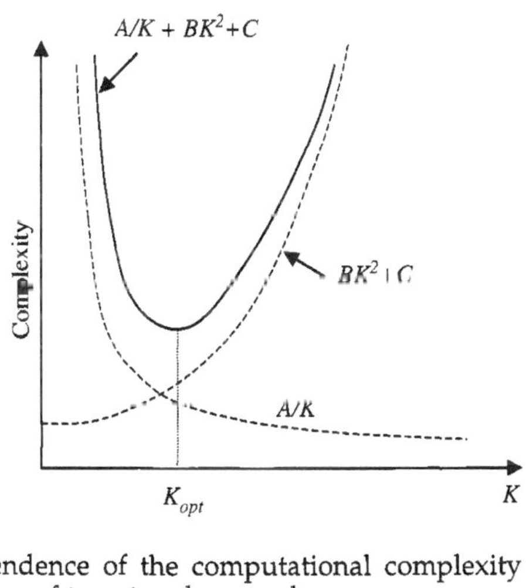

FIGURE 5.6 Dependence of the computational complexity of the single level FMM on the number of target and source boxes.

The latter is a general requirement for any method using the "Middleman" or factorization trick. So we can assume that this holds always. Then we can see that any translation method with

$$\text{TranslationCost}(p) = O(p^4) \tag{5.1.29}$$

provides a speed up for the computation, since in this case

$$O(N^{4/3}\,\text{TranslationCost}^{1/3}(p)) = O(N^{4/3}p^{4/3})$$

$$= o(N^{4/3}N^{2/3}) = o(N^2). \tag{5.1.30}$$

Note that the complexity in Eq. (5.1.29) can be achieved with a straightforward algorithm of translation based on matrix–vector multiplication (5.1.20), since a recursive computation of all entries of the truncated translation matrix requires $O(p^4)$ operations and the matrix–vector product for a $p^2 \times p^2$ matrix takes $O(p^4)$ operations as well.

In the following chapters we consider translation methods that are faster than straightforward multiplication of the translation matrix by a vector (say, using matrix decompositions or diagonal forms). While it may depend on $\log p$, the dependence of the translation cost on p can be usually modeled for evaluations as

$$\text{TranslationCost}(p) = O(p^{2\nu}), \qquad \nu \geq 1. \tag{5.1.31}$$

In this case the first term in the sum (5.1.27) is the principal term if

$$N^{1/3} \gg p^{2-2\nu/3}. \tag{5.1.32}$$

The complexity of the method is then

$$\text{Complexity}_{\text{opt}} = O(N^{4/3}p^{2\nu/3}). \tag{5.1.33}$$

Respectively, the principal term is represented by the second term in sum (5.1.27), if the inequality (5.1.32) does not hold.

Figure 5.7 illustrates the SLFMM. It is seen that savings in the number of operations appear because pairwise interactions are computed only between groups of sources and evaluation points. As before, for clarity we do not show direct evaluations of the source potentials at evaluation points from the neighborhoods.

It is also interesting to note the mnemonic nature of our notation. The source boxes (S) are associated with the singular (S) expansions and the receiver boxes (R) are associated with the regular expansions (R). *SR* translation is a conversion of the S to R expansion and at the same time it is a transfer of expansions for source boxes to the receiver boxes (S to R).

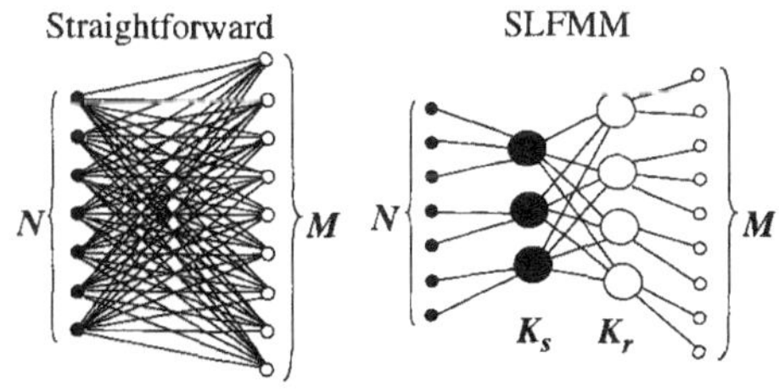

FIGURE 5.7 A scheme for straightforward computation of the field generated by N sources at M evaluation points is shown in the left picture. The picture on the right illustrates the computational scheme for the single level fast multipole method. Direct computations for the evaluation points in the neighborhood of sources are not shown in the chart.

One can keep in mind that the letter S is related to both "source" and "singular" and the letter R is related to both "receiver" and "regular".

5.1.4 Hierarchical space partitioning (MLFMM)

A further development of the idea for fast evaluation of a potential due to superposition of a large number of sources results in the MLFMM. As seen above, the grouping of sources and evaluation points reduces the number of operations, while the optimum number of boxes is large ($K \sim N^{2/3}$) and all pairwise interactions between the source and receiver boxes are taken into account. We can look for a method which provides a further level of space partitioning to replace the individual pairwise interactions with interactions between groups of boxes.

The idea is therefore to create a computational structure as shown in Fig. 5.8. Here, at the finest level the sources and evaluation points are grouped based on their spatial location in smaller sets (say, points inside boxes). At a coarser level smaller boxes are grouped

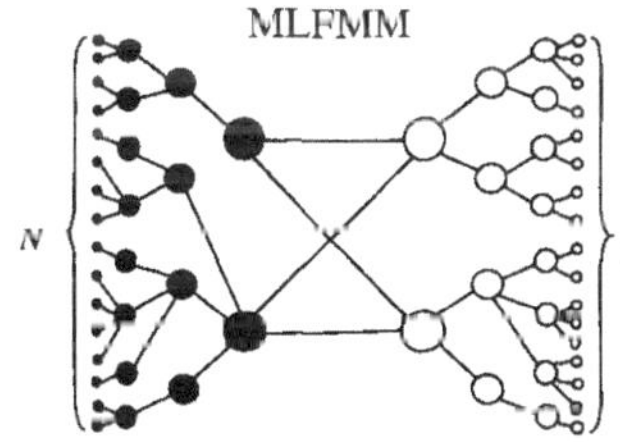

FIGURE 5.8 A scheme for the multilevel fast multipole method. Interactions between finer levels of the source and receiver hierarchies are not shown.

into larger boxes, and so on up to the coarsest level, which contains just a few boxes. The interaction between source and receiver boxes can occur at any level. However, since the neighborhood for each source or receiver box is finite, the number of the neighboring source boxes for each receiver box at the same level is relatively small. The effect of the sources outside the neighborhood can be computed by translations of expansions from the coarser level.

Since this algorithm requires more detailed explanation and there are many issues which should be considered carefully, we allocated special sections for the description of the algorithm and the design of data structures. Here we just state the core idea that the data should be organized hierarchically and they should employ so-called *fractal* structures, i.e., structures that repeat themselves at all levels of the hierarchy, to enable organization of the computations into loops over the levels.

We can preliminarily evaluate the complexity of the MLFMM method, using Fig. 5.8. It is clear that there should be a computational cost for the generation and evaluation of expansions, which results in the complexity of the Middleman method $O(p^2(N+M))$ (note that this term appeared also in the SLFMM complexity). Other operations are related to the cost of translation. Translations in the MLFMM should include all types of translations, including multipole-to-multipole $(S|S)$ when we go up with respect to the source hierarchy (upward pass), local-to-local $(R|R)$ when we go down with respect to the receiver hierarchy (downward pass) and multipole-to-local $(S|R)$ whenever we go from the source to the receiver hierarchy. Since the number of boxes in the hierarchy reduces from the finest to the coarsest level as a geometrical progression, the total number of source boxes is of the order $O(N)$ and the receiver boxes $O(M)$. So we can expect $O((M+N)\text{Translation Cost}(p))$ operations and the total complexity:

$$\text{Complexity} = O((M+N)(\text{TranslationCost}\,(p) + p^2)). \tag{5.1.34}$$

These estimates show that the expected complexity of the MLFMM for $N \sim M$ and p independent on N and M is $O(N)$, which is better than the $O(N^{4/3})$ complexity of the SLFMM. In the next chapter dedicated to the optimization of the complexity of the MLFMM, we provide more accurate estimations and consider some optimization issues as for the modified Middleman method and the SLFMM.

5.1.5 Truncation number dependence

While performing evaluations of the computational complexity of different methods, we treated the truncation number p as an independent fixed parameter. In practice, this number is not independent of N. This is especially so for the Helmholtz equation, where it depends on the wavenumber k, the size of the domain, and the size of boxes used for space partitioning. These parameters are interrelated, and p should be found based on the error ϵ, which is acceptable for a particular problem and is a user-specified input parameter. This analysis is substantially more involved and may be confusing if introduced at this stage. It is described in a separate chapter dedicated to error bounds.

5.1.5.1 *Geometrically decaying error*

For example, consider a case when the error of approximation decays geometrically with p for $p > p_{\min}$. In this case the absolute error of the far-field expansion of function Φ entering Eq. (5.1.1) can be represented as

$$\epsilon_p = \left| \Phi(\mathbf{y} - \mathbf{x}_\alpha) - \sum_{n'=0}^{p-1} \sum_{m'=-n'}^{n'} C_{n'}^{(n)m'} S_{n'}^{m'}(\mathbf{y} - \mathbf{x}_{n*}^{(S)}) \right|$$

$$= C\eta^p, \qquad 0 < \eta < 1. \tag{5.1.35}$$

For $\epsilon_p < \epsilon$

$$p > \frac{\log(C/\epsilon)}{\log(1/\eta)} \sim \log\left(\frac{1}{\epsilon}\right). \tag{5.1.36}$$

This is true for a single expansion. If we have N sources, the error of function approximation of the field of N sources will be $N\epsilon_p$ and we need the stronger condition $N\epsilon_p < \epsilon$ which ensures the accuracy of this function approximation. This yields

$$p > \frac{\log(CN/\epsilon)}{\log(1/\eta)} \sim \log\left(\frac{N}{\epsilon}\right) = O(\log N). \tag{5.1.37}$$

Thus, in this case p depends on N as $\log N$ and for the Middleman method for $M \sim N$ we will have complexity:

$$\text{ComplexityMiddleman} = O(Np^2) = O(N \log^2 N). \tag{5.1.38}$$

Evaluations of the translation cost of the type in Eq. (5.1.31) make the complexity of the SLFMM (Eq. (5.1.27)):

$$\text{ComplexitySLFMM} = (N^{4/3} \log^{2\nu/3} N) \tag{5.1.39}$$

and the complexity of the MLFMM (Eq. (5.1.34)):

$$\text{ComplexityMLFMM} = (N \log^{4\nu} N). \tag{5.1.40}$$

This shows that all approximate algorithms for a specified error ϵ, in fact, have powers of $\log N$ in the asymptotic complexity estimates, simply due to the error of truncation. In the literature dedicated to FMM one can find claims that the algorithms scale as $O(N)$ or some powers of N. Such estimates neglect the dependence of p on N. Note that $\log N$ terms also come from another source, namely from queries of data sets and organization of data structures. We discuss this in more detail later in this chapter.

5.1.5.2 Dependence of the truncation number on the box size

Another cause for the dependence of p on N for the Helmholtz equation appears from the relation between the approximation error and the size of the expansion domain, a. As a first example let us consider computation of the sum of sources which appear in the boundary element method coupled with the SLFMM. Assume that we want to solve a boundary value problem for a sphere of radius R, and for real $k > 0$. Here, the sources and evaluation points are distributed over the surface of this sphere, and we have $M \sim N$. If these points are distributed more or less evenly, the characteristic distance between neighboring sources or evaluation points is

$$d_{\min} \sim \left(\frac{4\pi R^2}{N}\right)^{1/2} \sim RN^{-1/2}. \tag{5.1.41}$$

A requirement for stable discretization (the discretization must not affect the result of computation) is that the wavelength be much larger than the discretization distance, $2\pi/k \gg d_{\min}$. In practice it is used in the form $2\pi/k \gtrsim Ad_{\min}$, where A is some constant. This leads to the following estimate for the number of surface elements:

$$N^{1/2} \gtrsim kR. \tag{5.1.42}$$

In the SLFMM we subdivide the space containing the source and evaluation points with K boxes. Therefore, the characteristic size of a box is of the order of the size of the expansion domain a and we have

$$a \sim \left(\frac{4\pi R^2}{K}\right)^{1/2} \sim RK^{-1/2}, \tag{5.1.43}$$

and each box contains of order N/K points. Assume now that the function $p = p_c(ka, \epsilon)$ specifies the minimum truncation number which provides a uniform approximation of the Helmholtz equation solution inside or outside a sphere of radius a for a given error ϵ. For large ka ($ka \gg 1$) the truncation number p_c grows with ka, while for $ka \lesssim 1$ a constant p_c can be used (in practice for simulations with $ka \lesssim 10$ there is no need to vary p with ka).

Thus for $ka \gg 1$, for larger boxes we need a larger p to achieve the required accuracy. As a simple model for evaluations we can use the following estimate:

$$p_c(ka, \epsilon) \gtrsim p_{\min} \sim ka. \tag{5.1.44}$$

Here we dropped the dependence on ϵ understanding that for $p_c \gtrsim p_{\min}$ we expect a geometrical decay of the error as considered in Section 5.1.5.1. If so, we find from Eqs. (5.1.42)–(5.1.44) that for limiting cases in the inequalities,

$$p \sim (N/K)^{1/2}. \tag{5.1.45}$$

With this estimate (Eqs. (5.1.23) and (5.1.31)) for $M \sim N$ and $\mathrm{Pow}(d) \sim 1$ (for large N and K the number of neighbors is as for $d = 2$, since we consider only non-empty boxes on the surface), we obtain

$$\begin{aligned} &\text{ComplexitySLFMM} = O(F(K; N)), \\ &F = \frac{N^2}{K} + K^2\left(\frac{N}{K}\right)^{\nu}, \qquad \nu \leq 2. \end{aligned} \tag{5.1.46}$$

The optimal number of boxes can then be found using Eq. (5.1.24), which yields

$$K_{\text{opt}} \sim \left(\frac{N^{4-2\nu}}{4-2\nu}\right)^{1/(8-2\nu)}. \tag{5.1.47}$$

It is interesting that as $\nu \to 2$ the optimal number of boxes

$$K_{\text{opt}}\Big|_{\nu \to 2} \sim \log N, \tag{5.1.48}$$

and the asymptotic complexity of the SLFMM according to Eq. (5.1.46) is $O(N^2)$, so there is no advantage of using SLFMM compared to the straightforward method in the case $ka \gg 1$! This is the reason why there is significant research on translation operators for the FMM. The complexity of the optimized SLFMM can be written using Eqs. (5.1.46) and (5.1.47) as

$$\text{ComplexitySLFMM}_{\text{opt}}(\nu) = O(N^{4/(4-\nu)}). \tag{5.1.49}$$

This formula works for $\nu \leq 2$. In fact we have $1 \leq \nu \leq 2$ ($\nu = 1$ is the minimum possible, since one operation per computation of one expansion coefficient is needed), so

$$\begin{aligned} \text{ComplexitySLFMM}_{\text{opt}}(1.5) &= O(N^{8/5}), \\ \text{ComplexitySLFMM}_{\text{opt}}(1) &= O(N^{4/3}). \end{aligned} \tag{5.1.50}$$

This shows that for $\nu < 2$ the SLFMM is asymptotically faster than straightforward evaluation. Dependence of the truncation number on N will introduce some logarithmic multipliers into complexity as mentioned earlier. However, we note that for optimum K we will have the following truncation numbers from Eq. (5.1.45):

$$p = O(N^{1/(4-\nu)}). \tag{5.1.51}$$

Compared to logarithmic terms depending on ϵ this is the principal term for $N \gg 1$.

Note that the optimum complexity of the modified Middleman method is $O(N^2)$ in the considered case and there is no advantage of using this method compared to the straightforward computations.

For the MLFMM we can also obtain complexity results for high frequency computations, $ka \gg 1$. Since the size of boxes in the MLFMM changes from level to level, we can change the truncation number with the level of space subdivision. Rigorous evaluation takes a bit more algebra and we consider this in more detail later. Here we can very roughly estimate that if we subdivide 3D space hierarchically with an octree, and at the finest level $l_{\max}$ we need $p_{l_{\max}}^2$ terms (one can assume $p_{l_{\max}} \sim 1$ if $ka \sim 1$ at this level), then at level $l_{\max} - 1$ we can keep $p_{l_{\max}-1}^2 \sim (2p_{l_{\max}})^2$ terms for the same accuracy due to the increase of the size of the domain two times. The truncation number for level l will be then $p_l \sim p_{l_{\max}} 2^{l_{\max}-l}$.

If the MLFMM is applied to speed up the BEM, then the number of non-empty boxes at level l will be $N_l \sim 2^{2l}$ with $N \sim 2^{2l_{\max}}$. Therefore, the total number of operations at level l due to translations with cost (5.1.31) can be estimated as

$$\begin{aligned} \text{ComplexityMLFMM}_l &= O(N_l p_l^{2\nu}) \\ &= O(2^{2l} 2^{2(l_{\max}-l)\nu} p_{l_{\max}}^{2\nu}) = O(2^{2l} 2^{2(l_{\max}-l)\nu}). \end{aligned} \tag{5.1.52}$$

The total complexity of the MLFMM should be of the order:

$$\text{ComplexityMLFMM}(\nu) = O\left(\sum_{l=0}^{l_{\max}} 2^{2l} 2^{2(l_{\max}-l)\nu}\right)$$

$$= O\left(2^{2l_{\max}\nu} \sum_{l=0}^{l_{\max}} 2^{2l(1-\nu)}\right) \sim 2^{2l_{\max}\nu} \frac{1 - 2^{2l_{\max}(1-\nu)}}{1 - 2^{2(1-\nu)}}. \tag{5.1.53}$$

This shows that

$$\text{ComplexityMLFMM}(1) = O(N \log N), \qquad N = 2^{2l_{\max}}, \tag{5.1.54}$$

$$\text{ComplexityMLFMM}(\nu) = O(N^{\nu}), \qquad \nu > 1, \; N = 2^{2l_{\max}}. \tag{5.1.55}$$

Thus, the translation cost heavily influences the complexity of the MLFMM and for minimum or critical exponent $\nu = 1$ we can expect $O(N \log N)$ asymptotic performance of the MLFMM used for speed up of the BEM. Note that the critical exponent can be larger for different distribution of the sources and receivers, and in a separate section we provide more detailed analysis. This preliminary estimate also shows that for speed up of the straightforward method we should have $\nu < 2$ at high frequencies.

5.1.6 Multipole summations

Equation (5.1.1) is written in the form which, for given Φ, determines ψ as a function of the coefficients Q_α, $\alpha = 1, \ldots, N$. This form seems appropriate for monopole sources, but in the case of multipole sources we have more coefficients and so rewrite it as:

$$\psi(\mathbf{y}_\beta) = \sum_{\alpha=1}^{N} \sum_{n=0}^{p-1} \sum_{m=-n}^{n} Q_n^{(\alpha)m} S_n^m(\mathbf{y}_\beta - \mathbf{x}_\alpha), \qquad \beta = 1, \ldots, M. \tag{5.1.56}$$

In the FMM we perform expansions of functions of this type about various expansion centers followed by translations. This requires the same amount of operations as expansions of monopole sources, since in any case we sum up expansion coefficients for given n and m and all $\mathbf{x}_\alpha$ from a particular box. The FMM for multipole sources can be efficiently used for multiple scattering problems, when each scatterer is represented by a sum of a number of multipoles (see the T-matrix method in the previous chapter). Details related to application of the FMM to this problem are provided in Chapter 10.

5.1.7 Function representations

The FMM was introduced originally as a method for summation of multipoles (or coefficients of function expansion over some basis). It can be extended by generalizing the function representations used and by specifying certain operations on these representations. To understand this point, let us analyze the FMM procedure from the computational point of view.

5.1.7.1 Concept

First, we encounter the question how one should represent a function in an algorithm. One of the ways to do this is to provide a table, which contains values of the function for certain values of the argument. The value of the function for arguments which are not listed in the table can be found approximately using interpolation or extrapolation. Another way to represent the function is by its expansion coefficients over some basis, e.g. spherical basis functions. This method can then be referred to as a spectral method. Spectral transforms can include expansions over special bases (solutions of the Helmholtz equation), more or less general bases independent of the equation (e.g. the Taylor series for arbitrary differentiable functions in the vicinity of some point), or use techniques such as the Fourier transform. The problems can involve scalar functions, vector functions (e.g. solutions of the Maxwell equation), or more generally tensors of various ranks, which depend again on scalars, vectors, or tensors of the same or some other ranks. In any case computationally we come to *function representations* using a finite set of some numbers or bits of information. These representing numbers or bits can be aligned in *strings* which are treated computationally as *vectors* of finite size P. Even if we have an infinite dimensional case (say, infinite series representing functions), these vectors are finite as we are concerned with computations.

Let us consider a function (scalar, vector, or tensor) f of a variable $\mathbf{y} \in \mathbb{R}^d$, where d is the space dimensionality (in this book we consider $d = 1, 2, 3$). We consider *approximations* of $f(\mathbf{y})$ inside or outside a sphere $\Omega_a(\mathbf{x}_*)$ of radius a centered at $\mathbf{y} = \mathbf{x}_*$. We say that function $R_P(\mathbf{y}; \mathbf{x}_*)$ *uniformly* approximate $f(\mathbf{y})$ inside a sphere $\Omega_a(\mathbf{x}_*)$ if

$$\exists \epsilon > 0, \qquad \forall \mathbf{y} \in \Omega_a(\mathbf{x}_*) \subset \mathbb{R}^d, \qquad |f(\mathbf{y}) - R_P(\mathbf{y}; \mathbf{x}_*)| < \epsilon, \qquad (5.1.57)$$

and function $S_P(\mathbf{y})$ uniformly approximate $f(\mathbf{y})$ outside a sphere $\Omega_a(\mathbf{x}_*)$ if

$$\exists \epsilon > 0, \qquad \forall \mathbf{y} \notin \Omega_a(\mathbf{x}_*), \qquad |f(\mathbf{y}) - S_P(\mathbf{y}; \mathbf{x}_*)| < \epsilon. \qquad (5.1.58)$$

The subscript P in $R_P(\mathbf{y};\mathbf{x}_*)$ and $S_P(\mathbf{y};\mathbf{x}_*)$ means that these functions can be determined by specification of the vector $\mathbf{C}$ in the complex P-dimensional space $\mathbb{C}^P$, which we call *representing vector.*

The dimensionality of this vector space P depends on ϵ and the approximation domain $\Omega_a(\mathbf{x}_*)$. Once the functions $R_P(\mathbf{y};\mathbf{x}_*)$ or $S_P(\mathbf{y};\mathbf{x}_*)$ are given, we can find uniquely corresponding $\mathbf{C}(\mathbf{x}_*)$ and once $\mathbf{C}(\mathbf{x}_*)$ is given we can uniquely determine $R_P(\mathbf{y};\mathbf{x}_*)$ and $S_P(\mathbf{y};\mathbf{x}_*)$. So we have a one-to-one mapping of the space of functions $R_P(\mathbf{y};\mathbf{x}_*)$ or $S_P(\mathbf{y};\mathbf{x}_*)$ to $\mathbf{C}(\mathbf{x}_*)$:

$$R_P(\mathbf{y},\mathbf{x}_*) \rightleftarrows \mathbf{C}(\mathbf{x}_*) = (c_1,\ldots,c_P), \qquad \mathbf{C}\in\mathbb{C}^P,$$
$$S_P(\mathbf{y},\mathbf{x}_*) \rightleftarrows \mathbf{C}(\mathbf{x}_*) = (c_1,\ldots,c_P), \qquad \mathbf{C}\in\mathbb{C}^P. \tag{5.1.59}$$

The representing vector $\mathbf{C}(\mathbf{x}_*)$ for $R_P(\mathbf{y};\mathbf{x}_*)$ is identified as the *local representation*. In the case when $\mathbf{C}(\mathbf{x}_*)$ corresponds to $S_P(\mathbf{y};\mathbf{x}_*)$ we call it the *far-field* representation. Note that the functions $R_P(\mathbf{y};\mathbf{x}_*)$ and $S_P(\mathbf{y};\mathbf{x}_*)$ can be the same or different.

For example, a function regular inside a sphere of radius a can be approximated as

$$R_P(\mathbf{y};\mathbf{x}_*) = \sum_{\alpha=1}^{N} \frac{Q_\alpha\, e^{ik|\mathbf{y}-\mathbf{x}_\alpha|}}{4\pi|\mathbf{y}-\mathbf{x}_\alpha|}, \qquad |\mathbf{x}_\alpha - \mathbf{x}_*| > a,$$
$$\mathbf{C} = (x_{11}, x_{12}, x_{13}, Q_1, \ldots, x_{N1}, x_{N2}, x_{N3}, Q_N), \qquad P = 4N. \tag{5.1.60}$$

So a $4N$-dimensional vector characterizing the source coordinates and intensities is a local representation. Similarly, a radiating solution of the Helmholtz equation regular outside a sphere of radius a can be represented with vector $\mathbf{C}$ as well:

$$S_P(\mathbf{y};\mathbf{x}_*) = \sum_{\alpha=1}^{N} \frac{Q_\alpha\, e^{ik|\mathbf{y}-\mathbf{x}_\alpha|}}{4\pi|\mathbf{y}-\mathbf{x}_\alpha|}, \qquad |\mathbf{x}_\alpha - \mathbf{x}_*| < a,$$
$$\mathbf{C} = (x_{11}, x_{12}, x_{13}, Q_1, \ldots, x_{N1}, x_{N2}, x_{N3}, Q_N), \qquad P = 4N. \tag{5.1.61}$$

The case of different $R_P(\mathbf{y};\mathbf{x}_*)$ and $S_P(\mathbf{y};\mathbf{x}_*)$ is clear from the previous consideration of multipole expansions. In this case

$$F_P(\mathbf{y};\mathbf{x}_*) = \sum_{n=0}^{p-1}\sum_{m=-n}^{n} C_n^m(\mathbf{x}_*) F_n^m(\mathbf{y}), \qquad F = S, R;\ \ P = p^2, \tag{5.1.62}$$

$$\mathbf{C} = \left(C_0^0, C_1^{-1}, C_1^0, C_1^1, C_2^{-2}, \ldots, C_{p-1}^{p-1}\right). \tag{5.1.63}$$

As an example of an alternate representation, in Chapter 7 we introduce the concept of the *far-field signature function*. In this case to any solution of the Helmholtz equation $\psi(\mathbf{y})$ inside or outside some sphere we correspond a surface function $\Psi(\mathbf{s})$ on the unit sphere, $|\mathbf{s}| = 1$. $\Psi(\mathbf{s})$ can be treated as action of some linear operator $\mathcal{F}$ on $\psi(\mathbf{y})$. This surface function can be *sampled* at, say, P points $\mathbf{s}_1, \mathbf{s}_2, \ldots, \mathbf{s}_P$. So, e.g. the far-field representation in this case will be as follows:

$$\Psi(\mathbf{s}) = \mathcal{F}[\psi(\mathbf{y})], \qquad \mathbf{C} = (\Psi(\mathbf{s}_1), \Psi(\mathbf{s}_2), \ldots, \Psi(\mathbf{s}_P)). \tag{5.1.64}$$

5.1.7.2 *FMM operations*

From the above examples of the FMM, the method consists of operations of generation of function representations, summations, translations (or conversion of one representation to some other representation) and evaluations of functions based on known representations. The operations of generation and evaluation of function representations (5.1.59) we refer as basic operations.

Consolidation

Summation of functions in the space of representing vectors is an easy and computationally cheap operation if the *operator* that maps the space of functions into the space of representing vectors (5.1.59) is *linear*. In this case, the sum of the functions is mapped into the sum of the representing vectors, so if $\mathbf{C}_1(\mathbf{x}_*)$ and $\mathbf{C}_2(\mathbf{x}_*)$ are representing vectors for functions $R_{P_1}(\mathbf{y};\mathbf{x}_*)$ and $R_{P_2}(\mathbf{y};\mathbf{x}_*)$ we have

$$R_{P_1}(\mathbf{y};\mathbf{x}_*) + R_{P_2}(\mathbf{y};\mathbf{x}_*) \rightleftarrows \mathbf{C}_1(\mathbf{x}_*) + \mathbf{C}_2(\mathbf{x}_*). \tag{5.1.65}$$

Of course, a similar relation can be written for the far-field functions and their representations. Examples of the linear mapping operator include expansions over the basis functions (5.1.62) and representations generated by linear integral transforms, such as given by Eq. (5.1.64). Note also that in this case the size of vectors $\mathbf{C}_1$ and $\mathbf{C}_2$ is the same.

Summation can be a difficult (or computationally expensive operation) if the mapping is *non-linear*. In this case the sum of functions is mapped to some representing vector $\mathbf{C}(\mathbf{x}_*)$, which should be produced from given vectors $\mathbf{C}_1$ and $\mathbf{C}_2$ with the aid of some operation, which we denote as $[+]$ and generally call *consolidation*:

$$R_{P_1}(\mathbf{y};\mathbf{x}_*) + R_{P_2}(\mathbf{y};\mathbf{x}_*) \rightleftarrows \mathbf{C}(\mathbf{x}_*) = \mathbf{C}_1(\mathbf{x}_*)[+]\mathbf{C}_2(\mathbf{x}_*). \tag{5.1.66}$$

For example, if we fix the length of the representing vectors $\mathbf{C}_1$, $\mathbf{C}_2$, and $\mathbf{C}$ as $P = 4N$ and try to produce $\mathbf{C} = \mathbf{C}_1[+]\mathbf{C}_2$ for representations of the type (5.1.60) the consolidation operation is non-linear and expensive. General consolidation operation $[+]$ turns into summation " $+$ " in the linear case.

Translations

There are three types of translations employed in the MLFMM. $R|R$, $S|R$, and $S|S$.

- *Local-to-local.* The $R|R$ translation operator in functional space acts on the function $R_{P_1}(\mathbf{y};\mathbf{x}_{*1})$ and produces a new function $R_{P_2}(\mathbf{y};\mathbf{x}_{*2})$ centered at a different point. Respectively, it takes the vector $\mathbf{C}_1(\mathbf{x}_{*1})$ and produces a new vector $\mathbf{C}_2(\mathbf{x}_{*2})$:

$$(R|R)(\mathbf{x}_{*2} - \mathbf{x}_{*1})[\mathbf{C}_1(\mathbf{x}_{*1})] = \mathbf{C}_2(\mathbf{x}_{*2}). \tag{5.1.67}$$

The length of vector $\mathbf{C}_1$ can differ from the length of $\mathbf{C}_2$, so if the length of the former vector is P_1, the length of the latter vector is P_2, not necessarily P_1. The translation operator can be linear or nonlinear, depending on whether the mapping (5.1.59) is linear or not. For linear mappings $(R|R)(\mathbf{x}_{*2} - \mathbf{x}_{*1})$ can be thought as a matrix. The computational cost of translation, however, can be cheaper than P_1P_2 required for straightforward matrix–vector multiplication as we stated above and provide algorithms in Chapter 7. It is important to note that in the FMM we require that the domain where $R_{P_1}(\mathbf{y};\mathbf{x}_{*1})$ approximates $f(\mathbf{y})$ *completely covers* the domain where $R_{P_2}(\mathbf{y};\mathbf{x}_{*1})$ approximates $f(\mathbf{y})$. Figure 5.9 illustrates this. Also if $f(\mathbf{y})$ has singularities located at $\mathbf{x}_\alpha$, these should be located outside the domain of validity of approximations.
- *Multipole-to-local.* Similarly, we can consider the $S|R$ translation operator, which transforms the representing vector $\mathbf{C}_1(\mathbf{x}_{*1})$ for

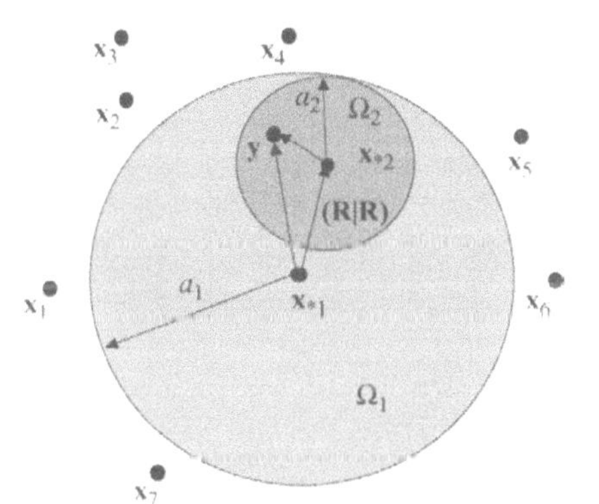

FIGURE 5.9 Local-to-local translation. The function centered at $\mathbf{x}_{*1}$ approximates the solution in the lighter gray domain Ω_1 and the translated function centered at $\mathbf{x}_{*2}$ approximates the solution in the darker gray domain Ω_2. Sources are shown as filled circles outside Ω_1.

function $S_{P_1}(\mathbf{y};\mathbf{x}_{*1})$ to the representing vector $\mathbf{C}_2(\mathbf{x}_{*2})$ for function $R_{P_2}(\mathbf{y};\mathbf{x}_{*2})$:

$$(S|R)(\mathbf{x}_{*2}-\mathbf{x}_{*1})[\mathbf{C}_1(\mathbf{x}_{*1})]=\mathbf{C}_2(\mathbf{x}_{*2}). \tag{5.1.68}$$

Again this operator can be linear or non-linear. The requirement that domain over which $f(\mathbf{y})$ is approximated by function $S_{P_1}(\mathbf{y};\mathbf{x}_{*1})$ is completely covered the domain of approximation of the same function $R_{P_2}(\mathbf{y};\mathbf{x}_{*2})$ leads to the picture shown in Fig. 5.10. All singularities of $f(\mathbf{y})$ should be located inside the sphere of radius a_1 centered at $\mathbf{x}_{*1}$.

- *Multipole-to-multipole.* In this case $\mathbf{C}_1(\mathbf{x}_{*1})$ and $\mathbf{C}_2(\mathbf{x}_{*2})$ are representing vectors for functions $S_{P_1}(\mathbf{y};\mathbf{x}_{*1})$ and $S_{P_2}(\mathbf{y};\mathbf{x}_{*2})$. The translation equation is similar to Eq. (5.1.69) and Fig. 5.11 illustrates location of the domains $\bar{\Omega}_2\subset\bar{\Omega}_1$ and possible singularities of $f(\mathbf{y})$.

5.1.7.3 SLFMM

As an example of the use of representing vectors, we formulate the SLFMM algorithm for three-dimensional space:

- Subdivide the computational domain into N_b boxes of size $b\times b\times b$.
- For each source $\mathbf{x}_\alpha$ obtain vectors $\mathbf{C}_\alpha(\mathbf{x}_{n*}^{(s)})$ of length P_1 corresponding to the function $S_{P_1}(\mathbf{y};\mathbf{x}_{n*}^{(s)})$ approximating the function $Q_\alpha\Phi(\mathbf{y}-\mathbf{x}_\alpha)$ in the domain outer to the sphere $\Omega_a(\mathbf{x}_{n*}^{(s)})$, where $a=\frac{3}{2}b$ (the sphere includes the box but enclosed into the box neighborhood) and $\mathbf{x}_{n*}^{(s)}$ is the center of the box containing $\mathbf{x}_\alpha$.

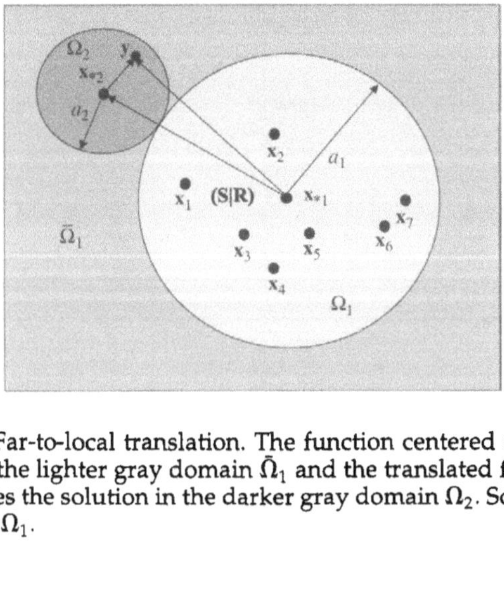

FIGURE 5.10 Far-to-local translation. The function centered at $\mathbf{x}_{*1}$ approximates the solution in the lighter gray domain $\bar{\Omega}_1$ and the translated function centered at $\mathbf{x}_{*2}$ approximates the solution in the darker gray domain Ω_2. Sources are shown as filled circles in Ω_1.

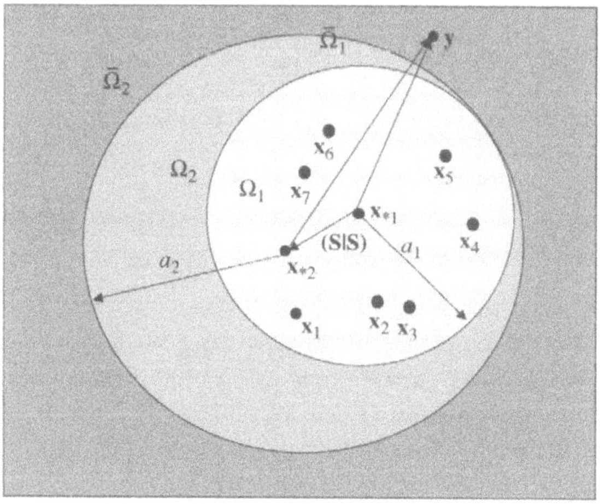

FIGURE 5.11 Far-to-far translation. The function centered at $\mathbf{x}_{*1}$ approximates the solution in the lighter gray domain $\bar{\Omega}_1$ and the translated function centered at $\mathbf{x}_{*2}$ approximates the solution in the darker gray domain $\bar{\Omega}_2 \subset \bar{\Omega}_1$. Sources are shown as filled circles in Ω_1.

- For each source box S_n containing q_n sources $\mathbf{x}_\alpha$, $\alpha = \alpha_1, \ldots, \alpha_{q_n}$, obtain vector of length $P_2^{(n)}$ (consolidation of all sources inside the source box):

$$\mathbf{C}(\mathbf{x}_{n*}^{(s)}) = \mathbf{C}_{\alpha_1}(\mathbf{x}_{n*}^{(s)})[+]\mathbf{C}_{\alpha_2}(\mathbf{x}_{n*}^{(s)})[+]\cdots[+]\mathbf{C}_{\alpha_{q_n}}(\mathbf{x}_{n*}^{(s)}). \tag{5.1.69}$$

  This vector represents the potential due to all the sources inside the box in the domain outside the neighborhood of this box S_n^-.
- $S|R$ translate each $\mathbf{C}(\mathbf{x}_{n*}^{(s)})$ from $\mathbf{x}_{n*}^{(s)}$ to the center $\mathbf{y}_{m*}^{(r)}$ of each receiver box R_m, such that the neighborhood of R_m does not contain S_n:

$$(S|R)(\mathbf{y}_{m*}^{(r)} - \mathbf{x}_{n*}^{(s)})[\mathbf{C}(\mathbf{x}_{n*}^{(s)})] = \mathbf{D}_n(\mathbf{y}_{m*}^{(r)}), \qquad R_m^+ \cap S_n = \emptyset, \tag{5.1.70}$$

  where $\mathbf{D}_n$ is the vector of length $P_3^{(n)}$ representing function in the domain inner to the sphere of radius a centered at $\mathbf{y}_{m*}^{(r)}$ due to sources in box S_n.
- For each receiver box R_m obtain the vector (consolidation of all sources outside the receiver neighborhood):

$$\mathbf{D}(\mathbf{y}_{m*}^{(r)}) = \mathbf{D}_{n_1}(\mathbf{y}_{m*}^{(r)})[+]\mathbf{D}_{n_2}(\mathbf{y}_{m*}^{(r)})[+]\cdots[+]\mathbf{D}_{n_{p_m}}(\mathbf{y}_{m*}^{(r)}),$$
$$R_m^+ \cap S_{n_1}, \ldots, S_{n_{p_m}} = \emptyset. \tag{5.1.71}$$

- For each receiver box evaluate the sum:

$$\psi(\mathbf{y}_\beta) = \sum_{\mathbf{x}_\alpha \in R_m^+} Q_\alpha \Phi(\mathbf{y}_\beta - \mathbf{x}_\alpha) + R_{P_2}(\mathbf{y}_\beta; \mathbf{y}_{m*}^{(r)}), \quad \mathbf{y}_\beta \in R_m, \tag{5.1.72}$$

  where $R_{P_4}(\mathbf{y};\mathbf{y}_{m*}^{(r)})$ is the local function represented by $\mathbf{D}(\mathbf{y}_{m*}^{(r)})$ and R_m is the mth receiver box.

The algorithm written in this form does not specify the functions S_P and R_P used and the sizes of the representing vectors and their sums. It just requires definition (subroutines) for consolidation and translation operations. For linear operators the operation $[+]$ turns into the usual vector summation $+$.

5.2 MULTILEVEL FAST MULTIPOLE METHOD

In this section we consider the MLFMM in generalized setting. As was mentioned in the introduction the basic ideas here are due to Rokhlin and Greengard's papers published in the late 1980s. While the description of the method in this book is sufficient to understand and implement the method, the interested reader is advised to review these publications as well. The MLFMM consists of several basic steps, which include setting up the hierarchical data structure, upward and downward passes, and final summation. The first step for given locations of the source and evaluation points can be performed once, while the other steps can be performed many times (say, in iterative solvers) for the same source and evaluation point locations. So the complexity of the algorithm consists of two parts. The first part is related to some precomputing (setting) and the second to the computation (FMM solver). We will consider both parts in detail. In Section 5.3 we also consider an efficient way of implementing the data structures ("bookkeeping") based on the bit-interleaving technique.

5.2.1 Setting up the hierarchical data structure

5.2.1.1 Generalized octrees (2^d trees)

One of the most important properties of d-dimensional Euclidean space ($\mathbb{R}^d$) is that it can be subdivided into rectangular boxes (we mostly operate with cubes). In practice, the problems we are concerned with are posed in finite domains, which can then be enclosed in a bounding box. We assign this bounding box to level 0, in a hierarchical division scheme. The level 0 box can be subdivided into 2^d smaller boxes of equal size by dividing each side in half. All boxes of this size are assigned to level 1. Repeating this procedure, we produce a sequence of boxes at level 2, level 3, and so on. While this process of subdivision could continue forever, in practice we would stop at some finite level $l_{\max}$, which is determined by some criterion (e.g. that there are at least s particles in a box at the finest level). Figure 5.12, left, illustrates this for $d = 2$. By the process of division we obtain a 2^d-tree (see Fig. 5.13), in which each node corresponds to a box. Any two nodes at different levels are connected in the tree if the box corresponding to the first node at the finer level is obtained by subdivision of the box corresponding to the second node at the coarser level. At level l of a 2^d-tree we have 2^{ld} boxes, with each node having the index as n, with n ranging from 0 to $2^{ld} - 1$. Therefore, any box in a 2^d-tree can be characterized by the pair (n, l).

A 2^d-tree graph clearly displays "parent–child" relationships, where the "children" boxes at level $l + 1$ are obtained by subdivision of a "parent" box at level l. For a 2^d-tree with $l_{\max}$ levels, any box at level $l \geq 1$ has exactly one parent, and any box at level $l \leq l_{\max} - 1$ has exactly 2^d children. So we can define operations Parent(n, l), which returns the index of the parent box, and ChildrenAll(n, l) that returns the indexes of the children boxes. The children of the same parent are called "siblings". Each box at a given level $l \geq 1$ has $2^d - 1$ siblings.

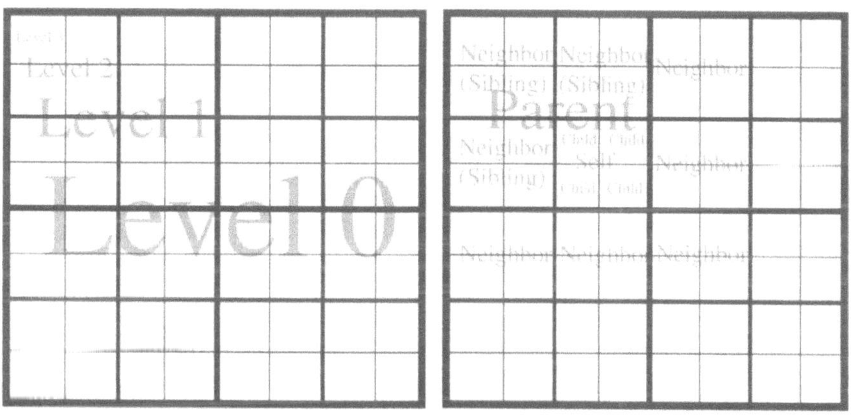

FIGURE 5.12 The left graph shows levels in quadtree space subdivision. The right graph shows children, parent, siblings, and neighbors of the box marked as "self".

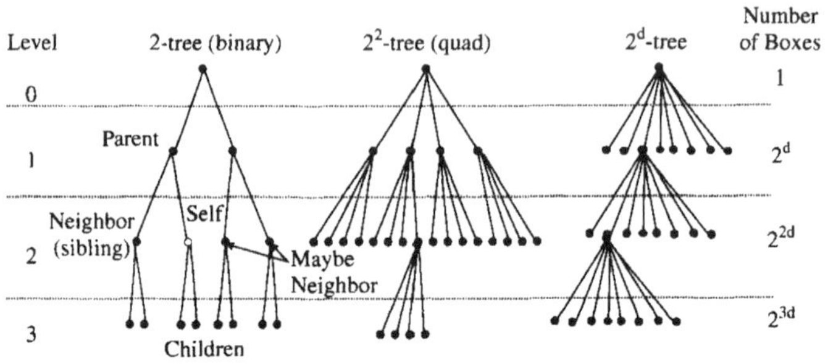

FIGURE 5.13 2^d-trees and terminology.

In the FMM we are also interested in neighbor relationships between boxes. These are determined exclusively by the relative spatial locations of the boxes, and not by their locations in the tree. We call two different boxes "neighbors" if their boundaries have at least one common point. We also use the terminology "power of a set" and "power of a neighborhood" to denote the number of boxes in a particular set and number of boxes in a particular neighborhood, respectively. Note that the concept of neighborhood can be extended, for example, to include into consideration neighbors of neighbors. In the present book, for simplicity, we will not consider this issue, which, however, cannot be avoided for dimensionality $d \geq 4$ (we are confined to $d = 1, 2, 3$) and refer the interested reader for extensions of the concept to our technical report [GDB03].

The number of neighbors that a given box has in a finite 2^d-tree space subdivision depends on its location relative to the boundary of the domain (the boundaries of the box at level 0). For example, a box at level $l \geq 1$ in a quadtree situated at the corner of the largest box has only three neighbors, while a box situated far from the boundaries (indicated as "self" in Fig. 5.12, right) has eight neighbors. The number of neighbors depends on the dimension d. In the general d-dimensional case, the minimum and maximum numbers of neighbors are

$$N_{\min}^{(\text{Neighbors})}(d) = 2^d - 1, \qquad N_{\max}^{(\text{Neighbors})}(d) = 3^d - 1. \qquad (5.2.1)$$

The minimum number of neighbors is achieved for a box in the corner, for which all neighbors are children of the same parent (siblings). Since the number of siblings is $2^d - 1$ this provides the minimum number of neighbors. The maximum number of neighbors is for a box located far from the boundary. Consider a box not on the boundary at a higher level. It has right and left neighbors in each dimension, and can be considered the central box of a cube divided into $3 \times \cdots \times 3 = 3^d$ sub-boxes, which is the power of the neighborhood. Excluding the box itself from this count we obtain the number of its neighbors as in Eq. (5.2.1).

Equations (5.2.1) show that the number of neighbors for large d far exceeds the number of siblings. The neighbor relationships are not easily determined by position on the 2^d-tree graph (see Fig. 5.13) and potentially any two boxes at the same level could be neighbors. On this graph the neighbors can be close to each other (siblings), or very far apart, so that a connecting path between them may have to go through a higher node (even through the node at level 0). For further consideration we can introduce operation NeighborsAll(n, l) which returns indexes of all neighbors of the box (n, l).

The above example shows that the power of neighborhood (i.e., the number of boxes it contains) depends on the level and on the location of the box. The maximum value of the power of neighborhood is

$$\text{MaxPower(Neighborhood)} = 3^d. \tag{5.2.2}$$

This shows that it depends exponentially on the dimension d.

5.2.1.2 *Data hierarchies*

The 2^d-trees provide a space subdivision without any consideration for the distribution of the source, $\mathbb{X}$, and evaluation, $\mathbb{Y}$, data points:

$$\mathbb{X} = \{\mathbf{x}_1, \mathbf{x}_2, \ldots, \mathbf{x}_N\}, \qquad \mathbf{x}_\alpha \in \mathbb{R}^d, \ \alpha = 1, \ldots, N, \tag{5.2.3}$$

$$\mathbb{Y} = \{\mathbf{y}_1, \mathbf{y}_2, \ldots, \mathbf{y}_M\}, \qquad \mathbf{y}_\beta \in \mathbb{R}^d, \ \beta = 1, \ldots, M. \tag{5.2.4}$$

These data can be structured with 2^d-trees and organized in the X data hierarchy (or source hierarchy) and the Y data hierarchy (or evaluation/target hierarchy) according to the coordinates of the source and evaluation points. We prescribe to each source or evaluation point the index of the box (n, l) to which it belongs, so that X and Y are sets of indices (n, l).

For each data hierarchy we define the operations Parent(n, l), Children(n, l), and Neighbors(n, l). The operation Parent(n, l) is the same for both hierarchies, since the parent of each box already contains points of the hierarchy. The other two operations return the sets of children and neighbor boxes at levels $l + 1$ and l, respectively, that contain the points from the particular hierarchies. To discriminate between the two sets, we denote them as Children$(X; n, l)$, and Neighbors$(X; n, l)$ for the X hierarchy and Children$(Y; n, l)$, and Neighbors$(Y; n, l)$ for the Y hierarchy.

5.2.1.3 Hierarchical spatial domains

We define notation here that will permit a succinct description of the FMM algorithm, and further allow its optimization. By optimization we mean the selection of parameters, e.g. one of the parameters to be chosen is the number of points, s, that are contained at the finest level ($l_{\max}$) in a non-empty box.

We define the following four spatial domains that are used in the FMM. These can be defined for each box with index $n = 0, \ldots, 2^{ld} - 1$ at level $l = 0, \ldots, l_{\max}$, and have fractal structure:[1]

- $E_1(n,l) \subset \mathbb{R}^d$ denotes spatial points *inside* the box (n,l).
- $E_2(n,l) \subset \mathbb{R}^d$ denotes spatial points in the neighborhood of box (n,l).
- $E_3(n,l) = E_1(0,0) \backslash E_2(n,l)$ denotes spatial points *outside* the neighborhood of box (n,l).
- $E_4(n,l) = E_2(\text{Parent}(n,l), l-1) \backslash E_2(n,l)$ denotes spatial points in the neighborhood of the parent box $(\text{Parent}(n,l), l-1)$, which do not belong to the neighborhood of the parent box itself.

We denote the sets of indices for boxes that constitute each of the domains $E_j(n,l)$, $j = 1, \ldots, 4$, as $I_j(n,l)$. Boxes $I_j(n,l) \cap X$ and $I_j(n,l) \cap Y$ belong to the X and Y hierarchies, respectively. Figure 5.14 illustrates these domains in the case $d = 2$.

To choose $l_{\min}$, the level from which we start the FMM we note that at $l_{\min}$ there should be at least one non-empty box outside the $E_2(n, l_{\min})$ domain, while at level $l_{\min} - 1$ the domain $E_1(0,0)$ resides completely inside the $E_2(n, l_{\min} - 1)$ domain. This happens if $2^{l_{\min}-1} < 2$. Thus,

$$l_{\min} = 2. \tag{5.2.5}$$

Based on these domains, the following functions can be defined for each box:

$$\psi_{n,l}^{(m)}(\mathbf{y}) = \sum_{\mathbf{x}_\alpha \in E_m(n,l)} Q_\alpha \Phi(\mathbf{y} - \mathbf{x}_\alpha), \qquad m = 1, \ldots, 4. \tag{5.2.6}$$

Note that since the domains $E_2(n,l)$ and $E_3(n,l)$ are complementary, we have from Eqs. (5.1.1) and (5.2.6)

$$\psi(\mathbf{y}) = \psi_{n,l}^{(2)}(\mathbf{y}) + \psi_{n,l}^{(3)}(\mathbf{y}), \tag{5.2.7}$$

for arbitrary l and n.

---

[1] By fractal we mean that these structures have the same shape at different levels of the hierarchy.

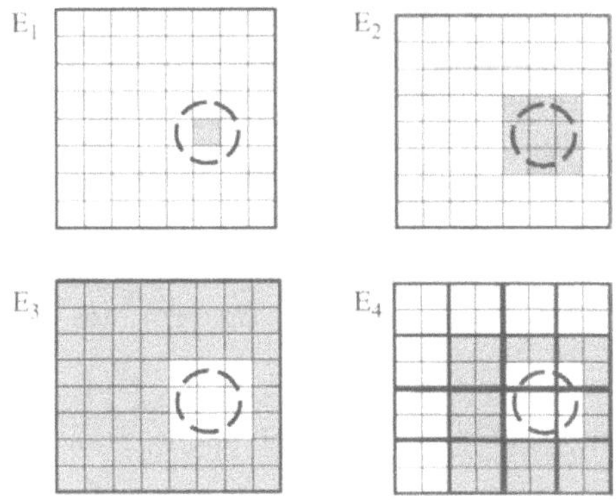

FIGURE 5.14 The domains used for construction of hierarchical reexpansion procedure in FMM ($d = 2$). A circle separates the box for which domains are drawn.

5.2.1.4 Spatial scaling and size of neighborhood

For the Helmholtz equation with real and positive k, which has the intrinsic length scale k^{-1}, the length of the representing vectors P (Eq. (5.1.59)) is a function of the actual size of the domain of approximation (the size of the sphere a inside which we approximate the solution). P depends on the dimensionless size of the box kD_l where D_l is the length of the box at level l, so we have for a given error ϵ:

$$P = P(kD_l, \epsilon), \qquad D_l = 2^{-l}D_0, \;\; l = 0, 1, \ldots, l_{\max}. \tag{5.2.8}$$

This is different, say, from the Laplace equation, which does not have an intrinsic length scale, and so P for fixed error ϵ does not depend on the box size (if we note that Laplace equation is the Helmholtz equation at $k = 0$, relation (5.2.8) also shows independence of P on the level). In the multilevel algorithm we encounter $S|R$ translation at every level $l = l_{\min}, \ldots, l_{\max}$. So for S-type representations at every level we will need approximately the same length vector as for R-type representations at the same level. That is why P should increase for both the S_p and R_p functions with ka.

It is also important to note that in the FMM, the length of the representing vector P depends on the size of the neighborhood q. Indeed, the number of terms in the expansion that approximate $\psi(\mathbf{y})$ in some domain depends on how far from the boundaries of this domain the singularities $\mathbf{x}_\alpha$ of $\psi(\mathbf{y})$ are located. At the singularities the expansions blow up, and we can treat this as $P \to \infty$ for $\mathbf{y} \to \mathbf{x}_\alpha$. To better understand the dependence of P on q, let us consider all situations that appear in the MLFMM.

(1) The singular type of approximation, S_P, near the center of the nth box at level l for $\mathbf{x}_\alpha \in E_1(n, l)$ is valid for any $\mathbf{y}$ in the domain $E_3(n, l)$. In d-dimensional space the maximum distance from the center of the unit box to its boundary is $d^{1/2}/2$ and the minimum distance from the center to the boundary of domain ($E_3(n, l)$) is $3/2$. The latter distance should be always larger than the former distance (which always holds for $d = 1, 2, 3$). The dimensionless measure for the length of expansion P is then the following quantity:

$$\eta_S = \tfrac{1}{3} d^{1/2}. \tag{5.2.9}$$

The closer this quantity is to 1, the larger P should be, and as $\eta_S \to 1$ we have $P \to \infty$ as stated above, so

$$P = P(kD_l, \eta_S, \epsilon), \qquad P(kD_l, 1, \epsilon) = \infty. \tag{5.2.10}$$

(2) The local approximation, R_P, near the center of the nth box at level l for $\mathbf{x}_\alpha \in E_3(n, l)$ is valid for any $\mathbf{y}$ from the domain $E_1(n, l)$. A similar calculation as that led to Eq. (5.2.9) shows that, in this case, the same ratio

$$\eta_R = \eta_S = \tfrac{1}{3} d^{1/2} < 1 \tag{5.2.11}$$

influences P.

(3) The $S|S$ translation of the S_P function from the center of the nth box at level l for $\mathbf{x}_\alpha \in E_1(n, l)$ to the center of its parent box preserves the validity of the S approximation for any $\mathbf{y}$ from the domain $E_3(\text{Parent}(n, l), (l-1))$. Parameter η_S is also useful for evaluation of the error of the $S|S$ translation.

(4) The $R|R$ translation of the R_P function from the center of the nth box at level l for $\mathbf{x}_\alpha \in E_1(n, l)$ to the centers of its children preserves the validity of the R approximation for any $\mathbf{y}$ from the domain $E_1(\text{ChildrenAll}(n, l), l+1)$. Parameter η_R is then useful for evaluation of the error of the $R|R$ translation.

(5) The $S|R$ translation of the S approximation from the center of the mth box at level l which belongs to $E_4(n, l)$ for $\mathbf{x}_\alpha \in E_1(m, l)$ to the center of the box $E_1(n, l)$ provides a valid R approximation for any $\mathbf{y}$ from the domain $E_1(n, l)$. If the size of the box at level l is 1, then the minimum distance between the centers of boxes (m, l) (say, $\mathbf{x}_{*1}$) and (n, l) (say, $\mathbf{x}_{*2}$) is 2. The maximum $|\mathbf{x}_\alpha - \mathbf{x}_{*1}|$ is $d^{1/2}/2$ the minimum $|\mathbf{x}_\alpha - \mathbf{x}_{*2}|$ is $3/2$, and the maximum $|\mathbf{y} - \mathbf{x}_{*2}|$ is $d^{1/2}/2$. For dimensionality $d = 1, 2, 3$ since $4 - d^{1/2} \leq 3$ this leads to the

following measure:

$$\eta_{SR} = \frac{d^{1/2}}{4 - d^{1/2}} < 1. \tag{5.2.12}$$

When $\eta_{SR} \to 1$ we have $P \to \infty$:

$$P = P(kD_l, \eta_{SR}, \epsilon), \qquad P(kD_l, 1, \epsilon) = \infty. \tag{5.2.13}$$

Since $\eta_{SR} \geq \eta_S$, namely, this criterion determines the length of the expansions at level l.

As it is seen η_{SR} is below the unity for $d \leq 3$ and so finite P can be expected for approximations. On the other hand, this parameter is related to the convergence of series and accuracy of approximations (geometrical convergence of the "tails").

5.2.2 MLFMM procedure

According to the Rokhlin–Greengard terminology, the FMM procedure consists of an upward pass, which is performed for each box at level l_{max} up to level l_{min} of the X hierarchy and uses the multipole S representations for these boxes, in the two-step downward pass, which is performed for each box from level l_{min} down to level l_{max} of the Y hierarchy and uses local R representations for boxes of this hierarchy, and the final summation. Translations within the X hierarchy are $S|S$ translations, within the Y hierarchy are $R|R$ translations, and translations from the boxes of the X hierarchy to the boxes of the Y hierarchy are the $S|R$ translations.

Here we confine ourselves with linear representations and linear translation and consolidation operators. This is dictated by a desire to use simpler notation. A reader who will be interested in development of methods based on non-linear consolidation should understand the consolidation process and modify algorithms for this case (which, in fact is not difficult, since algorithmically all it requires is a special routine for "summation", and we provided an example of such generalized algorithm for the SLFMM, see Eqs. (5.1.69)–(5.1.72)).

5.2.2.1 *Upward pass*

Step 1. For each source $\mathbf{x}_\alpha$ obtain the vectors $\mathbf{B}^{(\alpha)}(\mathbf{x}_*^{(n,l_{max})})$ of length $P_{l_{max}}$ corresponding to function $S_{P*}^{(\alpha)}(\mathbf{y}; \mathbf{x}_*^{(n,l_{max})})$, which approximates

$Q_\alpha \Phi(\mathbf{y} - \mathbf{x}_\alpha)$ in the domain outer to the sphere $\Omega_a(\mathbf{x}_*^{(n,l_{\max})})$:

$$S^{(\alpha)}_{P_{l_{\max}}}(\mathbf{y};\mathbf{x}_*^{(n,l_{\max})}) \rightleftarrows \mathbf{B}^{(\alpha)}(\mathbf{x}_*^{(n,l_{\max})}),$$

$$[\forall \mathbf{y}, |\mathbf{y}-\mathbf{x}_*^{(n,l_{\max})}| \geq a, |S^{(\alpha)}_{P_{l_{\max}}}(\mathbf{y};\mathbf{x}_*^{(n,l_{\max})}) - Q_\alpha \Phi(\mathbf{y}-\mathbf{x}_\alpha)| < \epsilon_*, a = \tfrac{3}{2} D_{l_{\max}}], \tag{5.2.14}$$

where $\mathbf{x}_*^{(n,l_{\max})}$ is the center of the box containing $\mathbf{x}_\alpha$ and ϵ_* is some specified error of approximation. Note that the selection of a is motivated by the requirement that the S approximation (5.2.14) for the nth box is valid in the domain $E_3(n,l_{\max})$ (see Fig. 5.14).

Step 2. For each box $(n, l_{\max})$ in the X hierarchy, generate the representing vector $\mathbf{C}(\mathbf{x}_*^{(n,l_{\max})})$ of length $P_{l_{\max}}$ corresponding to function $S_{P_{l_{\max}}}(\mathbf{y}; \mathbf{x}_*^{(n,l_{\max})})$ which approximates $\psi^{(1)}_{n,l_{\max}}(\mathbf{y})$:

$$\begin{aligned} &\mathbf{C}^{(n,l_{\max})} = \sum_{\mathbf{x}_\alpha \in E_1(n,l_{\max})} \mathbf{B}^{(\alpha)}(\mathbf{x}_*^{(n,l_{\max})}), \qquad (n, l_{\max}) \in X, \\ &[S_{P_{l_{\max}}}(\mathbf{y}; \mathbf{x}_*^{(n,l_{\max})}) \rightleftarrows \mathbf{C}^{(n,l_{\max})}, \quad \forall \mathbf{y} \in E_3(n, l_{\max}); \\ &\quad |S_{P_{l_{\max}}}(\mathbf{y}; \mathbf{x}_*^{(n,l_{\max})}) - \psi^{(1)}_{n,l_{\max}}(\mathbf{y})| < \epsilon_{l_{\max}}]. \end{aligned} \tag{5.2.15}$$

Here $\epsilon_{l_{\max}}$ is some specified error of approximation, which can be linked further to the overall error of the method.

Step 3. Repeat for $l = l_{\max} - 1, \ldots, l_{\min}$. For each box (n, l) in the X hierarchy recursively determine the representing coefficients $\mathbf{C}^{(n,l)}$ of length P_l corresponding to function $S_{P_l}(\mathbf{y}; \mathbf{x}_*^{(n,l)})$ which approximates $\psi^{(1)}_{n,l}(\mathbf{y})$ by translation the representing coefficients from its children boxes followed by summation:

$$\begin{aligned} &\mathbf{C}^{(n,l)} = \sum_{n' \in \text{Children}(X;n,l)} (S|S)(\mathbf{x}_*^{(n,l)} - \mathbf{x}_*^{(n',l+1)})[\mathbf{C}^{(n',l+1)}], \qquad (n, l) \in X, \\ &[S_{P_l}(\mathbf{y}; \mathbf{x}_*^{(n,l)}) \rightleftarrows \mathbf{C}^{(n,l)}, \quad \forall \mathbf{y} \in E_3(n, l), \quad |S_{P_l}(\mathbf{y}; \mathbf{x}_*^{(n,l)}) - \psi^{(1)}_{n,l}(\mathbf{y})| < \epsilon_l]. \end{aligned} \tag{5.2.16}$$

For the nth box, this expansion is valid in the domain $E_3(n, l)$ which is a subdomain of $E_3(\text{Children}(X; n, l), l + 1)$, and the far-to-far translation is applicable. Figure 5.15 illustrates this for $d = 2$. Indeed, the spheres that enclose each child box are themselves enclosed by the larger sphere around the parent box. Thus, for the domain $E_3(n, l)$ shaded in dark gray (see Fig. 5.14), the $S|S$ translation is applicable.

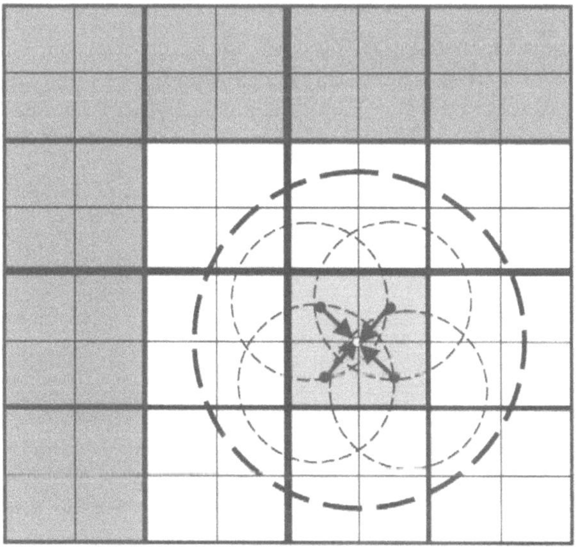

FIGURE 5.15 Step 3 of the FMM upward pass. The S-approximation near the center of each box can be obtained by $S|S$-translations of the S-approximations centered at centers of its children.

5.2.2.2 *Downward pass*

The downward pass applies Steps 1 and 2 below for each of the levels $l = l_{\min}, \ldots, l_{\max}$.

Step 1. In this step we form the representing vectors, $\tilde{\mathbf{D}}^{(n,l)}$, of length P_l for the regular function $\tilde{R}_{P_l}(\mathbf{y}; \mathbf{y}_*^{(n,l)})$ which approximate the function $\psi_{n,l}^{(4)}(\mathbf{y})$ about the center $\mathbf{y}_*^{(n,l)}$ of box $(n,l) \in Y$. To build the local approximation near the center of each box at level l, the coefficients $\mathbf{C}^{(n',l)}, n' \in I_4(n,l) \cap X$ should be $S|R$-translated to the center of the box.

Thus, we have

$$\tilde{\mathbf{D}}^{(n,l)} = \sum_{n' \in I_4(n,l) \cap X} (S|R)(\mathbf{y}_*^{(n,l)} - \mathbf{y}_*^{(n',l)})[\mathbf{C}^{(n',l)}], \qquad (n,l) \in Y,$$
$$[\tilde{R}_{P_l}(\mathbf{y}; \mathbf{y}_*^{(n,l)}) \rightleftarrows \tilde{\mathbf{D}}^{(n,l)}, \quad \forall \mathbf{y} \in E_1(n,l), \quad |\tilde{R}_{P_l}(\mathbf{y}; \mathbf{y}_*^{(n,l)}) - \psi_{n,l}^{(4)}(\mathbf{y})| < \epsilon_l]. \tag{5.2.17}$$

Condition (5.2.12) ensures that the far-to-local translation is applicable. Figure 5.16 illustrates this step for $d = 2$.

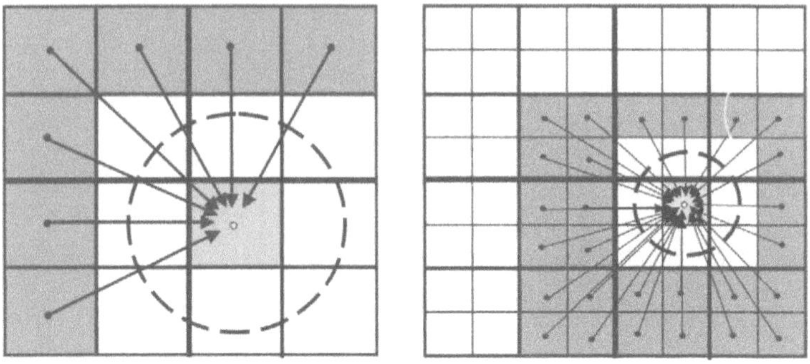

FIGURE 5.16 Step 1 of the downward pass of the FMM. The representing vectors for singular approximation corresponding to dark gray boxes are $S|R$-translated to the center of the light gray box. Figures illustrate this step for quadtree at levels 2 and 3.

Step 2. Assuming that for $l = l_{\min}$

$$\mathbf{D}^{(n,l_{\min})} = \tilde{\mathbf{D}}^{(n,l_{\min})}, \qquad (n, l_{\min}) \in Y, \qquad [\psi^{(3)}_{n,l_{\min}}(\mathbf{y}) = \psi^{(4)}_{n,l_{\min}}(\mathbf{y})], \quad (5.2.18)$$

we form the representing vectors $\mathbf{D}^{(n,l)}$ of length P_l for the regular function $R_{P_l}(\mathbf{y};\mathbf{y}^{(n,l)}_*)$ which approximate the function $\psi^{(3)}_{n,l}(\mathbf{y})$ about the box center $(n, l) \in Y$. We do this by adding $\tilde{\mathbf{D}}^{(n,l)}$ to the vectors obtained by the $R|R$-translation of representing vectors for $\psi^{(3)}_{\mathrm{Parent}(n,l),l-1}(\mathbf{y})$ from the parent box to the center of the child box (n, l):

$$\mathbf{D}^{(n,l)} = \tilde{\mathbf{D}}^{(n,l)} + (R|R)(\mathbf{y}^{(n,l)}_* - \mathbf{y}^{(n',l-1)}_*)[\mathbf{D}^{(n',l-1)}],$$
$$n' = \mathrm{Parent}(n,l), \ (n,l) \in Y, \qquad (5.2.19)$$
$$[R_{P_l}(\mathbf{y};\mathbf{y}^{(n,l)}_*) \rightleftarrows \mathbf{D}^{(n,l)}, \ \forall \mathbf{y} \in E_1(n,l), \ |R_{P_l}(\mathbf{y};\mathbf{y}^{(n,l)}_*) - \psi^{(3)}_{n,l}(\mathbf{y})| < \epsilon_l].$$

For the nth box, this approximation is valid in the domain $E_1(n,l)$ which is a subdomain of $E_1(\mathrm{Parent}(n,l), l-1)$, and local-to-local translation is allowed. Figure 5.17 illustrates this for $d = 2$. Indeed, here the smaller sphere is located completely inside the larger sphere, and union of domains $E_3(n,l)$ and $E_4(n,l+1)$ produces $E_3(n,l+1)$:

$$E_3(n,l+1) = E_3(n,l) \cup E_4(n,l+1). \qquad (5.2.20)$$

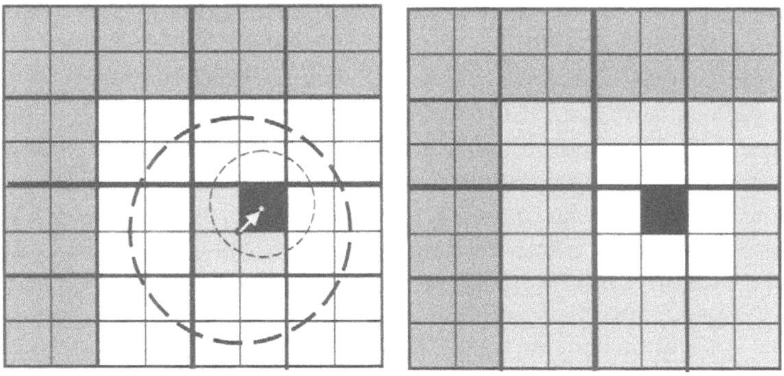

FIGURE 5.17 Step 2 of the downward pass of the FMM. On the left figure the representing vectors of the parent box (light gray) are locally translated to the center of the black box. On the right figure contribution of the light gray boxes is added to the sum of the dark boxes to repeat the structure at the finer hierarchical level.

5.2.2.3 *Final summation*

As soon as coefficients $\mathbf{D}^{(n,l_{\max})}$ are determined, the total sum $\psi(\mathbf{y}_\beta)$ can be computed for any point $\mathbf{y}_\beta \in E_1(0,0)$ using Eq. (5.2.7), where $\psi_{n,l}^{(2)}(\mathbf{y})$ can be computed in a straightforward way, using Eq. (5.2.6). Thus,

$$\psi(\mathbf{y}_\beta) = \psi_{n,l_{\max}}^{(2)}(\mathbf{y}_\beta) + R_{p_{l_{\max}}}(\mathbf{y}_\beta; \mathbf{y}_*^{(n,l_{\max})}), \qquad \mathbf{y}_\beta \in E_1(n, l_{\max}). \quad (5.2.21)$$

5.3 DATA STRUCTURES AND EFFICIENT IMPLEMENTATION

Our goal is to achieve the matrix–vector multiplication, or sum, in Eq. (5.1.1) in $O(N)$ or $O(N \log N)$ operations (assuming $M \sim N$). Accordingly, all methods used to perform indexing and searching for parent, children, and neighbors in the data hierarchies should be consistent with this. It is obvious that methods based on naive traversal algorithms have asymptotic complexity $O(N^2)$ and are not allowed since this would defeat the purpose of the FMM. There is also the issue of memory complexity. If the size of the problem is small enough, then the neighbor search procedures on trees can be $O(1)$ and the complexity of the MLFMM procedure can be $O(N)$. More restrictive memory requirements bring the complexity of such operations to $O(\log N)$ and the complexity of the MLFMM procedure to $O(N \log N)$. Reduction of this complexity to $O(N)$ can be achieved in some cases using hashing (this technique, however, depends on the properties of the source and evaluation data sets and may not always result in savings) [CLR90].

From the preceding discussion, 2^d-tree data structures are rather natural for use with the MLFMM. Since the regions of expansion validity are specified in terms of Euclidean distance, subdivision of space into d-dimensional cubes is convenient for range evaluation between points. We note that in the spatial data-structure literature, the data structures used most often for higher dimensional spaces are $k-d$ trees (e.g. see [Sam90, Sam94]). Such structures could also be employed in the MLFMM, especially for cases when expansions are tensor products of expansions with respect to each coordinate, however, no such attempts have been reported to our knowledge. We also remark that 2^d-tree data structures can be easily generated from $k-d$ data structures, so methods based on $k-d$ trees can be used for the MLFMM. The relative merits of these and other spatial data structures for the FMM remain a subject for investigation.

The main technique for working with 2^d-trees (and $k-d$ trees) is the bit interleaving technique (perhaps, first mentioned by Peano in 1890 [Pea90], for more details see Refs. [Sam90, Sam94]) which we apply in d dimensions. This technique enables $O(1)$ or constant, algorithms for parent and sibling search and $O(\log N)$ algorithms for neighbor and children search. Using the bit interleaving technique the MLFMM setting step will achieve $O(N \log N)$ time complexity in case we wish to minimize the amount of memory used. If we are able to store the occupancy maps for the given data sets we can obtain $O(N)$ complexity.

While these algorithms are well known in the spatial data-structures community, they have not been described in the context of the FMM before, and it is a lack of such a clear exposition that has held back their wider use.

5.3.1 Indexing

To index the boxes in a more efficient way we can, for example, do the following. Each box in the tree can be identified by assigning it a unique index among the 2^d children of its parent box, and by knowing the index of its parent in the set of its grandparent's children, and so on. For reasons that will be clear in the next section we index the 2^d children of a particular parent box using the numbers $0, 1, \ldots, 2^d - 1$. Then the index of a box can be written as the string:

$$\text{String}(n, l) = (N_1, N_2, \ldots, N_l), \qquad N_j = 0, \ldots, 2^d - 1, \; j = 1, \ldots, l, \quad (5.3.1)$$

where l is the level at which the indexed box is located and N_j is the index of a box at level j containing that box. We drop N_0 from the indexing, since it is the only box at level 0, and has no parents. We can assign the index 0 to this box. For example, in two dimensions for the quadtree we have the numbering shown in Fig. 5.18. The smaller black box will have the index string (3,1,2) and the larger black box will have index string (2,3). From the construction it is clear that each box can be described by such a string, and each string uniquely determines a box.

The index string can be converted to a single number as follows:

$$n = (2^d)^{l-1}N_1 + (2^d)^{l-2}N_2 + \cdots + 2^d N_{l-1} + N_l. \tag{5.3.2}$$

Note that this index depends on the level l at which the box is considered and unless this information is included, different boxes could be described by the same index. For example, boxes with strings (0,2,3) and (2,3) map to the same index, $n = 11$, but are different. The box (0,2,3) is the small gray box and (2,3) is the larger black box in Fig. 5.18. The unique index of any box can be represented by the pair:

$$\text{UniversalIndex} = (n, l). \tag{5.3.3}$$

We could instead say "box 11 at level 2" (this is the larger black box in Fig. 5.18) or "box 11 at level 3" (this is the smaller gray box in Fig. 5.18). We also have the box with index 0 at each level, which is located in the left bottom corner. "Box 0 at level 0" refers to the largest box in the 2^d-tree.

The string could be mapped in a different way, so that all boxes map to a unique index, instead of a pair. However, storing the level number does not increase the memory or time complexity, since anyway the

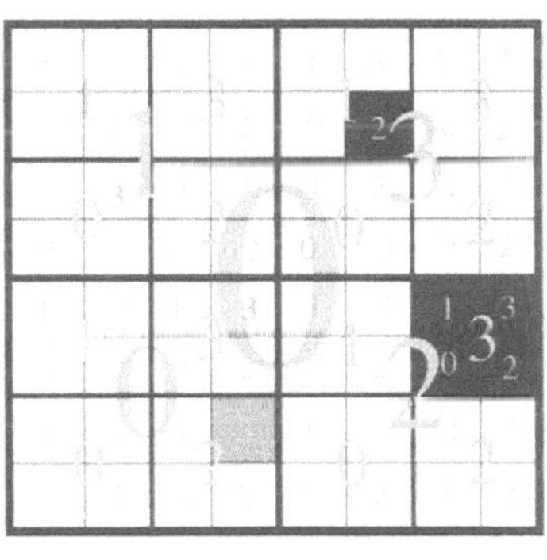

FIGURE 5.18 Hierarchical numbering in quadtree.

MLFMM loops go through the level hierarchy, and one always has the level value.

If the indexing at each level is performed in a consistent way (for example, in Fig. 5.18 we always assign 0 to the child at the bottom left corner of the parent, 1 to the child in the left upper corner, 2 to the child in the right bottom corner, and 3 to the child in the right upper corner; for quadtrees this can also be called "Z order" following [OM84]), then we call such a indexing scheme "hierarchical". A consistent hierarchical scheme has the following desirable properties.

(1) *Determining the parent.* Consider a box at level l of the 2^d-tree, whose index is given by Eq. (5.3.2). The parent of this box is

$$\text{Parent}(n) = (2^d)^{l-2}N_1 + (2^d)^{l-3}N_2 + \cdots + N_{l-1}. \qquad (5.3.4)$$

To obtain this index there is no need to know whether N_1, N_2, and so on are zeros or not. We also do not need to know l, since this index is produced from string simply by dropping the last element:

$$\text{Parent}(N_1, N_2, \ldots, N_{l-1}, N_l) = (N_1, N_2, \ldots, N_{l-1}). \qquad (5.3.5)$$

This means that the function "Parent" in such a numbering system is simple and level independent. For example, at $d = 2$ for box index 11 the parent always will be Parent(11) = 2 independent of the level being considered. Obtaining the parent's index in the universal numbering system in Eq. (5.3.3) is also simple, since the level of the parent is $l - 1$. Therefore,

$$\text{Parent}(n, l) = (\text{Parent}(n), l - 1). \qquad (5.3.6)$$

(2) *Determining the children.* For the function "ChildrenAll" as well we do not need to know the level. Indeed, to get the indices of all 2^d children of a box represented by the string (5.3.1), we need to simply add one more element to the string, which runs from 0 to $2^d - 1$, to list all the children:

$$\text{ChildrenAll}(N_1, N_2, \ldots, N_l) = \{(N_1, N_2, \ldots, N_l, N_{l+1})\},$$
$$N_{l+1} = 0, \ldots, 2^d - 1, \qquad (5.3.7)$$

or

$$\text{ChildrenAll}(n) = \{(2^d)^l N_1 + (2^d)^{l-1} N_2 + \cdots + (2^d) N_l + N_{l+1}\},$$
$$N_{l+1} = 0, \ldots, 2^d - 1. \qquad (5.3.8)$$

For the universal numbering system (5.3.3), the operation of finding the children is simply the calculation of the children numbers and assigning their level to $l+1$:

$$\mathrm{ChildrenAll}(n, l) = (\mathrm{ChildrenAll}(n), l+1). \tag{5.3.9}$$

Note that

$$\mathrm{Parent}(n) = [n/2^d], \tag{5.3.10}$$

$$\mathrm{ChildrenAll}(n) = \{2^d n + j\}, \qquad j = 0, \ldots, 2^d - 1, \tag{5.3.11}$$

where [ ] means integer part.

The use of 2^d-trees makes obtaining parent and children indices very convenient. Indeed, the above operations are nothing but shift operations in the *bit representation* of n. Performing a right bit-shift operation on n by d bits, one can obtain the index of the parent. One can list all indices of the children boxes of n by a left bit-shift operation on n by d bits and adding all possible combinations of d bits.

5.3.2 Spatial ordering

The above method of indexing provides a simple and natural way for representing a 2^d-tree graph structure and easy $O(1)$ algorithms to determine parent–children (and therefore sibling) relationships. However, we still do not have a way for determining neighbors. Further the MLFMM algorithm requires finding of the box center for a given index (n, l) and the box index to which a given spatial point $\mathbf{x}$ belongs. To do this a spatial ordering in d-dimensional space should be introduced. We provide below such an ordering and $O(1)$ algorithms for these operations.

5.3.2.1 *Scaling*

As assumed above, the part of the d-dimensional space we are interested in can be enclosed within a bounding box with dimensions $L_1 \times L_2 \times \cdots \times L_d$. In problems in physical space of dimensions ($d = 1$, 2, 3) we usually have isotropy of directions and can enclose that box in a cube of size $D_0 \times \cdots \times D_0$, where

$$D_0 = \max_d L_d, \tag{5.3.12}$$

with one corner assigned the minimum values of Cartesian coordinates:

$$\mathbf{x}_{\min} = (x_{1,\min}, \ldots, x_{d,\min}). \tag{5.3.13}$$

This cube can then be mapped to the unit cube $[0,1]\times\cdots\times[0,1]$ by a shift of the origin and scaling:

$$\bar{\mathbf{x}} = \frac{\mathbf{x}-\mathbf{x}_{\min}}{D_0}, \qquad (5.3.14)$$

where $\mathbf{x}$ are the true Cartesian coordinates of any point in the cube and $\bar{\mathbf{x}}$ are normalized coordinates of the point.

In the following, we will work only with the unit cube, assuming that, if necessary, such a scaling has already been performed, and that the point $\mathbf{x}$ in the original d-dimensional space can be found given $\bar{\mathbf{x}} \in [0,1]\times\cdots\times[0,1]$.

5.3.2.2 *Ordering in one dimension (binary ordering)*

Let us first consider the case $d=1$, where our 2^d-tree becomes a binary tree (see Fig. 5.13). In the one-dimensional case all the points $\bar{x} \in [0,1]$ are naturally ordered and can be represented in the decimal system as

$$\bar{x} = (0.a_1a_2a_3\ldots)_{10}, \qquad a_j = 0,\ldots,9, \;\; j=1,2,\ldots. \qquad (5.3.15)$$

Note that the point $\bar{x}=1$ can also be written as

$$\bar{x} = 1 = (0.999999\ldots)_{10}, \qquad (5.3.16)$$

which we consider to be two equivalent representations. The latter representation reflects the fact that $\bar{x}=1$ is a limiting point of sequence $0, 0.9, 0.99, 0.999, \ldots$.

We can also represent any point $\bar{x} \in [0,1]$ in the binary system as

$$\bar{x} = (0.b_1b_2b_3\ldots)_2, \qquad b_j = 0,1, \;\; j=1,2,\ldots, \qquad (5.3.17)$$

and we can write the point $\bar{x}=1$ in the binary system as

$$\bar{x} = 1 = (0.111111\ldots)_2. \qquad (5.3.18)$$

Even though the introduced indexing system for the boxes in the case $d=1$ results in a rather trivial result, since all the boxes are already ordered by their indices at a given level l from 0 to 2^{l-1}, and there is a straightforward correspondence between box indices and coordinates of points, we still consider the derivation of the neighbor, parent and children and other relationships in detail, so as to conveniently extend them to the general d-dimensional case.

Finding the index of the box containing a given point

Consider the relation between the coordinate of a point and the index of the box where the point is located. We note that the size of a box at each level is 1 placed at the position equal to the level number after the decimal in its binary record as shown in the table below.

Level	Box size (dec)	Box size (bin)
0	1	1
1	0.5	0.1
2	0.25	0.01
3	0.125	0.001
⋮	⋮	⋮

If we consider level 1 where there are two boxes:

$$(0.0b_1b_2b_3\ldots)_2 \in \mathrm{Box}((0)), \qquad (0.1b_1b_2b_3\ldots)_2 \in \mathrm{Box}((1)),$$
$$\forall b_j = 0, 1, \quad j = 1, 2, \ldots, \tag{5.3.19}$$

where Box((0)) and Box((1)) denote sets of spatial points that belong to boxes with indices (0) and (1), respectively (at this point we will use binary strings for indexing). At level 2 we will have

$$(0.00b_1b_2b_3\ldots)_2 \in \mathrm{Box}((0,0)), \qquad (0.01b_1b_2b_3\ldots)_2 \in \mathrm{Box}((0,1)),$$
$$(0.10b_1b_2b_3\ldots)_2 \in \mathrm{Box}((1,0)), \qquad (0.11b_1b_2b_3\ldots)_2 \in \mathrm{Box}((1,1)), \tag{5.3.20}$$
$$\forall b_j = 0, 1, \quad j = 1, 2, \ldots.$$

This process can be continued. At the lth level we obtain

$$(0.N_1, N_2\ldots N_l b_1b_2b_3\ldots)_2 \in \mathrm{Box}((N_1, N_2, \ldots, N_l)),$$
$$\forall b_j = 0, 1, \quad j = 1, 2, \ldots. \tag{5.3.21}$$

Therefore, to find the index of the box at level l to which the given point belongs, we need to simply shift the binary number representing this point by l positions and take the integer part of this number:

$$(0.N_1N_2\ldots N_l b_1b_2b_3\ldots)_2 \rightarrow (N_1N_2\ldots N_l.b_1b_2b_3\ldots)_2,$$
$$N_1N_2\ldots N_l = [(N_1N_2\ldots N_l.b_1b_2b_3\ldots)_2]. \tag{5.3.22}$$

This procedure can also be written as

$$(n, l) = [2^l \bar{x}]. \tag{5.3.23}$$

Finding the center of a given box

The relation between the coordinate of the point and the box index can also be used to find the coordinate of the center for given box index. Indeed, if the box index is $N_1N_2...N_l$ then at level l we use l bit shift to obtain

$$n = (N_1N_2...N_l)_2 \rightarrow (0.N_1N_2...N_l)_2,$$

then we add 1 as extra digit, so we have for the center of the box at level l:

$$\bar{x}_c(n, l) = (0.N_1N_2...N_l1)_2. \tag{5.3.24}$$

Indeed, any point with coordinates $(0.N_1N_2...N_l)_2 \leq \bar{x} \leq (0.N_1N_2...N_l111111...)_2$ belongs to this box. This procedure can also be written in the form:

$$\bar{x}_c(n, l) = 2^{-l}(n + 2^{-1}), \tag{5.3.25}$$

since addition of one at position $l + 1$ after the point in the binary system is the same as addition of 2^{-l-1}.

Finding neighbors

In a binary tree each box has two neighbors, except the boxes adjacent to the boundaries. Since at level l all boxes are ordered, we can find indices for all neighbors using the function:

$$\text{NeighborAll}(n, l) = \{(n - 1, l), (n + 1, l)\}. \tag{5.3.26}$$

For the binary tree the neighbors have indices that are ± 1 of the given box index. If the neighbor index at level l computes to a value larger than $2^l - 1$ or smaller than 0, we drop this box from the neighbor list.

5.3.2.3 Ordering in d dimensions

Coordinates of a point $\bar{\mathbf{x}} = (\bar{x}_1, \ldots, \bar{x}_d)$ in the d-dimensional unit cube can be represented in the binary form:

$$\bar{x}_m = (0.b_{m1}b_{m2}b_{m3}\ldots)_2, \qquad b_{mj} = 0, 1, \;\; j = 1, 2, \ldots, \;\; m = 1, \ldots, d. \tag{5.3.27}$$

Instead of having d indices characterizing each point, we can form a single binary index that represents the same point by an ordered mixing

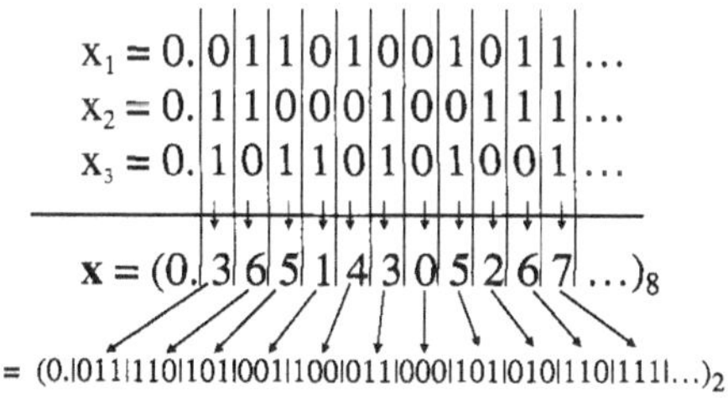

FIGURE 5.19 Example of converting of coordinates in three dimensions to a single octal or binary number.

of the digits in the above binary representation (this is also called *bit interleaving*), so we can write:

$$\bar{\mathbf{x}} = (0.b_{11}b_{21}\ldots b_{d1}b_{12}b_{22}\ldots b_{d2}\ldots b_{1j}b_{2j}\ldots b_{dj}\ldots)_2. \tag{5.3.28}$$

This can be rewritten in the system with base 2^d:

$$\bar{\mathbf{x}} = (0.N_1 N_2 N_3 \ldots N_j \ldots)_{2^d},$$
$$N_j = (b_{1j}b_{2j}\ldots b_{dj})_2, \quad j = 1, 2, \ldots, \quad N_j = 0, \ldots, 2^d - 1. \tag{5.3.29}$$

An example of converting three-dimensional coordinates to octal and binary indices is shown in Fig. 5.19.

Finding the index of the box containing a given point

Consider a relation between the coordinate of point, which is now a single number $x \in [0, 1]$ and the index of the box in the 2^d-tree where this point is located. We use the convention of ordered indexing of boxes in the hierarchical structure. 2^d children of any box will be indexed according to coordinate order. Since the children boxes are obtained by division of each side of the parent box in 2, we assign 0 to the box with the smaller center coordinate and 1 to the other box. In d dimensions, 2^d combinations are produced by d binary coordinates. So any set of d coordinates can be interpreted as a binary string, which can then be converted to a single index in the binary or some counting system, e.g. with the base 2^d:

$$(b_1, b_2, \ldots, b_d) \rightarrow (b_1 b_2 \ldots b_d)_2 = N_{2^d}. \tag{5.3.30}$$

Examples of such an ordering for $d = 2$ and 3 are shown in Fig. 5.20.

Obviously, such an ordering is consistent for all levels of the hierarchical structure, since it can be performed for children of each

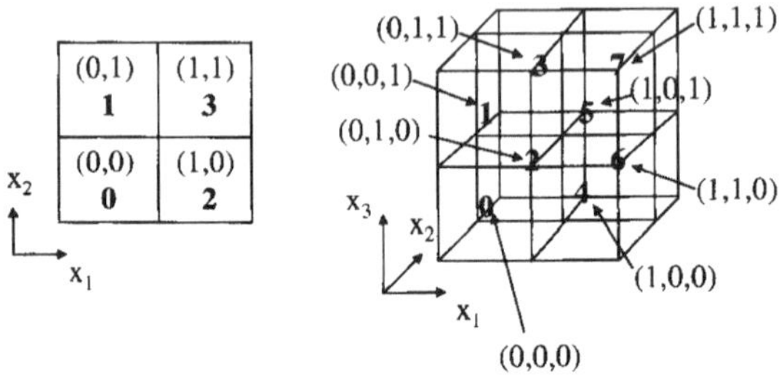

FIGURE 5.20 Ordering of children boxes in quadtree and in octree.

box. Therefore, the functions "Parent" and "ChildrenAll" introduced above can be used, and they are not level dependent.

Now we can show that the box index that contains a given spatial point can be found using the same method as for the binary tree with slight modification. The size of the boxes at each level is nothing but 1 placed at the position equal to the level number after the point in its binary record as shown in the table for binary tree. At level 1, where we have 2^d boxes, the binary record determines

$$\bar{\mathbf{x}} = (0.b_{11}b_{21}\ldots b_{d1}b_{12}b_{22}\ldots b_{d2}\ldots b_{1j}b_{2j}\ldots b_{dj}\ldots)_2 \in \text{Box}((b_{11}b_{21}\ldots b_{d1})_2)$$
$$= \text{Box}((N_1)_{2^d}). \quad (5.3.31)$$

Indeed, for each coordinate the first digit only determines the box at level 1, which is exactly equal to the mixed coordinate by the convention of ordered indexing of children. At level 2 the same happens with the second digit of each coordinate. At level l we have using a 2^d-based system and string representation of the box index:

$$(0.N_1N_2\ldots N_lc_1c_2c_3\ldots)_{2^d} \in \text{Box}((N_1, N_2, \ldots, N_l)_{2^d}),$$
$$\forall c_j = 0, \ldots, 2^d - 1, \;\; j = 1, 2, \ldots. \quad (5.3.32)$$

Therefore, to find the index of the box at level l, to which the given point belongs, we need to shift the 2^d-index representing this point by l positions and take the integer part of this index:

$$(0.N_1N_2\ldots N_lc_1c_2c_3\ldots)_{2^d} \rightarrow (N_1N_2\ldots N_l.c_1c_2c_3\ldots)_{2^d},$$
$$N_1N_2\ldots N_l = [(N_1N_2\ldots N_l.b_1b_2b_3\ldots)_{2^d}]. \quad (5.3.33)$$

This procedure can also be performed in the binary system by a dl left bit shift:

$$(0.b_{11}b_{21}\ldots b_{d1}b_{12}b_{22}\ldots b_{d2}\ldots b_{1l}b_{2l}\ldots b_{dl}b\ldots)_2$$
$$\rightarrow (b_{11}b_{21}\ldots b_{d1}b_{12}b_{22}\ldots b_{d2}\ldots b_{1l}b_{2l}\ldots b_{dl}.b\ldots)_2, \tag{5.3.34}$$
$$\text{index} = (b_{11}b_{21}\ldots b_{d1}b_{12}b_{22}\ldots b_{d2}\ldots b_{1l}b_{2l}\ldots b_{dl})_2.$$

In another counting system this can be obtained by multiplication of the coordinate of the point by 2^{dl} and taking the integer part. So

$$(n, l) = [2^{dl}\bar{\mathbf{x}}]. \tag{5.3.35}$$

This example shows that in contrast to the one-dimensional case, in the d-dimensional case, the advantage in conversion of the index to binary form is substantial, since this enables use of bit interleaving to produce a single multi-index consistent with the 2^d-tree data structure. This procedure is natural in computations, since, anyway all indices are represented finally in binary format. So the algorithm of finding the box index for a given spatial point is $O(1)$ for any dimension. We also note that the algorithm does not require conversion to octal or other 2^d-based system, as a binary representation of coordinates and bit-shift procedures are available.

Finding the center of a given box

The relation between the coordinate of a point and its box index enables easy finding of the coordinates of the center for a given box index. To do this we first convert the box index at level l into binary form:

$$n = (b_{11}b_{21}\ldots b_{d1}b_{12}b_{22}\ldots b_{d2}\ldots b_{1l}b_{2l}\ldots b_{dl})_2. \tag{5.3.36}$$

Then we decompose this index to d coordinate indices (this is also called *bit deinterleaving*):

$$n_1 = (b_{11}b_{12}\ldots b_{1l})_2,$$
$$n_2 = (b_{21}b_{22}\ldots b_{2l})_2,$$
$$\vdots \tag{5.3.37}$$
$$n_d = (b_{d1}b_{d2}\ldots b_{dl})_2.$$

This is a simple operation since the bit string (5.3.36) should be rewritten in the form of a matrix $d \times l$ column by column. Note that because some values of b can be zero we need to check the length of the bit string n and complete it by adding zeros before the first non-zero b to achieve a length dl, or we can simply fill the matrix starting with the last element b_{dl} then putting $b_{d-1,l}$ in the same column above b_{dl} and so on.

Further conversion from the indices to the coordinate values is similar to the procedures in binary tree (Eqs. (5.3.24) and (5.3.25)). The coordinates of the box center in binary form are

$$\bar{x}_m(n,l) = (0.b_{m1}b_{m2}\ldots b_{ml}1)_2, \qquad m = 1,\ldots,d, \tag{5.3.38}$$

or in a form that does not depend on the counting system:

$$\bar{x}_m(n,l) = 2^{-l}(n_l + \tfrac{1}{2}), \qquad l = 1,\ldots,d. \tag{5.3.39}$$

Finding the neighbors of a given box

The procedure for finding neighbors of a given box at level l in the 2^d-tree with ordered hierarchical indexing can be reduced to the procedure of finding neighbors with respect to each dimension. This procedure is described above for binary tree. In the general case we need to modify it just slightly. First we perform bit deinterleaving according to Eqs. (5.3.36) and (5.3.37). Then for each coordinate index we generate the indices:

$$n_m^+ = n_m + 1, \qquad n_m^- = n_m - 1, \qquad m = 1,\ldots,d, \tag{5.3.40}$$

and check if any of these indices is out of range $[0, 2^l - 1]$. If so, the index should be dropped from the list of indices generating the neighbor list. So for each dimension m of d-dimensional space we will have a set, Neighbor_m, consisting of 3 or 2 (if one of the indices $\text{index}_m^\pm$ is dropped) indices:

$$\text{Neighbor}_m = \begin{cases} \{n_m^-, n_m, n_m^+\}, & n_m \neq 0, 2^l - 1, \\ \{n_m, n_m^+\}, & n_m = 0, \\ \{n_m^-, n_m\}, & n_m = 2^l - 1, \end{cases} \qquad m = 1,\ldots,d. \tag{5.3.41}$$

The set of neighbor generating indices is then

$$\nu = (\nu_1,\ldots,\nu_d), \qquad \nu_m \in \text{Neighbor}_m, \qquad m = 1,\ldots,d. \tag{5.3.42}$$

where each ν_m can be any element of Neighbor_m (Eq. (5.3.41)), except for the case when all $\nu_m = n_m$ simultaneously for all $q = 1,\ldots,d$, since this case corresponds to the box itself. For a box situated far from the boundary of the domain we therefore have $3^d - 1$ possible combinations of $(\nu_1,\ldots,\nu_d)$, and each of them corresponds to a neighbor.

Note that n_m is obtained from the bit deinterleaving procedure in binary form. Thus, the operations of finding $n_m^{\pm}$ are also convenient to perform in the binary form to obtain each n_m, $m = 1, \ldots, d$, in the binary format. This yields

$$\nu_1 = (\nu_{11}\nu_{12}\ldots\nu_{1l})_2, \nu_2 = (\nu_{21}\nu_{22}\ldots\nu_{2l})_2, \ldots, \nu_d = (\nu_{d1}\nu_{d2}\ldots\nu_{dl})_2, \quad (5.3.43)$$

where $\nu_{mj} = 0, 1$ are the bits of ν_m. The interleaved bit strings produce the neighbor indices:

$$\begin{aligned} &\text{NeighborAll}(n, l) = \{(\nu_{11}\nu_{21}\ldots\nu_{d1}\nu_{12}\nu_{22}\ldots\nu_{d2}\ldots\nu_{1l}\nu_{2l}\ldots\nu_{dl})_2\}, \\ &\nu_m \in \text{Neighbor}_m, \qquad m = 1, \ldots, d. \end{aligned} \quad (5.3.44)$$

Note that the lengths of bit strings $(\nu_{m1}\nu_{m2}\ldots\nu_{ml})_2$ for different m can be different because the several first bits can be zero, $\nu_{m1} = 0$, $\nu_{m2} = 0, \ldots$. In this case either each string should be completed with zeros to length l, or the formation of the neighbor index can start from the last digit ν_{dl} assuming that 0 corresponds to the absent bits.

5.3.3 Structuring data sets

The 2^d-tree hierarchical space subdivision enables organization of infinite (continuum) data sets. However, in practice even large data sets are finite. Scaling and mapping finite d-dimensional data sets into a unit d-dimensional cube yield a set $\mathbb{W}$ of N *different* points inside the cube:

$$\begin{aligned} &\mathbb{W} = \{\bar{\mathbf{x}}_1, \bar{\mathbf{x}}_2, \ldots, \bar{\mathbf{x}}_N\}, \\ &\bar{\mathbf{x}}_\alpha \in (0, 1) \times \cdots \times (0, 1) \subset \mathbb{R}^d, \ \ \bar{\mathbf{x}}_\alpha \neq \bar{\mathbf{x}}_\beta, \ \ \alpha \neq \beta, \ \ \alpha, \beta = 1, \ldots, N, \end{aligned} \quad (5.3.45)$$

where the scaling can be made in such way that no data points are located on the cube boundaries.

Because the number of points we consider is finite, there always exists a level of space subdivision L_1, at which some boxes do not contain points from $\mathbb{W}$. Indeed, for $2^{Ld} > N$, or

$$L_1 > \frac{1}{d}\log_2 N, \quad (5.3.46)$$

the number of boxes is larger then the number of points. There also exists a finite L_2, such that at levels $L_2, L_2 + 1, L_2 + 2, \ldots$, all boxes will contain not more than one point. This is easy to prove, since if we consider the minimum distance between the points from $\mathbb{W}$:

$$\bar{D}_{\min} = \min_{\alpha \neq \beta} |\bar{\mathbf{x}}_\alpha - \bar{\mathbf{x}}_\beta|, \qquad \alpha, \beta = 1, \ldots, N, \quad (5.3.47)$$

where $|\bar{\mathbf{x}}_\alpha - \bar{\mathbf{x}}_\beta|$ is the Euclidean distance, then L_2 can be determined from the requirement that the main diagonal of the box at this level $d^{1/2}2^{-L_2}$ be smaller than $\bar{D}_{\min}$, or

$$L_2 > \log_2 \frac{d^{1/2}}{\bar{D}_{\min}}. \tag{5.3.48}$$

At some $l_{\max}$ we will have a situation, where each box at such a level will contain not more than s data points $(1 \leq s \leq N)$, while at level $l = l_{\max} - 1$ there exists at least one box containing more than s data points (assuming that the total number of data points $N \geq 2$). We call s the *grouping* or *clustering parameter*, and $l_{\max}$ as the *threshold level* and will provide $O(N)$ algorithm for its determination. Note that for the MLFMM procedure with q neighborhoods another clustering parameter, q, might be more appropriate for determination of $l_{\max}$. This parameter is the number of source points in the neighborhood of the evaluation point. So at level $l = l_{\max} - 1$ there exists at least one box containing an evaluation point, whose neighborhood contains more than q sources, while at level $l = l_{\max}$ there are no such boxes. Determination of q requires both the data sets and a more complicated procedure than determination of s, while it can be performed for $O(N \log N)$ operations.

5.3.3.1 *Ordering of d-dimensional data*

We can introduce a "correspondence vector", which is a vector of length N where the αth component is equal to the index of the box at level $l \geq l_{\max}$ of the 2^d-tree. We denote this vector as $\mathbf{v}$:

$$\begin{aligned} \mathbf{v} &= (v_1, v_2, \ldots, v_N), \qquad v_\alpha = \mathrm{Index}(\bar{\mathbf{x}}_\alpha, l), \\ \alpha &= 1, \ldots, N, \quad l \geq l_{\max}, \end{aligned} \tag{5.3.49}$$

where "Index" can be determined using the bit interleaving technique. The array $\mathbf{v}$ can then be sorted in non-descending order as

$$(v_1, v_2, \ldots, v_N) \to (v_{\alpha_1}, v_{\alpha_2}, \ldots, v_{\alpha_N}), \qquad v_{\alpha_1} \leq v_{\alpha_2} \leq \cdots \leq v_{\alpha_N}. \tag{5.3.50}$$

Such a sorting requires $O(N \log N)$ operations using standard sorting algorithms and provides the permutation index (or "permutation vector" or "pointer vector") of length N:

$$\mathrm{ind} = (\alpha_1, \alpha_2, \ldots, \alpha_N) \tag{5.3.51}$$

that can be stored in the memory. To save memory the array $\mathbf{v}$ should not be rewritten and stored again, since "ind" is a pointer and

$$\mathbf{v}(\alpha) = v_\alpha, \qquad \text{ind}(j) = \alpha_j, \qquad \mathbf{v}(\text{ind}(j)) = \mathbf{v}(\alpha_j) = v_{\alpha_j}, \\ \alpha, j = 1, \ldots, N, \tag{5.3.52}$$

so that

$$\mathbf{v}(\text{ind}) = (v_{\alpha_1}, v_{\alpha_2}, \ldots, v_{\alpha_N}). \tag{5.3.53}$$

At level $l \geq l_{\text{max}}$ there may exist $\alpha \neq j$ such that $v_\alpha = v_j$ and the order of these elements in the sorted list can be arbitrary. We will fix this order once for all time, in other words, we assume that a permutation index exists and does not change even though two subsequent elements in the list can be identical.

To machine precision each coordinate of the data point is represented with Bit_{max} bits. This means that there is no sense in using more than Bit_{max} levels of space subdivision—if two points have identical d coordinates in terms of Bit_{max}-truncation then they can be considered as identical. We assume that $l_{\text{max}} \leq \text{Bit}_{\text{max}}$. Note that the operation "Parent" in the present hierarchical indexing system preserves the non-descending order, so once data points are sorted at the maximum resolution level Bit_{max} and the permutation index is fixed, this operation should not be repeated and can be performed once *before* the level l_{max} for a given set is determined.

5.3.3.2 *Determination of the threshold level*

To determine the threshold level l_{max}, for example, the following $O(N)$ algorithm can be used:

```
α = 0, m = s, while m < n
    α = α + 1, m = m + 1;
    a = Interleaved(v(ind(α)));
    b = Interleaved(v(ind(m)));
    j = Bit_max + 1
    while a ≠ b
        j = j − 1;
        a = Parent(a);
        b = Parent(b);
        l_max = max(l_max, j);
    end;
end;
```

The idea of this algorithm is rather simple and it exploits the fact that the array $\{\text{Index}(\bar{\mathbf{x}}_{\text{ind}(\alpha)}, \text{Bit}_{\max}), \alpha = 1, \ldots, N\}$ is sorted (ordered). At level $l_{\max}$ only s subsequent data points may have the same bit strings. The level independent operation "Parent" can be performed several times to find the level at which two points differ.

5.3.3.3 *Search procedures and operations on point sets*

We also assume that some standard functions for working with the sets, such as the difference of two sets, $\mathbb{C} = \mathbb{A} \backslash \mathbb{B}$, intersection, $\mathbb{C} = \mathbb{A} \cap \mathbb{B}$, and union, $\mathbb{C} = \mathbb{A} \cup \mathbb{B}$ are available as library programs. Note that for ordered sets such procedures are much faster than for arbitrary sets since they do not require a step for sorting each set preceding an operation on that set. As a result of the initial data sorting, we also have fast standard search procedures in sorted lists with complexity $O(\log N)$.

We also mention that the complexity of the set intersection procedure of a small set of power p_w and large set of power P_w is $O(p_w \log P_w)$, since one can look for each element of the smaller set in the larger set, and such search has $O(\log P_w)$ complexity. This yields in $O(\log N)$ complexity for Neighbors$(W; n, l)$ and Children$(W; n, l)$ procedures, for given W data hierarchy ($W = X, Y$). Indeed, we have

$$\begin{aligned} \text{Neighbors}(W; n, l) &= \text{NeighborsAll}(n, l) \cap W, \qquad W = X, Y, \\ \text{Children}(W; n, l) &= \text{ChildrenAll}(n, l) \cap W, \qquad W = X, Y, \end{aligned} \tag{5.3.54}$$

and because the procedures NeighborsAll(n, l) and ChildrenAll(n, l) have $O(1)$ complexity and return $O(1)$ elements, the complexity of the above operations is $O(\log N)$ for $W = X$, and $O(\log M)$ for $W = Y$. As was mentioned earlier, if the memory is sufficient for the problem to be solved, the search procedure can be $O(1)$, which makes the functions Neighbors$(W; n, l)$ and Children$(W; n, l)$, $O(1)$ procedures.

5.4 SUMMARY

In this chapter we considered basic ideas of the fast multipole method. We identified several methods such as the "Middleman", "modified Middleman (with space partitioning)", "SLFMM", and "MLFMM". We evaluated roughly the complexities of these methods and found that they depend on the problems being solved and on the cost of translation methods used. Then we discussed function representation issues and described the SLFMM and MLFMM algorithms in more or less formal, problem (kernel) independent terms. This makes these algorithms applicable to solution of various problems in d dimensions. Further we discussed data structures, and how to build fast procedures for children and neighbor search that are based on spatial ordering and bit interleaving.

CHAPTER 6

Complexity and Optimization of the MLFMM

Strictly speaking, the complexity of the MLFMM is at least $O(N \log N)$ if the initial step for the ordering of spatial data is required. However, in practice the step that must be repeated over and over is not the step setting the data structure, but the MLFMM core procedure. Further, the latter step is usually much more costly and despite it potentially being performed for $O(N + M)$ operations, the constant in this asymptotic estimate is normally much larger than $\log N$ for practical values of N. Other issues of complexity of a real algorithm are related to dependence of the error on the length of representing vectors and to dependences of the search procedures on N. To obtain an understanding of the algorithm complexity, and as a basis for the development of optimization schemes, we consider the complexity of each step of the MLFMM.

6.1 MODEL FOR LEVEL-DEPENDENT TRANSLATION PARAMETERS

To proceed further with complexity evaluations, we need to accept some *models* of behavior of the length of the representing vectors P_l as functions of l and for the $S|S$, $S|R$, and $R|R$ translation costs, $\mathrm{CostTrans}(P_l)$. The model described below is appropriate for real (positive) k, while one can replace k with its positive real part for the case of complex k with $\Re(k) > 0$. For purely imaginary k, which appears, e.g. in the Klein–Gordan equation, the solutions decay exponentially at infinity, and this case can be covered with a simpler model, when P does not depend on the level of space subdivision. This is also the case for "low frequency" limit of the complexity of the algorithm. As mentioned earlier, there is also the one special linearly unstable case ($k_r > 0, k_i < 0$) with explosive growth of solutions. The results of this chapter are not applicable to this case, and the reader interested in this case must develop the theory for it perhaps following the methods presented here.

We assume that these costs for different types of translations are of the same order. The latter quantity can be modeled as

$$\text{CostTrans}(P_l) = CP_l^{\nu}, \qquad \nu \geq 1. \tag{6.1.1}$$

Indeed $\nu = 1$ represents the lowest possible translation cost, when $O(1)$ operations are spent to obtain each component of the translated vector. While translation algorithms can have complexity $O(P_l \log P_l)$, their cost can be overestimated with function (6.1.1) with some ν close to 1. For translation algorithms with polynomial complexity, ν can be any real number, $\nu \geq 1$ (particularly for translations of p truncated sums of multipole expansions $P_l = p_l^2$, and for $O(p^4)$ and $O(p^3)$ algorithms we have $\nu = 2$ and 1.5, respectively). Further, we use the terminology "translation exponent" to denote ν.

For the dependence of P_l on l we also accept a model based on the ideas of spatial scaling discussed above (see Eqs. (5.1.44), (5.2.8), (5.2.13) and associated discussion). These dependences are heavily influenced by the wavenumber of the problem considered. The character of dependences is also different for low and high frequencies (or wavenumbers). As a model we can introduce some breakdown level l_{b*}, determined by the given wavenumber k, and some dimensionless threshold κ_b:

$$kD_{l_{b*}} = \kappa_b, \tag{6.1.2}$$

such that for course levels we can use the "high frequency" model,

$$P_l = A(kD_l)^{d-1}, \qquad l \leq l_{b*}, \ d = 2, 3, \tag{6.1.3}$$

and for fine levels we can use the "low frequency" model,

$$P_l = A\kappa_b^{d-1} = P_*, \qquad l \geq l_{b*}, \ d = 2, 3. \tag{6.1.4}$$

Here P_*, l_{b*}, and P_l are integers and we should understand that we take integer part of these expressions. The power $d - 1$ in Eq. (6.1.3) is dictated by estimations based on approximation of solutions with a series of multipoles, where at high frequencies one can assume $p \sim ka$ (this is the principal term of asymptotics of the truncation number, see Chapter 9). Since the total number of terms in the expansion in the two-dimensional case is $P = p$, and in the three-dimensional case is $P = p^2$, we can write this as $P = p^{d-1}$, $d = 2, 3$. The constant of the model A depends on the specified error ϵ and the size of the neighborhood, or the parameter η_{SR} (Eq. (5.2.12)). This constant is of order of 1 if high frequency asymptotics $p \sim ka$ hold. This model says that for finer levels (if they are present) P_l

does not depend on l, while for coarse levels (if present) it grows as some power of the box size.

Equations (6.2.8) and (6.1.2) yield the following relation between the breakdown level and parameters of the problem:

$$l_b = \begin{cases} l_{min} = 2 & l_{b*} \leq 2 \\ l_{b*}, & l_{min} < l_{b*} < l_{max}, \\ l_{max}, & l_{b*} \geq l_{max} \end{cases} \quad l_{b*} = \left[\log_2 \frac{kD_0}{\kappa_b}\right]. \tag{6.1.5}$$

Also we can determine for the dependence of P_l on l explicitly:

$$P_l = \begin{cases} P_{min} 2^{(d-1)(l_b - l)}, & l \leq l_b \\ P_{min}, & l \geq l_b \end{cases}, \tag{6.1.6}$$

$$P_{min} = A(kD_{l_b})^{d-1} = P_* 2^{(d-1)(l_{b*} - l_b)}, \quad d = 2, 3,$$

where P_{min} is the minimum size of the representing vector which provides the required accuracy at the finest level and D_{l_b} is the size of boxes at level l_b defined according to Eq. (6.1.5). With this equation the translation cost (6.1.1) can be rewritten as

$$\text{CostTrans}(P_l) = C_{min} \begin{cases} 2^{\nu(l_b - l)(d-1)}, & l \leq l_b \\ 1, & l \geq l_b \end{cases}, \tag{6.1.7}$$

$$C_{min} = CP^{\nu}_{min}, \quad d = 2, 3.$$

6.2 SPATIALLY UNIFORM DATA

The actual number of operations in the FMM depends on the particular distributions of the sources and evaluation points. The distribution can be more or less uniform (for example, if we solve the problem of scattering from N scatterers distributed in the volume or problems for the non-uniform Helmholtz equation with sources and targets distributed in volume) or substantially non-uniform (sources distributed over a surface, while the evaluation points can be either on the surface (integral equations) or in the volume). To evaluate the number of operations, we consider "the worst" case when at the finest level of subdivision l_{max} each box contains s source points. Then, there will be no empty boxes and we have no savings by skipping such boxes. In this case the number of boxes at the l_{max}th level will be $2^{l_{max}d}$ and since each box contains s sources, the total number of sources will be

$$N = 2^{l_{\max}d}s. \tag{6.2.1}$$

We also assume that M is large enough and the target points are distributed more or less evenly, so we need to build local expansions for each box at the finest level. Let us analyze the MLFMM procedure described above step by step.

6.2.1 Upward pass

6.2.1.1 *Step 1*

Since the length of the representing vectors $\mathbf{B}^{(\alpha)}(\mathbf{x}_*^{(n,l_{\max})})$ is $P_{\min}$, the cost of obtaining each such vector can be referred as $O(P_{\min})$ and often this can be done analytically (e.g. we have analytical expressions for expansion of monopole sources into a sum of multipoles centered at a different point). However, in some cases such as those involving integral transforms, the representing vector is the vector of coefficients of the function image, and can require P_*^2 operations. Fast transforms of functions sampled at $P_{\min}$ points to $P_{\min}$ coefficients (such as FFT or fast spherical transform) can require $O(P_{\min} \log P_{\min})$ or $O(P_{\min} \log^2 P_{\min})$ operations. So in general, the computational cost for the generation of representing vectors can be denoted as $\mathrm{CostRep}(P_{\min})$ and the cost of generation of the representing vectors for all source points constitutes the cost of the first step of the upward pass:

$$\mathrm{CostUpward}_1 = N\,\mathrm{CostRep}(P_{\min}). \tag{6.2.2}$$

6.2.1.2 *Step 2*

The cost of the second step of the upward pass is related to the complexity of the consolidation operation (5.1.66). Since we confine ourselves to linear operations, this cost is the cost of summation, which usually is dropped from the count of operations compared to the much more costly multiplication operations. Everywhere below we consider such a simplified case, which neglects the cost of consolidation compared with other operations, such as translations and function evaluations and representations. So we set

$$\mathrm{CostUpward}_2 = 0. \tag{6.2.3}$$

6.2.1.3 *Step 3*

Let $\mathrm{CostTranslation}(P_l)$ be the computational complexity of a single $S|S$ translation of the representing vector of length P_l to a representing vector of length P_{l-1}. Since each box contains 2^d children and there are 2^{ld} boxes at level l, the number of operations to translate all vectors $\mathbf{C}^{(n,l)}$ from level $l = l_{\max} - 1$ to level $l = l_{\min}$, i.e. the cost of Step 3 of the upward pass (6.2.16) will be

$$\text{CostUpward}_3 = \sum_{l=l_{\min}+1}^{l_{\max}} 2^{ld}\,\text{CostTrans}(P_l). \qquad (6.2.4)$$

To evaluate this sum we will use the model (6.1.7). Assume that $l_{\min} + 1 \leq l_b \leq l_{\max}$. We then have

$$\text{CostUpward}_3(l_b) = \sum_{l=l_{\min}+1}^{l_b-1} 2^{ld}\,\text{CostTrans}(P_l) + \sum_{l=l_b}^{l_{\max}} 2^{ld}\,\text{CostTrans}(P_l)$$

$$= C_{\min}\left[\sum_{l=l_{\min}+1}^{l_b-1} 2^{ld} 2^{\nu(d-1)(l_b-l)} + \sum_{l=l_b}^{l_{\max}} 2^{ld}\right]$$

$$= C_{\min}\left[2^{\nu l_b(d-1)} \sum_{l=l_{\min}+1}^{l_b-1} 2^{l[d-(d-1)\nu]} + \sum_{l=l_b}^{l_{\max}} 2^{ld}\right]. \qquad (6.2.5)$$

The sum of a geometric progression can be computed as

$$\sum_{l=l_1}^{l_2} \alpha^l = \frac{\alpha^{l_2+1} - \alpha^{l_1}}{\alpha - 1} = \begin{cases} O(\alpha^{l_2}), & \alpha > 1, \\ O(\alpha^{l_1}), & \alpha < 1. \end{cases} \qquad (6.2.6)$$

The case $\nu = d/(d-1)$ is an important case. For two-dimensional problems it is realized for $\nu = 2$, or $O(p^2)$ complexity of translations ($P = p$), while for three-dimensional problems $\nu = 1.5$, which corresponds to $O(p^3)$ complexity of translations ($P = p^2$). In this case all terms in the first sum are equal, and we have

$$\text{CostUpward}_3(l_b) = O(C_{\min}[2^{dl_b}(l_b - l_{\min} - 1) + 2^{dl_{\max}}]),$$
$$\nu = \frac{d}{d-1}. \qquad (6.2.7)$$

This result has a simple meaning. The terms in the first sum, $2^{l[d-(d-1)\nu]}$, are produced by two competitive effects. First, the number of boxes for finer levels increases, due to space subdivision, and second, the translation cost at finer levels decreases, due to the decrease of the translation domain size. These two effects are completely balanced for $\nu = d/(d-1)$, so the number of operations at each level for $l_{\min} < l < l_b$

is the same. We can identify the value $\nu = d/(d-1)$ as "critical translation exponent". If we have faster translations than this critical case, $\nu < d/(d-1)$, then the first sum is a geometric progression with $\alpha > 1$ and using Eq. (6.2.6), we can evaluate the cost of the Step 3 in the upward pass as

$$\text{CostUpward}_3(l_b) = O(C_{\min}(2^{dl_b} + 2^{dl_{\max}})) = O(C_{\min}2^{dl_{\max}}), \quad \nu < \frac{d}{d-1}. \tag{6.2.8}$$

This can be referred to as the case of "fast translations". As we can see, for fast translations the asymptotic complexity is of the same order as for translations with constant truncation number, determined by the finest level of space subdivision. Correspondingly, we can identify the case $\nu > d/(d-1)$ as "slow translations". Here the geometric progression has $\alpha < 1$ and, hence

$$\text{CostUpward}_3(l_b) = O(C_{\min}(2^{\nu l_b(d-1)}2^{-(l_{\min}+1)[(d-1)\nu-d]} + 2^{dl_{\max}})), \quad \nu > \frac{d}{d-1}. \tag{6.2.9}$$

This case requires a bit more attention to understand which exponent is dominating here. The latter term prevails or both terms are of the same order, if

$$dl_{\max} \geq \nu l_b(d-1) - (l_{\min}+1)[(d-1)\nu - d]. \tag{6.2.10}$$

This can be rewritten as

$$\nu \leq \frac{d}{d-1}\frac{l_{\max} - l_{\min} - 1}{l_b - l_{\min} - 1}. \tag{6.2.11}$$

Note that for $l_b < l_{\max}$ this provides a weaker condition for "fast translations", than $\nu < d/(d-1)$, which includes the critical case $\nu = d/(d-1)$. In other words, the complexity of the third step of the upward pass is of the same order as for translations with constant $P = P_{\min}$ if the condition (6.2.11) holds and conditions for Eq. (6.2.8) can be relaxed as

$$\text{CostUpward}_3(l_b) = O(C_{\min}2^{dl_{\max}}),$$
$$\nu \leq \begin{cases} \dfrac{d}{d-1}\dfrac{l_{\max} - l_{\min} - 1}{l_b - l_{\min} - 1}, & l_{\max} > l_b > l_{\min} + 1, \\ \infty, & l_b \leq l_{\min} + 1. \end{cases} \tag{6.2.12}$$

If condition (6.2.11) does not hold, we can neglect the latter term in the parenthesis of Eq. (6.2.9). This yields

$$\begin{aligned}\text{CostUpward}_3(l_b) &= O(C_{\min}(2^{\nu l_b(d-1)}2^{-(l_{\min}+1)[(d-1)\nu-d]})) \\ &= O(C_{\min}2^{\nu l_b(d-1)}), \qquad \nu > \frac{d}{d-1}\frac{l_{\max}-l_{\min}-1}{l_b-l_{\min}-1}.\end{aligned} \tag{6.2.13}$$

The only case, which is not covered by Eqs. (6.2.11) and (6.2.12), is the special case (6.2.7) for $l_b = l_{\max}$, where we have

$$\text{CostUpward}_3(l_b) = O(C_{\min}2^{dl_{\max}}(l_{\max}-l_{\min})), \quad \nu = \frac{d}{d-1}, \quad l_b = l_{\max}. \tag{6.2.14}$$

6.2.2 Downward pass

6.2.2.1 *Step 1*

In the downward pass for purposes of estimation we will take the number of boxes in the $E_4(n,l)$ neighborhood of a given box (n,l) as if it is not a box near the boundary. If we denote $\text{Pow}_4(d)$ the maximum number of boxes from which centers $S|R$ translation is performed to the center of a given box for dimensionality d, then according to Eq. (6.2.2) we have

$$\text{Pow}_4(d) = (2^d - 1)3^d. \tag{6.2.15}$$

(Indeed, the number of boxes of level l in parent's neighborhood is 2^d3^d, from which we need to subtract the power of the box neighborhood, i.e. 3^d.) In this case the cost of the first step in the downward pass (see Eq. (6.2.17)) will be

$$\text{CostDownward}_1 < (2^d - 1)3^d \sum_{l=l_{\min}}^{l_{\max}} 2^{ld}\,\text{CostTrans}(P_l). \tag{6.2.16}$$

This sum is similar to Eq. (6.2.2), except the summation is taken from $l_{\min}$ instead of $l_{\min}+1$. So in the framework of the model of dependence of the length of the representing vectors on the level the evaluation of the asymptotic complexity (Eqs. (6.2.12)–(6.2.14)) can be used with corrected

$l_{\min}$. This yields

$$\text{CostDownward}_1(l_b) = (2^d-1)3^dO\left(C_{\min}\begin{cases}2^{dl_{\max}}, & \nu<\nu_*,\\ 2^{dl_{\max}}, & \nu=\nu_*, l_b<l_{\max},\\ 2^{dl_{\max}}(l_{\max}-l_{\min}+1), & \nu=\nu_*, l_b=l_{\max},\\ 2^{\nu l_b(d-1)}, & \nu>\nu_*\end{cases}\right), \quad (6.2.17)$$

$$\nu_* = \frac{d}{d-1}\frac{l_{\max}-l_{\min}}{l_b-l_{\min}}.$$

According to relations (6.1.5)–(6.1.7) $C_{\min}$ in this formula depends on l_b and l_{b*}. More explicit expression for the complexity can be obtained by inserting these relations to the obtained cost:

$$\text{CostDownward}_1(l_b) = (2^d-1)3^dO\left(C_*\begin{cases}2^{\nu(d-1)(l_{b*}-l_b)+dl_{\max}}, & \nu<\nu_*,\\ 2^{dl_{\max}}, & \nu=\nu_*, l_{b*}<l_{\max},\\ 2^{\nu(d-1)(l_{b*}-l_{\max})+dl_{\max}}(l_{\max}-l_{\min}+1), & \nu=\nu_*, l_{b*}\geq l_{\max},\\ 2^{\nu l_{b*}(d-1)}, & \nu>\nu_*\end{cases}\right),$$

$$C_*=CP_*^{\nu},\ \nu_*=\frac{d}{d-1}\frac{l_{\max}-l_{\min}}{l_b-l_{\min}}. \quad (6.2.18)$$

6.2.2.2 *Step 2*

Since each box has only one parent the cost of the second step in the downward pass (6.2.19) that is performed for levels $l_{\min}+1, \ldots, l_{\max}$ is

$$\text{CostDownward}_2 = \sum_{l=l_{\min}+1}^{l_{\max}} 2^{ld}\,\text{CostTrans}(P_l). \quad (6.2.19)$$

This coincides with the complexity of the third step in the upward pass (6.2.4):

$$\text{CostDownward}_2(l_b) = \text{CostUpward}_3(l_b). \quad (6.2.20)$$

6.2.3 Final summation

To evaluate the sum at M points that in the worst case occupy different boxes, we need to sum up to $3^d s$ sources in the neighborhood of each point (the first term in the right-hand side of Eq. (6.2.21), see also Eq. (6.2.2)) and compute the approximating functions $R_{P_{l_{\max}}}$ corresponding to the representing vectors of length $P_{l_{\max}} = P_{\min}$. This yields

$$\text{CostEvaluation} = M[s3^d\,\text{CostFunc} + \text{CostEval}(P_{\min})]. \tag{6.2.21}$$

where CostFunc is the cost of direct evaluation of the function $Q_\alpha \Phi(\mathbf{y} - \mathbf{x}_\alpha)$ at one point $\mathbf{y}$ and CostEval is the cost of evaluation of $R_{P_{l_{\max}}}$ at one evaluation point for a given representing vector. Note that the computation of the latter quantity may be of the order $P_{\min}$ (e.g. for function representation via series) or larger if integral transforms are used for building the function representations. This cost is similar to CostRep($P_{\min}$) which we introduced in the first step of the upward pass and related to the representations used and the availability of fast transform procedures.

As follows from Eq. (6.2.1) the parameter s in Eq. (6.2.21) can be expressed via N and the maximum level of space subdivision. This yields

$$\text{CostEvaluation} = M[3^d N 2^{-dl_{\max}}\,\text{CostFunc} + \text{CostEval}(P_{\min})]. \tag{6.2.22}$$

6.2.4 Total complexity of the MLFMM

Summing up the above estimates and taking into account that $(2^d - 1)3^d \gg 1$, the total complexity of the MLFMM procedure on the preset data within the assumptions and models taken can be estimated as

$$\text{CostMLFMM} = O(\text{CostDownward}_1(l_b) + 3^d M N 2^{-dl_{\max}}\,\text{CostFunc}$$
$$+ M\,\text{CostEval}(P_{\min}) + N\,\text{CostRep}(P_{\min})). \tag{6.2.23}$$

Since $(2^d - 1)3^d \gg 1$, Step 1 of the downward pass is much more expensive than Step 3 in the upward pass and Step 2 in the downward pass. The translation cost is then mostly due to the $S|R$ translation operations.

To make the model simpler for analysis, let us assume that CostEval and CostRep are linear functions of their arguments. Moreover, assume that they are of the same order. These assumptions are acceptable in many cases, while some special situations should be considered with respective corrections. Then, using expression (6.1.6) we can express the contribution of the representation and evaluation operations in the total cost as

$$M\text{CostEval}(P_{\min})+N\text{CostRep}(P_{\min})\sim B_*P_*(M+N)2^{(d-1)(l_{b*}-l_b)}, \quad (6.2.24)$$

where B_* is some constant.

Note that for a given problem (N and M) and fixed error (C_* and l_{b*}) all parameters in Eq. (6.2.23) with cost functions (6.2.18) and (6.2.24) are functions of l_{max} only. This quantity should be determined from solution of the optimization problem, which we discuss in a section below. If we set roughly $2^{dl_{max}} \sim N$ (i.e. each box at the finest level contains one source point), then these equations provide the following complexity estimates:

$$\text{CostMLFMM} = \begin{cases} O(M + N\log N), & l_b = l_{max}, \quad \nu = \dfrac{d}{d-1}, \\ O(M+N), & \text{else}. \end{cases} \quad (6.2.25)$$

It may look strange that for $\nu > d/(d-1)$ the asymptotics are formally smaller than for $\nu = d/(d-1)$. The explanation here is rather simple. Indeed, estimate (6.2.18) shows that for $\nu > \nu_*$ the cost of the downward pass does not depend on l_{max} or N and is some constant. However, this constant can be huge! Indeed, if $l_{b*} > l_{max}$ and $\nu > d/(d-1)$ it is larger than N. For such cases the asymptotic complexity can be achieved at really large N or M, when the terms growing with N and M prevail. For lower N and M, in this case, we can even write

$$\text{CostMLFMM} = \text{const}, \qquad N, M \ll N_*, M_*, \quad (6.2.26)$$

where N_* and M_* are large enough numbers related to the parameters of the problem. To understand better when the constant or other intermediate asymptotics can be realized, we can consider the following limits, assuming $2^{dl_{max}} \sim N \sim M$, since in this case the second term in sum (6.2.23) is scaled linearly with N.

- $l_{b*} < l_{max}$. In this case according to definition (6.1.5) we have $l_b = l_{b*}$. Therefore, costs of evaluation and representation (6.2.24), in this case, are $O(M+N)$, while the cost of the first step in the downward pass (Eq. (6.2.18)) is

$$\text{CostDownward}_1(l_b) = O\left(\begin{cases} N, & \nu \leq \nu_*, \\ 2^{\nu l_{b*}(d-1)}, & \nu > \nu_* \end{cases}\right), \quad (6.2.27)$$

$$\nu_* = \frac{d}{d-1}\frac{l_{max}-l_{min}}{l_{b*}-l_{min}}.$$

The constant here is large, since

$$\nu l_{b*}(d-1) > dl_{\max}\frac{l_{b*}}{l_{\max}}\frac{l_{\max}-l_{\min}}{l_{b*}-l_{\min}} = \gamma_* l_{\max} d,$$
$$\gamma_* = \frac{1-l_{\min}/l_{\max}}{1-l_{\min}/l_{b*}} > 1. \qquad (6.2.28)$$

This shows that

$$2^{\nu l_{b*}(d-1)} > (2^{dl_{\max}})^{\gamma_*} = N^{\gamma_*}, \qquad \gamma_* > 1, \qquad (6.2.29)$$

and, finally

$$\text{CostMLFMM} = \begin{cases} O(M+N), & \nu \leq \nu_*,\ l_{b*} < l_{\max}, \\ \text{const} > O(M+N^{\gamma_*}), & \nu > \nu_*,\ l_{b*} < l_{\max},\ \gamma_* > 1. \end{cases} \qquad (6.2.30)$$

- $l_{b*} = l_{\max}$. Here, again we have $l_b = l_{b*}$ and the same analysis can be performed, except for special case $\nu = d/(d-1)$. So we have

$$\text{CostMLFMM} = \begin{cases} O(M+N), & \nu < \dfrac{d}{d-1},\ l_{b*} = l_{\max}, \\ O(M+N\log N), & \nu = \dfrac{d}{d-1},\ l_{b*} = l_{\max}, \\ \text{const} > O(M+N^{\gamma}), & \nu > \dfrac{d}{d-1},\ l_{b*} < l_{\max},\ \gamma = \dfrac{\nu(d-1)}{d} > 1. \end{cases} \qquad (6.2.31)$$

- $l_{b*} > l_{\max}$. In this case according to definition (6.1.5) we have $l_b = l_{\max}$. The costs of evaluation and representation (Eq. (6.2.24)) in this case are $O(M+N)$ with factor $2^{(d-1)(l_{b*}-l_{\max})}$, which presumably is large. However, factor $2^{\nu(d-1)(l_{b*}-l_{\max})} \geq 2^{(d-1)(l_{b*}-l_{\max})}$ appears in the asymptotics of the downward pass (Eq. (6.2.18)) and, therefore, this step determines the total complexity. We can then write

$$\text{CostMLFMM} \sim \text{CostDownward}_1$$
$$= 2^{\nu(d-1)(l_{b*}-l_{\max})} \begin{cases} O(M+N), & \nu \le \dfrac{d}{d-1}, \\ \text{const} > O(M+N^{\gamma}), & \nu > \dfrac{d}{d-1}, \gamma = \dfrac{\nu(d-1)}{d} > 1, \end{cases} \tag{6.2.32}$$

and $l_{b*} > l_{\max}$. Thus, while formally this is $O(N+M)$ algorithm, it should perform much slower than the algorithm for which $l_{\max} \le l_{b^*}$.

Since $l_{\max} \sim d^{-1} \log_2 N$ we can relate these cases with $N \sim M$. Condition $l_{b*} > l_{\max}$ can be rewritten in the form:

$$\log_2 \frac{kD_0}{\kappa_b} > \frac{1}{d} \log_2 N. \tag{6.2.33}$$

Therefore, as l_{b*} and N are independent prescribed parameters, we determine condition for the above asymptotic cases:

$$\left(\frac{kD_0}{\kappa_b}\right)^d > N \Longrightarrow l_{b*} > l_{\max}, \qquad \left(\frac{kD_0}{\kappa_b}\right)^d < N \Longrightarrow l_{b*} < l_{\max}. \tag{6.2.34}$$

We should keep in mind that the above evaluations are obtained in assumption that we partitioned the space until the level where each box contains just 1 point. Such subdivision is far from optimal, and we will discuss this in corresponding section below. Nevertheless, the above estimates provide an idea on the complexity of the algorithm and its dependence on various parameters.

6.3 ERROR OF MLFMM

We note that the asymptotic behavior (6.2.25) takes place assuming that $P_* = \text{const}$. To provide estimates of the algorithm complexity for computations within specified error bound, ϵ, we need to introduce some dependence $P_*(N, \epsilon)$ into the model.

To determine this dependence, we can consider the error due to approximation of the field of a single source at the location of a single evaluation point. Then, we can sum up the errors of the sources to determine the maximum absolute error for the given receiver. Of course, different sources will make different contributions. For example, the sources from the neighborhood of the receiver at the finest level do not contribute into the error at all since we sum them up directly (assuming that roundoff errors for machine precision arithmetic are negligible). The computation of the contribution of the sources into the field at a given evaluation point via many translations can cause more error than the error of a source for which we need just one translation, etc. To put the count of errors in more or less regular way, we can state that the actual total absolute error, ϵ_{tot}, is smaller than

$$\epsilon_{tot} \lesssim N\epsilon, \qquad \epsilon = \max_l(\epsilon_l), \qquad l = 1, \ldots, N, \tag{6.3.1}$$

where ϵ_l is the error of the MLFMM for a single source and ϵ is the maximum of these errors.

We now assume that each translation causes some error. So the worst situation will be realized for locations of the source and receiver points, when the maximum number of the $S|S$ and $R|R$ translations is required. We also should add to this count one $S|R$ translation. Hence, for space partitioning from level l_{max} to l_{min} we will have the maximum number of single source translations:

$$\text{MaxNumberTranslations} = 2(l_{max} - l_{min} - 1) + 1. \tag{6.3.2}$$

We can also add to this count 2 if we include to the error of approximation of a single source and the errors of representation and evaluation of function, which are usually errors of the same order as translation errors. Now we should remind the reader that the variation of P with the level in the model is to provide the *same* error of function representation for each level. This also means that this error is determined by P_{min}, the length of the representing vector at the finest level. Let us denote this error as ϵ_P. Then we have

$$\epsilon = [2(l_{max} - l_{min}) + 1]\epsilon_P \lesssim 2l_{max}\epsilon_P, \qquad \epsilon_{tot} \lesssim N\epsilon \lesssim 2l_{max}N\epsilon_P. \tag{6.3.3}$$

While more or less accurate evaluations of ϵ_P can be obtained similar to that presented in Chapter 9, for error modeling purposes we can accept that ϵ_P is the truncation error of a series which converges absolutely and uniformly, and the truncated "tail" can be majorated by some geometric progression (exponential error decay). In this case the number of terms in expansion for the finest level will be

$$p = P_{min}^{1/(d-1)} = B \log \frac{1}{\epsilon_P}, \tag{6.3.4}$$

where B is some model constant, which is related to the exponent of the decay of terms in the series. Replacing here ϵ_P with its relation to other model parameters (6.3.3), we obtain

$$P_{\min} \sim B^{d-1} \log^{d-1} \frac{2 l_{\max} N}{\epsilon_{\text{tot}}} = O\left(\log^{d-1} \frac{N}{\epsilon_{\text{tot}}}\right). \tag{6.3.5}$$

Here we took into account that in the MLFMM $l_{\max} = O(\log N)$ and, hence $l_{\max} \ll N$.

Equation (6.1.7) now shows that

$$C_{\min} = O\left(\log^{\nu(d-1)} \frac{N}{\epsilon_{\text{tot}}}\right). \tag{6.3.6}$$

This affects the complexity of the MLFMM given by Eq. (6.2.25) since it introduces powers of logarithmic function due to the dependence of translation cost on $C_{\min}$ (Eq. (6.2.17)). For example, in the three-dimensional case and $\nu = 1.5$ we will have the following scaling of the MLFMM with N:

$$\text{CostMLFMM} = \begin{cases} O(M + N \log^4 N), & l_b = l_{\max},\ \nu = 1.5, \\ O(M + N \log^3 N), & l_b < l_{\max},\ \nu = 1.5. \end{cases} \tag{6.3.7}$$

Numerical examples in Chapter 9 show, that the error bound can be substantially overestimated. Nevertheless, at this point we can just bound the error with some upper limit to show that the error constraints may introduce logarithmic factors.

6.4 OPTIMIZATION

For a given problem and translation method, when the parameters N, M, d, ν, l_{b*}, and ϵ_{tot} are fixed, the only parameter that remains unknown is $l_{\max}$. This parameter substantially influences the performance of the algorithm and should be selected properly. This can be done by minimization of function CostMLFMM considered as a function of the single variable $l_{\max}$. Even in the model case, considered here, this function is complex enough to provide analytical expressions. In practice, this can be done numerically by solving and scaling some test problems and varying $l_{\max}$ to determine the optimum maximum level of space subdivision.

Having a goal to derive and analyze basic qualitative dependences, we note that this parameter enters the dependences in two forms. First, in the form of exponents of type $2^{l_{\max}}$ and second in the form of polynomial dependences $l_{\max}^{\alpha}$ where α is 1 or of order of 1. Obviously that the first quantity proportional to some power of N is much larger than the second quantity proportional to some power of $\log N$, since we consider $N \to \infty$.

By this reason to simplify the formula we can consider that parameter

$$\zeta = 2^{dl_{\max}} \tag{6.4.1}$$

is varying (large), while $l_{\max}$ as it appears in polynomial dependences stays almost constant. This is a valid approach to obtain the principal terms of dependences and optimize the problem in this approximation.

6.4.1 Lower frequencies or larger number of sources and receivers

First, consider the case of translations, when we assume that the translation exponent is small enough and the optimum $l_{\max}$ is located in the range:

$$l_{\max} > l_{b*}, \qquad \nu \leq \nu_* = \frac{d}{d-1} \frac{l_{\max} - l_{\min}}{l_{b*} - l_{\min}}. \tag{6.4.2}$$

In this case from definition (6.1.5) we have $l_b = l_{b*}$ and the complexity of the algorithm does not depend on the wavenumber, or l_{b*}, and we have from Eqs. (6.2.18) and (6.2.24):

$$\begin{aligned} \text{CostMLFMM} = O((2^d - 1)3^d C_* \zeta + 3^d MN\zeta^{-1} \,\text{CostFunc} \\ + M\,\text{CostEval}(P_*) + N\,\text{CostRep}(P_*)). \end{aligned} \tag{6.4.3}$$

Neglecting possible dependence of P_* on N due to error bounds, we can see that the cost function is a sum of linear function on ζ and a hyperbola. This sum has the only minimum for

$$\zeta^{(\text{opt})} = \left(\frac{MN}{2^d - 1} \frac{\text{CostFunc}}{C_*}\right)^{1/2}, \qquad \nu \leq \nu_*, \tag{6.4.4}$$

which can be easily found by taking the first derivative of the cost function with respect to ζ.

According to relation (6.4.1) this can be rewritten in terms of optimum maximum level of space partitioning:

$$l_{\max}^{(\text{opt})} = \frac{1}{d} \log_2 \zeta^{(\text{opt})} = \frac{1}{2d} \log_2 \left(\frac{MN}{2^d - 1} \frac{\text{CostFunc}}{C_*}\right). \tag{6.4.5}$$

To get an idea how large $l_{\max}^{(\text{opt})}$ can be, we consider the example of the three-dimensional case, with the following parameters $d = 3$, $M = N = 10^6$, $\nu = 1$, $P_* = p_*^2 \sim 10^2$, and $\text{CostFunc}/C \sim 10^{-2}$, where C_* can be calculated using Eq. (6.2.18). This formula then yields $l_{\max}^{(\text{opt})} \approx 4$ (here and below we round off $l_{\max}$ to the closest integer). For the same parameters, but $\text{CostFunc}/C \sim 1$ we obtain $l_{\max}^{(\text{opt})} \approx 5$. As we can see these numbers are not very large even for a million sources and receivers. Let us compare this result with a non-optimized case, when we subdivide space till the level when each box has one (or order of 1 point). In this case we have $l_{\max} = d^{-1}\log_2 N \approx 7$. Note that in three-dimensional case the number of boxes $N_{l_{\max}}$ is $8^{l_{\max}}$. So if we have two space partitionings with some $l_{\max}^{(1)}$ and $l_{\max}^{(2)}$ such that $l_{\max}^{(2)} = l_{\max}^{(1)} + 1$, then the ratio of corresponding $N_{l_{\max}}$ is 8 ($N_{l_{\max}}^{(2)}/N_{l_{\max}}^{(1)} = 8$) or about one order of magnitude.

Substituting the optimum value of ζ given by Eq. (6.4.4) into Eq. (6.4.3), we obtain the following evaluation for the optimized complexity of the algorithm:

$$\text{CostMLFMM}^{(\text{opt})}$$
$$= O\left(2(2^d-1)3^d\left(\frac{MN}{2^d-1}C_*\text{CostFunc}\right)^{1/2} + M\text{CostEval}(P_*) + N\text{CostRep}(P_*)\right). \tag{6.4.6}$$

It is noticeable that since $C_* = CP_*^{\nu}$ we can write for a simplified case $M \sim N$, $\text{CostEval}(P_*) \sim \text{CostRep}(P_*) \sim P_*$ and constants of order of 1:

$$\text{CostMLFMM}^{(\text{opt})} = O\left(NP_*^{\nu/2} + NP_*\right). \tag{6.4.7}$$

This is an interesting result that shows that for $\nu = 2$ the contribution of the translations and direct summation can be of the same order as the contribution of the representation and evaluation operations. Moreover, if P_* is large enough and $\nu < 2$, the latter operations determine the complexity of the algorithm which is scaled optimally as $O(NP_*)$ for *any* $\nu \leq 2$ if this ν satisfies condition (6.4.2) and the condition

$$\left(\frac{MN}{2^d-1}\frac{\text{CostFunc}}{C_*}\right)^{1/2} > 2^{dl_{b*}} = \left(\frac{kD_0}{\kappa_b}\right)^d \tag{6.4.8}$$

holds. This follows from substitution of the optimum level determined by relation (6.4.5) into the first condition (6.4.2) where we also inserted the explicit expression (6.1.5) for l_{b*}. This constraint relates to the whole case considered in this section and shows that it can be realized either at sufficiently low frequency, or sufficiently large MN.

Note the second constraint in condition (6.4.2) which can be rewritten in the form:

$$l_{b*} - l_{\min} = \log_2 \frac{kD_0}{2^{l_{\min}}\kappa_b} < \frac{d}{d-1}\frac{l_{\max} - l_{\min}}{\nu}. \tag{6.4.9}$$

Substituting here the $l_{\max}^{(\mathrm{opt})}$ determined by Eq. (6.4.5), we obtain

$$\log_2\left(\frac{kD_0}{2^{l_{\min}}\kappa_b}\right)^{\nu} < \frac{1}{2(d-1)}\log_2\left[\frac{MN}{2^d-1}\frac{\mathrm{CostFunc}}{C_* 2^{2(d-1)l_{\min}}}\right], \tag{6.4.10}$$

or, neglecting factors of type $2^{l_{\min}}$:

$$\left(\frac{MN}{2^d-1}\frac{\mathrm{CostFunc}}{C_*}\right)^{1/2} > \left(\frac{kD_0}{\kappa_b}\right)^{\nu(d-1)}. \tag{6.4.11}$$

This result being compared with Eq. (6.4.8) shows that for $\nu > d/(d-1)$ it is more restrictive, while no additional restrictions apply for $\nu \leq d/(d-1)$.

While the optimized and non-optimized algorithms have the same asymptotic complexity, the constants in these asymptotics can differ substantially. To show this let us return to our example and evaluate the ratio of the costs for the non-optimized and optimized algorithms. As we could see that for example cases the difference in levels for optimized and non-optimized algorithms was 2 or 3. Since $\zeta = 2^{3l_{\max}}$ this results in ratios $\zeta/\zeta^{(\mathrm{opt})}$ of 64 and 512, respectively. Usually, the costs of function representation and evaluation are negligible compared to translation costs. If so, the ratio of the costs of non-optimized and optimized algorithms can be estimated approximately as

$$\frac{\mathrm{CostMLFMM}}{\mathrm{CostMLFMM}^{(\mathrm{opt})}} \sim \frac{1}{2}\left(\frac{\zeta}{\zeta^{(\mathrm{opt})}} + \frac{\zeta^{(\mathrm{opt})}}{\zeta}\right)$$

$$= \frac{1}{2}\left[2^{d\left(l_{\max}-l_{\max}^{(\mathrm{opt})}\right)} + 2^{d\left(l_{\max}^{(\mathrm{opt})}-l_{\max}\right)}\right]. \tag{6.4.12}$$

This shows that the optimized algorithm should be 32 and 256 times faster than the non-optimized one for $\mathrm{CostFunc}/C \sim 10^{-2}$ and $\mathrm{CostFunc}/C \sim 1$, respectively.

6.4.2 Higher frequencies or smaller number of sources and receivers

In the case of higher frequencies or relatively small number of sources and receivers, condition (6.4.8) may not hold. In this case the optimum level found for the previous case is not valid and we should consider this case separately. As before, we assume that the translation exponent is small enough, while the optimum l_{max} is located in the following range:

$$l_{max} \leq l_{b*}, \qquad \nu < \frac{d}{d-1}. \tag{6.4.13}$$

As shown during the derivation of relation (6.2.32) the costs of evaluation and representation can be neglected for $\nu > 1$, while for $\nu = 1$ they can be of the same order as the cost of the downward pass. The latter case adds some constants to the expression for the optimum level, which we neglect for simplicity, since the purpose of this theoretical consideration is to establish qualitative rather than quantitative relations. The latter relations should anyway be found from numerical tests for particular problems. So, neglecting the evaluation and representation costs, and taking into account that for $l_{max} \leq l_{b*}$ we have $l_b = l_{max}$ (see Eq. (6.1.5)), we can write the total complexity of the method in the form:

$$\text{CostMLFMM} = O((2^d - 1)3^d 2^{\nu(d-1)l_{b*}} C_* \zeta^{\gamma} + 3^d MN\zeta^{-1}\,\text{CostFunc}),$$
$$\gamma = 1 - \frac{\nu(d-1)}{d} > 0. \tag{6.4.14}$$

Here we used Eq. (6.4.1) and the expression for the cost of the first step of the downward pass (Eq. (6.2.18)). The function to be minimized is then a sum of growing and decaying functions, which has the only minimum. As above this minimum can be found by taking the derivative, which provides the following optimum value:

$$\zeta^{(\text{opt})} = \left(2^{-\nu(d-1)l_{b*}} \frac{MN}{2^d - 1} \frac{\text{CostFunc}}{\gamma C_*}\right)^{1/(\gamma+1)}, \qquad \nu < \frac{d}{d-1}. \tag{6.4.15}$$

Substituting this expression into Eq. (6.4.14) we obtain

$$\text{CostMLFMM}^{(\text{opt})}$$
$$= O\left(\left[(2^d - 1)3^d 2^{\nu(d-1)l_{b*}} C_*\right]^{1/(\gamma+1)} \left[3^d MN\,\text{CostFunc}\right]^{\gamma/(\gamma+1)}\right). \tag{6.4.16}$$

Particularly, for $M \sim N$ we have

$$\text{CostMLFMM}^{(\text{opt})} = O(N^{2\gamma/(\gamma+1)}). \tag{6.4.17}$$

For $\nu = 1$ and $d = 3$ this yields

$$\text{CostMLFMM}^{(\text{opt})} = O(N^{1/2}). \tag{6.4.18}$$

The optimum level can be found from Eqs. (6.4.1) and (6.4.15)

$$l_{max}^{(opt)} = \frac{1}{d}\log_2 \zeta^{(opt)}$$

$$= \frac{1}{2d - \nu(d-1)}\left[\log_2\left(\frac{MN}{2^d - 1}\frac{\text{CostFunc}}{C_*}\right) - \nu(d-1)l_{b*}\right]. \quad (6.4.19)$$

The condition, for which this optimum can be applied, follows from constraints (6.4.1) and (6.4.13):

$$\left(\frac{MN}{2^d - 1}\frac{\text{CostFunc}}{C_*}\right)^{1/2} \leq 2^{dl_{b*}} = \left(\frac{kD_0}{\kappa_b}\right)^d. \quad (6.4.20)$$

Comparing this with relation (6.4.8), we can see that the cases are complementary, and therefore, the entire range of parameters is covered by these two cases for translation exponents $\nu < d/(d-1)$.

The question now is what happens in the case $\nu \geq d/(d-1)$ at higher frequencies. If $l_{max} = l_{b*}$ and $\nu = \nu_* = d/(d-1)$ we already have l_{max} defined and there is no optimization problem. In this case the complexity of the algorithm is $O(N \log N)$ as follows from Eq. (6.2.31). For the same critical $\nu = \nu_* = d/(d-1)$ and $l_{max} < l_{b*}$ the cost of the first step of the downward pass (Eq. (6.2.18)) grows weakly (linearly) with l_{max}, while parameter ζ grows exponentially with respect to l_{max}. The minimum exists in this case, and can be determined from an equation of type (6.4.3), where for simplicity we can neglect the evaluation and representation costs. This results in the following equation for the optimum level or related parameter ζ (Eq. (6.4.1)):

$$\frac{d}{d\zeta}\left[\frac{2^d - 1}{d}C_* 2^{dl_{b*}} \log_2 \zeta + \zeta^{-1} MN \,\text{CostFunc}\right] = 0. \quad (6.4.21)$$

The solution of this equation is

$$\zeta^{(opt)} = \ln 2 \frac{d}{2^d - 1}\frac{\text{CostFunc}}{C_*} MN 2^{-dl_{b*}}. \quad (6.4.22)$$

The cost of the optimized FMM is then

$$\text{CostMLFMM} = O\left(\left(\frac{kD_0}{\kappa_b}\right)^d (\log M + \log N)\right), \quad (6.4.23)$$

with substantially large constant $2^{dl_{b*}} > 2^{dl_{max}} = \zeta$. This shows that the complexity in this case grows slowly (logarithmically) with N and the optimum level is

$$l_{\max}^{(\text{opt})} = \frac{1}{d}\log_2 \zeta^{(\text{opt})} = \frac{1}{d}\log_2\left(\ln 2 \frac{d}{2^d - 1} \frac{\text{CostFunc}}{C_*} MN2^{-dl_{b*}}\right). \quad (6.4.24)$$

This optimum level grows with N and therefore at some N the constraint $l_{\max} < l_{b*}$ will be violated. The condition $l_{\max}^{(\text{opt})} < l_{b*}$ can be written using definition (6.1.5) of l_{b*} as

$$(NM)^{1/2} < [d \ln 2]^{1/2}\left[\frac{(2^d - 1)C_*}{\text{CostFunc}}\right]^{1/2}\left(\frac{kD_0}{\kappa_b}\right)^d. \quad (6.4.25)$$

Comparing this result with the lower frequency relation (6.5.20) valid for $\nu = 1$ we can see that the bounds for N are almost the same, since the constant $[d \ln 2]^{1/2}$ is close to 1. Moreover, when we did the estimates of the orders, we dropped some constant factors, and if were more precise these two cases should match. In this analysis it is important for us to identify asymptotic regions for different algorithm behavior depending on the problem controlling parameters.

If $\nu > d/(d-1)$ the cost of the downward pass in this range does not depend on $l_{\max}$ (large constant, as we discussed above, see Eq. (6.2.32)). At fixed N the neglected term due to function evaluation and representation decays at growing $l_{\max}$ as the term due to direct summation does. So there is no extremum. In fact, the balance between these terms has a minor effect and so the cost of the FMM is almost constant, since it is determined by the largest constant term. In this case according to Eq. (6.2.18) we can write:

$$\text{CostMLFMM} = O\left(\left(\frac{kD_0}{\kappa_b}\right)^{\nu(d-1)/d}\right). \quad (6.4.26)$$

6.4.2.1 Volume element methods

The estimations obtained above are valid for arbitrary N, M, and kD_0. While these parameters can be treated independently in some methods, they are closely related. For example, in computations of volume integrals the source points can be distributed in such a way that there are a certain number of points per wavelength determined by some accuracy of computations. Another example, which has the same type of relation between the number of sources and the computational domain size, is provided in Chapter 10, where we considered application of the FMM to multiple scattering problems. In these problems we can assume that the number of sources N_s per wave of length $\lambda = 2\pi/k$ is fixed. Since there are D_0/λ waves in the computational domain, we should have $N_s D_0/\lambda$

sources along the side of the box, and since the problem is d dimensional, the total number of sources will be

$$N = \left(\frac{N_s}{2\pi} kD_0\right)^d. \tag{6.4.27}$$

Assume now that all the constants of the method are of the order of unity and $M \sim N$. Then from the above estimations we obtain the following different types of the algorithm behavior:

$$\text{CostMLFMM} = \begin{cases} O(N), & \nu < d/(d-1), \\ O(N \log N), & \nu = d/(d-1), \\ O(N^{\nu(d-1)/d}), & \nu > d/(d-1). \end{cases} \tag{6.4.28}$$

Here the lowest complexity holds for any high frequency or low frequency regime. The case of critical translation exponent also applies to all cases, while the algorithm is scaled as $O(N)$ for the values of constants corresponding to the low frequency asymptotics. The low frequency range, however, is limited and at larger N namely $O(N \log N)$ realizes. The scaling for the latter case follows from Eq. (6.4.26) since at larger N condition (6.4.20) will be violated for any $\nu > d/(d-1)$ and the high frequency case applies.

This shows that the critical translation exponent provides the boundary between the "fast" and "slow" translation methods. For three-dimensional problems we have $\nu = 1.5$ and, thus $O(p^3)$ or faster translation methods are of particular interest, since they result in total complexities of the MLFMM $O(N \log^{\beta} N)$ (which includes *log* complexities from error bounds and data structure management) for volume data distributions.

6.4.2.2 *Some numerical tests*

Figure 6.1 shows the number of multiplication operations in a typical FMM run. To combine the model with the actual code and obtain results for different ν, we introduced counters for calls of each subroutine into our program. Then we assumed that the complexity of a single translation is given by Eq. (6.1.1) with $C = 1$ and the cost of each direct function evaluation is 1. Based on the actual number of calls, we sum up the results to obtain the total cost (which does not include the cost of operations with the data structure). This figure also shows that the major contribution to the total number of operations is due to translations and direct summations, while the cost of evaluation and representation ($\text{CostEval}(P_*) = \text{CostRep}(P_*) = P_*$) did not exceed a few percent of the total cost in any case.

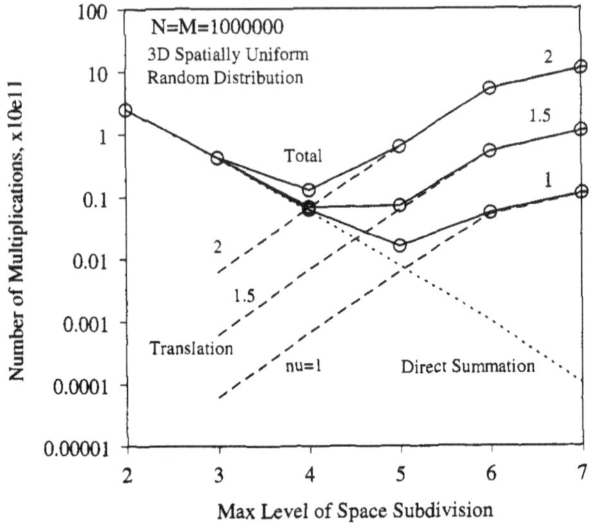

FIGURE 6.1 The number of floating point multiplication operations required by the multilevel FMM for computation of the potential due to a million sources at a million evaluation points. Sources and targets are distributed randomly uniformly, $l_{b*} = 2$, $C = 1$, $P_* = 100$, the translation exponent, ν, is indicated near the curves. The dotted line shows the number of operations spent for direct summation (CostFunc = 1), and the dashed lines show the number of operations required for translation. Solid lines marked with circles show the total number of operations. The number of operations was obtained during an actual run of the MLFMM code by introducing counters for multiplications into translation and summation subroutines.

Figure 6.2 displays the complexity of computations in the MLFMM, where as before we set parameters to investigate the theoretical model and count the number of multiplications. The maximum level here was optimized for each data point on the graph, by variation of $l_{\max}$ and determination at which value the number of operations reaches its minimum. Two sets of data corresponding to the "low frequency case" ($l_{b*} = 2$) and "high frequency case" ($l_{b*} = 5$) are distinguishable. In the low frequency case, computations with different translation exponents show an almost linear scaling of the algorithm with N. Of course, the constant in this scaling is such that the MLFMM is faster than the straightforward matrix–vector multiplication method (which scaling is shown in the same figure) only for N larger than $N \sim 10^3 - 10^4$. The "breakeven" point at which it makes sense to use the FMM depends on the translation cost and is different for different ν. For large problems,

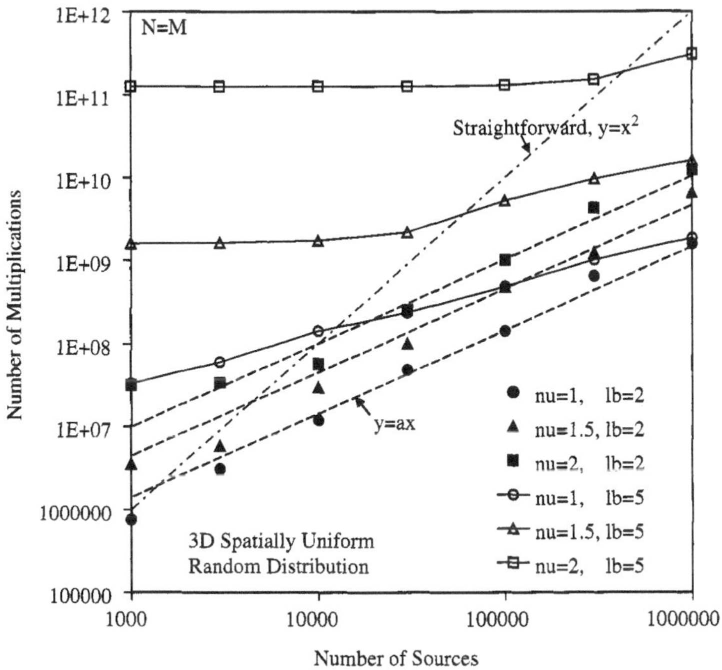

FIGURE 6.2 The dependencies of the number of floating point multiplications on the number of sources, N. Sources and evaluation points ($M = N$) are distributed uniformly randomly. Computations are performed with different translation exponents, ν, and breakdown levels, l_{b*}, as shown in the chart legend. So each curve is plotted at constant kD_0. The maximum level is optimized by finding the minimum of total number of multiplications for each given N, ν, and l_{b*} numerically. The dash-dotted line shows the number of multiplications when using the straightforward method, and the dashed lines show linear dependency. $P_* = 100$, $C = 1$.

say $N \sim 10^6$ the MLFMM is faster than the straightforward method by several orders of magnitude.

Qualitatively different behavior is observed in the "high frequency" case. We note that the optimum level in any case displayed in Fig. 6.2 did not exceed 5 (see also Fig. 6.1). So for a problem with $l_{b*} = 5$ all computations were performed with $l_{max} \leq l_{b*}$. In this region of parameters the scaling of the algorithm complexity with N is $O(N^{\alpha})$, $\alpha < 1$. For example, this α is about 0.5 for $\nu = 1$ and close to 0 for $\nu = 2$. This is consistent with evaluations (6.2.32) and (6.4.18). As mentioned above, an asymptotic behavior $O(N^{\alpha})$ does not mean that the method becomes faster in this case, due to a large asymptotic constant (in fact, any computation with $l_{b*} = 5$ is more expensive than with $l_{b*} = 2$ if other parameters are fixed). The asymptotic constants are so large, that the MLFMM can outperform the straightforward method only for $N > 10^4$. The translation exponent is a very important parameter here. For example, if $\nu = 2$ and the size of the problem is $N < 500{,}000$, the straightforward method is faster than the MLFMM. We also see that due to complexity $O(N^{\alpha})$, $\alpha < 1$, the "high frequency" curves approach the "low frequency" curves for larger N. In fact, the number of operations for $N = 10^6$ for $\nu = 1$ and $l_{b*} = 5$ and 2 is almost the same. When a "low frequency curve" reaches the "high frequency curve" corresponding to the same ν (which happens for $l_{\max}^{(\mathrm{opt})} = l_{b*}$), the character of the latter curve should change (since any computation with larger l_{b*} is more expensive), and these two curves should almost coincide for larger N. This effect is seen in computations with data distributed uniformly over a sphere surface, which is illustrated below.

6.5 NON-UNIFORM DATA

6.5.1 Use of data hierarchies

In the case of a regular mesh the MLFMM algorithm goes through all the boxes and there is no need to search for neighbors or children in the data hierarchies. Thus, for this case we do not even need to perform the initial step of setting up the data structures necessary and can use the functions NeighborsAll and ChildrenAll to get the neighbors and the children. As pointed out in the previous section, these procedures have constant, or $O(1)$ complexity. These provide an $O(N + M)$ algorithm for the case of a regular mesh if P is constant. Such a complexity also applies for the case of arbitrary data sets, since one simply can assign zero representing vectors $\mathbf{C}^{(n,l)}$ to boxes that do not contain source points and not build $\tilde{\mathbf{D}}^{(n,l)}$ and $\mathbf{D}^{(n,l)}$ vectors for boxes without evaluation points.

The first simple step to algorithm adaptivity is skipping boxes that do not contain sources (in the upward pass and for the $S|R$ translations from such boxes) or do not contain evaluation points (in the downward pass), in other words the use of X and Y data hierarchies. The use of these hierarchies increases the complexity of operations Neighbors and Children to $O(\log N)$ or $O(\log M)$. Usually, we assume that

$$\log N, \log M \ll \mathrm{CostTrans}(P_{l_{\min}}), \tag{6.5.1}$$

which allows us to exclude the costs of operations with data hierarchies compared to translations. However, if the search operations are performed as intersections (say, sets Neighbors for a given box and $\mathbb{X}$) condition (6.5.1) should be modified at least as

$$3^d(2^d - 1)\log N \ll \text{CostTrans}(P_{l_{\min}}), \tag{6.5.2}$$

since we need to check for each neighbor box whether it is in the list of boxes. In this case the cost in the right-hand side can be substantial. For example, in the three-dimensional case the multiplier near $\log N$ will be approximately 200, and if $N \sim 10^6$ we have 2000 in the right-hand side. So if a single translation involves of order 1000 operations, the costs of search operations can be comparable with translation costs and $O(N \log N)$ complexity of the algorithm will be just due to the search operations. Reduction in the costs of search operations overhead can be achieved by extension of memory, where one can keep all precomputed lists of neighbors and children for each box. In this case large constant factors, like $3^d(2^d - 1)$ can be avoided. Of course, this makes sense only for small enough translation costs, when inequality (6.5.2) does not hold.

6.5.2 Surface distributions of sources and receivers: simple objects

Solution of integral equations involving single and double layer potentials requires evaluation of sums of sources (or dipoles) at points located on the same surface. It is typical for these problems that the number of receivers is of the same order as the number of sources, $M \sim N$ and for simplicity of complexity estimations we assume that $M = N$.

The surface in d-dimensional space is a $d - 1$-dimensional manyfold. Once the computational domain space is partitioned to K boxes (cubes) with sufficiently large K only a fraction of these boxes is non-empty. As follows from the example case with the sphere (see Eq. (5.1.41) and below) the number of non-empty boxes should be of order $K^{(d-1)/d}$. Consider the complexity of the MLFMM in this case, which we can specify as the case of simple objects (like a sphere). We define the simple objects as objects, for which reduction of dimensionality happens at low levels l. For surfaces of simple objects each box has the number of non-empty neighbors and children approximately the same as in $d - 1$-dimensional case, while for function representation and translation we need to use formulae for d dimensions.

If the surface area is S and it is sampled by N points distributed uniformly over the surface, then $N_s = N/S$ is the number of points per unit of surface area, which can be assumed constant. The characteristic distance between two neighbor points can be evaluated as $N_s^{-1/(d-1)}$ and since the box size at the finest level is $D_{l_{\max}} = 2^{-l_{\max}} D_0$ (Eq. (6.2.8)), we should have

$$s = \left[\frac{2^{-l_{\max}} D_0}{N_s^{-1/(d-1)}} \right]^{d-1} = 2^{-(d-1)l_{\max}} \frac{1}{\xi_S} N, \qquad N_s = \frac{N}{S}, \quad \xi_S = \frac{S}{D_0^{d-1}}. \tag{6.5.3}$$

Here ξ_S is the dimensionless surface area. The power $(d-1)$ comes from scaling, since for a simple object the increase of the level by 1 reduces the size of the box twice and so the number of points located in the box, which intersects the surface, should decrease by 2^{d-1} times.

This equation shows that the number of non-empty boxes at level $l_{\max}$ is N/s and thus at level $l \leq l_{\max}$ the number of non-empty boxes can be estimated as

$$N_l \sim 2^{(d-1)(l-l_{\max})} \frac{N}{s} = 2^{(d-1)l} \xi_S. \tag{6.5.4}$$

For simple objects we assume that this holds for $l = l_{\min}, \ldots, l_{\max}$.

6.5.2.1 *Complexity of MLFMM*

The costs of the first and the second steps in the upward pass will be the same as in general case, while for the third step we will have the following estimate, instead of Eq. (6.2.4):

$$\text{CostUpward}_3 = \xi_S \sum_{l=l_{\min}+1}^{l_{\max}} 2^{l(d-1)} \,\text{CostTrans}(P_l). \tag{6.5.5}$$

Proceeding further with the model for the translation cost (6.1.7), we obtain the following modification of Eq. (6.2.5):

$$\begin{aligned} &\text{CostUpward}_3(l_b) \\ &= \xi_S \sum_{l=l_{\min}+1}^{l_b-1} 2^{l(d-1)} \,\text{CostTrans}(P_l) + \xi_S \sum_{l=l_b}^{l_{\max}} 2^{l(d-1)} \,\text{CostTrans}(P_l) \\ &= \xi_S C_{\min} \left[2^{\nu l_b (d-1)} \sum_{l=l_{\min}+1}^{l_b-1} 2^{-l(d-1)(\nu-1)} + \sum_{l=l_b}^{l_{\max}} 2^{l(d-1)} \right]. \end{aligned} \tag{6.5.6}$$

Now we can see that in the case of surface distribution the critical value of the translation exponent for any dimensionality is $\nu = 1$, which corresponds to the fastest possible translation method. For this special case the cost of the upward pass is

$$\mathrm{CostUpward}_3(l_b) = O(\xi_S C_{\min}[2^{l_b(d-1)}(l_b - l_{\min} - 1) + 2^{l_{\max}(d-1)}])$$
$$= O(\xi_S C_{\min} l_{\max} 2^{l_{\max}(d-1)}), \quad \nu = 1. \tag{6.5.7}$$

Since this cost can be majorated by the last estimate for any l_{b*} in the critical case, the complexity of the problem complexity is bounded by some value, which does not depend on the frequencies or sizes of the domain considered.

It is interesting to note that the critical translation exponent is lower in the case of surface data distributions than in the case of spatial distribution. The reason for this is pretty clear. In the three-dimensional case when we go to the coarser level, the number of non-empty boxes reduces by eight times for uniform spatial data distributions and four times for data points uniformly distributed over the surface. In the high frequency limit the length of the representing vector P in both cases increases by four times, when we go to the coarser level. So the efficiency necessary for translation should be higher in the latter case (it should be proportional to P ($\nu = 1$)) to avoid the increase of the cost of operations at the coarser level compared to the finer level. For the spatial distribution the efficiency of the translation is not restricted so strongly, and $\nu = 1.5$ ($4^{1.5} = 8$) is sufficient to avoid the increase of the cost of operations at coarser levels.

Translation methods with exponent $\nu > 1$ can then be classified as "slow translation methods" for surface distributions. For these methods we can calculate the first sum in Eq. (6.5.6) as a geometric progression and obtain

$$\mathrm{CostUpward}_3(l_b) = O\left(\xi_S C_{\min}[2^{\nu l_b(d-1)} 2^{-(l_{\min}+1)(d-1)(\nu-1)} + 2^{l_{\max}(d-1)}]\right), \quad \nu > 1. \tag{6.5.8}$$

Similar to the case of spatially uniform data distributions we can introduce an "effective" critical translation exponent, ν_*, which depends on l_b. This exponent is defined as some value of ν for which the contribution of the second term in the sum (6.5.8) is larger or equal to the contribution of the first term in the sum. This yields the following expression for ν_*:

$$\nu_* = \frac{l_{\max} - l_{\min} - 1}{l_b - l_{\min} - 1}. \tag{6.5.9}$$

Hence, we have

$$\text{CostUpward}_3(l_\text{b}) = O\left(\xi_S C_{\min}\begin{cases} 2^{l_{\max}(d-1)}, & \nu < \nu_*, \\ 2^{l_{\max}(d-1)}, & \nu = \nu_*, l_\text{b} < l_{\max}, \\ 2^{l_{\max}(d-1)}(l_{\max} - l_{\min}), & \nu = \nu_*, l_\text{b} = l_{\max}, \\ 2^{\nu l_\text{b}(d-1)}, & \nu > \nu_* \end{cases}\right). \tag{6.5.10}$$

In the downward pass, we should modify these estimates by putting $l_{\min}$ instead of $l_{\min}+1$ and the power of the E_4 neighborhood as a factor, similar to what we did to derive Eq. (6.2.17). We need, however, to modify the geometric factor $(2^d-1)3^d$ to $(2^{d-1}-1)3^{d-1}$, which is the number of non-empty boxes in the E_4-domain for the surface data distribution. Thus, we have

$$\text{CostDownward}_1(l_\text{b}) = (2^{d-1}-1)3^{d-1} \times O\left(\xi_S C_{\min}\begin{cases} 2^{l_{\max}(d-1)}, & \nu < \nu_*, \\ 2^{l_{\max}(d-1)}, & \nu = \nu_*, l_\text{b} < l_{\max}, \\ 2^{l_{\max}(d-1)}(l_{\max} - l_{\min}+1), & \nu = \nu_*, l_\text{b} = l_{\max}, \\ 2^{\nu l_\text{b}(d-1)}, & \nu > \nu_* \end{cases}\right), \tag{6.5.11}$$

where the "effective" critical exponent is modified as

$$\nu_* = \frac{l_{\max} - l_{\min}}{l_\text{b} - l_{\min}}. \tag{6.5.12}$$

This estimate contains $C_{\min}$ which depends on l_b and it is more convenient to introduce this dependence explicitly, as we did deriving Eq. (6.2.18):

$$\text{CostDownward}_1(l_\text{b}) = (2^{d-1}-1)3^{d-1} \times O\left(\xi_S C_*\begin{cases} 2^{(d-1)[\nu(l_{\text{b}*}-l_\text{b})+l_{\max}]}, & \nu < \nu_*, \\ 2^{l_{\max}(d-1)}, & \nu = \nu_*, l_{\text{b}*} < l_{\max}, \\ 2^{(d-1)[\nu(l_{\text{b}*}-l_{\max})+l_{\max}]}(l_{\max} - l_{\min}+1), & \nu = \nu_*, l_{\text{b}*} \geq l_{\max}, \\ 2^{\nu l_{\text{b}*}(d-1)}, & \nu > \nu_* \end{cases}\right). \tag{6.5.13}$$

Consider now the final summation step. Equation (6.2.21) should be modified for $M=N$ as

$$\text{CostEvaluation} = N[s3^{d-1}\text{CostFunc} + \text{CostEval}(P_{\min})], \tag{6.5.14}$$

since only of order 3^{d-1} non-empty boxes are in the neighborhood of a box containing a surface evaluation point. Parameter s is here specified by relation (6.5.3).

Equation (6.5.11) then provides an expression for the term related to translations in the total cost of the algorithm, if we neglect $\text{CostUpward}_3(l_b)$ and $\text{CostDownward}_2(l_b)$ compared to this term. The total cost can be written similar to Eq. (6.2.23), where we set $M = N$, as

$$\text{CostMLFMM} = O\Big(\text{CostDownward}_1(l_b) + 3^{d-1}2^{-(d-1)l_{\max}}\frac{1}{\xi_S}N^2\,\text{CostFunc} + N[\text{CostEval}(P_{\min}) + \text{CostRep}(P_{\min})]\Big). \tag{6.5.15}$$

Preliminary evaluations of the algorithm complexity can be done by neglecting the evaluation and representation costs (or considering them as term of the same order of magnitude as the other terms in the sum) and assuming $N \sim \xi_S 2^{l_{\max}(d-1)}$. In this case we will obtain

$$\text{CostMLFMM} = \begin{cases} O(N\log N), & l_{b*} = l_{\max},\ \nu = 1, \\ O(N), & \text{otherwise}. \end{cases} \tag{6.5.16}$$

As in the case of spatially uniform distributions, we note that the formal asymptotics in the latter case do not mean that it is faster than the former case, due to possibility that the asymptotic constant is huge. We will make a more detailed analysis of the cases in the optimization section.

6.5.2.2 *Error of the MLFMM*

The non-uniformity of distributions of sources and receivers does not affect Eq. (6.3.6) or estimations based on other error models, once they are performed in the way as Eq. (6.3.6) was obtained. This equation introduces logarithmic factors $\log^{\nu(d-1)}N$ into complexity estimates (Eq. (6.5.16)), similar to Eq. (6.3.7).

6.5.2.3 *Optimization for lower frequencies or larger number of sources and receivers*

Selection of the appropriate maximum level of space subdivision in the case

$$l_{\max} > l_{b*}, \qquad \nu \le \nu_* = \frac{l_{\max} - l_{\min}}{l_{b*} - l_{\min}} \tag{6.5.17}$$

can be done similar to the case of spatially uniform distributions with corrections to the factors depending on the dimensionality of the problem. Let us denote

$$\zeta = 2^{(d-1)l_{\max}}. \tag{6.5.18}$$

The optimum level can be found then as

$$l_{\max}^{(\text{opt})} = \frac{1}{d-1}\log_2 \zeta^{(\text{opt})} = \frac{1}{2(d-1)}\log_2\left(\frac{N^2}{2^{d-1}-1}\,\frac{\text{CostFunc}}{\xi_S^2 C_*}\right). \tag{6.5.19}$$

This is similar to relation (6.4.5) and so we can consider the example next to that equation ($d = 3$, $N \sim 10^6$, $\nu = 1$, $P_* = p_*^2 \sim 10^2$, $\text{CostFunc}/C \sim 10^{-2}$, and $\xi_S \sim 1$) to see the effect of non-uniformity of data distribution. This yields $l_{\max}^{(\text{opt})} \approx 6$. For the same parameters, but $\text{CostFunc}/C \sim 1$ we obtain $l_{\max}^{(\text{opt})} \approx 8$. As we can see the optimum levels here are higher than in the case of spatially uniform distribution of the same amount of points, when we had $l_{\max}^{(\text{opt})} = 4$ and 5, respectively. The optimum level required to handle surface data distribution efficiently is approximately 1.5 times larger than the optimum level for spatial distributions. Obviously, this is due to factor $1/(d-1)$ in Eq. (6.5.19) which appears instead of factor $1/d$ in relation (6.4.5). The ratio of these factors is $d/(d-1) = 1.5$. For non-uniform distributions the maximum level corresponding to the space subdivision until the state when each box contains not more than one source also increases compared to uniform distributions. Indeed, in the present example for surface data points we have $l_{\max} = (d-1)^{-1}\log_2 N \approx 10$. The number that approximately 1.5 times larger than $l_{\max} = 7$ is obtained for spatially uniform distributions.

As we are concerned with the efficiency of optimization, the effect of non-uniformity is rather weak, and this efficiency is almost the same as for uniform and non-uniform data. Indeed, for evaluation of the optimization efficiency (6.4.12) with ζ defined by relation (6.5.18) instead of ζ defined as Eq. (6.4.1) can be used. For optimum levels we have $(d-1)l_{\max}^{(\text{opt})(\text{surf})} \approx d l_{\max}^{(\text{opt})(\text{vol})}$, where $l_{\max}^{(\text{opt})(\text{surf})}$ and $l_{\max}^{(\text{opt})(\text{vol})}$ as the optimum maximum levels for surface and volume distribution of the same amount of points. Therefore, the parameters ζ are of the same order of magnitude for these two distributions. We can rephrase this fact as that for surface (non-uniform) distributions optimization is as important as for spatially uniform distributions. As we could see from the examples, optimization of $l_{\max}$ can speed up the algorithm 100 times or more.

We now note that the optimum found is valid if it satisfies the first constraint (6.5.17) that can be rewritten taking into account the explicit expression (6.1.5) for l_{b*}:

$$N > \xi_S \left[\frac{(2^{d-1}-1)C_*}{\text{CostFunc}} \right]^{1/2} \left(\frac{kD_0}{\kappa_b} \right)^{d-1}. \tag{6.5.20}$$

This shows that the optimum is realized either at larger N or smaller frequencies. The second constraint (6.5.17) is always satisfied for $\nu = 1$. However, if l_{max} is close enough to l_{b*} and $\nu > 1$, then this constraint may limit the range of applicability of the obtained results.

To see what additional restriction is, we rewrite the latter inequality (6.5.17) in the form:

$$l_{b*} - l_{min} = \log_2 \frac{kD_0}{2^{l_{min}} \kappa_b} < \frac{l_{max} - l_{min}}{\nu}. \tag{6.5.21}$$

Substituting here $l_{max}^{(opt)}$ determined by Eq. (6.5.19) we obtain

$$\log_2 \left(\frac{kD_0}{2^{l_{min}} \kappa_b} \right)^{\nu} < \frac{1}{2(d-1)} \log_2 \left[\frac{N^2}{(2^{d-1}-1)2^{2(d-1)l_{min}}} \frac{\text{CostFunc}}{\xi_S^2 C_*} \right], \tag{6.5.22}$$

or

$$\left(\frac{kD_0}{2^{l_{min}} \kappa_b} \right)^{\nu} < \frac{1}{2^{l_{min}}} \left[\frac{N^2}{(2^{d-1}-1)} \frac{\text{CostFunc}}{\xi_S^2 C_{min}} \right]^{1/[2(d-1)]}, \tag{6.5.23}$$

$$N > \xi_S \left[\frac{(2^{d-1}-1)C_*}{\text{CostFunc}} \right]^{1/2} 2^{(1-\nu)(d-1)l_{min}} \left(\frac{kD_0}{\kappa_b} \right)^{\nu(d-1)}. \tag{6.5.24}$$

Comparing this restriction with that specified by inequality (6.5.20), we can see that at large frequencies the latter restriction specifies higher N and so it limits the range of N.

6.5.2.4 *Optimization for higher frequencies or smaller number of sources and receivers*

For higher frequencies or relatively small number of sources and receivers conditions (6.5.20) and (6.5.24) may not hold. The cases to consider here is $l_{max} < l_{b*}$ and $l_{max} = l_{b*}$. According to definition (6.1.5) both these cases specify $l_b = l_{b*}$ and, therefore, the critical translation exponent is $\nu_* = 1$ (see its definition (6.5.12)). Since ν cannot be less than 1, we have only cases $\nu > \nu_*$ and $\nu = \nu_*$.

If $l_{max} = l_{b*}$ and $\nu = \nu_* = 1$ we have nothing to optimize, since this already prescribes l_{max} and the complexity of the algorithm is $O(N \log N)$ as prescribed by Eq. (6.5.16). For the same critical $\nu = 1$ and $l_{max} < l_{b*}$ the cost of the first step of the downward pass (Eq. (6.5.13)) grows linearly with $l_{\max}$, while parameter ζ grows exponentially with respect to l_{max}. The minimum in this case exists, and despite not being sharp it can be determined from Eq. (6.5.15) where we neglect the evaluation and representation costs for simplicity, as we did earlier. This results in the following equation for the optimum level or related parameter ζ (Eq. (6.5.18)):

$$\frac{d}{d\zeta}\left[\frac{2^{d-1}-1}{d-1}C_*\xi_S^2 2^{(d-1)l_{b*}}\log_2\zeta+\zeta^{-1}N^2\,\mathrm{CostFunc}\right]=0. \qquad (6.5.25)$$

Solution of this equation is

$$\zeta^{(\mathrm{opt})}=\ln 2\frac{d-1}{2^{d-1}-1}\frac{\mathrm{CostFunc}}{C_*\xi_S^2}N^2 2^{-(d-1)l_{b*}}. \qquad (6.5.26)$$

The cost of the optimized FMM can then be found as

$$\mathrm{CostMLFMM}=O\left(\left(\frac{kD_0}{\kappa_b}\right)^{d-1}\log N\right), \qquad (6.5.27)$$

with a large constant. This shows that the complexity grows slowly with N in this case and the optimum level is

$$l_{\max}^{(\mathrm{opt})}=\frac{1}{d-1}\log_2\zeta^{(\mathrm{opt})}$$
$$=\frac{1}{d-1}\log_2\left(\ln 2\frac{d-1}{2^{d-1}-1}\frac{\mathrm{CostFunc}}{C_*\xi_S^2}N^2 2^{-(d-1)l_{b*}}\right). \qquad (6.5.28)$$

This optimum level grow with N and therefore at some N the constraint $l_{\max} < l_{b*}$ will be violated. Condition $l_{\max}^{(\mathrm{opt})} < l_{b*}$ can be written using definition (6.1.5) of l_{b*} as

$$N<[(d-1)\ln 2]^{1/2}\xi_S\left[\frac{(2^{d-1}-1)C_*}{\mathrm{CostFunc}}\right]^{1/2}\left(\frac{kD_0}{\kappa_b}\right)^{d-1}. \qquad (6.5.29)$$

Comparing this result with the lower frequency relation (6.5.20) valid for $\nu=1$ we can see that the bounds for N almost coincide, since the constant $[(d-1)\ln 2]^{1/2}$ is close to 1. Here we can apply the same reasoning concerning the accuracy of the analysis and constants as in the case of spatially uniform data distributions.

The last case to consider is the case $\nu > \nu_* = 1$, $l_{\max} \leq l_{b*}$. Estimate (6.5.13) shows that, in this case, the complexity of the downward pass does not depend on N and is bounded by some large frequency-dependent constant. There is no extremum for this case, and the algorithm scales approximately as

$$\text{CostMLFMM} = O\left(\left(\frac{kD_0}{\kappa_b}\right)^{\nu(d-1)}\right), \qquad \nu > 1. \tag{6.5.30}$$

In the region where Eq. (6.5.24) holds, the optimum exists and we have CostMLFMM $= O(N)$. Comparing this result with the latter relation, we can see that the region at which constant asymptotic (6.5.30) is valid is bounded as

$$N < \xi_S \left[\frac{(2^{d-1}-1)C_*}{\text{CostFunc}}\right]^{1/2} 2^{(1-\nu)(d-1)l_{\min}} \left(\frac{kD_0}{\kappa_b}\right)^{\nu(d-1)}. \tag{6.5.31}$$

6.5.2.5 *Boundary element methods*

As in the case of spatial data distributions, we can estimate the complexity of the MLFMM for the boundary element methods. In these problems we can again assume that the number of sources N_s per wave of length $\lambda = 2\pi/k$ is fixed and is determined by the error of surface discretization. Correlation between parameters N and kD_0, in this case, is of the same type as Eq. (6.4.27) where we should modify the exponent:

$$N = \left(\frac{N_s}{2\pi} kD_0\right)^{d-1}. \tag{6.5.32}$$

Let us now assume that all the constants of the method are of the order of unity. Then, the analysis of the high and low frequency cases results in the following types of the algorithm behavior:

$$\text{CostMLFMM} = \begin{cases} O(N \log N), & \nu = 1, \\ O(N^{\nu}), & \nu > 1. \end{cases} \tag{6.5.33}$$

The critical case here is $\nu = 1$, while for other translation exponents the algorithm scales polynomially with N. For the three-dimensional problems with surface data distribution the "fast" translations should be really fast, since $\nu = 1$ is the theoretical minimum for translation exponent. In Chapter 8 we will consider translation methods of complexity $O(p^2 \log^{\alpha} p)$. This case is asymptotically close to $\nu = 1$ and provides MLFMM algorithms of total asymptotic complexity $O(N \log^{\beta} N)$, where β includes different factors, which we mentioned above.

Some numerical results

For the surface distributions of sources and receivers, $N = M$, we performed numerical tests using the actual MLFMM code, which was combined with the model to analyze performance of the MLFMM for various parameters and to get some insight into the method. All settings were the same as for the case of spatially uniform distributions. First, we performed optimization computations to determine $l_{\max}^{(\text{opt})}$. Results of this optimization task are shown in Figs. 6.3 and 6.4. As in the case of spatially uniform distributions we discriminated against two cases: "low frequencies" ($l_{b*} = 2$) and "high frequencies" ($l_{b*} = 5$). In all cases the minimum of $l_{\max}$ for $l_{\max}^{(\text{opt})}$ was strong and unique (i.e. the number of multiplications for computations with $l_{\max} > l_{\max}^{(\text{opt})}$ and $l_{\max} < l_{\max}^{(\text{opt})}$ was larger, then the number of multiplications for computations with $l_{\max} = l_{\max}^{(\text{opt})}$). The figure shows that in the "low frequency" and case the optimum level grows $\sim \log N$. The same type of behavior is observed for the "high frequency" case only for $\nu = 1$. The dependencies for $\nu = 1.5$ and 2 are qualitatively different. The optimum level is 2 (the minimum possible level for the MLFMM) for variation of N until some value N_* depending on ν and then the $l_{\max}^{(\text{opt})}$ "jumps" to the value, which is the optimum for the "low frequency" case. We also note that for surface distributions $l_{\max}^{(\text{opt})}$ are larger than for the spatial distributions. This is consistent with the above theory, which predicts 1.5 times larger $l_{\max}$ for surface distributions.

Figure 6.5 demonstrates the complexity of the MLFMM measured in terms of float multiplication operations as a function of N. Computations were performed using $l_{\max}^{(\text{opt})}$ shown in Figs. 6.3 and 6.4. The dependencies shown are qualitatively similar to the dependencies shown in Fig. 6.2. We can note only that the dependencies on N for the "high frequency" case and $\nu > 1$ are even more flat than that for spatially uniform distributions. In fact at high frequencies, $l_{\max} < l_{b*}$, and $\nu > 1$ the complexity of the MLFMM algorithm depends weakly on N, which follows from the theory (see Eq. (6.5.30)).

6.5.3 Surface distributions of sources and receivers: complex objects

Comparison of complexities and optimizations of the MLFMM for spatial and surface distributions of data points shows that the dimensionality of the problem plays an important role and related with the efficiency of translations and overall complexity of the method.

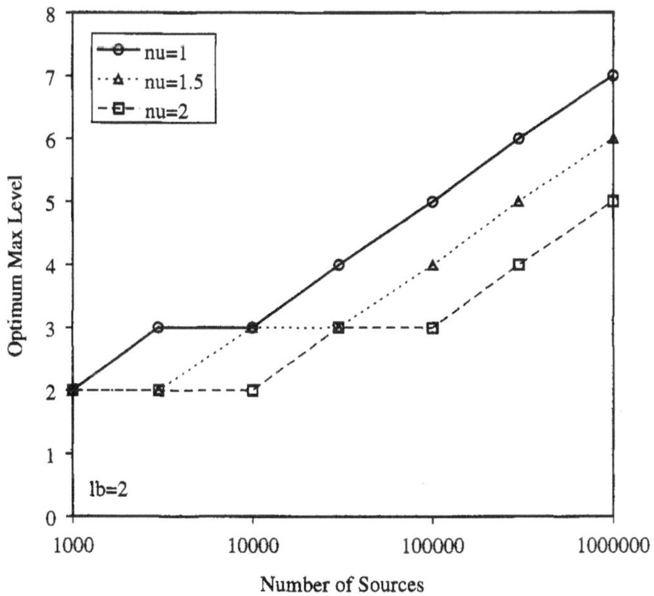

FIGURE 6.3 Dependence of the optimum maximum level of space subdivision on the number of sources for uniformly random distributions of sources and receivers over a sphere surface for different translation exponents ν for the "low frequency" case, $l_{b*} = 2$.

A point we want to make in this context is that even for a distribution of data points over a surface of some object the "effective" dimensionality of the problem can be d, not $d - 1$. Indeed, assume that a surface of some object passes through all the boxes in the computational domain at some level l. If the sampling of the surface is dense enough each box at this level contains sources and evaluation points. The same holds for level $l - 1$ and any level coarser than l. Therefore, in the MLFMM we need to perform all the translations for children–parent and neighbor boxes and this is the case for spatially uniform distribution of points. Therefore, this object behaves as a d-dimensional object for levels coarser than l. On the other hand, if we subdivide the space and go to levels finer than l, we can always reach some level that the surface intersects of order $O(2^{l(d-1)})$ boxes, since it is a $(d - 1)$-dimensional manifold. So in the case of a complex-shaped object or many objects in the domain, surface distributions of sources and receivers efficiently behave as d-dimensional or $(d - 1)$-dimensional manifolds. Moreover, the transition from one dimensionality to the

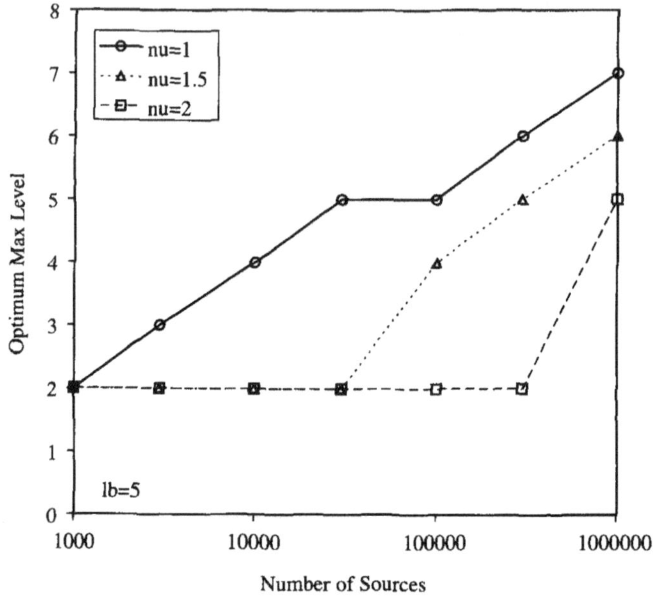

FIGURE 6.4 Dependence of the optimum maximum level of space subdivision on the number of sources for uniformly random distributions of sources and receivers over a sphere surface for different translation exponents ν for the "high frequency" case, $l_{b*} = 5$.

other does not happen as a "jump" and, in fact, one can consider some effective dimensionality, which can be even *fractional* (!).

Let us consider the following simple example to get a better understanding on what is going on when we divide the space hierarchically. Assume that all data points are distributed over a surface of a cube, which either coincide with the computational domain, or are just a bit smaller. Then we consider the number of non-empty boxes at each level. At level 0 all the points are located inside computational domain, and so the number of non-empty boxes is 1. At level 1 again all the boxes contain surface points, and the number of non-empty boxes is 8 ($d = 3$). At level 2 there are 56 non-empty boxes, while the total number of boxes is 64. A formula for the number of non-empty boxes at level $l \geq 2$ is

$$N_{\text{non-empty}}(l) = (2^l)^3 - (2^l - 2)^3 = 6 \times 2^{2l} - 12 \times 2^l + 8, \qquad l \geq 2. \tag{6.5.34}$$

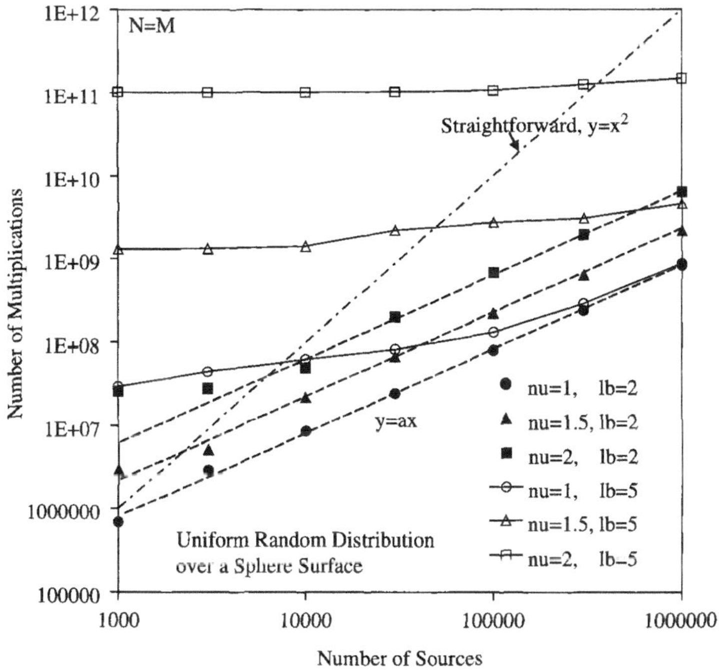

FIGURE 6.5 Dependences of the number of float multiplications on the number of sources, N, for uniform distribution of sources and receivers over a sphere surface ($M = N$). For each curve kD_0 is constant (determined by l_{b*}). Notations and parameters are the same as for Fig. 6.2.

This can be easily derived if we notice that only boxes adjacent to the boundary of the computational domain are non-empty and their number is a difference between the total number of boxes and boxes located inside a cube whose size is smaller than the size of the computational domain by two boxes.

Now if we consider two subsequent levels $l-1$ and l the ratio of the numbers of non-empty boxes at these levels determines an "effective" dimensionality of the problem, defined as

$$d_{\text{eff}}(l) = \log_2 \frac{N_{\text{non-empty}}(l)}{N_{\text{non-empty}}(l-1)}, \qquad l = 1, 2, \ldots. \tag{6.5.35}$$

Indeed, in the case when all the boxes are occupied, the case which is realized for spatially uniform distributions, we have $d = 3$, and in the case when we approach scaling $N_{\text{non-empty}}(l) \sim 2^{(d-1)l}$ we

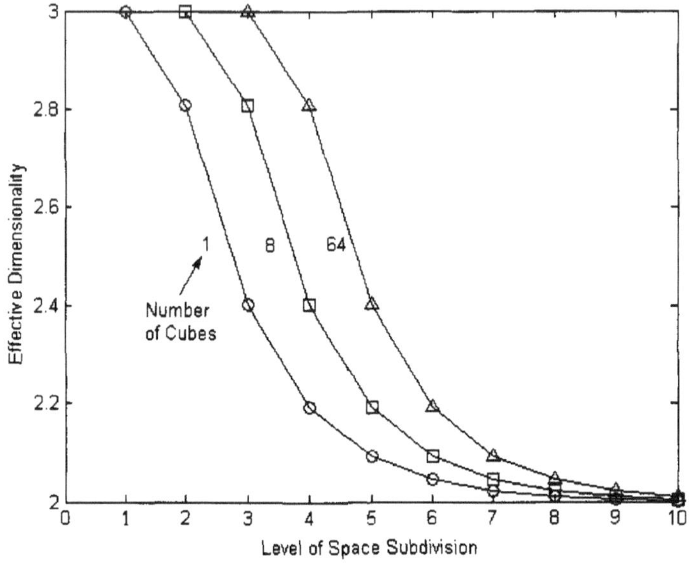

FIGURE 6.6 Dependence of the effective dimensionality of the problem on the maximum level of space subdivision for distribution of data points over the surfaces of several cubes (number indicated next to the curves).

have $d_{\mathrm{eff}}(l) = d - 1 = 2$. A dependence of $d_{\mathrm{eff}}(l)$ for this example is marked by circles in Fig. 6.6. We can see that for levels $l \geq 4$ the effective dimensionality is close to 2, while for levels $l \leq 2$ it is close to 3. For level 3 it is difficult to say whether we should apply the results of dimensionality 2 or 3 to evaluate the complexity of the algorithm.

We can make the case a bit complex, but easy for evaluation if we consider, say, eight cubes with surface point distributions, such that each cube has approximately half of size of the computational domain and their sides are close to the sides of boxes of level 1. In this case due to the hierarchical nature of space subdivision, we have simply a shift of the graph for effective dimensionality to the right (see Fig. 6.6). We can go further and consider computations, when the sources and receivers are distributed over the surfaces of 64 cubes, and so on. So, for more complex surfaces the effective dimensionality of the problem increases and, in fact, it depends on $l_{\max}$ which we select for computations whether we should treat the surface distribution as two dimensional or three dimensional. This also clarifies what we meant by "simple" objects. Now we can define this more accurately by saying that for simple objects the effective dimensionality is close to 2 for any $l \gtrsim l_{\min}$. Of course, this is an idealization, since even for the simplest object as a cube this condition does not hold at level 3.

Pointing out this issue, we will not go too far with complexity models and optimizations. Several qualitative conclusions, however, can be made immediately.

First, complexity estimates and relation of the translation exponent ν to dimensionality (especially critical value of this parameter) are obtained within some idealization and are of a rather qualitative character. So this should be considered in relation to particular cases and one of the factors that influence the complexity is the complexity of the object boundary. Second, for complex-shaped objects, or multiple objects, surface distributions of sources and targets can behave as three-dimensional manifolds or manifolds with fractional dimensionality. Since we obtained the values for critical translation exponents $\nu = d/(d-1)$ and $\nu = 1$ for two limiting cases, the actual complexity of the MLFMM can either be $O(N)$ or $O(N \log N)$ for some intermediate ν, $d/(d-1) > \nu > 1$. Finally, we can note that in the high frequency limit when the maximum level increases with frequency and is large enough, the effective dimensionality of the problem should be close to $d - 1$ for fine levels. For $\nu > 1$ computations for coarser levels are more expensive than for finer levels, and the change of the effective dimensionality can substantially influence the speed of the algorithm.

6.5.4 Other distributions

The MLFMM can be applied for different non-uniform distributions of sources and receivers. For example, an important problem can be to determine three-dimensional field scattered by an object. In this case the BEM provides a solution of the problem, where the sources are located on the surface of the scatterer, while the receivers are distributed in space. Distribution of this type is very different from spatially uniform distribution of both sets and from distribution of both sets of data points over the surface. While currently under active research and development, we obtained some preliminary results, which show that in this case the use of a version of the MLFMM, which is fully adaptive with respect to both sets (we call it "adaptive FMM" opposed to "regular FMM"), provides better results, speed up of algorithms and may not require optimization procedures. We describe this method in a separate section below.

6.6 ADAPTIVE MLFMM

The MLFMM algorithm described above is adaptive in a sense that it does not go over all the boxes of space subdivision, but only over boxes which are in the source and target hierarchies. This data structure, strictly speaking, is not a *tree* data structure as usually accepted in computer science, but a *pyramid* structure. The difference between the tree and pyramid hierarchies is that the leaves (end points) of a tree can be located at any level of space subdivision, while all end points of the pyramid are located at the finest level. So a pyramid is a particular kind of tree, all leaves of which are located at the finest level. Computations utilizing pyramid structures may not be optimal, since the result may be obtained at the coarser level, while the algorithms proceeds to the finest level and performs some unnecessary computations. The purpose of a fully adaptive method is to avoid such unnecessary operations using tree structures.

There exist in literature several versions of adaptive FMM algorithms. Some researchers call the MLFMM "adaptive" if it skips non-empty boxes, i.e. operates with data hierarchy, not with all the boxes. That is the method described above, which we call "regular MLFMM". On the other hand, we can find adaptive algorithms, which utilize tree structures. For example, in Ref. [CGR99] one can find a description of such an algorithm applied for solution of the Laplace equation. In this algorithm the neighbors of a given box can be boxes of different sizes (different levels). The authors suggest to use different evaluation strategies (either to use translations, or just apply expansions near the centers of boxes) depending on the number of points, size of the boxes and distance from a given box. A deficiency of the described method is that it is developed for the case when the target and source sets are the same ($\mathbb{X} = \mathbb{Y}$), so it should be modified for the case when these are different (say, in the case when sources are distributed over some surface, while the receivers occupy some volume in space). In addition, introduction of the concept of neighborhoods with different box sizes makes data structures more complex and the complexity of creating and operating with such data structures is not investigated yet.

Below we describe some other method developed and tested in our lab, which we call "adaptive MLFMM". While it was developed and tested for "low frequency" problems (constant length of the representing vectors), and its efficiency for "high frequency" problems should be investigated more carefully, we present the basic ideas of the method here to show how the data structure can be modified. This method can operate with arbitrary data sets $\mathbb{X}$ and $\mathbb{Y}$, fully adaptive with respect to *both* sets, does not use concepts of neighborhood with different box sizes, and operates with data organized in trees and *forests* (sets of trees). Similar to the regular MLFMM the adaptive MLFMM has some arbitrary parameter, which controls the data structure and can be a subject for optimization. If in the regular MLFMM the maximum level of space subdivision, $l_{\max}$, is a subject for optimization, in the adaptive MLFMM we introduce some parameter q which is the maximum number of sources in the neighborhood of a receiver, which is acceptable for direct evaluation of their contribution to the field at the location of the evaluation point.

6.6.1 Setting up the hierarchical data structure

6.6.1.1 General idea

Before providing the details of the adaptive algorithm, let us describe its general idea. It is based on decomposition (5.2.7) that is valid at any level. We note that the potential at any evaluation point of set $\mathbb{Y}$ can be found in the form similar to Eq. (5.2.21):

$$\psi(\mathbf{y}_\beta) = \psi^{(2)}_{n,l}(\mathbf{y}_\beta) + R_{P_l}(\mathbf{y}_\beta; \mathbf{y}^{(n,l)}_*), \qquad \mathbf{y}_\beta \in E_1(n,l), \ \forall l, \tag{6.6.1}$$

which is valid for any level of the hierarchical space subdivision. This formula consists of two terms. The first term is related to the direct summation of sources in the neighborhood of the evaluation point, and the second term represents the contribution of all the sources outside the neighborhood. Therefore, for each point $\mathbf{y}_\beta$ we can determine a level l_β of the space subdivision at which the number of sources in the neighborhood does not exceed some prescribed number q and the cost of direct computation of the field of sources in the neighborhood is $O(q)$. To determine the second term in the right-hand side of Eq. (6.6.1), we need to compute representing vectors $\mathbf{D}^{(n,l)}$ only for level $l = l_\beta < l_{\max}$ that is specific for each point $\mathbf{y}_\beta$ and is determined by the number of points of the set $\mathbb{X}$ in the neighborhood of that point. The set of box numbers and levels (n_β, l_β), $\beta = 1, \ldots, M$, for which we need to evaluate $\mathbf{y}_\beta$ will provide a set of *target* boxes for which we need to obtain $\mathbf{D}^{(n_\beta, l_\beta)}$.

To compute vectors $\mathbf{D}^{(n_\beta, l_\beta)}$ we will construct a tree which will show what other vectors $\mathbf{D}^{(n,l)}$, $l < l_\beta$ are needed to obtain the value of $\mathbf{D}^{(n_\beta, l_\beta)}$. We will call such tree as *D-tree*. Since $\mathbf{D}^{(n_\beta, l_\beta)}$ depends not only on $\mathbf{D}^{(n,l)}$, but also on $\tilde{\mathbf{D}}^{(n,l)}$, we will also construct a *$\tilde{D}$-tree*, which shows the entire set of vectors $\tilde{\mathbf{D}}^{(n,l)}$ that is needed for the computation of $\mathbf{D}^{(n_\beta, l_\beta)}$. A more detailed look at the formulae for computation of representing vectors shows that the *$\tilde{D}$-tree* is the same as the *D-tree*. Further, we will build a *C-forest* (which may be a single tree or a union of several trees) that will contain the numbers and levels of boxes, (n, l), for which we need to have vectors $\mathbf{C}^{(n,l)}$ to obtain $\tilde{\mathbf{D}}^{(n,l)}$ and further $\mathbf{D}^{(n,l)}$ belonging to the *D-tree* and *$\tilde{D}$-tree*.

It is clear that the described scheme is:

- Adaptive, since it provides the level of the box specific for each point $\mathbf{y}_\beta$, which depends on both sets $\mathbb{X}$ and $\mathbb{Y}$.
- Can be more economic than the regular FMM, since it is seeking for computation of the representing vectors that are necessary to achieve the objective and does not need computation of unnecessary representing vectors.

6.6.1.2 *Determination of the target box levels/numbers*

The first step of the present adaptive algorithm includes determination of all target boxes and an appropriate number of levels of the hierarchical space subdivision, $l_{\max}$. This depends on the prescribed number q that is the maximum allowed number of sources which can be summed directly to evaluate the potential due to these sources. All these tasks can be performed within one-step procedure that algorithmically can be described as follows. In our notation we denote the set of target boxes as $\mathbb{T}$ and the number of levels in the hierarchical space subdivision as $l_{\max}$. The set of target boxes at level l will be denoted as $\mathbb{T}_l$. So $\mathbb{T} = \mathbb{T}_2 \cup \cdots \cup \mathbb{T}_L$.

(1) Set the level of consideration $l = l_{\min} = 2$ and the remaining set of the evaluation points $\mathbb{Y}_{\text{rem}} = \mathbb{Y}$.
(2) For each box (n, l) that contains a point $\mathbf{y}_\beta \in \mathbb{Y}_{\text{rem}}$ find the number N_n of points $\mathbf{x}_\alpha \in E_2(n, l)$.
(3) If $N_n \leq q$ refer the box (n, l) to the set of target boxes $\mathbb{T}_l$ and exclude all $\mathbf{y}_\beta \in E_1(n, l)$ from the set $\mathbb{Y}_{\text{rem}}$.
(4) If $\mathbb{Y}_{\text{rem}} = \emptyset$, set the maximum number of levels $l_{\max} = l$ and stop the procedure, else increase the level, $l = l + 1$, and repeat steps 2–4 of this algorithm.

EXAMPLE 1 In Fig. 6.7 an example of the output of the above algorithm for $d = 2$ and particular distribution of the source and evaluation points is shown. It is seen that for $q = 3$ it is sufficient to have $l_{\max} = 4$ levels of the hierarchical space subdivision. The number of target boxes where the potential should be evaluated is much smaller than the total number of boxes if the regular FMM is used.

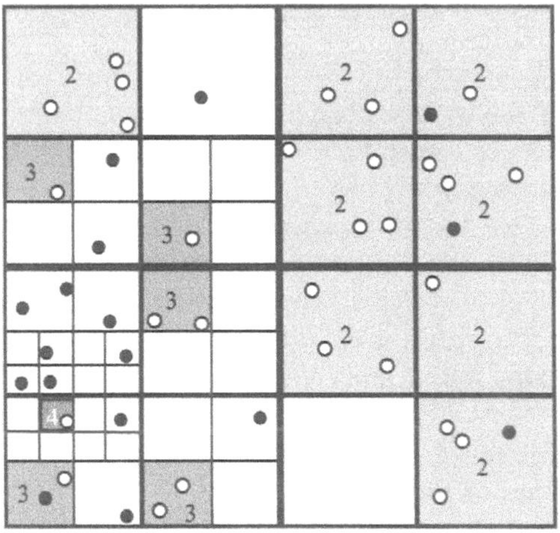

FIGURE 6.7 Example of determination of the target boxes and space subdivision level in the present adaptive FMM. These are boxes that contain evaluation points (the open circles) and colored by gray. The level for each target box is shown by the number in the box. This level is determined from the requirement that $E_2(n, l)$ domain for each target box does not contain more than $q = 3$ source points (the filled circles).

6.6.1.3 Construction of the D-tree

Let us denote the set of the boxes in the *D*-tree as $\mathbb{D}$. We also denote through $\mathbb{D}_l$ sets of boxes in the *D*-tree at level $l \geq 2$, so $\mathbb{D} = \mathbb{D}_2 \cup \cdots \cup \mathbb{D}_{l_{\max}}$. According to Eq. (6.2.19), to find $\mathbf{D}^{(n,l)}$ we need to know $\mathbf{D}^{(n',l-1)}$, where $n' = \text{Parent}(n)$. So the tree can be generated from the target box set by applying function Parent until the level $l = l_{\min}$ is reached. In other words, if box $(n, l) \in \mathbb{D}$, $l > l_{\min}$, then box $(\text{Parent}(n), l - 1) \in \mathbb{D}$. The algorithm here is straightforward:

(1) Refer all target boxes at the finest level of space subdivision $l_{\max}$ to set $\mathbb{D}_{l_{\max}}$ and set $l = l_{\max}$.
(2) If $l = l_{\min}$ stop the procedure, else set $l = l - 1$ and perform steps 2–4 in a loop.
(3) Determine $(\text{Parent}(n), l)$ for all boxes from $\mathbb{D}_{l+1}$ and refer them to set $\mathbb{D}_l$.
(4) Make a union of $\mathbb{D}_l$ and target boxes $\mathbb{T}_l$ and return the result back to $\mathbb{D}_l$.

Note that in the *D*-tree all target boxes represent *leaves* of the tree (in other words end points for each branch of the tree). So $\mathbb{T}$ is the set of all leaves in the D-tree and $\mathbb{T}_l$ is the set of leaves of this tree at level $l = l_{\min}, \ldots, l_{\max}$.

EXAMPLE 2 In Fig. 6.8 an example of the D-tree is shown for a quadtree ($d = 2$). The target boxes and the maximum level of the space subdivision ($l_{\max} = 4$) are determined by the algorithm described in the previous section. These target boxes represent leaves of the D-tree shown by filled circles. Function Parent applied to the target boxes generates nodes of the tree shown in gray. This function recursively applied to the gray boxes generates the entire tree up to level 0. However, we need to consider only levels $l \geq l_{\min}$ since vectors $\mathbf{D}^{(n,l)}$ are defined only for levels $l \geq l_{\min}$.

6.6.1.4 Construction of the $\tilde{D}$-tree

Equation (5.2.19) shows that for computation of $\mathbf{D}^{(n,l)}$ vectors we also need to know $\tilde{\mathbf{D}}^{(n,l)}$. Since vectors $\tilde{\mathbf{D}}^{(n,l)}$ should be computed exactly for the same boxes as vectors $\mathbf{D}^{(n,l)}$ the $\tilde{D}$-tree simply coincides with the D-tree, and further we will not make differences between these trees.

6.6.1.5 Construction of the C-forest

Now, when the D-tree is built, let us consider the boxes for which we need to know vectors $\mathbf{C}^{(n,l)}$ to compute $\tilde{\mathbf{D}}^{(n,l)}$ for all boxes belonging to $\mathbb{D}$. If $\mathbb{C}$ is the set of boxes for which we need to compute $\mathbf{C}^{(n,l)}$, this set can also be

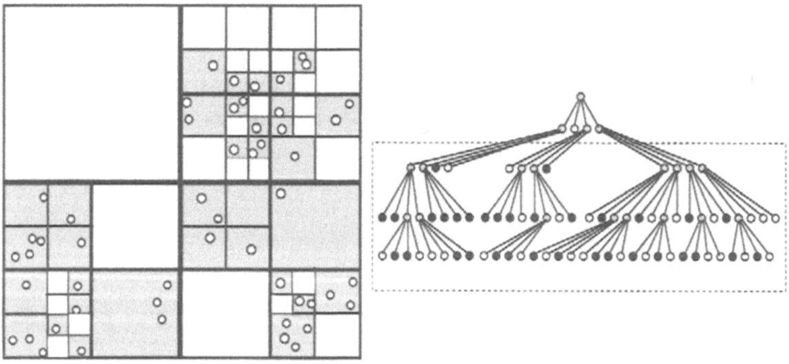

FIGURE 6.8 Example of the D-tree (on the right chart) corresponding to the target boxes shown in gray on the left chart. Each target box in the D-tree represent a leaf indicated by filled circle. Open circles in the tree are boxes that do not contain evaluation points (also shown in white in the left chart). The gray circles in the tree correspond to the parent boxes. The D-tree as well as target boxes are defined only for levels $l \geq 2$. This part of the tree is enclosed in a box with the dashed borders in the right chart. The open circles on the left chart show evaluation points $\mathbf{y}_j \in \mathbb{Y}$.

organized in a hierarchical multilevel structure, which we call C-forest. Equation (5.2.17) shows that for computation of $\tilde{\mathbf{D}}^{(n,l)}$ at some level l we need to know vectors $\mathbf{C}^{(n,l)}$ only for level l. Therefore, the finest level in the C-forest will be the same as in the $\tilde{D}$-tree and equal to $l_{\max}$. We also note that by definition $\mathbf{C}^{(n,l)} = 0$ if box (n,l) does not contain the source data points from set $\mathbb{X}$. So we will exclude all such boxes from the C-forest.

The C-forest has a different structure than D-tree. We consider the structure called "forest" as a union of K trees. The number K of these trees depends on the D-tree and the source data set $\mathbb{X}$, and we will determine it within the algorithm that constructs the C-forest. We denote these trees as $\mathbb{C}^1, \ldots, \mathbb{C}^K$. These trees are independent, so

$$\mathbb{C}^j \cap \mathbb{C}^{j'} = \emptyset, \qquad j \neq j', \qquad \mathbb{C} = \mathbb{C}^1 \cup \cdots \cup \mathbb{C}^K. \tag{6.6.2}$$

Each tree will have its own coarsest and finest level, which we denote as $l_{\min}^{(m)}$ and $l_{\max}^{(m)}$, respectively, $m = 1, \ldots, K$. All these levels are within the range $l_{\min} \leq l_{\min}^{(m)} \leq l_{\max}^{(m)} \leq l_{\max}$, and will be determined within the algorithm provided below. We also denote as $\mathbb{C}_l$ the sets of boxes in the C-forest at level $l \geq l_{\min}$, so $\mathbb{C} = \mathbb{C}_{l_{\min}} \cup \cdots \cup \mathbb{C}_{l_{\max}}$. The boxes of level l can belong to different trees. We denote the set of boxes at level l that belongs to the mth tree as $\mathbb{C}_l^m$. It follows from the definition that $\mathbb{C}_l^m = \mathbb{C}^m \cap \mathbb{C}_l$.

Construction of a C-forest can be performed using the following algorithm, which consists of two steps. At the first step we determine all $\mathbb{C}_l$ and at the second step we organize the data in the forest structure.

Step 1
Determine sets $\mathbb{C}_l$, $l = l_{\min}, \ldots, l_{\max}$.
(1) For levels $l = l_{\max}$ to $l = l_{\min}$ perform the following operation.
(2) For each $(n,l) \in \mathbb{D}_l$ refer boxes $\in I_4(n,l)$ which contain points of $\mathbb{X}$ to $\mathbb{C}_l$.

Step 2
Determine K, $l_{\min}^{(m)}$, $l_{\max}^{(m)}$ and sets $\mathbb{C}_l^m$, $l = l_{\min}, \ldots, l_{\max}$, $m = 1, \ldots, K$.
(1) Find the number of elements P_2 in the set $\mathbb{C}_2$. Set $K = P_2$, and $\mathbb{K} = [1, \ldots, K]$. For $m = 1, \ldots, K$ set $l_{\min}^{(m)} = l_{\min}$, and refer each element to $\mathbb{C}_{l_{\min}}^m$.
(2) For $l = l_{\min} + 1, \ldots, l_{\max}$ perform the following loop.
(3) For $m \in \mathbb{K}$ find the sets $\mathbb{G}_l^m = \text{Children}(\mathbb{C}_{l-1}^m) \cap \mathbb{C}_l$. If $\mathbb{G}_l^m \neq 0$, refer all elements of $\mathbb{G}_l^m$ to $\mathbb{C}_l^m$, else exclude m from the set $\mathbb{K}$ and set $l_{\max}^{(m)} = l - 1$.
(4) Find the number of elements P_l in the set $\mathbb{C}_l \backslash (\cup_{m \in \mathbb{K}} \mathbb{G}_l^m)$. Increase $K = K + P_l$, $\mathbb{K} = \mathbb{K} \cup [K+1, \ldots, K+P_l]$. For $m = K+1, \ldots, K+P_l$ set $l_{\min}^{(m)} = l$, and refer each element of $\mathbb{C}_l \backslash (\cup_{m \in \mathbb{K}} \mathbb{G}_l^m)$ to $\mathbb{C}_l^m$.

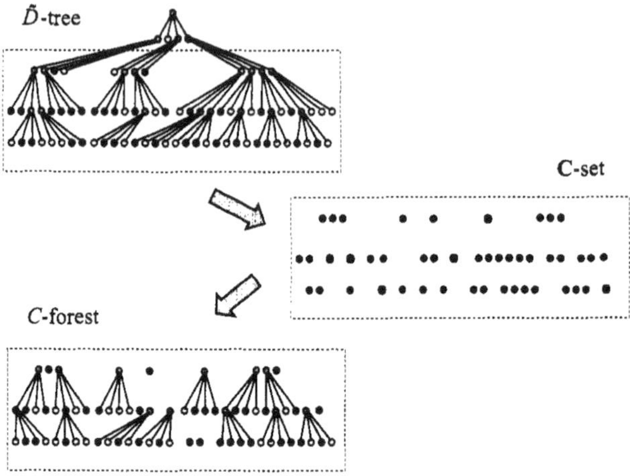

FIGURE 6.9 An example of construction of the C-forest, based on given $\tilde{D}$-tree (or D-tree) and the source data set $\mathbb{X}$. At first step we find all boxes that belong to this forest, by intersection of the I_4^1 and Parent(I_4^2) sets for each point of $\mathbb{D}$ with $\mathbb{X}$. At the second step we establish a forest structure for these data.

EXAMPLE 3 An example shown in Fig. 6.9 represents the two steps of the above algorithm for a quadtree ($d = 2$). As a result we obtain a C-forest, which in this example consists of $K = 16$ trees (single nodes in the graph are also considered as "trees"—such trees have a root and the only leaf, which coincides with the root).

6.6.2 Procedure

As in the regular multilevel FMM, in the adaptive FMM the evaluation procedure of the potential at all points of $\mathbb{Y}$ data sets consists of the upward and downward passes. As a result of the upward pass we will obtain all $\mathbf{C}^{(n,l)}$ vectors for the computation of necessary $\tilde{\mathbf{D}}^{(n,l)}$ vectors. At the end of the downward pass we will get $\tilde{\mathbf{D}}^{(n,l)}$ and $\mathbf{D}^{(n,l)}$ vectors sufficient for evaluation of the potential at the required points.

6.6.2.1 Upward pass

The upward pass in the present adaptive scheme is modified compared to the upward pass in the regular FMM to reduce the number of unnecessary translation operations. We note that structuring data in C-forest enables computation of representing vectors for each tree independently. Further, the representing vectors for the boxes that are leaves in the tree can be performed by direct summation of the far-field representations for all sources inside such a leaf box. The larger this number, the larger the savings in translations, since no additional subdivision of the leaf boxes is required. Finally, optimization occurs when we go to the parent boxes in a tree. In this case we will check whether sibling boxes that do not belong to the set $\mathbb{C}$ contain source points from $\mathbb{X}$. If this happens we can sum up directly the far-field representations of the points from that boxes near the center of the parent box to combine with translated representations for the children boxes that belong to $\mathbb{C}$. The algorithm for the upward pass can then be written as follows.

Step 1. For each source $\mathbf{x}_\alpha$ obtain vectors $\mathbf{B}^{(\alpha)}(\mathbf{x}_*^{(n,l)})$, $l = l_{\max}^{(m)}$, of length $P = P_{l_{\max}^{(m)}}$ corresponding to function $S_P^{(\alpha)}(\mathbf{y};\mathbf{x}_*^{(n,l)})$, which approximates $Q_\alpha\Phi(\mathbf{y}-\mathbf{x}_\alpha)$ in the domain outer to the sphere $\Omega_a(\mathbf{x}_*^{(n,l)})$:

$$S_P^{(\alpha)}(\mathbf{y};\mathbf{x}_*^{(n,l)}) \rightleftarrows \mathbf{B}^{(\alpha)}(\mathbf{x}_*^{(n,l)}), \qquad l = l_{\max}^{(m)},\ (n,l) \in \mathbb{C}_l^m,\ \mathbf{x}_\alpha \in E_1(n,l),\ P = P_{l_{\max}^{(m)}},$$
$$[\forall \mathbf{y},\ |\mathbf{y}-\mathbf{x}_*^{(n,l)}| \geq a^{(m)},\ |S_P^{(\alpha)}(\mathbf{y};\mathbf{x}_*^{(n,l)}) - Q_\alpha\Phi(\mathbf{y}-\mathbf{x}_\alpha)| < \epsilon_*,\ a^{(m)} = \tfrac{3}{2}D_l], \tag{6.6.3}$$

where $\mathbf{x}_*^{(n,l)}$ is the center of the leaf box of the mth tree ($m = 1,\ldots,K$) containing $\mathbf{x}_\alpha$ and ϵ_* is some specified error of approximation. Selection of $a^{(m)}$ is motivated by the requirement that the S approximation (5.2.14) for the nth box is valid in the domain $E_3(n, l_{\max}^{(m)})$. In contrast to the regular FMM the leaf boxes can be located at arbitrary levels, and therefore the length of the representing vector, P, can be different, if it is a function of the box size.

For $m = 1,\ldots,K$ (for each tree of the C-forest) perform the following steps.

Step 2. Generate $\mathbf{C}^{(n,l)} = \mathbf{C}(\mathbf{x}_*^{(n,l)})$ of length $P = P_l$ for the leaf boxes, $l = l_{\max}^{(m)}$, $(n,l) \in \mathbb{C}^m$. These vectors correspond to function $S_P(\mathbf{y};\mathbf{x}_*^{(n,l)})$ which approximates $\psi_{n,l}^{(1)}(\mathbf{y})$:

$$\mathbf{C}^{(n,l)} = \sum_{\mathbf{x}_\alpha \in E_1(n,l)} \mathbf{B}^{(\alpha)}(\mathbf{x}_*^{(n,l)}), \qquad l = l_{\max}^{(m)},\ (n,l) \in \mathbb{C}^m,\ P = P_{l_{\max}^{(m)}},$$
$$[S_P(\mathbf{y};\mathbf{x}_*^{(n,l)}) \rightleftarrows \mathbf{C}^{(n,l)},\ \forall \mathbf{y} \in E_3(n,l),\ |S_P(\mathbf{y};\mathbf{x}_*^{(n,l)}) - \psi_{n,l}^{(1)}(\mathbf{y})| < \epsilon_l]. \tag{6.6.4}$$

Here ϵ_l is some specified error of approximation, which can be linked to the overall error of the method. Particularly, it can be equal to ϵ_* of the Step 1.

Step 3. For $l = l_{\max}^{(m)} - 1, \ldots, l_{\min}^{(m)}$ recursively generate $\mathbf{C}^{(n,l)}$ for all other boxes in tree $\mathbb{C}^m$ (so $(n,l) \in \mathbb{C}^m$):

$$\mathbf{C}^{(n,l)} = \sum_{n' \in I^{(1)}} (S|S)(\mathbf{x}_*^{(n,l)} - \mathbf{x}_*^{(n',l+1)})[\mathbf{C}^{(n',l+1)}] + \sum_{\mathbf{x}_\alpha \in E_1(n',l+1), n' \in I^{(2)}} \mathbf{B}^{(\alpha)}(\mathbf{x}_*^{(n,l)}),$$

$$I^{(1)} = \text{Children}(X; n, l) \cap \mathbb{C}_{l+1}^m, \qquad I^{(2)} = \text{Children}(X; n, l) \cap \overline{\mathbb{C}_{l+1}^m},$$

$$\mathbb{C}_{l+1}^m \cap \overline{\mathbb{C}_{l+1}^m} = \emptyset,$$

$$[S_{P_l}(\mathbf{y}; \mathbf{x}_*^{(n,l)}) \rightleftarrows \mathbf{C}^{(n,l)}, \quad \forall \mathbf{y} \in E_3(n,l), \quad |S_{P_l}(\mathbf{y}; \mathbf{x}_*^{(n,l)}) - \psi_{n,l}^{(1)}(\mathbf{y})| < \epsilon_l]. \tag{6.6.5}$$

Here the first sum on the right-hand side represents translation of the already found representing vectors in the tree to the center of the parent box (n,l), and the second term represents a sum over the children boxes of (n,l) that are not boxes of the $\mathbb{C}^m$ tree, but contain sources $\mathbf{x}_\alpha \in \mathbb{X}$.

Figure 6.10 illustrates Step 3 of the algorithm. Box number 2 in this graph belongs to the $\mathbb{C}$ set and so $\mathbf{C}^{(n',l+1)}$ vectors for this box are computed. To obtain $\mathbf{C}^{(n,l)}$ for the parent box, $n = \text{Parent}(n')$ at the lower level, we translate the representing vector for box 2 to the center of the parent box. The contribution of sources located in the sibling boxes of $(n', l+1)$, which do not belong to the $\mathbb{C}$ set can be found directly by representation of the field due to these sources near the center of the parent box.

6.6.2.2 *Downward pass*

In the downward pass of the adaptive algorithm we perform the same operations as in the regular multilevel FMM with modifications that includes building $\tilde{\mathbf{D}}^{(n,l)}$ and $\mathbf{D}^{(n,l)}$ vectors only for nodes listed as nodes of the D-tree and stopping the recursive procedure when the leaves of these trees are reached. So, for completeness of the algorithm we rewrite the downward pass, which applies Steps 1 and 2 below for each level $l = l_{\min}, \ldots, l_{\max}$:

Step 1. In this step we form the representing vectors, $\tilde{\mathbf{D}}^{(n,l)}$, of length P_l for the regular function $\tilde{R}_{P_l}(\mathbf{y}; \mathbf{y}_*^{(n,l)})$ which approximates the function $\psi_{n,l}^{(4)}(\mathbf{y})$ about the center $\mathbf{y}_*^{(n,l)}$ of box $(n,l) \in \mathbb{D}$:

$$\tilde{\mathbf{D}}^{(n,l)} = \sum_{n' \in I_4(n,l) \cap X} (S|R)(\mathbf{y}_*^{(n,l)} - \mathbf{y}_*^{(n',l)})[\mathbf{C}^{(n',l)}], \qquad (n,l) \in \mathbb{D}, \tag{6.6.6}$$

$$[\tilde{R}_{P_l}(\mathbf{y}; \mathbf{y}_*^{(n,l)}) \rightleftarrows \tilde{\mathbf{D}}^{(n,l)}, \quad \forall \mathbf{y} \in E_1(n,l), \quad |\tilde{R}_{P_l}(\mathbf{y}; \mathbf{y}_*^{(n,l)}) - \psi_{n,l}^{(4)}(\mathbf{y})| < \epsilon_l].$$

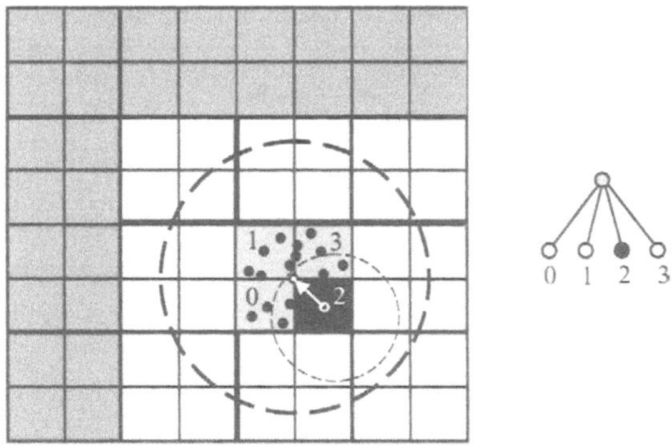

FIGURE 6.10 Illustration of the upward pass in the adaptive FMM. To obtain far-field expansion for the parent box we translate expansions from the children boxes which are in the tree (black) and directly reexpand and sum up all sources inside the other children boxes near the center of the parent box (gray). Such children (0,1,3) are shown by open circles in the tree graph.

Step 2. Assuming that for $l = l_{\min}$

$$\mathbf{D}^{(n,l_{\min})} = \tilde{\mathbf{D}}^{(n,l_{\min})}, \qquad (n, l_{\min}) \in \mathbb{D}, \qquad [\psi^{(3)}_{n,l_{\min}}(\mathbf{y}) = \psi^{(4)}_{n,l_{\min}}(\mathbf{y})], \quad (6.6.7)$$

we form the representing vectors $\mathbf{D}^{(n,l)}$ of length P_l for the regular function $R_{P_l}(\mathbf{y}; \mathbf{y}_*^{(n,l)})$ which approximates the function $\psi^{(3)}_{n,l}(\mathbf{y})$ about the box center $(n, l) \in \mathbb{D}$:

$$\mathbf{D}^{(n,l)} = \tilde{\mathbf{D}}^{(n,l)} + (R|R)(\mathbf{y}_*^{(n,l)} - \mathbf{y}_*^{(n',l-1)})\mathbf{D}^{(n',l-1)},$$

$$n' = \mathrm{Parent}(n, l), \quad (n, l) \in \mathbb{D}, \qquad (6.6.8)$$

$$[R_{P_l}(\mathbf{y}; \mathbf{y}_*^{(n,l)}) \rightleftarrows \mathbf{D}^{(n,l)}, \quad \forall \mathbf{y} \in E_1(n, l), \quad |R_{P_l}(\mathbf{y}; \mathbf{y}_*^{(n,l)}) - \psi^{(3)}_{n,l}(\mathbf{y})| < \epsilon_l].$$

6.6.2.3 Final summation

Evaluation of the potential at all required points $\mathbf{y}_\beta \in \mathbb{Y}$ can be made with the aid of Eq. (6.6.1) as soon as all necessary vectors $\mathbf{D}^{(n,l)}$ in the *D*-tree are obtained during the downward pass.

6.6.3 Complexity and optimization of the adaptive MLFMM

Investigation of complexity and optimization of the adaptive MLFMM described above in general case is beyond the scope of this book, since it is related to analysis of properties of various distributions of sources and receivers. Below we address just several qualitative issues concerning the use of this method in several typical cases.

6.6.3.1 *Data distributions*

The efficiency of the adaptive MLFMM (AMLFMM) compared to the regular MLFMM (RMLFMM) substantially depends on data (sources and receivers) distributions. We found that while in some cases there are no benefits or harm from the use of the adaptive method, there are some cases, when it performs better than the RMLFMM.

Spatially uniform distributions of sources and targets

In this case one should not expect any benefits of use the adaptive vs regular MLFMM. Indeed, the parameter q in the adaptive method is $q = 3^d s$, where s is the number of sources per box. As an optimum level $l_{\max}^{(\mathrm{opt})}$ is determined for the RMLFMM so that $s_{\mathrm{opt}} = 2^{-dl_{\max}^{(\mathrm{opt})}} N$, then the optimum q is

$$q_{\mathrm{opt}} = 3^d 2^{-dl_{\max}^{(\mathrm{opt})}} N. \tag{6.6.9}$$

In this case the C-forest coincides with the X hierarchy, and the D-tree with the Y hierarchy. The number of operations in the algorithm will be exactly the same, except that setting up the data structure is normally much more costly for the adaptive method (using the algorithms above). On the other hand, we can state that since the step that sets the data structure need be performed for given data distributions only once, and can be excluded from the evaluation of the algorithm itself, that there is no harm of using the adaptive method instead of regular one in this case.

Some difference between the regular and adaptive algorithm performances is expected if the number of sources per box at the finest level (for the regular algorithm) is low. In the case of random distributions there appear some source points which are located much closer to each other than the average distance between the sources. In this case the maximum level of space subdivision $l_{\max}$ in the RMLFMM can be large, if based on the maximum number of allowed sources per box (and this number is small). The adaptive algorithm should outperform the regular one in this case. However, such $l_{\max}$ seems is far from the optimum, since at the optimum $l_{\max}$ we have a balance of cost of all translations and direct evaluations. Therefore, for uniformly random distributions $l_{\max}^{(\mathrm{opt})}$ should be small enough to allow substantially large number of sources per box at the finest level. In this case the optimum performance of the regular FMM should be about the same as in the case of data distributed with a regular grid.

Uniform surface and line distributions

While this type of distribution can be classified as substantially non-uniform spatial distribution, one should not again expect any benefits or harm from using the adaptive version of the MLFMM. The reason for this is rather simple and is related to the fact that surface distributions behave as lower dimensional spatially uniform distributions (in terms of the number of children and neighbors). So the optimum level in the RMLFMM corresponds to some optimum q in the MLFMM, at which the number of operations in both methods should be approximately the same.

This reasoning applies to distribution over any manyfold of lower dimensionality. For example, if in the three-dimensional case both the source and target sets are distributed uniformly over some curve (line sources), then the change in the number of non-empty boxes from level to level (at sufficiently high l) will be similar to one-dimensional case with uniform distribution of both sets.

The same non-uniform distributions of sources and receivers

Some preliminary experiments with different types of distributions showed that when the sources and targets were distributed with the same probability density (e.g. three-dimensional Gaussian distributions), there were no advantages in the use of the adaptive method compared to the regular one. Particularly, this relates to the case when these sets are the same ($\mathbb{X} = \mathbb{Y}$). This issue requires some deeper insight and investigation and we are not going to go too far with conclusions, but some hint what may happen in this case is provided by the situation considered above when both sets are distributed uniformly over a low-dimensional manyfold.

Different effective dimensionality of source and receiver distributions

Our experiments with adaptive and regular MLFMM showed that the adaptive method can be a few times faster and/or stable to variation of q in a range, which is wide enough to avoid requiring optimization in the cases when the sources and receivers were distributed over manifolds of different dimensionality. This includes an important case, which is related to representation of the field via layer potentials. In this case the sources were distributed uniformly over the surface of scatterer, while the receivers were uniformly distributed over a substantially large three-dimensional domain outer to the scatterer to sample spatial distribution.

An example of such a distribution is shown in Fig. 6.11 where sources are distributed over a surface of a sphere, while the receivers are distributed uniformly inside the spherical shell between $r = 1$ and $r = r_{\max}$. When $r_{\max}$ is large enough, the adaptive method shows better performance (an example for $r_{\max} = 100$ is shown in Figs. 6.12 and 6.13). To study the method performance, we used a model where we set $l_{b*} = 2$ (so $kD_0 < 4\kappa_b$), and identified $C_{\min}$ as translation cost, since in this case the length of the representing vector does not vary. (The CPU time was measured on a 933 MHz Pentium III with 1 GB RAM.)

Two major effects are seen when comparing graphs in Figs. 6.12 and 6.13. First, the adaptive method shows minimum CPU time in a wide range of values for the clustering parameter, $q \ll N$. So no

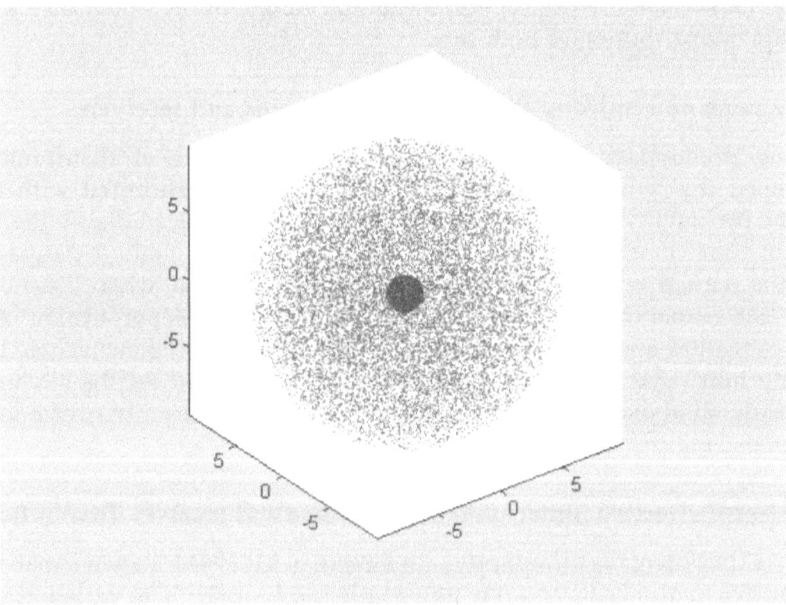

FIGURE 6.11 Three-dimensional uniformly random distributions of the targets and sources. The sources are distributed over the unit sphere, while the targets are distributed over a spherical shell bounded by two co-centered spheres of radii $r = 1$ and $r = r_{\max}$.

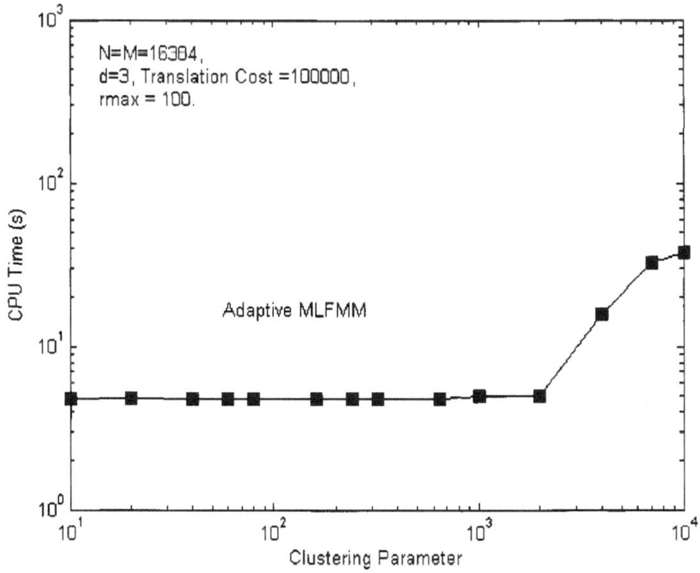

FIGURE 6.12 The CPU time for the adaptive MLFMM vs clustering parameter q for distribution shown in Fig. 6.11 with $r_{\text{max}} = 100$.

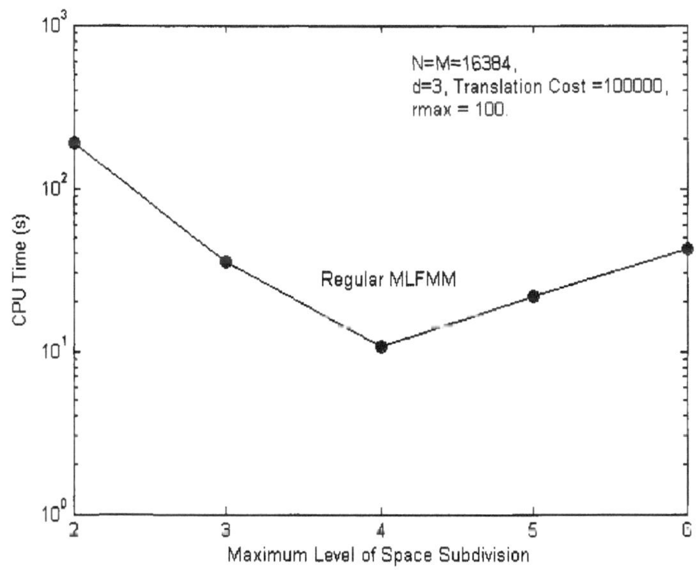

FIGURE 6.13 The CPU time for the regular MLFMM as a function of the maximum level of hierarchical space subdivision near the optimum l_{max} for the problem with the same parameters as in Fig. 6.12.

optimization is required for this range. On the other hand, the CPU time for the RMLFMM depends on l_{max} substantially and optimization is required. Second, we note that the minimum CPU time for the RMLFMM in this case exceeds two times the CPU time for the AMLFMM at $q \ll N$. We observed similar effect in two-dimensional case, when the sources were distributed over a circle, while the receivers occupied a large two-dimensional domain.

Another type of distribution that we tested was when the sources are distributed over a curve, while the receivers occupied some volume in space. An example, where the curve is a ring of unit radius and the domain occupied by the receivers is a spherical shell between $r = 1$ and $r = r_{max}$ as in the previous case, is shown in Fig. 6.14. Both sets had uniform random distributions over those domains. The dependencies of the CPU time for the adaptive and regular methods are shown in Figs. 6.15 and 6.16. They are very similar to those shown in Figs. 6.12 and 6.13 for distribution of the sources over a unit sphere surface. We noticed that for large $r_{max} \gg 1$

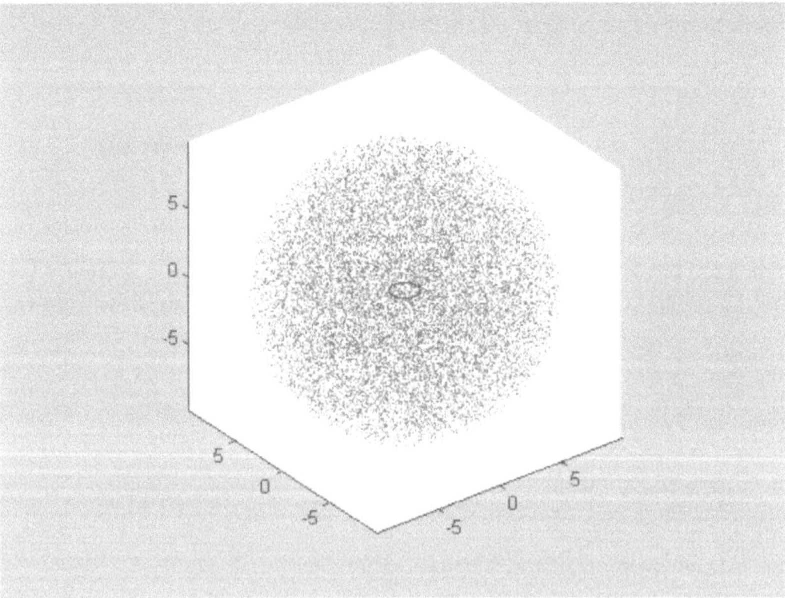

FIGURE 6.14 Three-dimensional uniformly random distributions of the targets and sources. The sources are distributed over the unit circle, while the targets are distributed over a spherical shell bounded by two co-centered spheres of radii $r = 1$ and $r = r_{max}$.

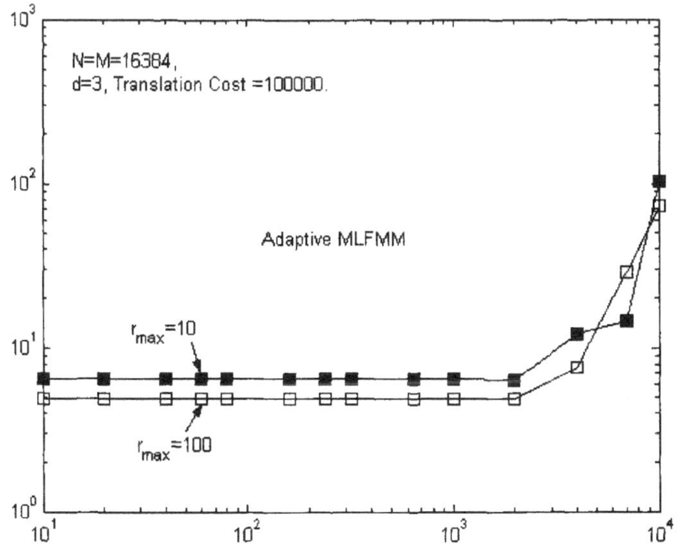

FIGURE 6.15 The CPU time for the adaptive MLFMM vs clustering parameter q (on the left) for distribution shown in Fig. 6.14. The open squares and circles correspond to $r_{\max} = 100$ and the filled squares and circles correspond to $r_{\max} = 10$.

the performance of the RMLFMM depends slightly on $r_{\max}$. For example, Fig. 6.12 shows that CPU times for the RMLFMM for $r_{\max} = 10$ and 100 are visibly hardly distinguishable from each other. The minimum CPU times for the AMLFMM in these two cases differ by about 30%. So we can see that while the effect of increase of the ratio of the sizes of the domains for targets and sources on the ratio of the CPU times holds in this case also and this is due to decrease of the CPU time of the AMLFMM.

Finally, we investigated three-dimensional distributions of type shown in Fig. 6.17. Here we distributed the targets uniformly inside a unit cube. The sources were distributed as clusters whose centers were also located inside the unit cube randomly with a uniform distribution. Each cluster was a three-dimensional normally distributed set with standard deviation σ. In the numerical experiments we fixed the total number of sources, while varying the number of clusters, so each cluster contained N/N_{clust} points, where N_{clust} is the number of clusters. The minimum CPU time was measured for the RMLFMM by varying the number of maximum levels, while for the AMLFMM we determined this quantity by finding a minimum by varying the clustering parameter q.

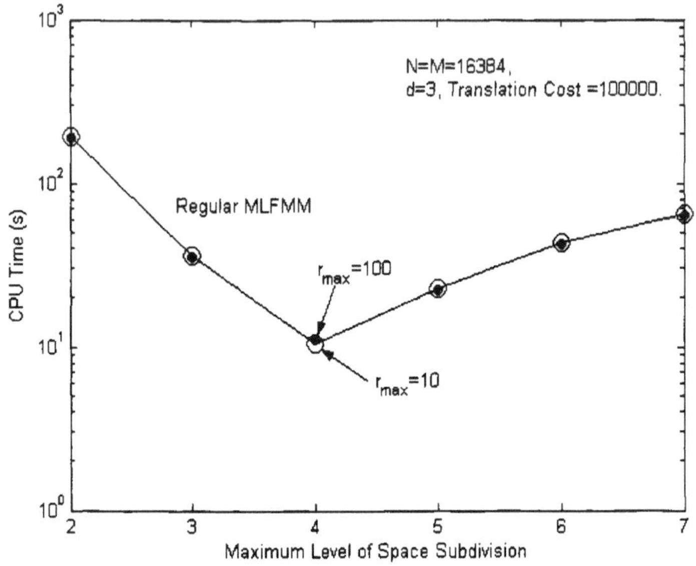

FIGURE 6.16 The CPU time for the regular MLFMM as a function of the maximum level of hierarchical space subdivision near the optimum l_{max}. The case and notations are the same as for Fig. 6.15.

The ratio of these CPU times is plotted in Fig. 6.18 as a function of N_{clust} for other parameters fixed.

While for more accurate conclusions more than one statistical realization of cluster center distribution is required, the tendency is clear: the increase of N_{clust} leads to the decrease of the ratio of the optimum CPU times, which for larger N_{clust} is expected to be close to 1. This can be explained as that the distribution becomes more and more uniform with the increase of N_{clust}, and at $N_{clust} = N$ we have the case when each cluster consists of one point and these points are distributed uniformly inside the cube. This is the case when there is no advantage of using the fully adaptive method, since for uniform distributions the optimum performance of both methods is approximately the same. When N_{clust} is relatively small, we also noticed that the dependences of the CPU time for the adaptive and regular methods had the same patterns as shown in Figs. 6.12, 6.13, 6.15, and 6.16. So the performance of the AMLFMM was stable to variation of q in a wide range, while the performance characteristics of the regular methods had a sharp extremum at which the CPU time was usually several times smaller than the CPU time at non-optimal l_{max}.

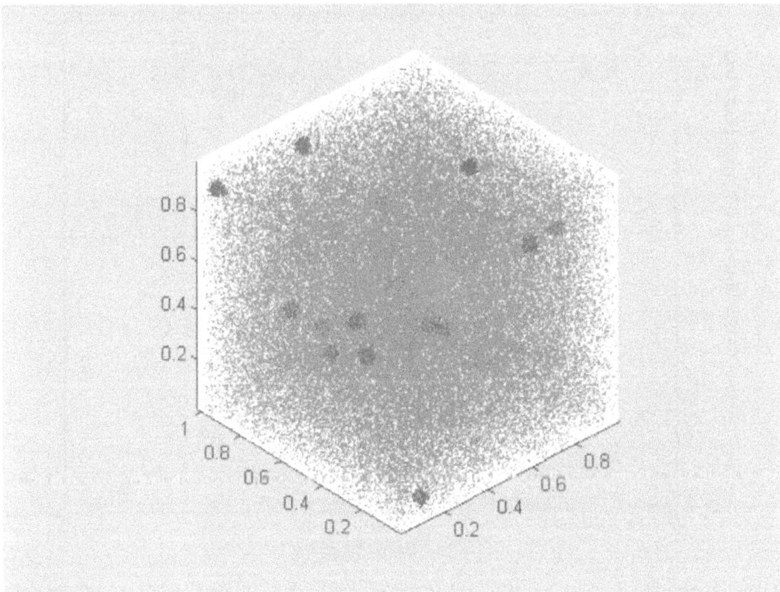

FIGURE 6.17 Three-dimensional uniformly random distributions of the targets and source cluster centers inside a unit cube. Inside each cluster the sources are distributed normally with standard deviation $\sigma = 0.0001$.

This indicates the necessity of optimization procedures for the regular method.

6.6.3.2 *High frequencies*

While the tree-based adaptive method, as described, may have advantages for certain distributions, its application to the high frequency case shows some problems, and the concept of adaptivity should be revised in this case.

Indeed, assume that some function representation of length P is obtained for a large enough domain occupied only by the receivers. Then we have two opportunities: either to compute the potential at q_r points in this domain using this representation, the procedure of complexity $\sim Pq_r$ (the way of the present adaptive FMM), or to subdivide this domain to, say, eight boxes (in three-dimensional case), translate the representing vector to the center of each box and evaluate the potential with lower $P_2 = P/4$ (the way how the regular FMM works if the box is not at the finest level). The number of points in each box will be then $q_r/8$ and the complexity of evaluation will be of order $8(q_r/8)P_2 = (1/4)Pq_r$, which is

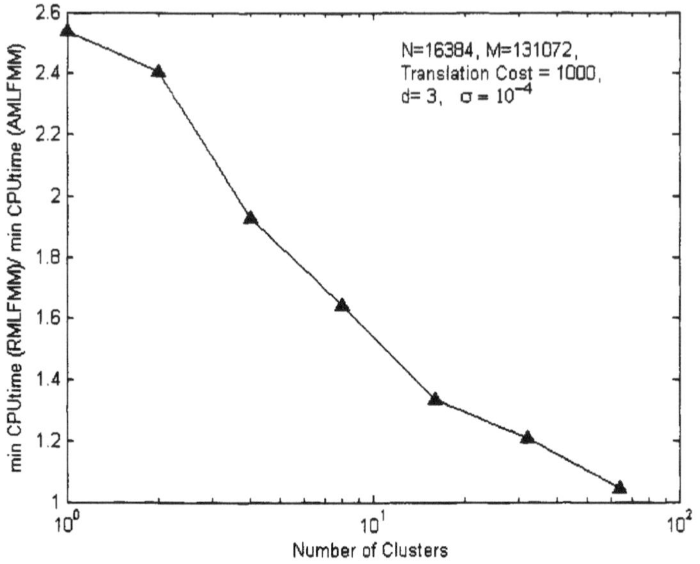

FIGURE 6.18 The ratio of the minimum CPU times required by the regular MLFMM and the adaptive MLFMM for computation of the same problem as a function of number of clusters in distribution shown in Fig. 6.17. The total number of sources is fixed.

four times faster than Pq_r. However, the price to pay for this cheaper evaluation is the cost of eight translations, which in the fastest case ($\nu = 1$) is $O(8P)$. The asymptotic constant here can be large enough, and one should decide whether the reduction in the evaluation cost worth additional translations. In any case, it is clear that for substantially large q_r a sourceless box should be subdivided, if this associated with a reduction of the representing vector length, the case for boxes at level $l < l_b$. Therefore, the criteria by which we constructed the data trees should be modified to include some criterion of type: "any leaf box of the D-tree at level $l < l_b$ cannot contain more than q_r receivers". It is likely that this criterion leads to data trees, which are close to data pyramids up to level l_b (so we can come back to the regular MLFMM!). Similar concerns relate to the upward pass of the AMLFMM. Here one should decide what is cheaper for large boxes in the C-forest: to compute contribution of qs sources immediately or perform translations from its children boxes. While it is possible to develop an adaptive method based on modified conditions, we leave this issue open for further research, since it should be considered in conjunction with translation complexities and particular problems to be solved.

6.7 SUMMARY

In this chapter we provided a deeper insight into the complexity of the MLFMM. We introduced a model of the MLFMM with level-dependent lengths of representing vectors. Based on this model we found some basic relations between the dimensionality of the problem, translation costs (translation exponents), and the dimensionless wavenumber to achieve $O(N \log^{\gamma} N)$ type of algorithms. In the FMM-related literature one can find a mistaken belief that this complexity for the 3D Helmholtz equation can be achieved only for translation exponents $\nu = 1$. As we showed $O(N\log^{\gamma} N)$ complexity holds for lager ν if the effective dimensionality of the problem is larger than two. We also found that the maximum level of space subdivision can be optimized and that optimization depends on the effective dimensionality of the problem and the translation exponent. Further, we considered a version of "fully adaptive MLFMM", which utilizes tree structures opposed to pyramid data structures of the regular MLFMM and discussed some situations when the adaptive method can be useful.

CHAPTER 7

Fast Translations: Basic Theory and $O(p^3)$ Methods

From the complexity analysis of the MLFMM, it follows that the cost of the translation operation plays a key role in determining the efficiency of the overall algorithm. Based on the model for the complexity of the FMM for the Helmholtz equation that was introduced and investigated in Chapter 6, we found that while $O(p^4)$ ($\nu = 2$) translation methods can be used for moderate and low frequencies, a translation cost of least $O(p^3)$ ($\nu = 1.5$) is required for higher frequencies to provide $O(N \log^\alpha N)$, $\alpha \geq 1$, complexity for spatial distributions required for computation of volume integrals or multiple scattering problems. $O(p^3)$ methods can also be applied for surface data distributions and result in $O(N^{1.5})$ complexity of the MLFMM applied for the BEM and makes it asymptotically faster than the $O(N^2)$ direct method. For high-frequency computations with surface distributions $O(p^4)$ translation algorithms are not efficient, since they provide $O(N^2)$ complexity of the MLFMM with larger asymptotic constant than the direct summation method.

In this chapter we focus on $O(p^3)$ translation methods. A proper question here is, why should one consider $O(p^3)$ techniques, if $O(p^2 \log^\alpha p)$ techniques exist? There are several reasons for this. First, as we shall show in this chapter, the $O(p^3)$ methods have a low asymptotic constant. So if we compare the actual number of operations in, say, an $O(p^3)$ method which requires p^3 operations and an $O(p^2)$ method, which requires, say, $25p^2$ operations, $O(p^3)$ can be *faster* than $O(p^2)$ for $p < 25$. In fact $p \sim 25$ is large enough to compute many practical problems in acoustics and electromagnetics, and this value allows computation of problems with $kD_0 = 4kD^2 \sim 4p = 100$, since translations are required only for levels $l \geq l_{\min} = 2$. Of course, at high frequencies $O(p^2 \log^\alpha p)$ would prove superior. However, since they normally require several FFTs (or other divide and conquer algorithms, such as those used in the *Fast Legendre Transform*) [DH94] or the use of the 1D FMM [ST02] (which themselves are several times slower than the FFT) per translation for stabilization and error control of the algorithms, the asymptotic constants there are large enough, and $O(p^3)$ methods can be the best available translation methods for low and moderate frequencies. Moreover, as shown in the following chapters, $O(p^3)$ matrix-based translation methods are stable to errors and do not require additional data filtering, which is an important stabilization step for high-frequency translation methods based on diagonal forms of translation operators. Finally, an insight into the $O(p^3)$ methods is required to understand some of the asymptotically faster algorithms. This motivates development and study of the fast $O(p^3)$ algorithms presented in this chapter.

7.1 REPRESENTATIONS OF TRANSLATION AND ROTATION OPERATORS

7.1.1 Functions and operators

The first thing that we need to proceed further is clarify the meaning of the terms *translation operator, rotation operator,* and any other *operator.* As is known from linear algebra and functional analysis the concept of an operator is related to the concept of the *space* of objects on which it acts.

7.1.1.1 Linear vector spaces

We are interested mainly with *linear spaces* or *linear vector spaces.* Such spaces consist of objects called vectors, and usual vectors in two and three dimensions provide an example of *finite-dimensional* spaces. More sophisticated examples are the *infinite-dimensional* spaces of functions, for example, continuous functions on a unit sphere, $f(\mathbf{s})$, where $\mathbf{s}$ is a unit vector, or infinitely differentiable solutions of the Helmholtz equation $\psi(\mathbf{r})$ in some spatial domain $\mathbf{r} \in \Omega$. Since, we will frequently speak about the latter space, it is denoted by the symbol $\mathcal{B}(\Omega)$. The fact that $\mathcal{B}(\Omega)$ is a *complex* linear space (complex means that the scalars we consider are complex numbers, $\alpha \in \mathbb{C}$) can be easily checked if we recall the definition of a linear space. This definition states that if $\mathbf{a}$, $\mathbf{b}$, and $\mathbf{c}$ are some elements of a linear vector space $\mathcal{U}$, then the following properties should hold:

- $\mathbf{a}+\mathbf{b} \in \mathcal{U}$ (indeed, the sum of two solutions of the Helmholtz equation again is a solution of the same equation);
- $(\mathbf{a}+\mathbf{b} = \mathbf{b}+\mathbf{a}$, $\mathbf{a}+(\mathbf{b}+\mathbf{c}) = (\mathbf{a}+\mathbf{b})+\mathbf{c}$ (simply the additivity and commutativity of the addition operation, which is obviously true for solutions of the Helmholtz equation);
- $\exists \mathbf{0} \in \mathcal{U}$, $\mathbf{a}+\mathbf{0} = \mathbf{a}$, $\mathbf{a}+(-\mathbf{a}) = \mathbf{a}-\mathbf{a} = \mathbf{0}$ (function $\psi(\mathbf{r}) \equiv 0$ is the

zero element of the space $\mathcal{B}(\Omega)$);

- $\forall \alpha \in \mathbb{C}$, $\alpha\mathbf{a} \in \mathcal{U}$ (indeed, if $\psi(\mathbf{r})$ is a solution of the Helmholtz equation, then $\alpha\psi(\mathbf{r})$ is a solution as well);
- $\forall \alpha, \beta \in \mathbb{C}$, $(\alpha\beta)\mathbf{a} = \alpha(\beta\mathbf{a})$, $1\mathbf{a} = \mathbf{a}$, $\alpha(\mathbf{a}+\mathbf{b}) = \alpha\mathbf{a} + \alpha\mathbf{b}$, $(\alpha+\beta)\mathbf{a} = \alpha\mathbf{a} + \beta\mathbf{a}$ (just properties of products of vectors by scalars, which are true for $\psi \in \mathcal{B}(\Omega)$).

7.1.1.2 *Linear operators*

Operators are defined as mappings from one linear vector space to another (both spaces can be the same). If we have a function $\psi \in \mathcal{B}(\Omega)$ and can correspond it to some other function $\psi' \in \mathcal{B}(\Omega')$, then such a mapping is an operator $\mathcal{A}$ acting from the space $\mathcal{B}(\Omega)$ to the space $\mathcal{B}(\Omega')$. This can be written as

$$\psi' = \mathcal{A}[\psi], \qquad \mathcal{A} : \mathcal{B}(\Omega) \to \mathcal{B}(\Omega'), \qquad \psi \in \mathcal{B}(\Omega), \; \psi' \in \mathcal{B}(\Omega'). \quad (7.1.1)$$

The operator is called linear if it satisfies the following property:

$$\mathcal{A}[\alpha\psi_1 + \beta\psi_2] = \alpha\mathcal{A}[\psi_1] + \beta\mathcal{A}[\psi_2], \qquad \alpha, \beta \in \mathbb{C}. \quad (7.1.2)$$

Important examples of linear operators are as follows.

Differential operators, $\mathcal{D}$. An example is $\partial/\partial z$ which performs partial differentiation with respect to z (we also used $\partial_z = k^{-1}\partial/\partial z$ and sometimes to stress that this operator acts in functional space, we use the notation $\mathcal{D}_z = \partial_z$). Differential operators include ∇, ∇^2, or differentiation in a direction $\mathbf{s}$, $\mathbf{s}\cdot\nabla$. These operators map $\mathcal{B}(\Omega) \to \mathcal{B}(\Omega)$, so that the domain of definition does not change and we have

$$\psi' = \mathcal{D}[\psi], \qquad \mathcal{D} : \mathcal{B}(\Omega) \to \mathcal{B}(\Omega). \quad (7.1.3)$$

Rotation operators, $\mathcal{R}$ot. These operators are generated by rotation of the basis vectors with matrix Q, so we can write this operator as $\mathcal{R}\mathrm{ot}(Q)$. If Ω is the interior or exterior of a sphere, then rotation does not change the domain of definition, and we have

$$\begin{aligned} &\psi' = \mathcal{R}\mathrm{ot}(Q)[\psi], \qquad \psi'(\mathbf{r}) = \psi(\hat{\mathbf{r}}), \qquad \hat{\mathbf{r}} = Q\mathbf{r}, \\ &\mathcal{R}\mathrm{ot}(Q) : \mathcal{B}(\Omega) \to \mathcal{B}(\Omega), \qquad \Omega : r \gtrless a. \end{aligned} \quad (7.1.4)$$

Translation operators, $\mathcal{T}$. These operators arise due to a shift in the function argument by a translation vector $\mathbf{t}$, so we can write them as $\mathcal{T}(\mathbf{t})$. The translation operator shifts the domain of definition of the transformed function $\psi'(\mathbf{r}) = \psi(\mathbf{r}+\mathbf{t})$ (indeed if the function $\psi(\mathbf{r})$ has only a singularity at $\mathbf{r} = -\mathbf{t}$, then $\psi'(\mathbf{r})$ has the same singularity at $\mathbf{r} = \mathbf{0}$). In general, we have

$$\psi' = \mathcal{T}(\mathbf{t})[\psi], \qquad \psi'(\mathbf{r}) = \psi(\mathbf{r}'), \qquad \mathbf{r}' = \mathbf{r} + \mathbf{t},$$
$$\mathcal{T}(\mathbf{t}) : \mathcal{B}(\Omega) \to \mathcal{B}(\Omega'), \tag{7.1.5}$$

where Ω' is obtained by the shift of the domain Ω.

7.1.1.3 Groups of transforms

We note that the rotation and translation operators appear as a result of the corresponding spatial transforms in the three-dimensional Euclidean space. Rotations and translations form a *group of Euclidean motion*. In the space of functions this group also generates a group of transforms, so for any spatial vector transform, we have a corresponding operation in the space of functions. This establishes a *homomorphism* between the group of Euclidean motion (denoted as $E(3)$ if rotations with reflections are included, and $SE(3)$ if we consider only "pure" rotations, which preserve the right-handed orientation of the orthonormal basis vectors, "3" denotes the space dimensionality, E means "Euclidean" and S means "special") and the group of transforms in the functional space (we will denote this group by adding F to the notation of 3D motion group, e.g. $FE(3)$ or $FSE(3)$). The term "homomorphism" means that for the group operation in 3D space (which include translations and rotations) we have a corresponding operation in the function space, so that a composition of operations in 3D space maps to the composition of operations in the space of functions. A formal definition of a group is the following:

DEFINITION 1 A set $\mathbb{G}$ of elements (objects) $a, b, c, \ldots$ is called a group if there is defined some binary operation $\odot$ called the "group operation", which for any pair of elements $a, b \in \mathbb{G}$ there corresponds some object $a \odot b$, such that for any $a, b, c \in \mathbb{G}$ the following properties hold:

$$\begin{aligned}
&(1)\ a\odot b \in \mathbb{G}, \qquad (\mathbb{G} \text{ is closed with respect to } \odot),\\
&(2)\ a\odot(b\odot c) = (a\odot b)\odot c, \qquad (\text{associativity}),\\
&(3)\ \exists e \in \mathbb{G},\ e\odot a = a, \qquad (\mathbb{G} \text{ contains the unity, } e),\\
&(4)\ \exists a^{-1} \in \mathbb{G},\ a^{-1}\odot a = e, \qquad (\mathbb{G} \text{ contains the inverse}\\
&\qquad \text{element for each } a \in \mathbb{G}).
\end{aligned} \tag{7.1.6}$$

The group of rigid body rotations (which is a *subgroup* of $E(3)$; subgroup is a subset of the group, which is also a group with respect to the group operation), is denoted as $O(3)$ for arbitrary rotations including reflections (letter O means here "orthogonal"), and $SO(3)$ for "pure" rotations which transforms vectors with rotation matrices. A homomorphism between $SO(3)$ and $FSO(3)$ or $O(3)$ and $FO(3)$ means

$$\mathcal{R}\text{ot}(Q_1Q_2) = \mathcal{R}\text{ot}(Q_1)\circ\mathcal{R}\text{ot}(Q_2), \tag{7.1.7}$$

$$\begin{aligned}\mathcal{R}\text{ot}(I) &= \mathcal{R}\text{ot}(Q^{-1}Q) = \mathcal{R}\text{ot}(Q^{-1})\circ\mathcal{R}\text{ot}(Q)\\ &= \mathcal{R}\text{ot}^{-1}(Q)\circ\mathcal{R}\text{ot}(Q) = \mathcal{I},\end{aligned} \tag{7.1.8}$$

where I is the 3×3 identity matrix and $\mathcal{I}$ means the identity rotation operator in the functional space, which does not change the function, and $\circ$ denotes the composition of operations. Groups $O(3)$ and $SO(3)$ are not commutative, so the same holds for the group of rotations in the functional space $FO(3)$ and $FSO(3)$.

The group of *rigid body translations* in 3D space, which we can denote as $T(3)$ is described by the translation transform $g(\mathbf{r}) = \mathbf{r} + \mathbf{t}$. So again due to the homomorphism we have

$$\mathcal{T}(g_1\circ g_2) = \mathcal{T}(g_1)\circ\mathcal{T}(g_2), \tag{7.1.9}$$

$$\mathcal{T}(I) = \mathcal{T}(g^{-1}\circ g) = \mathcal{T}(g^{-1})\circ\mathcal{T}(g) = \mathcal{T}^{-1}(g)\circ\mathcal{T}(g) = \mathcal{I}. \tag{7.1.10}$$

Note that the composition of two translations, with associated translation vectors $\mathbf{t}_1$ and $\mathbf{t}_2$, is a translation with the vector $\mathbf{t}_1 + \mathbf{t}_2 = \mathbf{t}_2 + \mathbf{t}_1$. This composition does not depend on the order these translations are performed. This shows that the group operation in case of translation is addition, which is the commutative operation in vector space. So $T(3)$ is a commutative or *Abelian* group. The same applies for the group of translations in the functional space, $FT(3)$ so

$$\mathcal{T}(\mathbf{t}_1)\circ\mathcal{T}(\mathbf{t}_2) = \mathcal{T}(\mathbf{t}_1 + \mathbf{t}_2) = \mathcal{T}(\mathbf{t}_2 + \mathbf{t}_1) = \mathcal{T}(\mathbf{t}_2)\circ\mathcal{T}(\mathbf{t}_1), \tag{7.1.11}$$

$$\mathcal{T}(0) = \mathcal{T}(-\mathbf{t} + \mathbf{t}) = \mathcal{T}(-\mathbf{t})\circ\mathcal{T}(\mathbf{t}) = \mathcal{T}^{-1}(\mathbf{t})\circ\mathcal{T}(\mathbf{t}) = \mathcal{I}. \tag{7.1.12}$$

Another example of the commutative group of spatial transforms is the special group of rotations in two dimensions, $SO(2)$. This means that both translation and rotation groups in case of two dimensions are commutative. Generally, translations and rotations do not commute.

7.1.1.4 Representations of groups

In practice linear transforms of functions in functional space are often related to transforms of the basis functions, as soon as such a basis $\{F_l\}$ exists; and any function from that functional space can be expanded as

$$\psi = \sum_n c_n F_n, \qquad c_n \in \mathbb{C}. \tag{7.1.13}$$

Acting with a linear operator on ψ we have

$$\psi' = \mathcal{A}[\psi] = \mathcal{A}\left[\sum_n c_n F_n\right] = \sum_n c_n \mathcal{A}[F_n]. \tag{7.1.14}$$

Assume now that the functions ψ' and $\mathcal{A}[F_n]$ can be expanded over some other basis $\{E_n\}$ as follows

$$\psi' = \sum_{n'} \hat{c}_{n'} E_{n'}, \qquad \mathcal{A}[F_n] = \sum_{n'} (F|E)_{n'n} E_{n'},$$
$$\hat{c}_n \in \mathbb{C}, \quad (F|E)_{n'n} \in \mathbb{C}, \tag{7.1.15}$$

where $(F|E)_{n'n}$ are the basis reexpansion coefficients. From Eqs. (7.1.14) and (7.1.15) we have

$$\sum_n c_n \sum_{n'} (F|E)_{n'n} E_{n'} = \sum_{n'} \hat{c}_{n'} E_{n'}. \tag{7.1.16}$$

If the infinite series converge absolutely and uniformly and expansion over the basis functions is unique (which they are from the definition of a basis), we obtain

$$\hat{c}_{n'} = \sum_n (F|E)_{n'n} c_n, \tag{7.1.17}$$

or in matrix–vector form:

$$\hat{\mathbf{c}} = (\mathbf{F}|\mathbf{E})\mathbf{c}. \tag{7.1.18}$$

We have already met such forms earlier in this book. Here we show the process in a more condensed form, to establish the correspondence between operators in a function space and reexpansion matrices. As soon as a basis is selected (functions centered at different points, are different functions !), the functions can be *represented* by infinite column-vectors, and linear operators can be *represented* by infinite matrices. Such representations are obviously related to the selected basis. As we can treat a vector as an invariant object, which has different coordinates in different bases, functions (which are also vectors in functional space) have different "coordinates", which are column-vectors of expansion coefficients that transform according to the function transform. An operator in functional space is also an invariant object, which has different representations in different bases. Hence, the same translation operator $\mathcal{T}(\mathbf{t})$ can be represented by, say, matrices $(\mathbf{R}|\mathbf{R})(\mathbf{t})$, $(\mathbf{S}|\mathbf{S})(\mathbf{t})$, and $(\mathbf{S}|\mathbf{R})(\mathbf{t})$. Respectively, we should realize the meaning of the expansion coefficients, since these coefficients can represent expansion of the translated function over the same type of basis as the original function, as $(\mathbf{R}|\mathbf{R})(\mathbf{t})$ and $(\mathbf{S}|\mathbf{S})(\mathbf{t})$ do, or over different types of bases, as $(\mathbf{S}|\mathbf{R})(\mathbf{t})$ does.

In any case as we come to the space of column-vectors and matrices, $(\mathbf{R}|\mathbf{R})(\mathbf{t})$, $(\mathbf{S}|\mathbf{S})(\mathbf{t})$, and $(\mathbf{S}|\mathbf{R})(\mathbf{t})$ become *linear operators* in this space, which can also be called the "space of expansion coefficients". Therefore, these matrices can also be called "translation operators". We already considered matrices such as $\mathbf{Rot}(Q)$ which represents operator $\mathcal{R}ot(Q)$ in the **R** or **S** basis. In the next section, we also deal with the differential operators in the space of coefficients, such as $\mathbf{D}_z$, which is a *matrix* representing operator, $\mathcal{D}_z = \partial_z$.

Rotations

To introduce some basic concepts of the *group representation theory*, let us consider first the group of rotation operators $FSO(3)$. It is not difficult to see that matrices $\mathbf{Rot}(Q)$ form a group with the group operation which is matrix multiplication, and the unit element which is the identity matrix. We can denote this group as $MSO(3)$, where M reminds that this is a group of rotation matrices. As soon as the homomorphism between $FSO(3)$ and $MSO(3)$ exists (which is more or less obvious) we call $MSO(3)$ a *representation* of the group $FSO(3)$. So the representation is a *group of matrices* which is homomorphic to the group that is represented. If the mapping between the two groups $FSO(3)$ and $MSO(3)$ is one-to-one then the representation is called *exact*. Such a one-to-one mapping or homomorphism is called an *isomorphism*. The concepts of *reducible* and *irreducible* representations are the key concepts in the theory of representations. These concepts can be perfectly illustrated for the rotation transforms. If we say that $MSO(3)$ is irreducible, this means that if we take *any* non-zero column-vector and rotate it with *all* matrices $\mathbf{Rot}(Q)$ we obtain *all* other column-vectors. This is not the case for the rotation matrix. Indeed, if we take a vector, which has non-zero entries only for components corresponding to harmonics of degree n, then any vector which is a product of any matrix $\mathbf{Rot}(Q)$ and this vector will have zero entries for components corresponding to harmonics of degree other than n. This is nothing, but stating the fact that the space of column-vectors has *invariant subspaces* (which are subspaces of degree n). We showed this in the chapter related to reexpansion matrices. If such a thing happens, the representation is called reducible. So $MSO(3)$ is a reducible representation of $FSO(3)$. The fact that a representation $\mathcal{R}$ is reducible (and so is a direct sum of subgroups $\mathcal{R}_0, \mathcal{R}_1, \ldots$, which represent the group of transforms for invariant subspaces) can be written in the form

$$\mathcal{R} = \mathcal{R}_0 \oplus \mathcal{R}_1 \oplus \cdots = \sum_{n=0}^{\infty} \oplus \mathcal{R}_n. \tag{7.1.19}$$

We note that rotation transforms can be considered as rotations of surface functions $F(\theta, \varphi)$. This means that upon rotation the interior and exterior of a sphere do not change. For application of multipole methods, such as the FMM, consideration of such domains is sufficient. If more complex domains are treated, then after rotation transforms the domain of definition of function $\psi(\mathbf{r})$ transforms as well and care should be taken to make sure that the transformed function $\psi'(\mathbf{r}) = \psi(\hat{\mathbf{r}})$ can be expanded in the entire domain over the same basis as $\psi(\mathbf{r})$.

Translations

Entire functions

First, let us consider a class of infinitely differentiable functions which satisfy the Helmholtz equation in the entire space. Examples of such functions are spherical basis functions $R_n^m(\mathbf{r})$, any of their derivatives, any finite sums with some coefficients, or infinite sums, which converge absolutely and uniformly for any finite $\mathbf{r}$. The last case includes, e.g. the plane wave solution $\psi(\mathbf{r}) = e^{i\mathbf{k}\cdot\mathbf{r}}$, which can be expanded over the basis functions $R_n^m(\mathbf{r})$ as given by Eq. (2.3.6). The domain of function definition Ω in this case is the entire space, and if we translate such a function then the translated domain is $\Omega' = \Omega$. We denote this class of functions as $\mathcal{B}(\mathbb{R}^3)$. These functions can be considered as an analog of the *entire* or *integral* functions in complex analysis (the entire functions are single-valued analytical functions for any finite value of the argument). We will call functions of class $\mathcal{B}(\mathbb{R}^3)$ also as entire or integral functions.

If we select $\mathbf{R}(\mathbf{r}) = \{R_n^m(\mathbf{r})\}$ as a basis for this class of functions, then translations in the space of infinite column-vectors of coefficients are described by the infinite matrix $(\mathbf{R}|\mathbf{R})(\mathbf{t})$, where $\mathbf{t}$ is an arbitrary translation vector. Obviously, for convergence of the series, which appear as products of the matrix rows and the column-vector, we should consider only a particular set of such columns, which we denote as $\mathbf{B}(\mathbb{R}^3)$. Indeed, generally, series over $\{R_n^m(\mathbf{r})\}$ can diverge (e.g. the expansion of the Green's function $G(\mathbf{r})$ converges only in the interior of a sphere, which does not contain $\mathbf{r} = \mathbf{0}$). However, the matrix $(\mathbf{R}|\mathbf{R})(\mathbf{t})$ "does not know" to which vector we are going to apply it and so its properties do not depend on a particular vector, and depend only on $\mathbf{t}$.

In the space of infinite matrices $(\mathbf{R}|\mathbf{R})(\mathbf{t})$ forms a group, with matrix multiplication as the group operation, the identity element $\mathbf{I} = (\mathbf{R}|\mathbf{R})(\mathbf{0})$, which is an infinite matrix (the only non-zero elements of which are the ones located on the main diagonal). The inverse element in such a group is a matrix

$$(\mathbf{R}|\mathbf{R})(-\mathbf{t}) = (\mathbf{R}|\mathbf{R})^{-1}(\mathbf{t}), \tag{7.1.20}$$

and, what is unusual for general matrices, this group is commutative, since

$$(\mathbf{R}|\mathbf{R})(\mathbf{t}_1)(\mathbf{R}|\mathbf{R})(\mathbf{t}_2) = (\mathbf{R}|\mathbf{R})(\mathbf{t}_1 + \mathbf{t}_2) = (\mathbf{R}|\mathbf{R})(\mathbf{t}_2 + \mathbf{t}_1) = (\mathbf{R}|\mathbf{R})(\mathbf{t}_2)(\mathbf{R}|\mathbf{R})(\mathbf{t}_1). \tag{7.1.21}$$

We can denote this group as $RR(3)$. It is isomorphic to the group of translations $FT(3)$ defined on functions from $\mathcal{B}(\mathbb{R}^3)$ and, therefore, $RR(3)$ is the representation of $FT(3)$.

It is noteworthy that $\mathcal{B}(\mathbb{R}^3)$ does not contain any radiating function (except the trivial function $\psi(\mathbf{r}) \equiv 0$). Indeed, any radiating function should have a singularity somewhere in $\mathbb{R}^3$, while functions of $\mathcal{B}(\mathbb{R}^3)$ are regular everywhere.

R|R translations

Consider now a superset of $\mathcal{B}(\mathbb{R}^3)$, which are functions regular *inside* a sphere Ω of radius a centered at $\mathbf{r} = \mathbf{0}$. We denote them as $\mathcal{B}(\Omega)$. The fact that functions of class $\mathcal{B}(\Omega)$ can be represented in Ω by absolutely and uniformly convergent series with respect to the basis $\mathbf{R}(\mathbf{r}) = \{R_n^m(\mathbf{r})\}$ means that for any ε there exists some truncation number $p(\varepsilon)$, such that some function $\psi(\mathbf{r}) \in \mathcal{B}(\Omega)$ can be approximated in the whole domain Ω by the truncated function $\psi^{(p)}(\mathbf{r})$ with the maximum absolute error ϵ. Function $\psi^{(p)}(\mathbf{r})$ given by Eq. (2.3.11) is of class $\mathcal{B}(\mathbb{R}^3)$, since it contains a finite number of terms in series over $R_n^m(\mathbf{r})$. As soon as $\psi^{(p)}(\mathbf{r})$ is an entire function its translations can be described using infinite matrices $(\mathbf{R}|\mathbf{R})(\mathbf{t})$, which have properties (7.1.20) and (7.1.21). However, we should notice that $\psi^{(p)}(\mathbf{r})$ is an approximation of $\psi(\mathbf{r})$ only for $\mathbf{r} \in \Omega$, and therefore, the translated function

$$\psi^{(p)\prime} = T(\mathbf{t})[\psi^{(p)}], \qquad \psi^{(p)\prime}(\mathbf{r}) = \psi^{(p)}(\mathbf{r}'), \qquad \mathbf{r}' = \mathbf{r} + \mathbf{t},$$
$$T(\mathbf{t}) : \mathcal{B}(\mathbb{R}^3) \to \mathcal{B}(\mathbb{R}^3), \tag{7.1.22}$$

can be considered as an approximation of the translated function $\psi(\mathbf{r}')$ only for $\mathbf{r}' \in \Omega$. If we limit ourselves to the translation vectors $|t| < a$ we ensure that the center of the new expansion is located inside Ω and the translated expansion approximates $\psi(\mathbf{r}')$ in a spherical domain $\Omega' \subset \Omega$. It is noteworthy that function $\psi^{(p)\prime}$ contains infinite number of terms in its expansion over the basis $\mathbf{R}(\mathbf{r}) = \{R_n^m(\mathbf{r})\}$, but this expansion converges for any finite $\mathbf{r}$, since $\psi^{(p)}(\mathbf{r})$ is an entire function. So the inverse operator

$(\mathbf{R}|\mathbf{R})^{-1}(\mathbf{t})$ is defined for any $\mathbf{t}$ and $\psi^{(p)\prime}$ can be translated back to the original center. If this translation is exact, then the expansion contains finite number of terms. It is nothing but the expansion $\psi^{(p)}(\mathbf{r})$ which approximates $\psi(\mathbf{r})$ in the domain Ω (since we return to the original expansion, and this expansion is unique).

S|S translations

The same trick of replacement of infinite series, which converges in some domain Ω, with an approximating function with finite term series (which can be expanded in a convergent series at any finite $\mathbf{r}$) can be used for singular or radiating solutions. Indeed, if Ω is the *exterior* of a sphere of radius a centered at $\mathbf{r} = \mathbf{0}$, where radiating function $\psi(\mathbf{r})$ is regular, $\psi(\mathbf{r}) \in \mathcal{B}(\Omega)$, this function can be represented by absolutely and uniformly convergent series in Ω with respect to the basis $\mathbf{S}(\mathbf{r}) = \{S_n^m(\mathbf{r})\}$. The truncated function $\psi^{(p)}(r)$ given by Eq. (2.3.11) is of class $\mathcal{B}(\mathbb{R}^3/\{\mathbf{0}\})$, since it contains a finite number of terms which are singular only at $\mathbf{r} = \mathbf{0}$. For $|\mathbf{t}| < a$ we can translate this truncated function. The translated function then will have its only singularity at $\mathbf{r} = -\mathbf{t}$ and can be expanded over the basis $\mathbf{S}(\mathbf{r}) = \{S_n^m(\mathbf{r})\}$ in absolutely and uniformly convergent series in Ω. This can be achieved by application of an infinite matrix $(\mathbf{S}|\mathbf{S})(\mathbf{t})$ to the vector of expansion coefficients. Obviously the inverse operation $(\mathbf{S}|\mathbf{S})^{-1}(\mathbf{t})$ is also defined and we can return to the original coefficients by the application of $(\mathbf{S}|\mathbf{S})^{-1}(\mathbf{t})$ to the vector of the translated coefficients. Again, since the expansion is unique, we obtain the same truncated function $\psi^{(p)}(\mathbf{r})$, which approximates $\psi(\mathbf{r})$ in Ω. Moreover, as shown earlier the matrix $(\mathbf{S}|\mathbf{S})(\mathbf{t})$ is nothing but matrix $(\mathbf{R}|\mathbf{R})(\mathbf{t})$. The latter matrices form a commutative group and, hence, translations with $|\mathbf{t}| < a$ in Ω for *radiating* functions of class $\mathcal{B}(\mathbb{R}^3/\{\mathbf{0}\})$ can be represented by the infinite matrices $(\mathbf{R}|\mathbf{R})(\mathbf{t})$ as well.

S|R translations

Finally, we consider the $(\mathbf{S}|\mathbf{R})(\mathbf{t})$ translation matrix. This case seems a bit more complicated, since we also change the type of expansion basis. Again we can consider the truncated function $\psi^{(p)}(\mathbf{r})$ of class $\mathcal{B}(\mathbb{R}^3/\{\mathbf{0}\})$. For any $|\mathbf{t}| > a$ there exists a sphere Ω'' of radius a' centered at $\mathbf{r} = -\mathbf{t}$ such that $\Omega'' \subset \Omega$, where Ω is the exterior of a sphere of radius a centered at $\mathbf{r} = \mathbf{0}$. Since $\psi^{(p)}(\mathbf{r})$ is regular in Ω'' the translated function $\psi^{(p)\prime}(\mathbf{r}) = \psi^{(p)}(\mathbf{r}+\mathbf{t})$ is regular in Ω', which is a sphere of radius a' centered at $\mathbf{r} = \mathbf{0}$. Therefore, it can be expanded into convergent series over the basis $\mathbf{R}(\mathbf{r}) = \{R_n^m(\mathbf{r})\}$. The relation between the coefficients of the original expansion over the basis $\mathbf{S}(\mathbf{r}) = \{S_n^m(\mathbf{r})\}$ and the coefficients of the translated function expansion over the basis $\mathbf{R}(\mathbf{r}) = \{R_n^m(\mathbf{r})\}$ are given by the infinite matrix $(\mathbf{S}|\mathbf{R})(\mathbf{t})$.

Note that the matrix $(\mathbf{S}|\mathbf{R})(\mathbf{t})$ is singular at $\mathbf{t} = \mathbf{0}$ and there is no operation such as the composition of two $(\mathbf{S}|\mathbf{R})$-translations similar to the composition property of the $(\mathbf{R}|\mathbf{R})(\mathbf{t})$ matrices (see Eq. (7.1.21)). However, since $(\mathbf{S}|\mathbf{R})(\mathbf{t})$ represents the translation operator (for particular type of translations) this type of matrices obey some properties, which involve the products of $(\mathbf{S}|\mathbf{R})$ and $(\mathbf{R}|\mathbf{R})$ matrices.

First, we note that once an $(\mathbf{S}|\mathbf{R})$-translation is performed to obtain a function $\psi^{(p)\prime}(\mathbf{r})$ regular in Ω', we can use further $(\mathbf{R}|\mathbf{R})$-translations for $|\mathbf{t}| < a'$ to get an R expansion near any point in Ω'. The same expansion can be obtained by a direct $(\mathbf{S}|\mathbf{R})$-translation to that center in Ω'. This yields

$$(\mathbf{S}|\mathbf{R})(\mathbf{t}_1 + \mathbf{t}_2) = (\mathbf{R}|\mathbf{R})(\mathbf{t}_2)(\mathbf{S}|\mathbf{R})(\mathbf{t}_1), \qquad |\mathbf{t}_1| > a, \quad |\mathbf{t}_2| < a' < a. \tag{7.1.23}$$

On the other hand, an $(\mathbf{S}|\mathbf{R})$ translation can be considered as a composition of an $(\mathbf{S}|\mathbf{S})$-translation with a translation vector of a smaller length followed by the $(\mathbf{S}|\mathbf{R})$-translation with a translation vector of a larger length. This yields

$$(\mathbf{S}|\mathbf{R})(\mathbf{t}_1 + \mathbf{t}_2) = (\mathbf{S}|\mathbf{R})(\mathbf{t}_1)(\mathbf{S}|\mathbf{S})(\mathbf{t}_2), \qquad |\mathbf{t}_1| > a, \quad |\mathbf{t}_2| < a' < a. \tag{7.1.24}$$

Because the matrices $(\mathbf{R}|\mathbf{R})$ and $(\mathbf{S}|\mathbf{S})$ for the Helmholtz equation are the same, we can see that the $(\mathbf{S}|\mathbf{R})$-translation matrix *commutes* with the $(\mathbf{R}|\mathbf{R})$-translation matrix. Note also that a and a' determine the radii of particular function convergence for particular expansion vectors, so they are not properties of matrices, which depend only on the translation vectors. Since these parameters can be selected arbitrarily, we can rewrite Eqs. (7.1.23) and (7.1.24) in the form

$$(\mathbf{S}|\mathbf{R})(\mathbf{t}_1)(\mathbf{R}|\mathbf{R})(\mathbf{t}_2) = (\mathbf{R}|\mathbf{R})(\mathbf{t}_2)(\mathbf{S}|\mathbf{R})(\mathbf{t}_1) = (\mathbf{S}|\mathbf{R})(\mathbf{t}_1 + \mathbf{t}_2), \qquad |\mathbf{t}_1| > |\mathbf{t}_2|. \tag{7.1.25}$$

7.1.2 Representations of translation operators using signature functions

Integral transforms on a unit sphere and integral representations of the translation operators play a key role in the theory of *diagonal forms* introduced by Rokhlin [Rok93]. In this section we derive integral representations and establish links between them and the matrix representations. In Chapter 8, we will elaborate the meaning of the word "diagonal" by exploiting relations between different representations of the translation operators. Integral representations can also be used for derivation of sparse matrix decomposition of the translation operators, which we consider later in this chapter.

7.1.2.1 (R|R) translation

We start the consideration of integral forms of the translation operators for a function $\psi(\mathbf{r}) \in \mathcal{B}(\Omega)$, where Ω is the interior of a sphere centered at $\mathbf{r} = \mathbf{0}$. This function can be represented by the following convergent series

$$\psi(\mathbf{r}) = \sum_{n=0}^{\infty} \sum_{m=-n}^{n} C_n^m R_n^m(\mathbf{r}). \tag{7.1.26}$$

Using the representation of the regular spherical basis functions (2.3.7), we can then write

$$\psi(\mathbf{r}) = \frac{1}{4\pi} \int_{S_u} e^{ik\mathbf{s}\cdot\mathbf{r}} \Psi(\mathbf{s}) dS(\mathbf{s}), \tag{7.1.27}$$

where the surface function

$$\Psi(\mathbf{s}) = \sum_{n=0}^{\infty} \sum_{m=-n}^{n} i^{-n} C_n^m Y_n^m(\mathbf{s}) \tag{7.1.28}$$

is the *far-field signature function* (compare with Eq. (2.3.39)) or simply *signature* function. The function $\psi(\mathbf{r})$ defined by Eq. (7.1.27) is called the Herglotz wave function. We assume that the series (7.1.28) converges. E.g. this is the case for entire functions or *band-limited* functions, where $C_n^m = 0$ for $n \geq p$.

Therefore, we can correspond some surface function $\Psi(\mathbf{s})$ one-to-one to any function $\psi(\mathbf{r})$. Indeed for a given set $\{C_n^m\}$ we can find $\Psi(\mathbf{s})$ and for given $\Psi(\mathbf{s})$ we can find the expansion coefficients

$$C_n^m = i^n \int_{S_u} \Psi(\mathbf{s}) Y_n^{-m}(\mathbf{s}) dS(\mathbf{s}). \tag{7.1.29}$$

The space of functions $\mathcal{B}(\Omega)$ or $\mathcal{B}(\mathbb{R}^3)$ then will be mapped into the space of functions on a unit sphere (say $L^2(S_u)$) and linear operators in $B(\Omega)$ can be represented by linear operators in $L^2(S_u)$. Then the translation operator $\mathcal{T}(\mathbf{t})$ in space $\mathcal{B}(\Omega)$ will be represented by operator $(\mathcal{R}|\mathcal{R})(\mathbf{t})$ in space $L^2(S_u)$. The symbol $(\mathcal{R}|\mathcal{R})(\mathbf{t})$ reminds that the signature function $\Psi(\mathbf{s})$ is built based on the expansion coefficients over the regular basis $\{R_n^m(\mathbf{r})\}$ and the translated signature function should also be referred to the regular basis.

If function $\psi(\mathbf{r})$ is an entire function (band-limited functions with respect to the regular basis are entire functions) it can be translated with an arbitrary translation vector $\mathbf{t}$. Using integral representation (7.1.27), we obtain

$$\begin{aligned}\hat{\psi}(\mathbf{r}) &= \mathcal{T}(\mathbf{t})[\psi(\mathbf{r})] = \psi(\mathbf{r}+\mathbf{t}) \\ &= \int_{S_u} \mathrm{e}^{\mathrm{i}k\mathbf{s}\cdot\mathbf{r}}\,\mathrm{e}^{\mathrm{i}k\mathbf{s}\cdot\mathbf{t}}\Psi(\mathbf{s})\mathrm{d}S(\mathbf{s}) = \int_{S_u} \mathrm{e}^{\mathrm{i}k\mathbf{s}\cdot\mathbf{r}}\hat{\Psi}(\mathbf{s})\mathrm{d}S(\mathbf{s}).\end{aligned} \tag{7.1.30}$$

This shows that upon translation of the entire function the corresponding far-field signature function gets multiplied by the factor $\mathrm{e}^{\mathrm{i}\mathbf{k}\cdot\mathbf{t}}$. So we can write the following representation of the translation operator

$$\hat{\Psi}(\mathbf{s}) = (\mathcal{R}|\mathcal{R})(\mathbf{t})[\Psi(\mathbf{s})] = \mathrm{e}^{\mathrm{i}k\mathbf{s}\cdot\mathbf{t}}\Psi(\mathbf{s}). \tag{7.1.31}$$

We also note that integral representation (7.1.27) yields an integral representation of the elements of the $(\mathbf{R}|\mathbf{R})$ translation matrix. Indeed, the expansion coefficients of the translated function can be expressed according to Eqs. (7.1.29) as

$$\begin{aligned}\hat{C}_n^m &= \mathrm{i}^n \int_{S_u} \Psi^{(p)\prime}(\mathbf{s})Y_n^{-m}(\mathbf{s})\mathrm{d}S(\mathbf{s}) = \mathrm{i}^n \int_{S_u} \mathrm{e}^{\mathrm{i}k\mathbf{s}\cdot\mathbf{t}}\Psi^{(p)}(\mathbf{s})Y_n^{-m}(\mathbf{s})\mathrm{d}S(\mathbf{s}) \\ &= \mathrm{i}^n \int_{S_u} \mathrm{e}^{\mathrm{i}k\mathbf{s}\cdot\mathbf{t}} \sum_{n'=0}^{\infty}\sum_{m'=-n'}^{n'} \mathrm{i}^{-n'} C_{n'}^{m'} Y_{n'}^{m'}(\mathbf{s})Y_n^{-m}(\mathbf{s})\mathrm{d}S(\mathbf{s}) \\ &= \sum_{n'=0}^{\infty}\sum_{m'=-n'}^{n'} C_{n'}^{m'}\mathrm{i}^{n-n'} \int_{S_u} \mathrm{e}^{\mathrm{i}k\mathbf{s}\cdot\mathbf{t}} Y_{n'}^{m'}(\mathbf{s})Y_n^{-m}(\mathbf{s})\mathrm{d}S(\mathbf{s}) \\ &= \sum_{n'=0}^{\infty}\sum_{m'=-n'}^{n'} (R|R)_{nn'}^{mm'}(\mathbf{t})C_{n'}^{m'}.\end{aligned} \tag{7.1.32}$$

so

$$(R|R)_{nn'}^{mm'}(\mathbf{t}) = \mathrm{i}^{n-n'} \int_{S_u} \mathrm{e}^{\mathrm{i}k\mathbf{s}\cdot\mathbf{t}} Y_{n'}^{m'}(\mathbf{s})Y_n^{-m}(\mathbf{s})\mathrm{d}S(\mathbf{s}), \tag{7.1.33}$$

$$n, n' = 0, 1, \ldots, \quad m = -n, \ldots, n, \quad m' = -n', \ldots, n'.$$

From a methodological point of view it is useful to represent Eq. (7.1.33) in matrix form. For this purpose we introduce the following vector notation

$$\mathbf{iY}(\mathbf{s}) = \{\mathrm{i}^n Y_n^{-m}(\mathbf{s})\}. \tag{7.1.34}$$

This vector is of the same type as the vector of expansion coefficients and can be aligned, e.g. as Eq. (7.2.3). We can also introduce the notation for the Kronecker product of two vectors of this type:

$$\mathbf{iY(s)} \otimes \overline{\mathbf{iY(s)}} = \{\mathrm{i}^n Y_n^{-m}(\mathbf{s}) \overline{\mathrm{i}^{n'} Y_{n'}^{-m'}(\mathbf{s})}\} = \{\mathrm{i}^n Y_n^{-m}(\mathbf{s}) \mathrm{i}^{-n'} Y_{n'}^{m'}(\mathbf{s})\}. \quad (7.1.35)$$

This product is a matrix of the same type as the $(\mathbf{R}|\mathbf{R})$-translation matrix, whose elements are terms in the curly brackets of Eq. (7.1.35). With this notation we have the following integral representation of the $(\mathbf{R}|\mathbf{R})$-translation matrix

$$(\mathbf{R}|\mathbf{R})(\mathbf{t}) = \int_{S_u} e^{ik\mathbf{s}\cdot\mathbf{t}} \mathbf{iY(s)} \otimes \overline{\mathbf{iY(s)}} dS(\mathbf{s}). \quad (7.1.36)$$

The vector notation (7.1.34) can be used to rewrite, Eqs. (7.1.28) and (7.1.29), describing transforms from the space of expansion coefficients to the space of signature functions in the following vector form:

$$\Psi(\mathbf{s}) = (\mathbf{iY(s)}, \mathbf{C}), \quad (7.1.37)$$

$$\mathbf{C} = \int_{S_u} \Psi(\mathbf{s}) \mathbf{iY(s)} dS(\mathbf{s}). \quad (7.1.38)$$

Here the scalar product in complex space can be defined as

$$(\mathbf{A}, \mathbf{B}) = \sum_{n=0}^{\infty} \sum_{m=-n}^{n} \overline{A_n^m} B_n^m, \quad (7.1.39)$$

assuming that the series converges.

7.1.2.2 (S|S) translation

Let us now turn to translations of a function $\psi(\mathbf{r}) \in \mathcal{B}(\Omega)$, where Ω is the exterior of a sphere centered at $\mathbf{r} = \mathbf{0}$. These functions can then be represented by the following series:

$$\psi(\mathbf{r}) = \sum_{n=0}^{\infty} \sum_{m=-n}^{n} C_n^m S_n^m(\mathbf{r}). \quad (7.1.40)$$

As in the case of the $S|S$ translations we will correspond to $\psi(\mathbf{r})$ the signature function given by Eq. (7.1.28) or (7.1.37). The expansion coefficient over the singular basis can then be found from Eq. (7.1.29) or (7.1.38). Using the addition theorem for spherical harmonics (2.1.72) we find the following integral representation for the function $\psi(\mathbf{r})$:

$$\psi(\mathbf{r}) = \sum_{n=0}^{\infty}\sum_{m=-n}^{n} C_n^m S_n^m(\mathbf{r}) = \sum_{n=0}^{\infty}\sum_{m=-n}^{n} \mathrm{i}^n \int_{S_u} \Psi(\mathbf{s}) S_n^m(\mathbf{r}) Y_n^{-m}(\mathbf{s}) dS(\mathbf{s})$$

$$= \sum_{n=0}^{\infty} \mathrm{i}^n h_n(kr) \int_{S_u} \left[\sum_{m=-n}^{n} Y_n^m\left(\frac{\mathbf{r}}{r}\right) Y_n^{-m}(\mathbf{s})\right] \Psi(\mathbf{s}) dS(\mathbf{s})$$

$$= \frac{1}{4\pi}\sum_{n=0}^{\infty} (2n+1)\mathrm{i}^n h_n(kr) \int_{S_u} P_n\left(\frac{\mathbf{r}\cdot\mathbf{s}}{r}\right) \Psi(\mathbf{s}) dS(\mathbf{s}). \quad (7.1.41)$$

We note that, in general, the operations of integration and summation are not interchangeable because of the *divergence* of the series

$$\Lambda_s(\mathbf{r};\mathbf{s}) = \sum_{n=0}^{\infty} (2n+1)\mathrm{i}^n h_n(kr) P_n\left(\frac{\mathbf{r}\cdot\mathbf{s}}{r}\right). \quad (7.1.42)$$

However, for band-limited functions $\psi^{(p)}(\mathbf{r})$ ($C_n^m = 0$ for $n \geq p$), for which the sum is finite, this procedure can be performed, and we have

$$\psi^{(p)}(\mathbf{r}) = \frac{1}{4\pi}\int_{S_u} \Lambda_s^{(p)}(\mathbf{r};\mathbf{s}) \Psi(\mathbf{s}) dS(\mathbf{s}), \quad (7.1.43)$$

Particularly, for the function $\psi^{(p)}(\mathbf{r}) = S_n^m(\mathbf{r})$ ($n \leq p-1$) we have, using Eqs. (7.1.28) and (7.1.40):

$$S_n^m(\mathbf{r}) = \frac{\mathrm{i}^{-n}}{4\pi}\int_{S_u} \Lambda_s^{(p)}(\mathbf{r};\mathbf{s}) Y_n^m(\mathbf{s}) dS(\mathbf{s}), \qquad n \leq p-1. \quad (7.1.44)$$

Here the kernel of the integral transform from the space of solutions of the Helmholtz equation to the space of surface functions is

$$\Lambda_s^{(p)}(\mathbf{r};\mathbf{s}) = \sum_{n=0}^{p-1} (2n+1)\mathrm{i}^n h_n(kr) P_n\left(\frac{\mathbf{r}\cdot\mathbf{s}}{r}\right), \quad (7.1.45)$$

and Eq. (7.1.41) can be rewritten as

$$\psi(\mathbf{r}) = \lim_{p\to\infty} \frac{1}{4\pi}\int_{S_u} \Lambda_s^{(p)}(\mathbf{r};\mathbf{s}) \Psi(\mathbf{s}) dS(\mathbf{s}). \quad (7.1.46)$$

As noted above, the kernel function for the S expansion is singular at $\mathbf{r} = \mathbf{0}$ and the series (7.1.45) diverges as $p \to \infty$. While for any prescribed error ϵ we can find a finite p that provides acceptable approximation of $\psi(\mathbf{r})$ with $\psi^{(p)}(\mathbf{r})$ in Ω, we are faced with the fact that the kernel is defined only for finite p and so it is a band-limited surface function. At the same time for any given p the integral entering Eq. (7.1.46) exists and the sequence of $\psi^{(p)}(\mathbf{r})$ as $p \to \infty$ converges, which allows us to put the limit sign *outside* the integral.

It is also useful to compare the singular and regular kernels. The latter band-limited kernel is given by

$$\Lambda_r^{(p)}(\mathbf{r};\mathbf{s}) = \sum_{n=0}^{p-1} (2n+1) \mathrm{i}^n j_n(kr) P_n\left(\frac{\mathbf{r}\cdot\mathbf{s}}{r}\right). \tag{7.1.47}$$

In contrast to $\Lambda_s^{(p)}(\mathbf{r};\mathbf{s})$, the regular kernel is an entire function and absolutely and uniformly converges as $p \to \infty$:

$$\lim_{p\to\infty} \Lambda_r^{(p)}(\mathbf{r};\mathbf{s}) = \Lambda_r^{(\infty)}(\mathbf{r};\mathbf{s}) = \mathrm{e}^{\mathrm{i}k\mathbf{s}\cdot\mathbf{r}}. \tag{7.1.48}$$

This is nothing but the plane wave expansion (2.3.5), which is the kernel of transform (7.1.30).

Consider now translation of the function $\psi(\mathbf{r})$ with some vector $\mathbf{t}$. We can use the fact that matrices $(\mathbf{S}|\mathbf{S})(\mathbf{t})$ and $(\mathbf{R}|\mathbf{R})(\mathbf{t})$ are the same and the elements of the latter matrix are given by Eq. (7.1.33). The expansion coefficients over the basis $\{S_n^m(\mathbf{r})\}$ for the translated function can then be found as

$$\begin{aligned}
\hat{C}_n^m &= \sum_{n'=0}^{\infty} \sum_{m'=-n'}^{n'} (S|S)_{nn'}^{mm'}(\mathbf{t}) C_{n'}^{m'} \\
&= \sum_{n'=0}^{\infty} \sum_{m'=-n'}^{n'} C_{n'}^{m'} \mathrm{i}^{n-n'} \int_{S_u} \mathrm{e}^{\mathrm{i}k\mathbf{s}\cdot\mathbf{t}} Y_{n'}^{m'}(\mathbf{s}) Y_n^{-m}(\mathbf{s}) dS(\mathbf{s}) \\
&= \mathrm{i}^n \int_{S_u} Y_n^{-m}(\mathbf{s}) \mathrm{e}^{\mathrm{i}k\mathbf{s}\cdot\mathbf{t}} \sum_{n'=0}^{\infty} \sum_{m'=-n'}^{n'} \mathrm{i}^{-n'} C_{n'}^{m'} Y_{n'}^{m'}(\mathbf{s}) dS(\mathbf{s}) \qquad (7.1.49) \\
&= \mathrm{i}^n \int_{S_u} \mathrm{e}^{\mathrm{i}k\mathbf{s}\cdot\mathbf{t}} \Psi(\mathbf{s}) Y_n^{-m}(\mathbf{s}) dS(\mathbf{s}) \\
&= \mathrm{i}^n \int_{S_u} \hat{\Psi}(\mathbf{s}) Y_n^{-m}(\mathbf{s}) dS(\mathbf{s}),
\end{aligned}$$

$$n = 0, 1, \ldots, \qquad m = -n, \ldots, n.$$

The latter equality expresses $\hat{C}_n^m$ via the translated signature function $\hat{\Psi}(\mathbf{s})$, (see Eq. (7.1.29)) which is related to the original signature function $\Psi(\mathbf{s})$ as

$$\hat{\Psi}(\mathbf{s}) = (\mathcal{S}|\mathcal{S})(\mathbf{t})[\Psi(\mathbf{s})] = e^{ik\mathbf{s}\cdot\mathbf{t}}\Psi(\mathbf{s}). \tag{7.1.50}$$

So the transform of the signature function in the $S|S$ translation occurs in the same way as the transform of the signature function in the $R|R$ translation (see Eq. (7.1.31)). As soon as the signature function is translated we can determine $\hat{\psi}(\mathbf{r})$ according to Eq. (7.1.41)

$$\begin{aligned}\hat{\psi}(\mathbf{r}) &= \frac{1}{4\pi}\sum_{n=0}^{\infty}(2n+1)i^n h_n(kr)\int_{S_u} P_n\left(\frac{\mathbf{r}\cdot\mathbf{s}}{r}\right)\hat{\Psi}(\mathbf{s})dS(\mathbf{s})\\ &= \lim_{p\to\infty}\frac{1}{4\pi}\int_{S_u}\Lambda_s^{(p)}(\mathbf{r};\mathbf{s})\hat{\Psi}(\mathbf{s})dS(\mathbf{s}).\end{aligned} \tag{7.1.51}$$

The translated function $\hat{\psi}(\mathbf{r})$ can also be represented as

$$\hat{\psi}(\mathbf{r}) = \mathcal{T}(\mathbf{t})[\psi(\mathbf{r})] = \psi(\mathbf{r}+\mathbf{t}) = \lim_{p\to\infty}\frac{1}{4\pi}\int_{S_u}\Lambda_s^{(p)}(\mathbf{r}+\mathbf{t};\mathbf{s})\Psi(\mathbf{s})dS(\mathbf{s}), \tag{7.1.52}$$

$|\mathbf{t}| < |\mathbf{r}|$.

Comparing Eqs. (7.1.51) and (7.1.52) and using Eq. (7.1.50) we can see that

$$\lim_{p\to\infty}\frac{1}{4\pi}\int_{S_u}[\Lambda_s^{(p)}(\mathbf{r}+\mathbf{t};\mathbf{s}) - e^{ik\mathbf{s}\cdot\mathbf{t}}\Lambda_s^{(p)}(\mathbf{r};\mathbf{s})]\Psi(\mathbf{s})dS(\mathbf{s}) = 0, \tag{7.1.53}$$

$|\mathbf{t}| < |\mathbf{r}|$.

Note that condition $|\mathbf{t}| < |\mathbf{r}|$ here is strictly necessary, since it is a requirement for the $S|S$ translations as discussed above.

7.1.2.3 *S|R translation*

Once we have defined the signature function both for the regular and the singular expansion, we can consider how it changes in the $S|R$ translation. In general, the translated function is described by the integral (7.1.52), where the shifted kernel can be decomposed as provided by Eq. (7.1.53). However, the latter equation is written in the form appropriate for the $S|S$ translation, where we have $|\mathbf{t}| < |\mathbf{r}|$.

In case of the $S|R$ translation we have the opposite situation, when we are looking for a local expansion with $|\mathbf{t}| > |\mathbf{r}|$. We can obtain the relation for this case simply by exchanging symbols $\mathbf{r}$ and $\mathbf{t}$ in Eq. (7.1.53):

$$\lim_{p\to\infty} \frac{1}{4\pi} \int_{S_u} [\Lambda_s^{(p)}(\mathbf{r}+\mathbf{t};\mathbf{s}) - e^{ik\mathbf{s}\cdot\mathbf{r}}\Lambda_s^{(p)}(\mathbf{t};\mathbf{s})]\Psi(\mathbf{s})dS(\mathbf{s}) = 0, \tag{7.1.54}$$

$|\mathbf{r}| < |\mathbf{t}|$.

Substituting this relation into Eq. (7.1.52) we obtain

$$\begin{aligned}\hat{\psi}(\mathbf{r}) = \mathcal{T}(\mathbf{t})[\psi(\mathbf{r})] &= \lim_{p\to\infty} \frac{1}{4\pi} \int_{S_u} \Lambda_s^{(p)}(\mathbf{r}+\mathbf{t};\mathbf{s})\Psi(\mathbf{s})dS(\mathbf{s}) \\ &= \lim_{p\to\infty} \frac{1}{4\pi} \int_{S_u} e^{ik\mathbf{s}\cdot\mathbf{r}}\Lambda_s^{(p)}(\mathbf{t};\mathbf{s})\Psi(\mathbf{s})dS(\mathbf{s}).\end{aligned} \tag{7.1.55}$$

This can also be rewritten as

$$\hat{\psi}(\mathbf{r}) = \lim_{p\to\infty} \hat{\psi}^{(p)}(\mathbf{r}), \qquad \hat{\psi}^{(p)}(\mathbf{r}) = \frac{1}{4\pi} \int_{S_u} e^{ik\mathbf{s}\cdot\mathbf{r}}\Lambda_s^{(p)}(\mathbf{t};\mathbf{s})\Psi_{(p)}(\mathbf{s})dS(\mathbf{s}). \tag{7.1.56}$$

Now we can see that Eq. (7.1.55) provides an integral representation of the function $\hat{\psi}^{(p)}(\mathbf{r})$ in its R form (Eq. (7.1.27)) with the signature function $\hat{\Psi}_{(p)}(\mathbf{s})$ related to $\Psi(\mathbf{s})$ as

$$\hat{\Psi}_{(p)}(\mathbf{s}) = (\mathcal{S}|\mathcal{R})(\mathbf{t})[\Psi(\mathbf{s})] = \Lambda_s^{(p)}(\mathbf{t};\mathbf{s})\Psi(\mathbf{s}). \tag{7.1.57}$$

So the transform of the signature function for the $S|R$ translation occurs with $\Lambda_s^{(p)}(\mathbf{t};\mathbf{s})$, used instead of $\Lambda_r(\mathbf{t};\mathbf{s}) = e^{ik\mathbf{s}\cdot\mathbf{t}}$ for the $R|R$ and $S|S$ translations (see Eqs. (7.1.31) and (7.1.50)). We also note that we understand the $S|R$ translation, $\hat{\psi}(\mathbf{r})$, of the function $\psi(\mathbf{r})$ as a limit of the convergent sequence of transforms $\psi(\mathbf{r}) \to \hat{\psi}^{(p)}(\mathbf{r})$ as $p \to \infty$.

Finally, we consider integral representations of the elements of the $(\mathbf{S}|\mathbf{R})$ translation matrix using the singular kernel function $\Lambda_s^{(p)}(\mathbf{t};\mathbf{s})$. While this can be done by several ways, we will derive the formula using the Clebsch–Gordan coefficients or related E-symbols. Equations (3.2.31), (3.2.33), and (3.2.36) then provide the following expression for the translation coefficients $(S|R)_{nn'}^{mm'}(\mathbf{t})$:

$$
\begin{aligned}
(S|R)_{nn'}^{mm'}(\mathbf{t}) &= \sum_{n''=|n-n'|}^{n+n'} (s|r)_{n''nn'}^{m'-m,mm'} S_{n''}^{m'-m}(\mathbf{t}) \\
&= \sum_{n''=0}^{n+n'} \sum_{m''=-n''}^{n''} (s|r)_{n''nn'}^{m''mm'} S_{n''}^{m''}(\mathbf{t}) \\
&= \sum_{n''=0}^{n+n'} \sum_{m''=-n''}^{n''} 4\pi \mathrm{i}^{n+n''-n'} S_{n''}^{m''}(\mathbf{t}) \int_{S_u} Y_{n'}^{m'}(\mathbf{s}) Y_n^{-m}(\mathbf{s}) Y_{n''}^{-m''}(\mathbf{s}) \mathrm{d}S \\
&= \int_{S_u} \sum_{n''=0}^{n+n'} 4\pi \mathrm{i}^{n+n''-n'} h_{n''}(kt) Y_{n'}^{m'}(\mathbf{s}) Y_n^{-m}(\mathbf{s}) \\
&\quad \times \left[\sum_{m''=-n''}^{n''} Y_{n''}^{m''}(\mathbf{t}) Y_{n''}^{-m''}(\mathbf{s}) \right] \mathrm{d}S \\
&= \mathrm{i}^{n-n'} \int_{S_u} \Lambda_s^{(n+n'+1)}(\mathbf{t};\mathbf{s}) Y_{n'}^{m'}(\mathbf{s}) Y_n^{-m}(\mathbf{s}) \mathrm{d}S.
\end{aligned} \tag{7.1.58}
$$

Here $\Lambda_s^{(n+n'+1)}$ can be replaced with $\Lambda_s^{(q)}$ with any $q \geq n+n'+1$. Particularly, we can consider elements of a rectangular truncated matrix, $n \leq p-1$, $n' \leq p'-1$. In this case we can replace $\Lambda_s^{(n+n'+1)}$ with $\Lambda_s^{(p+p'-1)}$ which is sufficient to represent any element of the truncated matrix. So if $(\mathbf{S}|\mathbf{R})^{(p,p')}(\mathbf{t})$ is a truncated translation matrix of size $p^2 \times p'^2$ it can be represented in the following form:

$$
(\mathbf{S}|\mathbf{R})^{(p,p')}(\mathbf{t}) = \int_{S_u} \Lambda_s^{(p+p'-1)}(\mathbf{t};\mathbf{s}) \mathbf{iY}^{(p)}(\mathbf{s}) \otimes \overline{\mathbf{iY}^{(p')}(\mathbf{s})} \mathrm{d}S(\mathbf{s}), \tag{7.1.59}
$$

where $\mathbf{iY}^{(p)}(\mathbf{s})$ is a truncated vector of total length p^2.

Another representation of the $(\mathbf{S}|\mathbf{R})(\mathbf{t})$ matrix is:

$$
\begin{aligned}
(S|R)_{nn'}^{mm'}(\mathbf{t}) = \sum_{n''=0}^{\infty} (2n''+1) \mathrm{i}^{n''} h_{n''}(kt) \mathrm{i}^{n-n'} \\
\int_{S_u} P_{n''}\left(\frac{\mathbf{t}\cdot\mathbf{s}}{t}\right) Y_{n'}^{m'}(\mathbf{s}) Y_n^{-m}(\mathbf{s}) \mathrm{d}S(\mathbf{s}).
\end{aligned} \tag{7.1.60}
$$

Here the integral is zero for $n'' > n' + n$. As a result we can write the following representation of the matrix $(\mathbf{S}|\mathbf{R})(\mathbf{t})$ through the infinite series:

$$(\mathbf{S}|\mathbf{R})(\mathbf{t}) = \sum_{n''=0}^{\infty} (2n''+1)\mathrm{i}^{n''} h_{n''}(kt) \int_{S_u} P_{n''}\left(\frac{\mathbf{t}\cdot\mathbf{s}}{t}\right) \mathbf{iY}(\mathbf{s}) \otimes \overline{\mathbf{iY}(\mathbf{s})} \mathrm{d}S(\mathbf{s}). \quad (7.1.61)$$

7.1.2.4 *Coaxial translations*

Similar integral representations can be obtained for the coaxial translation matrices. In this case the integrals involved will be one-dimensional, which simplifies the problem. Equation (7.1.33) yields the following representation of the $R|R$ coaxial translation coefficients (3.2.88):

$$(R|R)_{nn'}^{m}(t) = (R|R)_{nn'}^{mm}(t\mathbf{i}_z) = \mathrm{i}^{n-n'} \int_{S_u} \mathrm{e}^{\mathrm{i}kt\mathbf{s}\cdot\mathbf{i}_z} Y_{n'}^{m}(\mathbf{s}) Y_{n}^{-m}(\mathbf{s}) \mathrm{d}S(\mathbf{s})$$

$$= \mathrm{i}^{n-n'} \int_{-1}^{1} \mathrm{e}^{\mathrm{i}kt\mu} \underline{P}_{n'}^{m}(\mu) \underline{P}_{n}^{m}(\mu) \mathrm{d}\mu, \quad (7.1.62)$$

where the integration is performed over the spherical polar angle φ, and functions $\underline{P}_{n}^{m}(\mu)$ are related to the associated Legendre polynomials by the definition of the spherical harmonics (2.1.59) as

$$\underline{P}_{n}^{m}(\mu) = (-1)^{m} \sqrt{\frac{2n+1}{2} \frac{(n-|m|)!}{(n+|m|)!}} P_{n}^{|m|}(\mu) = \sqrt{2\pi} \mathrm{e}^{-\mathrm{i}m\varphi} Y_{n}^{m}(\mathbf{s})$$

$$= \sqrt{2\pi} Y_{n}^{\pm m}(\mathbf{s})|_{\varphi=0}, \qquad \mu = \mathbf{s}\cdot\mathbf{i}_z \quad (7.1.63)$$

Functions $\underline{P}_{n}^{m}(\mu)$ of given m form orthonormal basis in $L^2[-1,1]$ (see Eq. (2.1.58)):

$$(\underline{P}_{n}^{m}, \underline{P}_{n'}^{m}) = \int_{-1}^{1} \underline{P}_{n}^{m}(\mu) \underline{P}_{n'}^{m}(\mu) \mathrm{d}\mu = \delta_{nn'}. \quad (7.1.64)$$

These functions can also be arranged in vectors similar to Eq. (7.1.34)

$$\begin{aligned} \mathbf{i}\underline{\mathbf{P}}(\mu) &= \{\mathrm{i}^{n} \underline{P}_{n}^{m}(\mu)\}, \\ \mathbf{i}\underline{\mathbf{P}}^{m}(\mu) &= (\mathrm{i}^{|m|} \underline{P}_{|m|}^{m}(\mu), \mathrm{i}^{|m|+1} \underline{P}_{|m|+1}^{m}(\mu), \ldots)^{\mathrm{T}}. \end{aligned} \quad (7.1.65)$$

The coaxial translation matrix and its mth block can be represented using notation for the Kronecker product (7.1.35):

$$(\mathbf{R}|\mathbf{R})_{(\mathrm{coax})}(t) = \int_{-1}^{1} \mathrm{e}^{\mathrm{i}kt\mu} \mathbf{i}\underline{\mathbf{P}}(\mu) \otimes \overline{\mathbf{i}\underline{\mathbf{P}}(\mu)} \mathrm{d}\mu, \quad (7.1.66)$$

$$(\mathbf{R}|\mathbf{R})_{(\mathrm{coax})}^{(m)}(t) = \int_{-1}^{1} \mathrm{e}^{\mathrm{i}kt\mu} \mathbf{i}\underline{\mathbf{P}}^{m}(\mu) \otimes \overline{\mathbf{i}\underline{\mathbf{P}}^{m}(\mu)} \mathrm{d}\mu. \quad (7.1.67)$$

Similarly, we have a representation for the $S|R$ coaxial translation coefficients (3.2.86) that follows from Eq. (7.1.58):

$$(S|R)^{m}_{nn'}(t) = (S|R)^{mm}_{nn'}(t\mathbf{i}_z) = \mathrm{i}^{n-n'} \int_{-1}^{1} \Lambda_s^{(n+n'+1)}(kt;\mu) \underline{P}^{m}_{n'}(\mu)\underline{P}^{m}_{n}(\mu)\mathrm{d}\mu, \tag{7.1.68}$$

with the singular kernel common for all m:

$$\Lambda_s^{(p)}(kt;\mu) = \sum_{n=0}^{p-1} (2n+1)\mathrm{i}^n h_n(kt) P_n(\mu). \tag{7.1.69}$$

Corresponding truncated coaxial matrices can then be represented via the Kronecker products as

$$(\mathbf{S}|\mathbf{R})^{(p,p')}_{(\mathrm{coax})}(t) = \int_{-1}^{1} \Lambda_s^{(p+p'-1)}(kt;\mu)\mathbf{i}\underline{\mathbf{P}}^{(p)}(\mu)\otimes\overline{\mathbf{i}\underline{\mathbf{P}}^{(p')}(\mu)}\mathrm{d}\mu, \tag{7.1.70}$$

$$(\mathbf{S}|\mathbf{R})^{(p,p')(m)}_{(\mathrm{coax})}(t) = \int_{-1}^{1} \Lambda_s^{(p+p'-1)}(kt;\mu)\mathbf{i}\underline{\mathbf{P}}^{m(p)}(\mu)\otimes\overline{\mathbf{i}\underline{\mathbf{P}}^{m(p')}(\mu)}\mathrm{d}\mu, \tag{7.1.71}$$

where $\mathbf{i}\underline{\mathbf{P}}^{(p)}$ denotes p-truncated vector $\mathbf{i}\underline{\mathbf{P}}$.

7.1.2.5 Rotations

Here we mention that vectors $\mathbf{i}\underline{\mathbf{P}}(\mu)$ and $\mathbf{iY(s)}$ are related via a rotation transform. Let us perform a rotation of the reference frame with some rotation matrix $Q(\alpha,\beta,\gamma)$ which transforms vector $\mathbf{i}_z$ to $\mathbf{i}_{\hat{z}}$. The inverse rotation is described by matrix $Q^{-1} = Q(\gamma,\beta,\alpha)$ (see Eq. (3.3.14)). So we have from Eq. (3.3.37) and symmetry of $H_n^{m'm}(\beta)$ rotation coefficients (3.3.44):

$$T_n^{m'm}(Q^{-1}) = T_n^{m'm}(\gamma,\beta,\alpha) = \mathrm{e}^{\mathrm{i}m\gamma}\,\mathrm{e}^{-\mathrm{i}m'\alpha}H_n^{m'm}(\beta)$$
$$= \overline{\mathrm{e}^{-\mathrm{i}m'\alpha}\,\mathrm{e}^{\mathrm{i}m\gamma}H_n^{mm'}(\beta)} = \overline{T_n^{mm'}(\alpha,\beta,\gamma)}. \tag{7.1.72}$$

Now let $\mathbf{A}$ be a vector obtained by the application of the rotation matrix to the vector $\mathbf{iY(s)}$:

$$\mathbf{A} = \mathbf{Rot}(Q)\mathbf{iY(s)}. \tag{7.1.73}$$

Then the components of this vector are

$$A_n^m = \sum_{m'=-n}^{n} T_n^{mm'}(Q) i^n Y_n^{-m'}(\mathbf{s}) = i^n \overline{\sum_{m'=-n}^{n} \overline{T_n^{mm'}(Q)} Y_n^{m'}(\mathbf{s})}$$

$$= i^n \overline{\sum_{m'=-n}^{n} T_n^{m'm}(Q^{-1}) Y_n^{m'}(\mathbf{s})} = i^n \overline{Y_n^m(\hat{\mathbf{s}})} = i^n Y_n^{-m}(\hat{\mathbf{s}}). \qquad (7.1.74)$$

Here we used definition of rotation coefficients (3.3.22). If $\mathbf{s}$ are the coordinates of a spatial point M on a unit sphere, then $\hat{\mathbf{s}}$ are coordinates of the same point M in the rotated reference frame. These coordinates are related with the rotation matrix Q,

$$\hat{\mathbf{s}} = Q\mathbf{s} \qquad (7.1.75)$$

and so Eq. (7.1.74) proves the following relation:

$$\mathbf{iY}(Q\mathbf{s}) = \mathbf{Rot}(Q)\mathbf{iY}(\mathbf{s}). \qquad (7.1.76)$$

Now consider some point M that in the rotated reference frame is characterized by spherical polar angles $(\hat{\theta}, \hat{\varphi})$, such that $\mu = \cos\hat{\theta}$ and $\hat{\varphi} = 0$. In the original reference frame coordinates of this point are $\mathbf{s}$ and using Eq. (7.1.63) we obtain

$$\mathbf{i}\underline{\mathbf{P}}(\mu) = \frac{1}{\sqrt{2\pi}} \mathbf{Rot}(Q)\mathbf{iY}(\mathbf{s}). \qquad (7.1.77)$$

This can also be rewritten as

$$\mathbf{iY}(\mathbf{s}) = \sqrt{2\pi}\mathbf{Rot}^{-1}(Q)\mathbf{i}\underline{\mathbf{P}}(\mu). \qquad (7.1.78)$$

So for a given $\mathbf{i}\underline{\mathbf{P}}(\mu)$ we can produce an arbitrary $\mathbf{iY}(\mathbf{s})$ by some rotation transform. Note that if $\mathbf{s} = \mathbf{t}/t$ and rotation is performed to align the rotated axis $\mathbf{i}_{\hat{z}}$ with $\mathbf{t}/t$ we obtain using Eq. (7.1.63) and denoting the rotation matrix as $Q(\mathbf{t}/t)$:

$$\mathbf{i}\underline{\mathbf{P}}(1) = \frac{1}{\sqrt{2\pi}} \mathbf{Rot}\left(Q\left(\frac{\mathbf{t}}{t}\right)\right)\mathbf{iY}\left(\frac{\mathbf{t}}{t}\right). \qquad (7.1.79)$$

7.2 ROTATIONAL-COAXIAL TRANSLATION DECOMPOSITION

Consider translation of the expansion coefficients based on p-truncated translation matrices, which appear when infinite sums (3.1.6) are truncated to first p degrees of expansion.

$$C_n^m(\mathbf{r}_{*2}) = \sum_{n'=0}^{p-1} \sum_{m'=-n'}^{n'} (F|E)_{nn'}^{mm'}(\mathbf{t}) C_{n'}^{m'}(\mathbf{r}_{*1})$$

$$= \sum_{m'=-(p-1)}^{p-1} \sum_{n'=|m'|}^{p-1} (F|E)_{nn'}^{mm'}(\mathbf{t}) C_{n'}^{m'}(\mathbf{r}_{*1}), \tag{7.2.1}$$

where F, E denote any of S, R. For further compact notation we may write this in the following matrix–vector form:

$$\hat{\mathbf{C}} = (\mathbf{F}|\mathbf{E})(\mathbf{t})\mathbf{C}, \qquad \mathbf{F}, \mathbf{E} = \mathbf{S}, \mathbf{R}, \tag{7.2.2}$$

where $\mathbf{C}$ is the vector of expansion coefficients near the "old" center $\mathbf{r}_{*1}$, $\hat{\mathbf{C}}$ is the vector of expansion coefficients near the "new" center, and $(\mathbf{F}|\mathbf{E})(\mathbf{t})$ is the translation matrix. Note that in the form (7.2.2) elements of vectors $\mathbf{C}$ and $\hat{\mathbf{C}}$ can be aligned by various ways, e.g. as

$$\mathbf{C} = (C_0^0, C_1^{-1}, C_1^0, C_1^1, C_2^{-2}, C_2^{-1}, C_2^0, C_2^1, C_2^2, \ldots)^{\mathrm{T}}, \tag{7.2.3}$$

then for p-truncated series, the size of this vector will be p^2. Accordingly, the matrix $(\mathbf{F}|\mathbf{E})(\mathbf{t})$ should be aligned as

$$(\mathbf{F}|\mathbf{E}) = \begin{pmatrix} (F|E)_{00}^{00} & (F|E)_{01}^{0-1} & (F|E)_{01}^{00} & (F|E)_{01}^{01} & (F|E)_{02}^{0-2} & \cdots \\ (F|E)_{10}^{-10} & (F|E)_{11}^{-1-1} & (F|E)_{11}^{-10} & (F|E)_{11}^{-11} & (F|E)_{12}^{-1-2} & \cdots \\ (F|E)_{10}^{00} & (F|E)_{11}^{0-1} & (F|E)_{11}^{00} & (F|E)_{11}^{01} & (F|E)_{12}^{0-2} & \cdots \\ (F|E)_{10}^{10} & (F|E)_{11}^{1-1} & (F|E)_{11}^{10} & (F|E)_{11}^{11} & (F|E)_{12}^{1-2} & \cdots \\ (F|E)_{20}^{-20} & (F|E)_{21}^{-2-1} & (F|E)_{21}^{-20} & (F|E)_{21}^{-21} & (F|E)_{22}^{-2-2} & \cdots \\ \cdots & \cdots & \cdots & \cdots & \cdots & \cdots \end{pmatrix} \tag{7.2.4}$$

which for p-truncated series will be a matrix $p^2 \times p^2$.

Proceeding in this way for translation of the function expansion coefficients from one basis to another, we can compute all entries of the truncated translation matrix in the order of $O(p^4)$ operations using recurrences from Chapter 3, and then multiply the matrix by vector $\mathbf{C}$, which again is $O(p^4)$ procedure.

Another way to organize entries of the vectors of expansion coefficients is their representation as the direct sum of blocks. It is natural to consider the following two types of such blocky decompositions.

First, for each degree n we can consider the following vector

$$\mathbf{C}_n = (C_n^{-n}, \ldots, C_n^n)^{\mathrm{T}}, \qquad n = 0, 1, 2, \ldots \tag{7.2.5}$$

and use the following notation for the total vector of coefficients

$$\mathbf{C} = \mathbf{C}_0 \oplus \mathbf{C}_1 \oplus \cdots = \sum_{n=0}^{p-1} \oplus \mathbf{C}_n. \tag{7.2.6}$$

Second, for each order m we can consider the following vector:

$$\mathbf{C}^m = (C_{|m|}^m, C_{|m|+1}^m, C_{|m|+2}^m, \ldots)^{\mathrm{T}}, \qquad m = 0, \pm 1, \pm 2, \ldots \tag{7.2.7}$$

and represent the total vector of coefficients as

$$\mathbf{C} = \mathbf{C}^0 \oplus \mathbf{C}^{\pm 1} \oplus \cdots = \sum_{m=-(p-1)}^{p-1} \oplus \mathbf{C}^m. \tag{7.2.8}$$

We obviously know the total vector of the translated coefficients, $\hat{\mathbf{C}}$, if we know all its component blocks or smaller size vectors $\hat{\mathbf{C}}_0, \hat{\mathbf{C}}_1, \ldots$ or $\hat{\mathbf{C}}^0, \hat{\mathbf{C}}^{\pm 1}, \ldots$ Once one of the full sets of such smaller size vectors are known the other set can be found by rearranging entries. There is no advantage in using either arrangement (7.2.6) or (7.2.8) instead of arrangement (7.2.3) if we consider a general translation, since the translation matrix acts on all components and $\hat{\mathbf{C}}_n$ depends on all the $\mathbf{C}_n$ (or $\hat{\mathbf{C}}_m$ depends on all the $\mathbf{C}_m$). The situation changes if we deal with rotation or coaxial translation transforms.

7.2.1 Rotations

In the case of the rotation transform, the degree of the spherical harmonics, and so the degree of the expansion coefficients is conserved (see Eq. (3.1.15)). So with rotation we have

$$\hat{\mathbf{C}}_n = \mathbf{Rot}_n(Q)\mathbf{C}_n, \qquad n = 0, 1, 2, \ldots \tag{7.2.9}$$

where $\mathbf{Rot}_n(Q)$ is a $(2n+1) \times (2n+1)$ matrix for harmonics of degree n, consisting of the rotation coefficients $T_n^{mm'}$ defined by Eq. (3.3.22):

$$\mathbf{Rot}_n(Q) = \begin{pmatrix} T_n^{-n,-n} & T_n^{-n,-n+1} & \cdots & T_n^{-n,n} \\ T_n^{-n+1,-n} & T_n^{-n+1,-n+1} & \cdots & T_n^{-n+1,n} \\ \cdots & \cdots & \cdots & \cdots \\ T_n^{n,-n} & T_n^{n,-n+1} & \cdots & T_n^{nn} \end{pmatrix}, \quad (7.2.10)$$

$n = 0, 1, 2, \ldots$

This shows that the total rotation matrix, $\mathbf{Rot}(Q)$, applied to the total vector of coefficients can be represented as a direct sum of blocks:

$$\mathbf{Rot}(Q) = \mathbf{Rot}_0(Q) \oplus \mathbf{Rot}_1(Q) \oplus \cdots = \sum_{n=0}^{p-1} \oplus \mathbf{Rot}_n(Q). \quad (7.2.11)$$

Each of these blocks acts on a block $\mathbf{C}_n$ of the total vector $\mathbf{C}$ and produces the block $\hat{\mathbf{C}}_n$. In this case we can say that the rotation matrix $\mathbf{Rot}(Q)$ has *invariant subspaces*. Vectors $\mathbf{C}_n$ and $\hat{\mathbf{C}}_n$ are vectors of the same invariant subspace. So, if the operations on vectors are limited only by arbitrary rotations, each subspace is *closed* in the sense that the rotation of the basis will not produce vectors from other subspaces, and the intersection of these subspaces is empty, while their direct sum produces the whole space of the expansion coefficients.

Note that the rotation matrix is finite independently of the truncation number, so even for an infinite series for a given n, rotation transform is performed by multiplying a finite matrix by a finite vector. The size of matrices $\mathbf{Rot}_n(Q)$ depends on n and is limited by $(2p+1) \times (2p+1)$ as soon as we consider the truncated series. If multiplication of a matrix $\mathbf{Rot}_n(Q)$ by vector $\mathbf{C}_n$ of length $(2n+1)$ takes $(2n+1)^2$ operations, then the total number of operations to "rotate" the whole expansion will be

$$N_{\mathrm{Rot}}(p) = \sum_{n=0}^{p-1} (2n+1)^2 = \frac{1}{3} p(4p^2 - 1), \quad (7.2.12)$$

which is $O(p^3)$ in asymptotic sense. The same asymptotic complexity holds for computation of all rotation coefficients $T_n^{mm'}$ using the recurrence relations obtained in Chapter 3. So the total asymptotic complexity of "rotation" of the expansion is $O(p^3)$.

7.2.2 Coaxial translation

Another operation which has complexity $O(p^3)$ is the translation operation along the z-axis, which we called coaxial translation (see Eq. (3.2.85) and below). This operation acts on each vector $\mathbf{C}^m$ and produces a vector $\hat{\mathbf{C}}^m$ for a given order m. So we can write

$$\hat{\mathbf{C}}^m = (\mathbf{F}|\mathbf{E})^{(m)}_{(\text{coax})}(t)\mathbf{C}^m, \qquad m = 0, \pm 1, \pm 2, \ldots, \quad E, F = S, R, \qquad (7.2.13)$$

where $(\mathbf{F}|\mathbf{E})^{(m)}_{(\text{coax})}$ is the following $(p - |m|) \times (p - |m|)$ matrix, which depends only on the length of the translation vector:

$$(\mathbf{F}|\mathbf{E})^{(m)}_{(\text{coax})}(t) = \begin{pmatrix} (F|E)^m_{|m|,|m|} & (F|E)^m_{|m|,|m|+1} & \cdots & (F|E)^m_{|m|,p-1} \\ (F|E)^m_{|m|+1,|m|} & (F|E)^m_{|m|+1,|m|+1} & \cdots & (F|E)^m_{|m|+1,p-1} \\ \cdots & \cdots & \cdots & \cdots \\ (F|E)^m_{p-1,|m|} & (F|E)^m_{p-1,|m|+1} & \cdots & (F|E)^m_{p-1,p-1} \end{pmatrix}, \qquad (7.2.14)$$

where $m = 0, \pm 1, \pm 2, \ldots$ Due to the symmetry property (3.2.92) of the coaxial translation coefficients $(F|E)^m_{nn'}$ the coaxial translation matrix can be computed only for $m \geq 0$, since

$$(\mathbf{F}|\mathbf{E})^{(-m)}_{(\text{coax})}(t) = (\mathbf{F}|\mathbf{E})^{(m)}_{(\text{coax})}(t) = (\mathbf{F}|\mathbf{E})^{(|m|)}_{(\text{coax})}(t), \quad m = 0, \pm 1, \pm 2, \ldots \qquad (7.2.15)$$

The total coaxial translation matrix can then be represented as a direct sum of blocks

$$\begin{aligned}(\mathbf{F}|\mathbf{E})_{(\text{coax})}(t) &= (\mathbf{F}|\mathbf{E})^{(-p+1)}_{(\text{coax})}(t) \oplus (\mathbf{F}|\mathbf{E})^{(-p+2)}_{(\text{coax})}(t) \oplus \cdots \\ &= \sum_{m=-(p-1)}^{p-1} \oplus (\mathbf{F}|\mathbf{E})^{(m)}_{(\text{coax})}(t). \end{aligned} \qquad (7.2.16)$$

So the coaxial translation matrix also has invariant subspaces and vectors from each subspace translate independently.

In contrast to the rotation transform, the dimension of the invariant subspaces for translations is infinite, and p truncation means *projection* of these infinite-dimensional matrices and vectors into the first p dimensions. While we are limited with $(p - |m|) \times (p - |m|)$ matrices for each order m, the total number of operations due to all matrix–vector multiplications for the p-truncated coaxial translation is

$$N_{\mathrm{coax}}(p) = \sum_{m=-(p-1)}^{p-1} (p - |m|)^2 = p^2 + 2\sum_{m=1}^{p-1} m^2 = \frac{1}{3}p(2p^2 + 1). \quad (7.2.17)$$

With the recurrence formulae of Chapter 3, computation of entries for all coaxial translation matrices can be performed for $O(p^3)$ operations, which yields the total $O(p^3)$ asymptotic complexity for the coaxial translation.

7.2.3 Decomposition of translation

To perform an arbitrary translation we can then act as follows. First, we "rotate" the expansion. In other words, we find the expansion coefficients over basis functions in the system of coordinates rotated in such a way that the new axis $\hat{z}$ has the same direction as vector $\mathbf{t}$. So we obtain the vector of coefficients

$$\tilde{\mathbf{C}} = \mathbf{Rot}(Q)\mathbf{C}. \quad (7.2.18)$$

The computational cost of obtaining vector $\tilde{\mathbf{C}}$ from vector $\mathbf{C}$ is $O(p^3)$ as discussed above.

Second, we "translate coaxially" the expansion. This means that we find the coefficients of expansion of the same function near the new center. This center is located on axis $\hat{z}$ of the rotated coordinate system and so we can determine

$$\tilde{\tilde{\mathbf{C}}} = (\mathbf{F}|\mathbf{E})_{(\mathrm{coax})}(t)\tilde{\mathbf{C}}. \quad (7.2.19)$$

The cost of getting vector $\tilde{\tilde{\mathbf{C}}}$ from vector $\tilde{\mathbf{C}}$ is also $O(p^3)$.

Finally, we should return to the original orientation of the axes. So we need to rotate the expansion backward (rotate the frame of reference with matrix Q^{-1}). This is again an $O(p^3)$ operation which yields the translated coefficients $\hat{\mathbf{C}}$, which we are looking for. Formally, this can be written as

$$\hat{\mathbf{C}} = \mathbf{Rot}(Q^{-1})\tilde{\tilde{\mathbf{C}}}. \quad (7.2.20)$$

Comparing these results with Eq. (7.2.2), in terms of matrix multiplications this can be written as

$$(\mathbf{F}|\mathbf{E})(\mathbf{t})\mathbf{C} = \mathbf{Rot}(Q^{-1})(\mathbf{F}|\mathbf{E})_{(\mathrm{coax})}(t)\mathbf{Rot}(Q)\mathbf{C}, \qquad F, E = S, R, \quad (7.2.21)$$

or dropping symbol $\mathbf{C}$, since this applies to any vector of coefficients, as

$$(\mathbf{F}|\mathbf{E})(\mathbf{t}) = \mathbf{Rot}(Q^{-1})(\mathbf{F}|\mathbf{E})_{(\mathrm{coax})}(t)\mathbf{Rot}(Q), \qquad F, E = S, R. \quad (7.2.22)$$

This is a decomposition of the translation transform into rotation, coaxial translation, and inverse rotation transforms. The order of operations here is important, since the general rotation and coaxial translation operation do not commute. This equation can also be rewritten in the form

$$(\mathbf{F}|\mathbf{E})(\mathbf{t}) = \mathbf{Rot}^{-1}(Q)(\mathbf{F}|\mathbf{E})_{(\mathrm{coax})}(t)\mathbf{Rot}(Q), \qquad F, E = S, R, \tag{7.2.23}$$

since

$$\mathbf{Rot}^{-1}(Q) = \mathbf{Rot}(Q^{-1}). \tag{7.2.24}$$

The decomposition (7.2.22) results in reduction of the translation complexity from $O(p^4)$ to $O(p^3)$. This is illustrated schematically in Fig. 7.1.

We note that the left-hand side of Eq. (7.2.23) depends only on the vector $\mathbf{t}$, while the rotation operations depend on the rotation matrix, which is determined by the three rotation angles. In fact, the only requirement for this rotation is that the basis vector of the rotated axis is $\mathbf{i}_{\hat{z}} = \mathbf{t}/|\mathbf{t}|$. So the rotation angle γ (or the Euler angle γ_{E}, see Fig. 3.4) can be arbitrary, e.g. $\gamma = 0$. The rotation angles β and α are the spherical polar angles of $\mathbf{i}_{\hat{z}}$ in the original coordinate system (see Fig. 3.3). Therefore, they are the (θ_t, φ_t) angles of the unit vector $\mathbf{s}_t = \mathbf{t}/|\mathbf{t}|$ in the original frame of reference and we can write

$$\mathbf{Rot}(Q(\alpha, \beta, 0)) = \mathbf{Rot}(\mathbf{s}_t), \qquad \mathbf{s}_t = \frac{\mathbf{t}}{|\mathbf{t}|} = (\theta_t, \varphi_t) = (\beta, \alpha). \tag{7.2.25}$$

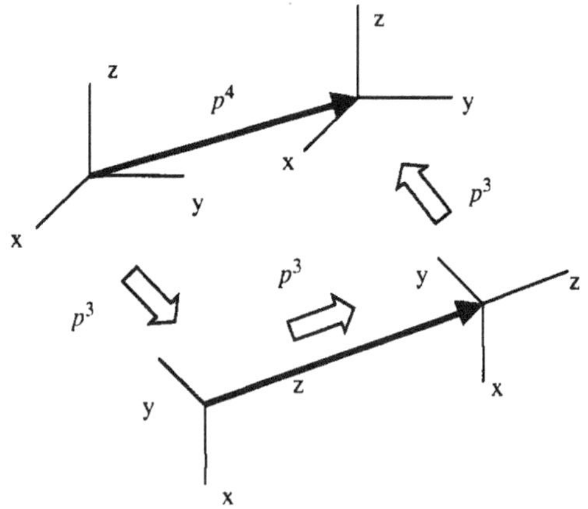

FIGURE 7.1 Illustration of decomposition of the general translation operation into rotation, coaxial translation, and inverse rotation operations. The total $O(p^4)$ complexity of the matrix–vector multiplication, reduces to $O(p^3)$ due to the rotation and coaxial translation operations have complexities $O(p^3)$.

The observation that decomposition (7.2.23) does not depend on the angle γ can also be found from the fact that the rotation matrix can be decomposed into three elementary rotations depending on the rotation angles (7.3.12). Rotation with angle γ obviously commute with the coaxial translation operation, so if we rotate axes with angle γ about the axis of translation $\mathbf{i}_{\hat{z}}$ and then rotate them back with angle $-\gamma$, the effect of γ disappears. We can also show this formally, using Eq. (7.2.24):

$$
\begin{aligned}
(\mathbf{F}|\mathbf{E})(\mathbf{t}) &= \mathbf{Rot}^{-1}(Q)(\mathbf{F}|\mathbf{E})_{(\mathrm{coax})}(t)\mathbf{Rot}(Q) \\
&= [\mathbf{Rot}(A(\gamma)B(\beta)A^{\mathrm{T}}(\alpha))]^{-1}(\mathbf{F}|\mathbf{E})_{(\mathrm{coax})}(t)[\mathbf{Rot}(A(\gamma)B(\beta)A^{\mathrm{T}}(\alpha))] \\
&= \mathbf{Rot}^{-1}(A^{\mathrm{T}}(\alpha))\mathbf{Rot}^{-1}(B(\beta))\mathbf{Rot}^{-1}(A(\gamma))(\mathbf{F}|\mathbf{E})_{(\mathrm{coax})}(t) \\
&\quad \times \mathbf{Rot}(A(\gamma))\mathbf{Rot}(B(\beta))\mathbf{Rot}(A^{\mathrm{T}}(\alpha)) \\
&= \mathbf{Rot}^{-1}(A^{\mathrm{T}}(\alpha))\mathbf{Rot}^{-1}(B(\beta))(\mathbf{F}|\mathbf{E})_{(\mathrm{coax})}(t) \\
&\quad \times \mathbf{Rot}(B(\beta))\mathbf{Rot}(A^{\mathrm{T}}(\alpha)).
\end{aligned} \tag{7.2.26}
$$

A remark about the error of this $O(p^3)$ method compared to the error of the $O(p^4)$ method is that we do not "trade" error for complexity in this case, and the error of the decomposed operation is the same as the error of the straightforward translation. Indeed, for any given degree n the rotation is performed *exactly* (machine precision), so the error comes only due to truncation of the translation operator and cannot increase for the $O(p^3)$ method as soon as p degrees of n are taken into account in the straightforward and in the fast methods. A comparison of the $O(p^4)$ and $O(p^3)$ translation methods is shown in Fig. 7.2.

Finally, we mention that rotational–coaxial translation decomposition of general translation is a universal technique, applicable for many other equations of physical nature in three dimensions. For example, it was mentioned in the paper of Cheng et al. (1999) [CGR99] concerning the solutions of the Laplace equation.

7.3 SPARSE MATRIX DECOMPOSITION OF TRANSLATION AND ROTATION OPERATORS

Differential operators can be considered as the limiting case of translation operators as $|\mathbf{t}| \to 0$. Thus they are related to *infinitesimal* translations. Such operators obey the important properties associated with translation operators, particularly, they commute with them. As translation operators

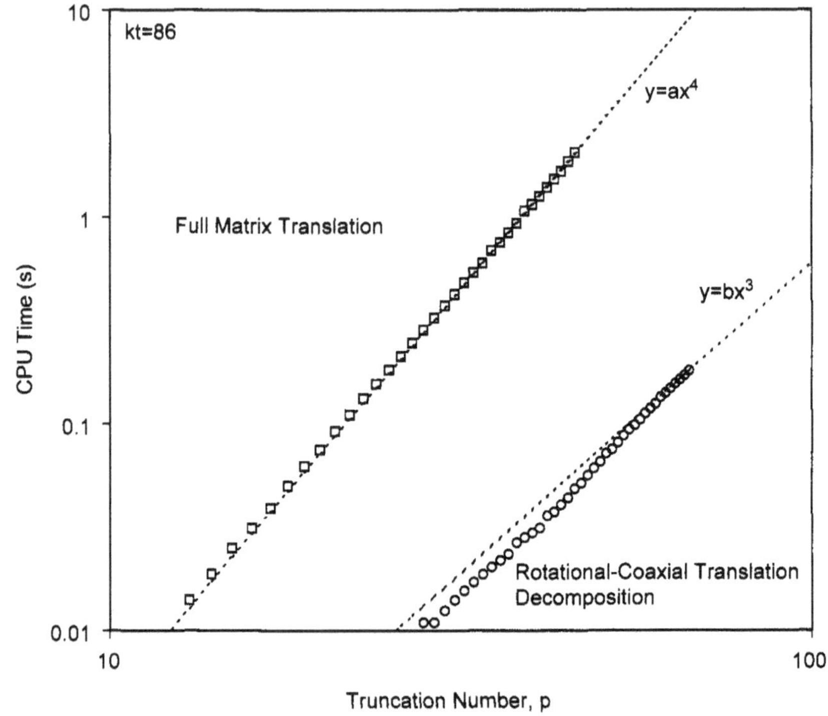

FIGURE 7.2 A comparison of CPU time required to perform an $S|R$-translation using matrix multiplication and rotational–coaxial translation decomposition. In both cases the matrix elements were computed recursively. Computations were performed with parameter $kt = 86$ (933 MHz Pentium III processor, 1 GB RAM).

can be represented by matrices, the infinitesimal translation operators can also be represented by matrices acting on the vectors of expansion coefficients. We show below that these matrices are very sparse, so infinitesimal translations can be performed very efficiently from the computational point of view. We will also show how arbitrary translations are related to infinitesimal translations and provide an algorithm for fast translation based on these decompositions.

7.3.1 Matrix representations of differential operators

In this section we consider differential operators acting in space of functions $\mathcal{B}(\Omega)$. We already considered such operators in Chapter 2, where we derived the differentiation theorems for elementary solutions of the Helmholtz equation in spherical coordinates. According to the notation adopted for operators in functional space we will use the following to denote differential operators in $\mathcal{B}(\Omega)$ for $k \neq 0$:

$$\mathcal{D}_z = \partial_z = \frac{1}{k}\frac{\partial}{\partial z}, \qquad \mathcal{D}_{x+iy} = \partial_{x+iy} = \frac{1}{k}\left(\frac{\partial}{\partial x} + \mathrm{i}\frac{\partial}{\partial y}\right),$$

$$\mathcal{D}_{x-iy} = \partial_{x-iy} = \frac{1}{k}\left(\frac{\partial}{\partial x} - \mathrm{i}\frac{\partial}{\partial y}\right), \qquad \mathcal{D}_t = \frac{\mathbf{t}}{k|\mathbf{t}|}\cdot\nabla. \tag{7.3.1}$$

Here $\mathcal{D}_{\mathbf{t}}$ denotes differentiation in the direction determined by vector $\mathbf{t} = (t_x, t_y, t_z)$. Such a differentiation can be represented as

$$\mathcal{D}_t = \frac{1}{2t}[(t_x + \mathrm{i}t_y)\mathcal{D}_{x-iy} + (t_x - \mathrm{i}t_y)\mathcal{D}_{x+iy}] + \frac{t_z}{t}\mathcal{D}_z,$$
$$t = |\mathbf{t}| = \sqrt{t_x^2 + t_y^2 + t_z^2}. \tag{7.3.2}$$

Differentiation of some function $\psi(\mathbf{r}) \in \mathcal{B}(\Omega)$ expanded into a series

$$\psi(\mathbf{r}) = \sum_{n=0}^{\infty}\sum_{m=-n}^{n} C_n^m F_n^m(\mathbf{r}), \qquad F = S, R, \tag{7.3.3}$$

can be written as

$$\psi'(\mathbf{r}) = \mathcal{D}_\alpha[\psi(\mathbf{r})], \qquad \mathcal{D}_\alpha = \mathcal{D}_z, \mathcal{D}_{x\pm iy}, \mathcal{D}_t, \tag{7.3.4}$$

where $\mathcal{D}_\alpha$ denotes any of the differential operators (7.3.1). The function $\psi'(\mathbf{r})$ satisfies the Helmholtz equation in Ω and so $\psi'(\mathbf{r}) \in \mathcal{B}(\Omega)$. Moreover, possible singularities of $\psi'(\mathbf{r})$ are located at the same points as singularities of $\psi(\mathbf{r})$. So this function can be expanded over the same basis as $\psi(\mathbf{r})$ where expansion (7.3.3) is valid

$$\psi'(\mathbf{r}) = \sum_{n=0}^{\infty}\sum_{m=-n}^{n} \widehat{C}_n^m F_n^m(\mathbf{r}), \qquad F = S, R. \tag{7.3.5}$$

Now we will consider infinite matrices $\mathbf{D}_\alpha$ which represent $\mathcal{D}_\alpha$ in the space of the expansion coefficients. Transform (7.3.4) can be written as

$$\hat{\mathbf{C}} = \mathbf{D}_\alpha \mathbf{C}, \qquad \mathbf{D}_\alpha = \mathbf{D}_z, \mathbf{D}_{x\pm iy}, \mathbf{D}_t. \tag{7.3.6}$$

7.3.1.1 Operator D_z

Due to the differentiation theorem (2.2.7) we have for $\mathcal{D}_\alpha = \mathcal{D}_z$:

$$\begin{aligned}
\mathcal{D}_z[\psi(\mathbf{r})] &= \sum_{n=0}^{\infty}\sum_{m=-n}^{n} C_n^m \mathcal{D}_z[F_n^m(\mathbf{r})] \\
&= \sum_{n=0}^{\infty}\sum_{m=-n}^{n} C_n^m [a_{n-1}^m F_{n-1}^m(\mathbf{r}) - a_n^m F_{n+1}^m(\mathbf{r})] \\
&= \sum_{n=0}^{\infty}\sum_{m=-n}^{n} [C_{n+1}^m a_n^m - C_{n-1}^m a_{n-1}^m] F_n^m(\mathbf{r}) = \sum_{n=0}^{\infty}\sum_{m=-n}^{n} \hat{C}_n^m F_n^m(\mathbf{r}) \\
&= \psi'(\mathbf{r}).
\end{aligned} \tag{7.3.7}$$

Therefore,

$$\hat{C}_n^m = a_n^m C_{n+1}^m - a_{n-1}^m C_{n-1}^m. \tag{7.3.8}$$

In matrix form this can be rewritten as

$$\hat{\mathbf{C}} = \mathbf{D}_z \mathbf{C}, \tag{7.3.9}$$

where $\mathbf{D}_z$ is a matrix of the same type as the $(\mathbf{R}|\mathbf{R})$ and $(\mathbf{S}|\mathbf{R})$ translation matrices considered above. The elements of this matrix are

$$(D_z)_{nn'}^{mm'} = \delta_{mm'}(a_n^m \delta_{n',n+1} - a_{n'}^m \delta_{n'+1,n}), \tag{7.3.10}$$

where δ is the Kronecker delta.

This operator does not change the order m of the expansion coefficients and, therefore, it can be represented as the direct sum of operators acting in invariant subspaces, similarly to representation of the coaxial translation operator (7.2.16):

$$\mathbf{D}_z = \mathbf{D}_z^{(0)} \oplus \mathbf{D}_z^{(\pm 1)} \oplus \cdots = \sum_{m=-\infty}^{\infty} \oplus \mathbf{D}_z^{(m)}. \tag{7.3.11}$$

Here each operator $\mathbf{D}_z^{(m)}$ acts only on the mth subspace and

$$\hat{\mathbf{C}}^m = \mathbf{D}_z^{(m)} \mathbf{C}^m, \qquad m = 0, \pm 1, \pm 2, \ldots \tag{7.3.12}$$

Due to the property of the recurrence coefficients $a_n^m = a_n^{|m|}$, (Eq. (2.2.8)) we have

$$\mathbf{D}_z^{(m)} = \mathbf{D}_z^{(-m)} = \mathbf{D}_z^{(|m|)}, \qquad m = 0, \pm 1, \pm 2, \ldots \tag{7.3.13}$$

The operator $\mathbf{D}_z^{(m)}$ is then the following infinite matrix:

$$\mathbf{D}_z^{(m)} = \begin{pmatrix} 0 & a_m^m & 0 & 0 & \cdots \\ -a_m^m & 0 & a_{m+1}^m & 0 & \cdots \\ 0 & -a_{m+1}^m & 0 & a_{m+2}^m & \cdots \\ 0 & 0 & -a_{m+2}^m & 0 & \cdots \\ \cdots & \cdots & \cdots & \cdots & \cdots \end{pmatrix}, \tag{7.3.14}$$

$m = 0, 1, 2, \ldots$

This matrix is real, tri-diagonal (or bidiagonal, if we count non-zero subdiagonals only) and skew-symmetric:

$$\mathbf{D}_z^{(m)\mathrm{T}} = -\mathbf{D}_z^{(m)}. \tag{7.3.15}$$

7.3.1.2 *Operator* D_{x+iy}

Similarly the operator $\mathcal{D}_\alpha = \mathcal{D}_{x+\mathrm{i}y}$ can be considered. Using theorem (2.2.9) we have:

$$\begin{aligned} \mathcal{D}_{x+\mathrm{i}y}[\psi(\mathbf{r})] &= \sum_{n=0}^{\infty} \sum_{m=-n}^{n} [b_{n+1}^{-m-1} F_{n+1}^{m+1}(\mathbf{r}) - b_n^m F_{n-1}^{m+1}(\mathbf{r})] C_n^m \\ &= \sum_{n=0}^{\infty} \sum_{m=-n}^{n} [C_{n-1}^{m-1} b_n^{-m} - C_{n+1}^{m-1} b_{n+1}^{m-1}] F_n^m(\mathbf{r}) \\ &= \sum_{n=0}^{\infty} \sum_{m=-n}^{n} \hat{C}_n^m F_n^m(\mathbf{r}) = \psi'(\mathbf{r}). \end{aligned} \tag{7.3.16}$$

This yields

$$\hat{C}_n^m = C_{n-1}^{m-1} b_n^{-m} - C_{n+1}^{m-1} b_{n+1}^{m-1}, \qquad m = 0, \pm 1, \pm 2, \ldots, \quad n = |m|, |m| + 1, \ldots \tag{7.3.17}$$

We can treat this relation then as the action of operator (matrix) $\mathbf{D}_{x+\mathrm{i}y}$ in the space of coefficients

$$\hat{\mathbf{C}} = \mathbf{D}_{x+\mathrm{i}y}\mathbf{C}. \tag{7.3.18}$$

The elements of matrix $\mathbf{D}_{x+iy}$ are given by

$$(\mathbf{D}_{x+iy})_{nn'}^{mm'} = \delta_{m',m-1}(b_n^{-m}\delta_{n',n-1} - b_{n'}^{m-1}\delta_{n',n+1}). \tag{7.3.19}$$

Note that this operator increases the order m of the expansion coefficients. Moreover, to produce the m-order subspace it acts only on the $(m-1)$-order subspace and can be represented as a direct sum of *non-invariant* operators acting on block-vectors $\mathbf{C}^m$ (see Eq. (7.2.7)). They map the $(m-1)$- order subspace to the m-order subspace. We can represent this as

$$\mathbf{D}_{x+iy} = \mathbf{D}_{x+iy}^{(0)} \oplus \mathbf{D}_{x+iy}^{(\pm 1)} \oplus \cdots = \sum_{m=-\infty}^{\infty} \oplus \mathbf{D}_{x+iy}^{(m)}, \tag{7.3.20}$$

so

$$\hat{\mathbf{C}}^m = \mathbf{D}_{x+iy}^{(m-1)}\mathbf{C}^{m-1}, \qquad m = 0, \pm 1, \pm 2, \ldots \tag{7.3.21}$$

where operator $\mathbf{D}_{x+iy}^{(m-1)}$ is the following infinite matrix:

$$\mathbf{D}_{x+iy}^{(m-1)} = \begin{pmatrix} b_m^{-m} & 0 & -b_{m+1}^{m-1} & 0 & \cdots \\ 0 & b_{m+1}^{-m} & 0 & -b_{m+2}^{m-1} & \cdots \\ 0 & 0 & b_{m+2}^{-m} & 0 & \cdots \\ 0 & 0 & 0 & b_{m+3}^{-m} & \cdots \\ \cdots & \cdots & \cdots & \cdots & \cdots \end{pmatrix}, \qquad m = 1, 2, \ldots,$$

$$\mathbf{D}_{x+iy}^{(m-1)} = \begin{pmatrix} -b_{|m|+1}^{m-1} & 0 & 0 & 0 & \cdots \\ 0 & -b_{|m|+2}^{m-1} & 0 & 0 & \cdots \\ b_{|m|+2}^{-m} & 0 & -b_{|m|+3}^{m-1} & 0 & \cdots \\ 0 & b_{|m|+3}^{-m} & 0 & -b_{|m|+3}^{m-1} & \cdots \\ \cdots & \cdots & \cdots & \cdots & \cdots \end{pmatrix},$$

$$m = 0, -1, -2, \ldots, \tag{7.3.22}$$

These matrices are real and have only two non-zero diagonals. We can also notice that

$$(\mathbf{D}^{(m-1)}_{x+\mathrm{i}y})^{\mathrm{T}} = -\mathbf{D}^{(-m)}_{x+\mathrm{i}y}. \tag{7.3.23}$$

7.3.1.3 *Operator* D_{x-iy}

For operator $\mathcal{D}_\alpha = \mathcal{D}_{x-\mathrm{i}y}$ using theorem (2.2.11) we have:

$$\begin{aligned}\mathcal{D}_{x-\mathrm{i}y}[\psi(\mathbf{r})] &= \sum_{n=0}^{\infty}\sum_{m=-n}^{n} [b^{m-1}_{n+1} F^{m-1}_{n+1}(\mathbf{r}) - b^{-m}_{n} F^{m-1}_{n-1}(\mathbf{r})] C^m_n \\ &= \sum_{n=0}^{\infty}\sum_{m=-n}^{n} [C^{m+1}_{n-1} b^m_n - C^{m+1}_{n+1} b^{-m-1}_{n+1}] F^m_n(\mathbf{r}) \\ &= \sum_{n=0}^{\infty}\sum_{m=-n}^{n} \hat{C}^m_n F^m_n(\mathbf{r}) = \psi'(\mathbf{r}).\end{aligned} \tag{7.3.24}$$

Therefore,

$$\begin{aligned}\hat{C}^m_n &= C^{m+1}_{n-1} b^m_n - C^{m+1}_{n+1} b^{-m-1}_{n+1}, \\ m &= 0, \pm 1, \pm 2, \ldots, \quad n = |m|, |m|+1, \ldots\end{aligned} \tag{7.3.25}$$

The matrix form of this relation is

$$\hat{\mathbf{C}} = \mathbf{D}_{x-\mathrm{i}y}\mathbf{C}, \tag{7.3.26}$$

where the elements of matrix $\mathbf{D}_{x-\mathrm{i}y}$ are given by

$$(\mathbf{D}_{x-\mathrm{i}y})^{mm'}_{nn'} = \delta_{m',m+1}(b^m_n \delta_{n',n-1} - b^{-m-1}_{n'} \delta_{n',n+1}). \tag{7.3.27}$$

Operator $\mathbf{D}_{x-\mathrm{i}y}$ is similar to $\mathbf{D}_{x+\mathrm{i}y}$, but in contrast to $\mathbf{D}_{x+\mathrm{i}y}$ it decreases the order m of the expansion coefficients. To produce the m-order subspace it acts only on the $(m+1)$-order subspace and can be represented as a direct sum of non-invariant operators acting on block-vectors $\mathbf{C}^m$ (see Eq. (7.2.7)). They map the $(m+1)$-order subspace to the m-order subspace. We can write this as

$$\mathbf{D}_{x-\mathrm{i}y} = \mathbf{D}^{(0)}_{x-\mathrm{i}y} \oplus \mathbf{D}^{(\pm 1)}_{x-\mathrm{i}y} \oplus \cdots = \sum_{m=-\infty}^{\infty} \oplus \mathbf{D}^{(m)}_{x-\mathrm{i}y}, \tag{7.3.28}$$

so

$$\hat{\mathbf{C}}^m = \mathbf{D}^{(m+1)}_{x-\mathrm{i}y} \mathbf{C}^{m+1}, \qquad m = 0, \pm 1, \pm 2, \ldots \tag{7.3.29}$$

where operator $\mathbf{D}_{x-iy}^{(m+1)}$ is the following infinite matrix:

$$\mathbf{D}_{x-iy}^{(m+1)} = \begin{pmatrix} -b_{m+1}^{-m-1} & 0 & 0 & 0 & \cdots \\ 0 & -b_{m+2}^{-m-1} & 0 & 0 & \cdots \\ b_{m+2}^{m} & 0 & -b_{m+3}^{-m-1} & 0 & \cdots \\ 0 & b_{m+3}^{m} & 0 & -b_{m+3}^{-m-1} & \cdots \\ \cdots & \cdots & \cdots & \cdots & \cdots \end{pmatrix},$$

$$m = 0, 1, 2, \ldots, \tag{7.3.30}$$

$$\mathbf{D}_{x-iy}^{(m+1)} = \begin{pmatrix} b_{|m|}^{m} & 0 & -b_{|m|+1}^{-m-1} & 0 & \cdots \\ 0 & b_{|m|+1}^{m} & 0 & -b_{|m|+2}^{-m-1} & \cdots \\ 0 & 0 & b_{|m|+2}^{m} & 0 & \cdots \\ 0 & 0 & 0 & b_{|m|+3}^{m} & \cdots \\ \cdots & \cdots & \cdots & \cdots & \cdots \end{pmatrix},$$

$$m = -1, -2, \ldots,$$

Comparing this result with Eq. (7.3.22) we find that

$$\mathbf{D}_{x-iy}^{(m+1)} = \mathbf{D}_{x+iy}^{(-m-1)}, \qquad m = 0, \pm 1, \pm 2, \ldots \tag{7.3.31}$$

This can also be derived from the fact that regular basis functions for real k have symmetry property $\overline{R_n^m(\mathbf{r})} = R_n^{-m}(\mathbf{r})$.

For the transposed matrices we then have

$$(\mathbf{D}_{x-iy}^{(m+1)})^{\mathrm{T}} = (\mathbf{D}_{x+iy}^{(-m-1)})^{\mathrm{T}} = -\mathbf{D}_{x+iy}^{(m)} = -\mathbf{D}_{x-iy}^{(-m)}, \tag{7.3.32}$$

$$m = 0, \pm 1, \pm 2, \ldots$$

7.3.1.4 Operator D_t

For operator $\mathcal{D}_\alpha = \mathcal{D}_t$ we can use theorem (2.2.12) to derive its matrix representation. We can also use Eq. (7.3.2) and representation of the operators considered above. In terms of matrix representations this equation turns to

$$\mathbf{D}_t = \frac{1}{2t}[(t_x + \mathrm{i}t_y)\mathbf{D}_{x-\mathrm{i}y} + (t_x - \mathrm{i}t_y)\mathbf{D}_{x+\mathrm{i}y}] + \frac{t_z}{t}\mathbf{D}_z. \tag{7.3.33}$$

as operators constituting $\mathbf{D}_t$ can be expressed via infinite sparse matrices. Since the mth-order subspaces of $\mathbf{D}_{x\pm\mathrm{i}y}$ are not invariant, the same holds for $\mathbf{D}_t$ (if $t_x^2 + t_y^2 \neq 0$, otherwise $\mathbf{D}_t = \mathbf{D}_z$). Therefore, for the transformed vector

$$\hat{\mathbf{C}} = \mathbf{D}_t\mathbf{C} \tag{7.3.34}$$

the mth-order block of $\hat{\mathbf{C}}$ is

$$\hat{\mathbf{C}}^m = \frac{1}{2t}[(t_x + \mathrm{i}t_y)\mathbf{D}_{x-\mathrm{i}y}^{(m-1)}\mathbf{C}^{m-1} + (t_x - \mathrm{i}t_y)\mathbf{D}_{x+\mathrm{i}y}^{(m+1)}\mathbf{C}^{m+1}] + \frac{t_z}{t}\mathbf{D}_z^{(m)}\mathbf{C}^m, \tag{7.3.35}$$

where $m = 0, \pm 1, \pm 2, \ldots$ We can also write the transform due to differentiation in an arbitrary direction as

$$\begin{aligned} \hat{C}_n^m &= \frac{1}{2t}[(t_x + \mathrm{i}t_y)(C_{n-1}^{m+1}b_n^m - C_{n+1}^{m+1}b_{n+1}^{-m-1}) + (t_x - \mathrm{i}t_y) \\ &\quad \times (C_{n-1}^{m-1}b_n^{-m} - C_{n+1}^{m-1}b_{n+1}^{m-1})] + \frac{t_z}{t}(a_n^m C_{n+1}^m - a_{n-1}^m C_{n-1}^m), \\ m &= 0, \pm 1, \pm 2, \ldots, \quad n = |m|, |m| + 1, \ldots \end{aligned} \tag{7.3.36}$$

7.3.1.5 Matrix form of the Helmholtz equation

Using definitions of differential operators (7.3.1) we can write the Helmholtz equation in the following form:

$$(\mathcal{D}_{x+\mathrm{i}y}\mathcal{D}_{x-\mathrm{i}y} + \mathcal{D}_z^2 + \mathcal{I})[\psi] = 0, \qquad \forall\psi \in \mathcal{B}(\Omega), \tag{7.3.37}$$

where $\mathcal{I}$ is the identity operator. As the differential operators are linear and are represented by matrix operators in the space of expansion coefficients we have

$$(\mathbf{D}_{x+\mathrm{i}y}\mathbf{D}_{x-\mathrm{i}y} + \mathbf{D}_z^2 + \mathbf{I})\mathbf{C} = 0, \tag{7.3.38}$$

where $\mathbf{I}$ is the identity matrix.

This can be checked directly using explicit expressions for matrices of differential operators. We can also notice that the composed operator

$$\mathbf{D}_{xy} = \mathbf{D}_{x+\mathrm{i}y}\mathbf{D}_{x-\mathrm{i}y} \tag{7.3.39}$$

has invariant subspaces with respect to the order m:

$$\mathbf{D}_{xy} = \mathbf{D}_{xy}^{(0)} \oplus \mathbf{D}_{xy}^{(\pm 1)} \oplus \cdots = \sum_{m=-\infty}^{\infty} \oplus \mathbf{D}_{xy}^{(m)}, \tag{7.3.40}$$

since operators $\mathbf{D}_{x+iy}$ and $\mathbf{D}_{x-iy}$ increase and decrease the order of expansion coefficients by 1, respectively. This is also clear from the operator form of the Helmholtz equation (7.3.38), since both $\mathbf{D}_z$ and $\mathbf{I}$ have invariant subspaces of order m and

$$\mathbf{D}_{xy} = -\mathbf{D}_z^2 - \mathbf{I}. \tag{7.3.41}$$

7.3.2 Spectra of differential and translation operators

7.3.2.1 Continuous spectra of differential operators

An eigen-function $\psi(\mathbf{r}; \lambda_\alpha)$ corresponding to the eigenvalue λ_α of a differential operator $\mathcal{D}_\alpha$ is defined as

$$\mathcal{D}_\alpha[\psi(\mathbf{r}; \lambda_\alpha)] = \lambda_\alpha \psi(\mathbf{r}; \lambda_\alpha), \qquad \mathcal{D}_\alpha = \mathcal{D}_z, \mathcal{D}_{x\pm iy}, \mathcal{D}_\mathbf{t}. \tag{7.3.42}$$

The set of all possible λ_α is called the spectrum of operator $\mathcal{D}_\alpha$. Since all operations are considered over the field of complex scalars, λ_α is also complex and the spectrum of operator $\mathcal{D}_\alpha$ is a domain in $\mathbb{C}$. Respectively, if the eigenfunction $\psi(\mathbf{r}; \lambda_\alpha)$ is represented in the space of expansion coefficients by vector $\mathbf{E}(\lambda_\alpha)$ this vector satisfies

$$\mathbf{D}_\alpha \mathbf{E}(\lambda_\alpha) = \lambda_\alpha \mathbf{E}(\lambda_\alpha), \qquad \mathbf{D}_\alpha = \mathbf{D}_z, \mathbf{D}_{x\pm iy}, \mathbf{D}_\mathbf{t}. \tag{7.3.43}$$

A set of eigenfunctions can be found more or less easily, if we notice that the plane wave solution, $\psi = e^{ik\mathbf{s}\cdot\mathbf{r}}$ for an arbitrary unit vector $\mathbf{s} = (s_x, s_y, s_z)$, $|\mathbf{s}| = 1$, is an eigenfunction for the following differential operators

$$\mathcal{D}_z[e^{ik\mathbf{s}\cdot\mathbf{r}}] = is_z\, e^{ik\mathbf{s}\cdot\mathbf{r}}, \tag{7.3.44}$$

$$\mathcal{D}_{x\pm iy}[e^{ik\mathbf{s}\cdot\mathbf{r}}] = i(s_x \pm is_y)e^{ik\mathbf{s}\cdot\mathbf{r}}, \tag{7.3.45}$$

$$\mathcal{D}_\mathbf{t}[e^{ik\mathbf{s}\cdot\mathbf{r}}] = i\left(\frac{\mathbf{s}\cdot\mathbf{t}}{|\mathbf{t}|}\right)e^{ik\mathbf{s}\cdot\mathbf{r}}. \tag{7.3.46}$$

Respectively, the eigenvalues of these operators are

$$\lambda_z = is_z, \qquad \lambda_{x\pm iy} = i(s_x \pm is_y), \qquad \lambda_\mathbf{t} = i\left(\frac{\mathbf{s}\cdot\mathbf{t}}{|\mathbf{t}|}\right). \tag{7.3.47}$$

As soon as the plane wave can be expanded over the basis of regular functions (2.3.6) with expansion coefficients $4\pi\{i^n Y_n^{-m}(\mathbf{s})\}$ we see that, in fact, vectors (7.1.34) are the eigenvectors for all the matrix differential operators in Eq. (7.3.43):

$$\lambda_\alpha(\mathbf{s})\mathrm{i}^n Y_n^{-m}(\mathbf{s}) = \sum_{n'=0}^{\infty} \sum_{m'=-n'}^{n'} (\mathbf{D}_\alpha)_{nn'}^{mm'} \mathrm{i}^{n'} Y_{n'}^{-m'}(\mathbf{s}), \qquad (7.3.48)$$
$$n = 0, 1, 2, \ldots, \qquad m = -n, \ldots, n,$$

or

$$\mathbf{D}_z \mathbf{iY}(\mathbf{s}) = \lambda_z \mathbf{iY}(\mathbf{s}), \qquad (7.3.49)$$

$$\mathbf{D}_{x\pm \mathrm{i}y} \mathbf{iY}(\mathbf{s}) = \lambda_{x\pm \mathrm{i}y} \mathbf{iY}(\mathbf{s}), \qquad (7.3.50)$$

$$\mathbf{D}_t \mathbf{iY}(\mathbf{s}) = \lambda_t \mathbf{iY}(\mathbf{s}). \qquad (7.3.51)$$

The eigenvalues (7.3.47) are complex (λ_z and λ_t are purely imaginary, while $\lambda_{x\pm iy}$ are arbitrary complex numbers) and bounded (owing to $\mathbf{s}$ being a unit vector):

$$|\lambda_z| \leq 1, \qquad |\lambda_{x\pm \mathrm{i}y}| \leq 1, \qquad |\lambda_t| \leq 1. \qquad (7.3.52)$$

7.3.2.2 *Continuous spectra of translation operators*

The translation operator $\mathcal{T}(\mathbf{t})$ commutes with any differential operator $\mathcal{D}_\alpha$:

$$\mathcal{T}(\mathbf{t})\mathcal{D}_\alpha = \mathcal{D}_\alpha \mathcal{T}(\mathbf{t}), \qquad \mathcal{D}_\alpha = \mathcal{D}_z, \mathcal{D}_{x\pm \mathrm{i}y}, \mathcal{D}_t. \qquad (7.3.53)$$

As a result the matrices representing translation and differential operators commute as well,

$$(\mathbf{R}|\mathbf{R})(\mathbf{t})\mathbf{D}_\alpha = \mathbf{D}_\alpha(\mathbf{R}|\mathbf{R})(\mathbf{t}), \qquad \mathbf{D}_\alpha = \mathbf{D}_z, \mathbf{D}_{x\pm \mathrm{i}y}, \mathbf{D}_t,$$
$$(\mathbf{S}|\mathbf{R})(\mathbf{t})\mathbf{D}_\alpha = \mathbf{D}_\alpha(\mathbf{S}|\mathbf{R})(\mathbf{t}), \qquad \mathbf{D}_\alpha = \mathbf{D}_z, \mathbf{D}_{x\pm \mathrm{i}y}, \mathbf{D}_t. \qquad (7.3.54)$$

Such commutation results in the fact that vectors $\mathbf{iY}(\mathbf{s}) = \{\mathrm{i}^n Y_n^{-m}(\mathbf{s})\}$ which are eigenvectors for the differential matrices also being eigenvectors for the translation matrices. This can be checked directly. We can also show this, e.g., for the $(\mathbf{R}|\mathbf{R})(\mathbf{t})$ matrix by consideration of the plane wave. Indeed, we have

$$\mathcal{T}(\mathbf{t})[\mathrm{e}^{\mathrm{i}k\mathbf{s}\cdot\mathbf{r}}] = \mathrm{e}^{\mathrm{i}k\mathbf{s}\cdot(\mathbf{r}+\mathbf{t})} = \mathrm{e}^{\mathrm{i}k\mathbf{s}\cdot\mathbf{t}}\, \mathrm{e}^{\mathrm{i}k\mathbf{s}\cdot\mathbf{r}}, \qquad (7.3.55)$$

which shows that the plane wave is the eigenfunction of operator $\mathcal{T}(\mathbf{t})$ with the eigenvalue

$$\Lambda_r^{(\infty)}(\mathbf{t};\mathbf{s}) = \mathrm{e}^{\mathrm{i}k\mathbf{s}\cdot\mathbf{t}}, \qquad (7.3.56)$$

and because the plane wave can be expanded over the basis of regular functions (2.3.6) with expansion coefficients $4\pi\{\mathrm{i}^n Y_n^{-m}(\mathbf{s})\}$ vectors (7.1.34) are eigenvectors for the translation operator, and

$$\Lambda_r^{(\infty)}(\mathbf{t};\mathbf{s})\mathrm{i}^n Y_n^{-m}(\mathbf{s}) = \sum_{n'=0}^{\infty}\sum_{m'=-n'}^{n'} (R|R)_{nn'}^{mm'}(\mathbf{t})\mathrm{i}^{n'} Y_{n'}^{-m'}(\mathbf{s}), \qquad (7.3.57)$$

$$n = 0,1,2,\ldots, \quad m = -n,\ldots,n.$$

This sheds new light on the function $\Lambda_r^{(\infty)}(\mathbf{t};\mathbf{s})$, which was introduced as the kernel of integral transform to the signature function (see Eq. (7.1.48)). We can now characterize this function also as the eigenvalue of the translation operator.

Another eigenvalue appears for the $(\mathbf{S}|\mathbf{R})(\mathbf{t})$ translation matrix. For such translations we limit ourselves with band-limited functions. So we can check directly that the singular kernel function $\Lambda_s^{(p)}(\mathbf{t};\mathbf{s})$ at large but finite p defined by Eq. (7.1.45) *approximates* the eigenvalue of the $(\mathbf{S}|\mathbf{R})(\mathbf{t})$ matrix. To see that, we consider expansion of surface function

$$\Psi^{(p)}(\mathbf{s}) = \Lambda_s^{(p)}(\mathbf{t};\mathbf{s})\mathrm{i}^n Y_n^{-m}(\mathbf{s}), \qquad n = 0,1,2,\ldots, \qquad (7.3.58)$$

$$m = -n,\ldots,n.$$

over the orthonormal basis $\{\mathrm{i}^{n'} Y_{n'}^{-m'}(\mathbf{s})\}$:

$$\Psi^{(p)}(\mathbf{s}) = \sum_{n'=0}^{\infty}\sum_{m'=-n'}^{n'} c_{n'}^{m'}\mathrm{i}^{n'} Y_{n'}^{-m'}(\mathbf{s}). \qquad (7.3.59)$$

The expansion coefficients can then be found similarly to Eq. (7.1.58):

$$\begin{aligned}
c_{n'}^{m'} &= \mathrm{i}^{-n'}\int_{S_u} \Psi^{(p)}(\mathbf{s}) Y_{n'}^{m'}(\mathbf{s})\mathrm{d}S(\mathbf{s}) \\
&= \mathrm{i}^{n-n'}\int_{S_u} \Lambda_s^{(p)}(\mathbf{t};\mathbf{s}) Y_n^{-m}(\mathbf{s}) Y_{n'}^{m'}(\mathbf{s})\mathrm{d}S(\mathbf{s}) \\
&= 4\pi \sum_{n''=0}^{p-1}\sum_{m''=-n''}^{n''} S_{n''}^{m''}(\mathbf{t})\mathrm{i}^{n-n'+n''}\int_{S_u} Y_n^{-m}(\mathbf{s}) Y_{n'}^{m'}(\mathbf{s}) Y_{n''}^{-m''}(\mathbf{s})\mathrm{d}S(\mathbf{s}) \\
&= \sum_{n''=0}^{p-1}\sum_{m''=-n''}^{n''} (s|r)_{n''nn'}^{m''mm'} S_{n''}^{m''}(\mathbf{t}) = (S|R)_{nn'}^{mm'}(\mathbf{t}), \quad p-1 \geq n+n'. \qquad (7.3.60)
\end{aligned}$$

So for any given n there exists some $q_n(n') = n + n'$, such that for any $p \geq q_n$ coefficients $c_{n'}^{m'}$ become the elements of the $(\mathbf{S}|\mathbf{R})(\mathbf{t})$ matrix and their values do not change as p increases.

7.3.3 Integral representations of differential operators

Once we have a correspondence between a solution of the Helmholtz equation and its signature function, we can investigate how the signature function changes under differentiation of this solution. Let $\{\hat{C}_n^m\}$ be a set of expansion coefficients for the function derivative, $\mathcal{D}_\alpha[\psi(\mathbf{r})]$, as stated in Eq. (7.3.5) ($\mathcal{D}_\alpha = \mathcal{D}_z, \mathcal{D}_{x\pm iy}, \mathcal{D}_t$). It is important to note that if $\{C_n^m\}$ is a band-limited vector,

$$C_n^m = 0, \qquad n \geq p, \tag{7.3.61}$$

then multiplication of any infinite matrix $\mathbf{D}_\alpha$ by such a vector produces again a band-limited vector $\hat{\mathbf{C}}_n^m$:

$$\hat{C}_n^m = \sum_{n'=0}^{\infty} \sum_{m'=-n'}^{n'} (\mathbf{D}_\alpha)_{nn'}^{mm'} C_{n'}^{m'} = 0, \qquad n \geq p+1, \tag{7.3.62}$$
$$\mathbf{D}_\alpha = \mathbf{D}_z, \mathbf{D}_{x\pm iy}, \mathbf{D}_t.$$

This is an important feature of the matrices $\mathbf{D}_\alpha$ which shows that differentiation of band-limited functions can be performed *exactly* as we increase the truncation number by 1 for each differentiation. So assuming that Eq. (7.3.61) holds, we have, using Eq. (7.1.29) and the fact that $\mathbf{iY(s)} = \{i^n Y_n^{-m}(\mathbf{s})\}$ is the eigenvector of $\mathbf{D}_\alpha$ corresponding to the eigenvalue λ_α, the following expression for any n:

$$\hat{C}_n^m - \sum_{n'=0}^{p-1} \sum_{m'=-n'}^{n'} (\mathbf{D}_\alpha)_{nn'}^{mm'} C_{n'}^{m'}$$
$$= \int_{S_u} \Psi^{(p)}(\mathbf{s}) \left[\sum_{n'=0}^{p-1} \sum_{m'=-n'}^{n'} (\mathbf{D}_\alpha)_{nn'}^{mm'} i^{n'} Y_{n'}^{-m'}(\mathbf{s}) \right] dS(\mathbf{s})$$
$$= i^n \int_{S_u} \lambda_\alpha(\mathbf{s}) \Psi^{(p)}(\mathbf{s}) Y_n^{-m}(\mathbf{s}) dS(\mathbf{s}). \tag{7.3.63}$$

On the other hand from the definition (7.1.29), we have

$$\hat{C}_n^m = i^n \int_{S_u} \Psi^{(p)'}(\mathbf{s}) Y_n^{-m}(\mathbf{s}) dS(\mathbf{s}). \tag{7.3.64}$$

Due to the orthonormality and completeness of the basis of spherical harmonics, we then have

$$\Psi^{(p)'}(\mathbf{s}) = \lambda_\alpha \Psi^{(p)}(\mathbf{s}), \qquad \lambda_\alpha = \lambda_z, \lambda_{x\pm iy}, \lambda_t. \tag{7.3.65}$$

As we noticed above, the transformed signature function $\Psi^{(p)'}(\mathbf{s})$ is band-limited and is *exactly*

$$\Psi^{(p)'}(\mathbf{s}) = \sum_{n=0}^{p} \sum_{m=-n}^{n} \mathrm{i}^{-n} \hat{C}_n^m Y_n^m(\mathbf{s}). \tag{7.3.66}$$

The integral representations of the elements of the differentiation matrices can be obtained immediately from Eq. (7.3.48) due to the fact that $(\mathbf{D}_\alpha)_{nn'}^{mm'}$ are the expansion coefficients of a surface function. So Eq. (7.1.68) yields

$$(\mathbf{D}_\alpha)_{nn'}^{mm'} = \mathrm{i}^{n-n'} \int_{S_u} \lambda_\alpha(\mathbf{s}) Y_n^{-m}(\mathbf{s}) Y_{n'}^{m'}(\mathbf{s}) dS. \tag{7.3.67}$$

In the matrix-vector form this can be rewritten using the Kronecker product as defined by Eq. (7.1.35)

$$\mathbf{D}_\alpha = \int_{S_u} \lambda_\alpha(\mathbf{s}) \mathbf{iY}(\mathbf{s}) \otimes \overline{\mathbf{iY}(\mathbf{s})} dS. \tag{7.3.68}$$

Particularly,

$$\mathbf{D}_z = \mathrm{i} \int_{S_u} s_z \mathbf{iY}(\mathbf{s}) \otimes \overline{\mathbf{iY}(\mathbf{s})} dS, \qquad s_z = \mathbf{s} \cdot \mathbf{i}_z \tag{7.3.69}$$

$$\mathbf{D}_{x \pm \mathrm{i}y} = \mathrm{i} \int_{S_u} (s_x \mp \mathrm{i} s_y) \mathbf{iY}(\mathbf{s}) \otimes \overline{\mathbf{iY}(\mathbf{s})} dS, \quad s_x \mp \mathrm{i} s_y = \mathbf{s} \cdot (\mathbf{i}_x \mp \mathrm{i} \mathbf{i}_y) \tag{7.3.70}$$

$$\mathbf{D}_t = \mathrm{i} \int_{S_u} \frac{\mathbf{s} \cdot \mathbf{t}}{|\mathbf{t}|} \mathbf{iY}(\mathbf{s}) \otimes \overline{\mathbf{iY}(\mathbf{s})} dS. \tag{7.3.71}$$

7.3.4 Sparse matrix decomposition of translation operators

7.3.4.1 Matrix exponential

Consider integral representation of the $(\mathbf{R}|\mathbf{R})$-translation matrix (7.1.36). The kernel of this integral can be expanded in Taylor series absolutely convergent at any finite $\mathbf{t}$. This shows that this matrix can be represented in the following form:

$$(\mathbf{R}|\mathbf{R})(\mathbf{t}) = \int_{S_u} e^{ik\mathbf{s}\cdot\mathbf{t}}\mathbf{iY(s)}\otimes\overline{\mathbf{iY(s)}}dS$$
$$= \int_{S_u} \sum_{n=0}^{\infty} \frac{(kt)^n}{n!}\left(\mathrm{i}\frac{\mathbf{s}\cdot\mathbf{t}}{t}\right)^n \mathbf{iY(s)}\otimes\overline{\mathbf{iY(s)}}dS$$
$$= \sum_{n=0}^{\infty} \frac{(kt)^n}{n!}\int_{S_u}\left(\mathrm{i}\frac{\mathbf{s}\cdot\mathbf{t}}{t}\right)^n \mathbf{iY(s)}\otimes\overline{\mathbf{iY(s)}}dS. \quad (7.3.72)$$

Now we can express the nth integral through the nth power of matrix $\mathbf{D_t}$. Indeed, from Eq. (7.3.71) we have

$$\int_{S_u} \mathrm{i}\frac{\mathbf{s}\cdot\mathbf{t}}{t}\mathbf{iY(s)}\otimes\overline{\mathbf{iY(s)}}dS = \mathbf{D_t},$$
$$\int_{S_u}\left(\mathrm{i}\frac{\mathbf{s}\cdot\mathbf{t}}{t}\right)^n \mathbf{iY(s)}\otimes\overline{\mathbf{iY(s)}}dS = \mathbf{D_t^n}, \quad (7.3.73)$$

since each multiplication of the matrix $\mathbf{D_t}$ by some vector results in the multiplication of the signature function corresponding to this vector by the eigenvalue of this differential operator. Therefore, Eqs. (7.3.72) and (7.3.73) result in the following expansion of the infinite matrix $(\mathbf{R}|\mathbf{R})(\mathbf{t})$ over the powers of matrix $\mathbf{D_t}$:

$$(\mathbf{R}|\mathbf{R})(\mathbf{t}) = \sum_{n=0}^{\infty} \frac{(kt)^n}{n!}\mathbf{D_t^n} = e^{kt\mathbf{D_t}}. \quad (7.3.74)$$

Here $e^{kt\mathbf{D_t}}$ is the matrix called the *matrix exponential*.

Note that at $\mathbf{t} = t\mathbf{i}_z$ this formula turns into the following expression for the coaxial translation matrix

$$(\mathbf{R}|\mathbf{R})_{(\mathrm{coax})}(t) = \sum_{n=0}^{\infty} \frac{(kt)^n}{n!}\mathbf{D}_z^n = e^{kt\mathbf{D}_z}. \quad (7.3.75)$$

Since both $(\mathbf{R}|\mathbf{R})_{(\mathrm{coax})}(t)$ and $\mathbf{D}_z$ have the same invariant subspaces (see Eqs. (7.2.16) and (7.3.11)), we also have

$$(\mathbf{R}|\mathbf{R})^{(m)}_{(\mathrm{coax})}(t) = \sum_{n=0}^{\infty} \frac{(kt)^n}{n!}[\mathbf{D}_z^{(m)}]^n = e^{kt\mathbf{D}_z^{(m)}}, \qquad m = 0, 1, 2, \ldots \quad (7.3.76)$$

Representation of the translation matrix via the sparse matrix exponential has several computational advantages. To demonstrate them let us consider the $R|R$ or $S|S$ translation of a band-limited vector of expansion coefficients $\mathbf{C}$ such that Eq. (7.3.61) holds.

The spectrum of operator $\mathbf{D}_t$ is limited (Eq. (7.3.52)), and we can also show that the *norm* of this operator is limited as well. The operator $\mathbf{D}_t$ is *bounded*. This means that the truncated series provide any required accuracy for some truncation number N_t, and we have for the translated vector

$$\hat{\mathbf{C}} = (\mathbf{R}|\mathbf{R})(\mathbf{t})\mathbf{C} = \sum_{n=0}^{N_t-1} \frac{(kt)^n}{n!} \mathbf{D}_t^n \mathbf{C} + \epsilon_{N_t}(\mathbf{t}), \tag{7.3.77}$$

where $\epsilon_{N_t}(\mathbf{t})$ is some truncation error which we assume is small. We will be interested here in computation of the finite sum. We state that this finite sum can be computed *exactly* using $O(N_t p^2)$ operations, where p^2 is the length of vector $\mathbf{C}$. Let us denote

$$\begin{aligned} &\mathbf{C}_0 = \mathbf{C}, \qquad \hat{\mathbf{C}}_0 = \mathbf{C}, \qquad \hat{\mathbf{C}}_{N_t-1} = \hat{\mathbf{C}}, \qquad \mathbf{C}_n = \mathbf{D}_t \mathbf{C}_{n-1}, \\ &\hat{\mathbf{C}}_n = \hat{\mathbf{C}}_{n-1} + \frac{(kt)^n}{n!} \mathbf{C}_n, \qquad n = 1, \ldots, N_t - 1. \end{aligned} \tag{7.3.78}$$

For an infinite matrix $\mathbf{D}_t$ and finite vector $\mathbf{C}_{n-1}$ we have a finite vector $\mathbf{C}_n$ of a larger length as stated in Eq. (7.3.62). So, in exact arithmetic, the length of vector $\mathbf{C}_0$ is p^2, of $\mathbf{C}_1$ is $(p+1)^2$, and so on until $\mathbf{C}_{N_t-1}$, which has length $(p + N_t - 1)^2$. The nth step of the recursive procedure (7.3.78), therefore, has complexity $(p+n)^2$, $n = 1, \ldots, N_t - 1$, since multiplication of the sparse matrix $\mathbf{D}_t$ by the finite length vector is the operation in the order of the length of this vector (see Eq. (7.3.36)). Thus the total number of operations to obtain $\hat{\mathbf{C}}$ will be

$$N_{\text{tot}} = \sum_{n=1}^{N_t-1} (p+n)^2 = (N_t - 1)[p^2 + pN_t + \tfrac{1}{6} N_t(2N_t - 1)]. \tag{7.3.79}$$

Particularly, if $N_t \sim p$ we have

$$N_{\text{tot}} \sim \tfrac{7}{3} p^3 = O(p^3). \tag{7.3.80}$$

We can see that in this case translation algorithm (7.3.78) has the same asymptotic complexity $O(p^3)$ as the algorithm using truncated matrix translation with rotational–coaxial translation decomposition considered earlier in this chapter. We also note that in the algorithm, which requires computation of the matrix elements we need to compute special functions, namely, spherical harmonics and the spherical Bessel functions. In the algorithm based on the computation of the matrix exponential this is not the case and so there is no need to control the accuracy of recursive computation of special functions. Another advantage is obvious: for relatively low $|kt|$ we can use N_t substantially smaller than p. Those N_t's come from the evaluation of the error bounds, and for low $|kt|$ just a few terms can provide sufficient accuracy. This brings the exponential algorithm at low $|kt|$ to complexity $O(p^2)$, which is the theoretical minimum for the asymptotic complexity of translation. We also note that for larger $|kt|$ a number of techniques for computation of the matrix exponentials are available. They can be computationally more efficient than direct summation of the power series [ML78]. What is important here is that $i\mathbf{D}_t$ is a *Hermitian* operator (as a simpler example one can consider matrices $\mathbf{D}_z^{(m)}$ (Eq. (7.3.13)), which are real skew-symmetric matrices, so $i\mathbf{D}_z^{(m)}$ are the Hermitian matrices[1]).

7.3.4.2 Legendre series

Another decomposition of the $(\mathbf{R}|\mathbf{R})$-translation matrix follows from the expansion of the plane wave solution over the Legendre polynomials (2.3.5). In this case we have

$$
\begin{aligned}
(\mathbf{R}|\mathbf{R})(\mathbf{t}) &= \int_{S_u} e^{ik\mathbf{s}\cdot\mathbf{t}} i\mathbf{Y}(\mathbf{s}) \otimes \overline{i\mathbf{Y}(\mathbf{s})} dS \\
&= \int_{S_u} \sum_{n=0}^{\infty} (2n+1) i^n j_n(kt) P_n\left(\frac{\mathbf{s}\cdot\mathbf{t}}{t}\right) i\mathbf{Y}(\mathbf{s}) \otimes \overline{i\mathbf{Y}(\mathbf{s})} dS \\
&= \sum_{n=0}^{\infty} (2n+1) i^n j_n(kt) \int_{S_u} P_n\left(\frac{\mathbf{s}\cdot\mathbf{t}}{t}\right) i\mathbf{Y}(\mathbf{s}) \otimes \overline{i\mathbf{Y}(\mathbf{s})} dS. \qquad (7.3.81)
\end{aligned}
$$

Here we note that P_n are polynomials and because the polynomials of eigenvalues can be related to polynomials of the matrix, as stated in Eq. (7.3.73) we have

$$
(\mathbf{R}|\mathbf{R})(\mathbf{t}) = \sum_{n=0}^{\infty} (2n+1) i^n j_n(kt) P_n(-i\mathbf{D}_t). \qquad (7.3.82)
$$

Similar decomposition of the $(\mathbf{S}|\mathbf{R})$-translation matrix follows from Eq. (7.1.61):

$$
(\mathbf{S}|\mathbf{R})(\mathbf{t}) = \sum_{n=0}^{\infty} (2n+1) i^n h_n(kt) P_n(-i\mathbf{D}_t). \qquad (7.3.83)
$$

Obviously, for the coaxial translation we have similar decomposition for

[1] The elements A_{mn} of a Hermitian matrix $\mathbf{A}$ obey the symmetry property $A_{mn} = \overline{A_{nm}}$. All eigenvalues of such matrices are real.

the invariant subspaces:

$$(\mathbf{R}|\mathbf{R})^{(m)}_{(\mathrm{coax})}(t) = \sum_{n=0}^{\infty} (2n+1)\mathrm{i}^n j_n(kt) P_n(-\mathrm{i}\mathbf{D}_z^{(m)}), \tag{7.3.84}$$

$$(\mathbf{S}|\mathbf{R})^{(m)}_{(\mathrm{coax})}(t) = \sum_{n=0}^{\infty} (2n+1)\mathrm{i}^n h_n(kt) P_n(-\mathrm{i}\mathbf{D}_z^{(m)}). \tag{7.3.85}$$

As in the case of the matrix exponential, translations with the truncated series can be computed efficiently using recursions for the Legendre polynomials (2.1.52):

$$\begin{aligned} &P_0(-\mathrm{i}\mathbf{D}_t) = \mathbf{I}, \qquad P_1(-\mathrm{i}\mathbf{D}_t) = -\mathrm{i}\mathbf{D}_t, \\ &P_{n+1}(-\mathrm{i}\mathbf{D}_t) = \frac{1}{n+1}[(2n+1)P_n(-\mathrm{i}\mathbf{D}_t)(-\mathrm{i}\mathbf{D}_t) - nP_{n-1}(-\mathrm{i}\mathbf{D}_t)]. \end{aligned} \tag{7.3.86}$$

In this case we have the following procedure for the $(R|R)$-translation of coefficients $\mathbf{C}$:

$$\begin{aligned} &\mathbf{C}_0 = \mathbf{C}, \qquad \mathbf{C}_1 = P_1(-\mathrm{i}\mathbf{D}_t)\mathbf{C} = -\mathrm{i}\mathbf{D}_t\mathbf{C}, \qquad \hat{\mathbf{C}}_0 = \mathbf{C}, \\ &\hat{\mathbf{C}}_{N_t-1} = \hat{\mathbf{C}}, \\ &\mathbf{C}_n = P_n(-\mathrm{i}\mathbf{D}_t)\mathbf{C} = \frac{1}{n}[-(2n-1)\mathrm{i}\mathbf{D}_t\mathbf{C}_{n-1} - (n-1)\mathbf{C}_{n-2}], \\ &n = 2, 3, \ldots, N_t - 1, \\ &\hat{\mathbf{C}}_n = \hat{\mathbf{C}}_{n-1} + (2n+1)\mathrm{i}^n j_n(kt)\mathbf{C}_n, \quad n = 1, \ldots, N_t - 1. \end{aligned} \tag{7.3.87}$$

For the $(S|R)$-translation the procedure is the same, but with the spherical Hankel function $h_n(kt)$ instead of the spherical Bessel function $j_n(kt)$.

The asymptotic complexity of this procedure obviously is the same as for the computation of the matrix exponent, since each step involves one sparse matrix multiplication $\mathbf{D}_t\mathbf{C}_{n-1}$ to produce $\mathbf{C}_n$.

Note that the expansion of the matrices over the Legendre polynomials of sparse matrices is applicable to both the translation matrices, $(\mathbf{R}|\mathbf{R})$ and $(\mathbf{S}|\mathbf{R})$, and so they are more general than the exponential expansion, which is valid only for the $(\mathbf{R}|\mathbf{R})$ translation matrix. There is also an issue about the convergence and error bounds of different series.

7.3.5 Sparse matrix decomposition of rotation operators

The sparse matrix decomposition for translation operators appeared from the consideration of infinitesimal translations described by sparse differentiation matrices. We can expand this idea and try to do the same for rotation operators. In this case we need to consider *infinitesimal rotation operators*.

In fact, to obtain computationally efficient sparse matrix decomposition we need to consider only infinitesimal rotations with the rotation angle β. To show this we note that an arbitrary rotation specified by the rotation angles α, β, and γ can be decomposed into three rotations (see Eq. (3.3.12)). According to Eq. (7.1.7) this results in the decomposition of the rotation operator corresponding to the rotation matrix Q as follows:

$$\begin{aligned}\mathcal{R}\mathrm{ot}(Q(\alpha,\beta,\gamma)) &= \mathcal{R}\mathrm{ot}(A(\gamma)B(\beta)A^{-1}(\alpha)) \\ &= \mathcal{R}\mathrm{ot}(A(\gamma))\circ\mathcal{R}\mathrm{ot}(B(\beta))\circ\mathcal{R}\mathrm{ot}^{-1}(A(\alpha)). \qquad (7.3.88)\end{aligned}$$

In terms of the rotation group representation this results in the decomposition of the matrix rotation operator as

$$\mathbf{Rot}(Q(\alpha,\beta,\gamma)) = \mathbf{Rot}(A(\gamma))\mathbf{Rot}(B(\beta))\mathbf{Rot}^{-1}(A(\alpha)). \qquad (7.3.89)$$

We also can see this directly. Indeed if $\hat{\mathbf{C}}$ is the vector of rotated coefficients $\mathbf{C}$,

$$\hat{\mathbf{C}} = \mathbf{Rot}(Q)\mathbf{C}, \qquad \hat{C}_n^m = \sum_{m'=-n}^{n} T_n^{mm'}(\alpha,\beta,\gamma)C_n^{m'}, \qquad (7.3.90)$$

then according to Eqs. (7.3.37) and (7.3.44) we have

$$\hat{C}_n^m = \mathrm{e}^{-im\gamma}\sum_{m'=-n}^{n} H_n^{mm'}(\beta)\mathrm{e}^{im'\alpha}C_n^{m'}. \qquad (7.3.91)$$

Let us denote

$$\tilde{\mathbf{C}} = \mathbf{Rot}^{-1}(A(\alpha))\mathbf{C}, \quad \tilde{\tilde{\mathbf{C}}} = \mathbf{Rot}(B(\beta))\tilde{\mathbf{C}}, \quad \hat{\mathbf{C}} = \mathbf{Rot}(A(\gamma))\tilde{\tilde{\mathbf{C}}}. \qquad (7.3.92)$$

Components of the first vector of coefficients are simply related to the original coefficients as

$$\tilde{C}_n^m = \mathrm{e}^{im\alpha}C_n^m, \qquad n = 0,1,2,\ldots, \qquad m = -n,\ldots,n. \qquad (7.3.93)$$

Obviously, this step is computationally efficient, since for a vector of total length p^2 $(n = 0,\ldots,p-1)$ we need only p^2 operations to rotate $\mathbf{C}$ with matrix $\mathbf{Rot}^{-1}(A(\alpha))$. This also shows that the nth invariant subspace of this operator is represented by a diagonal $(2n+1) \times (2n+1)$ matrix:

$$\mathbf{Rot}_n^{-1}(A(\alpha)) = \begin{pmatrix} e^{-in\alpha} & 0 & \cdots & 0 \\ 0 & e^{-i(n-1)\alpha} & \cdots & 0 \\ \cdots & \cdots & \cdots & \cdots \\ 0 & 0 & \cdots & e^{in\alpha} \end{pmatrix}, \tag{7.3.94}$$

$$\mathbf{Rot}^{-1}(A(\alpha)) = \sum_{n=0}^{\infty} \oplus \mathbf{Rot}_n^{-1}(A(\alpha)).$$

It easy to get the inverse of the diagonal matrix to see that

$$\mathbf{Rot}_n(A(\gamma)) = \begin{pmatrix} e^{in\gamma} & 0 & \cdots & 0 \\ 0 & e^{i(n-1)\gamma} & \cdots & 0 \\ \cdots & \cdots & \cdots & \cdots \\ 0 & 0 & \cdots & e^{-in\gamma} \end{pmatrix}, \tag{7.3.95}$$

$$\mathbf{Rot}(A(\gamma)) = \sum_{n=0}^{\infty} \oplus \mathbf{Rot}_n(A(\gamma)).$$

This can also be found directly from Eqs. (7.3.91) and (7.3.92).

By this way we showed that matrices $\mathbf{Rot}^{-1}(A(\alpha))$ and $\mathbf{Rot}(A(\gamma))$ are sparse (diagonal) for arbitrary angles α and γ and the complexity of the rotation operation comes from computation of $\tilde{\tilde{\mathbf{C}}}$, which requires multiplication of a dense real symmetric matrix $\{H_n^{mm'}(\beta)\}$. The nth invariant subspace of this matrix is the following $(2n+1) \times (2n+1)$ matrix:

$$\mathbf{Rot}_n(B(\beta)) = \begin{pmatrix} H_n^{-n,-n}(\beta) & H_n^{-n,-n+1}(\beta) & \cdots & H_n^{-n,n}(\beta) \\ H_n^{-n+1,-n}(\beta) & H_n^{-n+1,-n+1}(\beta) & \cdots & H_n^{-n+1,n}(\beta) \\ \cdots & \cdots & \cdots & \cdots \\ H_n^{n,-n}(\beta) & H_n^{n,-n+1}(\beta) & \cdots & H_n^{n,n}(\beta) \end{pmatrix}, \tag{7.3.96}$$

$$\mathbf{Rot}(B(\beta)) = \sum_{n=0}^{\infty} \oplus \mathbf{Rot}_n(B(\beta)).$$

We note now that the angles α, β, and γ are not the Euler rotation angles, while they are simply related to them by Eq. (3.3.15). To relate the rotation transform with these angles we need to make modification of the rotation matrices. So we have

$$\mathbf{Rot}_n^{-1}(A(\alpha)) = \mathbf{Rot}_n^{-1}(A(\pi - \alpha_E)) = \mathbf{\Delta}_n \mathbf{Rot}_n(A(\alpha_E)), \tag{7.3.97}$$

Here matrix $\mathbf{\Delta}_n$ is

$$\mathbf{\Delta}_n = \mathbf{Rot}_n(A(\pi)) = (-1)^n \begin{pmatrix} 1 & 0 & \cdots & 0 \\ 0 & -1 & \cdots & 0 \\ \cdots & \cdots & \cdots & \cdots \\ 0 & 0 & \cdots & 1 \end{pmatrix}, \tag{7.3.98}$$

where $\Delta_n^{mm'} = (-1)^m \delta_{mm'}$; $n = 0, 1, \ldots$; $m, m' = -n, \ldots, n$, and the direct sum of these blocks constitute the matrix $\mathbf{\Delta}$:

$$\mathbf{\Delta} = \sum_{n=0}^{\infty} \oplus \mathbf{\Delta}_n = \mathbf{Rot}(A(\pi)). \tag{7.3.99}$$

Note that

$$\mathbf{\Delta}_n^2 = \mathbf{I}_n, \quad \mathbf{\Delta}^2 = \mathbf{I}, \tag{7.3.100}$$

where $\mathbf{I}_n$ and $\mathbf{I}$ are identity matrices of the corresponding sizes.

This shows that

$$\mathbf{Rot}(Q) = \mathbf{Rot}(A(\gamma_E))\mathbf{Rot}(B(\beta_E))\Delta\mathbf{Rot}(A(\alpha_E)), \tag{7.3.101}$$

and we can represent this as

$$\begin{aligned} \mathbf{Rot}(Q) &= \mathbf{Rot}(A(\gamma_E))\mathbf{Rot}(C(\beta_E))\mathbf{Rot}(A(\alpha_E)), \\ \mathbf{Rot}(C(\beta_E)) &= \mathbf{Rot}(B(\beta_E))\mathbf{\Delta} = \mathbf{Rot}(B(\beta_E))\mathbf{Rot}(A(\pi)), \end{aligned} \tag{7.3.102}$$

where $C(\beta_E)$ is the rotation matrix, such that

$$Q = A(\gamma_E)C(\beta_E)A(\alpha_E), \qquad C(\beta) = B(\beta)A(\pi). \tag{7.3.103}$$

Representation through the Euler angles is important, since by the meaning of the Euler angles the rotation with zero angles is the identity rotation,

$$A(0) = C(0) = I. \tag{7.3.104}$$

Moreover, we can consider two subsequent rotations with the same axis of rotation, for which we have commutativity of rotations:

$$\begin{aligned} A(\alpha_{E1})A(\alpha_{E2}) &= A(\alpha_{E1} + \alpha_{E2}) = A(\alpha_{E2})A(\alpha_{E1}), \\ C(\beta_{E1})C(\beta_{E2}) &= C(\beta_{E1} + \beta_{E2}) = C(\beta_{E2})C(\beta_{E1}). \end{aligned} \tag{7.3.105}$$

This results in commutativity of the following rotation operators:

$$\begin{aligned}
&\mathbf{Rot}(A(\alpha_{E1}))\mathbf{Rot}(A(\alpha_{E2})) = \mathbf{Rot}(A(\alpha_{E1} + \alpha_{E2})) \\
&\quad = \mathbf{Rot}(A(\alpha_{E2}))\mathbf{Rot}(A(\alpha_{E1})), \\
&\mathbf{Rot}(C(\beta_{E1}))\mathbf{Rot}(C(\beta_{E2})) = \mathbf{Rot}(C(\beta_{E1} + \beta_{E2})) \\
&\quad = \mathbf{Rot}(C(\beta_{E2}))\mathbf{Rot}(C(\beta_{E1})).
\end{aligned} \tag{7.3.106}$$

While it is easy to check the first equality directly using Eq. (7.3.95), the second one is non-trivial and can be used for decomposition of the rotation operator as shown below. With the relation between the matrices C and B given by Eq. (7.3.103) we can rewrite the latter equality as

$$\begin{aligned}
&\mathbf{Rot}(B(\beta_1))\boldsymbol{\Delta}\mathbf{Rot}(B(\beta_2))\boldsymbol{\Delta} = \mathbf{Rot}(B(\beta_1 + \beta_2))\boldsymbol{\Delta} \\
&\quad = \mathbf{Rot}(B(\beta_2))\boldsymbol{\Delta}\mathbf{Rot}(B(\beta_1))\boldsymbol{\Delta}.
\end{aligned} \tag{7.3.107}$$

7.3.5.1 *Infinitesimal rotations*

First we note that for $\beta = 0$ we have from Eq. (3.3.39)

$$H_n^{mm'}(0) = (-1)^m \delta_{mm'} = \Delta_n^{mm'}, \qquad n = 0, 1, 2, \ldots \tag{7.3.108}$$

Using matrix notation we then have

$$\begin{aligned}
\boldsymbol{\Delta}_n &= \mathbf{Rot}_n(B(0)) \\
&= \begin{pmatrix} H_n^{-n,-n}(0) & H_n^{-n,-n+1}(0) & \cdots & H_n^{-n,n}(0) \\ H_n^{-n+1,-n}(0) & H_n^{-n+1,-n+1}(0) & \cdots & H_n^{-n+1,n}(0) \\ \cdots & \cdots & \cdots & \cdots \\ H_n^{n,-n}(0) & H_n^{n,-n+1}(0) & \cdots & H_n^{n,n}(0) \end{pmatrix}.
\end{aligned} \tag{7.3.109}$$

Since the rotation coefficients $H_n^{mm'}(\beta)$ are regular functions of β they can be expanded in convergent Taylor series for any β (remember $0 \leq \beta \leq \pi$)

$$H_n^{mm'}(\beta) = \sum_{l=0}^{\infty} \frac{1}{l!} \left. \frac{\mathrm{d}^l H_n^{mm'}}{\mathrm{d}\beta^l} \right|_{\beta=0} \beta^l. \tag{7.3.110}$$

This can also be written in matrix form as

$$\mathbf{Rot}_n(B(\beta)) = \sum_{l=0}^{\infty} \frac{\beta^l}{l!} \mathbf{Rot}_n^{(l)}(0) = \left[\mathbf{I} + \sum_{l=1}^{\infty} \frac{\beta^l}{l!} \mathbf{Rot}_n^{(l)}(0) \boldsymbol{\Delta}_n^{-1} \right] \boldsymbol{\Delta}_n, \tag{7.3.111}$$

where we use the following notation:

$$\mathbf{Rot}_n^{(l)}(\beta) = \frac{\mathrm{d}^l}{\mathrm{d}\beta^l}\mathbf{Rot}_n(B(\beta)). \tag{7.3.112}$$

From Eqs. (7.3.100) and (7.3.102) we can find also that

$$\mathbf{Rot}_n(C(\beta)) = \mathbf{I} + \sum_{l=1}^{\infty} \frac{\beta^l}{l!}\mathbf{Rot}_n^{(l)}(0)\mathbf{\Delta}_n. \tag{7.3.113}$$

The first-order term can be found from explicit formula for rotation coefficients (3.3.39):

$$\left.\frac{\mathrm{d}H_n^{mm'}}{\mathrm{d}\beta}\right|_{\beta=0} = (\delta_{m',m-1}c_n^{m-1} + \delta_{m',m+1}c_n^m), \qquad m', m = -n, \ldots, n, \tag{7.3.114}$$

where

$$c_n^m = \tfrac{1}{2}\mathrm{sgn}(m)(-1)^m[(n-m)(n+m+1)]^{1/2}, \qquad m = -n, \ldots, n, \tag{7.3.115}$$

and sgn(m) is defined as

$$\mathrm{sgn}(m) = \begin{cases} 1 & m \geq 0, \\ -1 & m < 0. \end{cases} \tag{7.3.116}$$

Let us introduce now the following real skew-symmetric matrix

$$\mathbf{D}_\beta = \left.\frac{\mathrm{d}}{\mathrm{d}\beta}\mathbf{Rot}(B(\beta))\right|_{\beta=0}\mathbf{\Delta}, \qquad \mathbf{D}_\beta = \sum_{n=0}^{\infty} \oplus \mathbf{D}_\beta^{(n)}, \tag{7.3.117}$$

where for the nth subspace we have

$$\mathbf{D}_\beta^{(n)} = (-1)^n \begin{pmatrix} 0 & -c_n^{-n} & 0 & \cdots & 0 & 0 \\ c_n^{-n} & 0 & c_n^{-n+1} & \cdots & 0 & 0 \\ 0 & -c_n^{-n+1} & 0 & \cdots & 0 & 0 \\ \cdots & \cdots & \cdots & \cdots & \cdots & \cdots \\ 0 & 0 & 0 & \cdots & 0 & c_n^{n-1} \\ 0 & 0 & 0 & \cdots & -c_n^{n-1} & 0 \end{pmatrix}. \tag{7.3.118}$$

and the elements of this matrix are

$$[\mathbf{D}_\beta^{(n)}]_{m'm} = (-1)^m(\delta_{m,m'-1}c_n^{m'-1} + \delta_{m,m'+1}c_n^{m'}),$$
$$m, m' = -n, \ldots, n. \tag{7.3.119}$$

With this notation we can rewrite series (7.3.113) as

$$\mathbf{Rot}_n(C(\beta)) = \mathbf{I} + \beta\mathbf{D}_\beta^{(n)} + \sum_{l=2}^{\infty} \frac{\beta^l}{l!}\mathbf{Rot}_n^{(l)}(0)\mathbf{\Delta}_n. \tag{7.3.120}$$

The fact that matrix $\mathbf{D}_\beta^{(n)}$ commutes with matrix $\mathbf{Rot}_n(B(\beta))\mathbf{\Delta}_n$

$$\mathbf{D}_\beta^{(n)}\mathbf{Rot}_n(B(\beta))\mathbf{\Delta}_n = \mathbf{Rot}_n(B(\beta))\mathbf{\Delta}_n\mathbf{D}_\beta^{(n)}, \tag{7.3.121}$$

yields the following relation for the real rotation coefficients $H_n^{mm'}(\beta)$:

$$d_n^{m-1}H_n^{m-1,m'} - d_n^m H_n^{m+1,m'} = d_n^{m'-1}H_n^{m,m'-1} - d_n^{m'}H_n^{m,m'+1}, \tag{7.3.122}$$

where coefficients d_n^m are related to coefficients c_n^m (Eq. (7.3.115)) as

$$d_n^m = (-1)^m c_n^m = \tfrac{1}{2}\operatorname{sgn}(m)[(n-m)(n+m+1)]^{1/2},$$
$$m = -n, \ldots, n. \tag{7.3.123}$$

This relation can be checked by writing Eq. (7.3.121) in component form and can be used for recursive computation of the rotation coefficients.

7.3.5.2 *Decomposition of the rotation operator for Euler angle β*

Consider now a small rotation with angle $\beta + \epsilon$. Keeping the terms with respect to ϵ only up to the first order, we have:

$$\begin{aligned}\mathbf{Rot}_n(C(\beta+\epsilon)) &= \mathbf{I} + (\beta+\epsilon)\mathbf{D}_\beta^{(n)} + \sum_{l=2}^{\infty} \frac{(\beta+\epsilon)^l}{l!}\mathbf{Rot}_n^{(l)}(0)\mathbf{\Delta}_n \\ &= \mathbf{I} + \beta\mathbf{D}_\beta^{(n)} + \sum_{l=2}^{\infty} \frac{\beta^l}{l!}\mathbf{Rot}_n^{(l)}(0)\mathbf{\Delta}_n \\ &\quad + \epsilon\left(\mathbf{D}_\beta^{(n)} + \sum_{l=2}^{\infty} \frac{\beta^{l-1}}{(l-1)!}\mathbf{Rot}_n^{(l)}(0)\mathbf{\Delta}_n\right) + O(\epsilon^2).\end{aligned} \tag{7.3.124}$$

On the other hand we can use the commutativity property of this matrix

(7.3.106) in the following form:

$$\mathbf{Rot}_n(C(\beta+\epsilon)) = \mathbf{Rot}_n(C(\beta))\mathbf{Rot}_n(C(\epsilon))$$

$$= \left(\mathbf{I} + \beta\mathbf{D}_\beta^{(n)} + \sum_{l=2}^{\infty} \frac{\beta^l}{l!}\mathbf{Rot}_n^{(l)}(0)\boldsymbol{\Delta}_n\right)[\mathbf{I} + \epsilon\mathbf{D}_\beta^{(n)} + O(\epsilon^2)]. \quad (7.3.125)$$

Comparing these series we find:

$$\left(\beta\mathbf{D}_\beta^{(n)} + \sum_{l=2}^{\infty} \frac{\beta^l}{l!}\mathbf{Rot}_n^{(l)}(0)\boldsymbol{\Delta}_n\right)\mathbf{D}_\beta^{(n)}$$

$$= \mathbf{Rot}_n^{(2)}(0)\boldsymbol{\Delta}_n + \sum_{l=2}^{\infty} \frac{\beta^l}{l!}\mathbf{Rot}_n^{(l+1)}(0)\boldsymbol{\Delta}_n. \quad (7.3.126)$$

Since this can also be considered as an asymptotic series as $\beta \to 0$ the coefficients of the same powers of β should be the same, which yields

$$\mathbf{Rot}_n^{(2)}(0)\boldsymbol{\Delta}_n = [\mathbf{D}_\beta^{(n)}]^2, \qquad \mathbf{Rot}_n^{(l+1)}(0)\boldsymbol{\Delta}_n = \mathbf{Rot}_n^{(l)}(0)\boldsymbol{\Delta}_n\mathbf{D}_\beta^{(n)}, \quad (7.3.127)$$

$l = 2, 3, \ldots$

The latter equation represents a recurrence, which shows that

$$\mathbf{Rot}_n^{(l)}(0)\boldsymbol{\Delta}_n = [\mathbf{D}_\beta^{(n)}]^l, \qquad l = 0, 1, 2, 3, \ldots, \quad (7.3.128)$$

since for $l = 0$ this is the unity matrix, and for $l = 1$ this follows from the definition of matrix $\mathbf{D}_\beta^{(n)}$ (Eq. (7.3.117)). Substituting these values into Eq. (7.3.113) we can see that $\mathbf{Rot}_n(C(\beta))$ is nothing but the matrix exponential of $\beta\mathbf{D}_\beta^{(n)}$

$$\mathbf{Rot}_n(C(\beta)) = \mathbf{I} + \sum_{l=1}^{\infty} \frac{\beta^l}{l!}[\mathbf{D}_\beta^{(n)}]^l = e^{\beta\mathbf{D}_\beta^{(n)}}, \qquad n = 0, 1, 2, \ldots \quad (7.3.129)$$

Because this holds for any subspace n, we can also write

$$\mathbf{Rot}(C(\beta)) = e^{\beta\mathbf{D}_\beta}, \qquad \mathbf{Rot}(B(\beta)) = e^{\beta\mathbf{D}_\beta}\boldsymbol{\Delta}. \quad (7.3.130)$$

So the rotation with angle β can be represented as the matrix exponential of a sparse matrix $\beta\mathbf{D}_\beta$. It is interesting to note that matrix $\mathbf{D}_\beta^{(n)}$ is skew-symmetric and only has two non-zero subdiagonals. So the structure of this matrix is the same as the structure of the differentiation matrix $\mathbf{D}_z^{(m)}$. While the exponent of $kt\mathbf{D}_z^{(m)}$ produces the mth invariant block of the coaxial translation operator, the exponent of the infinitesimal rotation matrix $\beta\mathbf{D}_\beta^{(n)}$ produces the nth invariant block of the rotation operator. Also the blocks $\mathbf{D}_\beta^{(n)}$ are always finite.

The value of this decomposition for computations depends on the capabilities for fast computation of the matrix exponentials. It is obvious that for relatively small angles β just a few terms of the expansion of the matrix exponential into the power series can be used and rotation of a vector of expansion coefficients can be performed in this case in the order of $O(p^2)$ operations. Even for larger β we can expect that this decomposition is useful, since the range of β is limited $0 \leq \beta \leq \pi$. Other methods for computation of the matrix exponential can also be exploited [ML78]. The fact that $i\mathbf{D}_\beta^{(n)}$ is the *Hermitian* matrix is important for these methods.

7.4 SUMMARY

In this chapter we considered some basics on representation of translation operators, including matrix and integral representations. We showed that rotational–coaxial translation decomposition with recursive computation of the rotation and translation matrix elements results in $O(p^3)$ translation algorithm with relatively small asymptotic constant. This method is faster than $O(p^4)$ method based on direct product of the translation matrix with the vector of expansion coefficients. We also introduced another $O(p^3)$ translation method based on sparse matrix decomposition of the translation operator, where the sparse matrices are differentiation matrices commutative with translation matrices. Similar type of decomposition of the rotation operator provides $O(p^3)$ method for rotation.

CHAPTER 8

Asymptotically Faster Translation Methods

In this chapter we consider translation and rotation methods which have complexity $o(p^3)$, more precisely $O(p^2)$, $O(p^2 \log p)$ and $O(p^2 \log^2 p)$. As noted earlier the practical value of fast translation methods depends on the problem (dimensionless wavenumber) and asymptotic constants of the method. In this context we note that to solve problems with large p, we need a method with translation exponent $\nu = 1$ or close to this value. Indeed, assume $p = 100$, then the length of the vector of expansion coefficients will be 10^4 and an $O(p^3)$ translation method requires in the order of million operations per translation (!).

Complexities $O(p^2) - O(p^2 \log^2 p)$ for a single translation can be achieved using different approaches. The most popular (and perhaps the only well known and developed) approach is the method based on *diagonal forms* of the translation operator due to Rokhlin [Rok93]. While this method potentially has a complexity $O(p^2)$, procedures such as those that increase or decrease the vector length with the FMM level and error control require additional $O(p^2) - O(p^2 \log^2 p)$ complexity. Another method, which we try to develop in the following are based on decompositions of rotation and translation matrices into products involving *structured matrices*. The Toeplitz and Hankel matrices are the well-known representatives of this class of matrices, which can be multiplied by an arbitrary vector of length p^2, using $O(p^2 \log p)$ operations due to the use of the FFT. Finally, we are interested in *asymptotic methods*, which may be applicable for high and low frequencies and achieve $O(p^2)$ complexity.

8.1 FAST ALGORITHMS BASED ON MATRIX DECOMPOSITIONS

8.1.1 Fast rotation transform

While a general rotation transform can be performed at a cost of $O(p^3)$ operations, it was found recently that this transform can be performed exactly at a cost of only $O(p^2 \log p)$ operations [Tan03]. We call this algorithm the *Fast rotation transform* (FRT). It is based on a further decomposition of the general rotation operator into a product of diagonaland structured (Toeplitz type) matrices. Below we provide details to this algorithm and some brief introduction to Toeplitz and Hankel matrices.

8.1.1.1 *Toeplitz and Hankel matrices*

Definitions

Consider an arbitrary $M \times N$ matrix $\mathbf{W}$ with entries W_{mn}, $m = 0, \ldots, M-1$, $n = 0, \ldots, N-1$. Let $\mathbf{U}$ be a vector of size $M + N - 1$ with entries U_l, $l = -N+1, \ldots, M-1$. By definition $\mathbf{W}$ is called the *Toeplitz matrix,* if

$$W_{mn} = U_{m-n}, \qquad m = 0, \ldots, M-1, \quad n = 0, \ldots, N-1. \tag{8.1.1}$$

A Toeplitz matrix has the following structure:

$$\mathbf{W} = \{W_{mn}\} = \begin{pmatrix} W_{00} & W_{01} & \cdots & W_{0,N-1} \\ W_{10} & W_{11} & \cdots & W_{1,N-1} \\ \vdots & \vdots & \ddots & \vdots \\ W_{M-1,0} & W_{M-1,1} & \cdots & W_{M-1,N-1} \end{pmatrix}$$

$$= \begin{pmatrix} U_0 & U_{-1} & \cdots & U_{-N+1} \\ U_1 & U_0 & \cdots & U_{1-N} \\ \vdots & \vdots & \ddots & \vdots \\ U_{M-1} & U_{M-1} & \cdots & U_{M-N} \end{pmatrix}. \tag{8.1.2}$$

One can call $\mathbf{U}$ the "generating vector", since all MN entries of the Toeplitz matrix can be found by specifying a single $M + N - 1$ vector.

Let us index the entries of the vector $\mathbf{U}$ as U_l, $l = 0, \ldots, M + N - 2$. In this case we define $\mathbf{W}$ as a *Hankel matrix,* if

$$W_{mn} = U_{m+n}, \qquad m = 0, \ldots, M-1, \qquad n = 0, \ldots, N-1. \tag{8.1.3}$$

So the structure of the Hankel matrix is

$$\mathbf{W} = \{W_{mn}\} = \begin{pmatrix} W_{00} & W_{01} & \cdots & W_{0,N-1} \\ W_{10} & W_{11} & \cdots & W_{1,n-1} \\ \vdots & \vdots & \ddots & \vdots \\ W_{M-1,0} & W_{M-1,1} & \cdots & W_{M-1,N-1} \end{pmatrix}$$

$$= \begin{pmatrix} U_0 & U_1 & \cdots & U_{N-1} \\ U_1 & U_2 & \cdots & U_N \\ \vdots & \vdots & \ddots & \vdots \\ U_{M-1} & U_M & \cdots & U_{N+M-1} \end{pmatrix}, \tag{8.1.4}$$

and this matrix, again, is completely determined by the generating vector $\mathbf{U}$.

Toeplitz and Hankel matrices obey a number of interesting properties. For example, the transpose of a Toeplitz matrix will be a Toeplitz matrix (the same is the case with Hankel matrices). While we refer the interested reader to a special literature [GL89, TS99], below we point out some properties, which make them valuable for fast computational algorithms.

Fast matrix–vector multiplication

Toeplitz matrices

A remarkable property of Toeplitz matrices is that they can be multiplied by a vector in $O(N \log N)$ operations (assuming $M \sim N$). Consider the matrix–vector product

$$B_m = \sum_{n=0}^{N-1} W_{mn} A_n = \sum_{n=0}^{N-1} U_{m-n} A_n, \qquad m = 0, \ldots, M-1, \tag{8.1.5}$$

which is a *convolution* operation of the vectors $\mathbf{U}$ and $\mathbf{A}$. This operation can be performed quickly using the FFT. Indeed, consider the following

2π-periodical functions

$$u(t) = \sum_{l=-N+1}^{M-1} U_l \, e^{ilt}, \qquad b(t) = \sum_{m=0}^{M-1} B_m \, e^{imt}, \qquad a(t) = \sum_{n=0}^{N-1} A_n \, e^{int}. \tag{8.1.6}$$

Then

$$\begin{aligned} a(t)u(t) &= \sum_{l=-N+1}^{M-1} \sum_{n=0}^{N-1} A_n U_l \, e^{i(n+l)t} = \sum_{m=-N+1}^{M+N-2} e^{imt} \sum_{n=0}^{N-1} A_n U_{m-n} \\ &= \sum_{m=-N+1}^{-1} \alpha_m \, e^{imt} + b(t) + \sum_{m=M}^{M+N-2} \beta_m \, e^{imt}, \end{aligned} \tag{8.1.7}$$

where α_m and β_m are some coefficients, and $U_l = 0$ for $l \leq -N$ or $l \geq M$. This shows that to determine vector $\mathbf{B}$ one can perform the inverse Fourier transforms of $\mathbf{U}$ and $\mathbf{A}$, to determine $u(t)$ and $a(t)$ sampled at appropriate points $\{t_j\}$ (operation of complexity $O((N+M)\log(N+M)))$, then multiply two vectors point-by-point (complexity $O(N+M)$), perform the forward Fourier transform (complexity $O((N+M)\log(N+M)))$, and take coefficients of harmonics from $m = 0$ to $M-1$, which are B_m. So the total complexity of the algorithm is $O((N+M)\log(N+M))$, if the FFT is used for the forward and inverse Fourier transforms.

Hankel matrices

A similar procedure can be applied to multiply a Hankel matrix by some vector, or compute a product

$$B_m = \sum_{n=0}^{N-1} W_{mn} A_n = \sum_{n=0}^{N-1} U_{m+n} A_n, \qquad m = 0, \ldots, M-1. \tag{8.1.8}$$

In this case we introduce the following 2π-periodical functions

$$u(t) = \sum_{l=0}^{M+N-2} U_l \, e^{ilt}, \qquad b(t) = \sum_{m=0}^{M-1} B_m \, e^{imt}, \qquad a(t) = \sum_{n=0}^{N-1} \overline{A_n} \, e^{int}. \tag{8.1.9}$$

Then

$$\overline{a(t)}u(t) = \sum_{l=0}^{M+N-2} \sum_{n=0}^{N-1} A_n U_l \, \mathrm{e}^{\mathrm{i}(l-n)t} = \sum_{m=-N+1}^{M+N-2} \mathrm{e}^{\mathrm{i}mt} \sum_{n=0}^{N-1} A_n U_{m+n}$$

$$= \sum_{m=-N+1}^{-1} \alpha_m \, \mathrm{e}^{\mathrm{i}mt} + b(t) + \sum_{m=M}^{M+N-2} \beta_m \, \mathrm{e}^{\mathrm{i}mt}. \quad (8.1.10)$$

So to get $\{B_m\}$ we should perform inverse FFT of $\{\overline{A_n}\}$ and $\{U_l\}$ multiply $\overline{a(t)}u(t)$, perform the forward FFT, and then take coefficients for harmonics from 0 to $M-1$.

8.1.1.2 Decomposition of rotation into product of Toeplitz and diagonal matrices

Axes flipping

As we could see from Chapter 7, the general rotation operator can be decomposed to a product of elementary rotations about the Cartesian axes with matrices A and B (or C) (Eq. (7.3.102)). Recall that rotation about $\mathbf{i}_z$ is computationally fast, while the complexity comes from rotation about $\mathbf{i}_y$-axis. We can consider a transform which exchanges the axes numeration (so $\mathbf{i}_y$ becomes $\mathbf{i}_z$) which allows us to make the rotation about the $\mathbf{i}_y$-axis as cheap as the rotation about $\mathbf{i}_z$. The transform which enables us to put $\mathbf{i}_y$ in the position of $\mathbf{i}_z$ can be understood from the definition of the Euler angles and Fig. 3.4. Consider the following two elementary rotations, which exchange the axes:

$$Q_1 = A\left(\frac{\pi}{2}\right): \quad (\mathbf{i}_x, \mathbf{i}_y, \mathbf{i}_z) \to (\tilde{\mathbf{i}}_x, \tilde{\mathbf{i}}_y, \mathbf{i}_z) = (\mathbf{i}_y, -\mathbf{i}_x, \mathbf{i}_z),$$
$$Q_2 = C\left(\frac{\pi}{2}\right): \quad (\mathbf{i}_x, \mathbf{i}_y, \mathbf{i}_z) \to (\tilde{\mathbf{i}}_x, \mathbf{i}_y, \tilde{\mathbf{i}}_z) = (\mathbf{i}_z, \mathbf{i}_y, -\mathbf{i}_x). \quad (8.1.11)$$

Composition of these rotations, when we first perform rotation with matrix A and then with matrix C results in the following transform of axes

$$Q_{21} = C\left(\frac{\pi}{2}\right)A\left(\frac{\pi}{2}\right): \quad (\mathbf{i}_x, \mathbf{i}_y, \mathbf{i}_z) \to (-\mathbf{i}_z, -\mathbf{i}_x, \mathbf{i}_y). \quad (8.1.12)$$

This is the type of transform we are looking for, since it puts $\mathbf{i}_y$ in the position of the z-axis. Now we can perform a rotation with the Euler angle β_E about the new z-axis (which coincides with the old y-axis) using matrix $A(\beta_E)$. To return to the original coordinate system we need to apply the matrix inverse to Q_{21} given by Eq. (8.1.12). Composition of all these operations provides an expression for elementary rotation about the $\mathbf{i}_y$-axis, which is a rotation with matrix $C(\beta_E)$. Therefore, we have the following decomposition of the elementary matrix $C(\beta_E)$:

$$C(\beta_E) = A^{-1}\left(\frac{\pi}{2}\right)C^{-1}\left(\frac{\pi}{2}\right)A(\beta_E)C\left(\frac{\pi}{2}\right)A\left(\frac{\pi}{2}\right)$$
$$= A\left(-\frac{\pi}{2}\right)C\left(-\frac{\pi}{2}\right)A(\beta_E)C\left(\frac{\pi}{2}\right)A\left(\frac{\pi}{2}\right). \tag{8.1.13}$$

Now we can put this result into general decomposition of the rotation matrix (7.3.103):

$$Q = A(\gamma_E)A\left(-\frac{\pi}{2}\right)C\left(-\frac{\pi}{2}\right)A(\beta_E)C\left(\frac{\pi}{2}\right)A\left(\frac{\pi}{2}\right)A(\alpha_E). \tag{8.1.14}$$

This can be simplified using the commutativity of the A matrices of different arguments (7.3.105):

$$Q = A\left(\gamma_E - \frac{\pi}{2}\right)C\left(-\frac{\pi}{2}\right)A(\beta_E)C\left(\frac{\pi}{2}\right)A\left(\frac{\pi}{2} + \alpha_E\right). \tag{8.1.15}$$

Correspondingly, a general rotation transform can be represented as

$$\mathbf{Rot}(Q) = \mathbf{Rot}\left(A\left(\gamma_E - \frac{\pi}{2}\right)\right)\mathbf{Rot}\left(C\left(-\frac{\pi}{2}\right)\right)\mathbf{Rot}(A(\beta_E))$$
$$\times\mathbf{Rot}\left(C\left(\frac{\pi}{2}\right)\right)\mathbf{Rot}\left(A\left(\frac{\pi}{2} + \alpha_E\right)\right). \tag{8.1.16}$$

Using Eqs. (7.3.102) and (7.3.105) we can rewrite this in terms of angles $(\alpha, \beta, \gamma) = (\pi - \alpha_E, \beta_E, \gamma_E)$ and the elementary rotation matrix B.

$$\mathbf{Rot}(Q(\alpha, \beta, \gamma)) = \mathbf{Rot}\left(A\left(\gamma - \frac{\pi}{2}\right)\right)\mathbf{Rot}\left(B\left(-\frac{\pi}{2}\right)\right)$$
$$\times\mathbf{Rot}(A(\beta + \pi))\mathbf{Rot}\left(B\left(\frac{\pi}{2}\right)\right)\mathbf{Rot}\left(A\left(\frac{\pi}{2} - \alpha\right)\right). \tag{8.1.17}$$

Note, that from symmetry relation (3.3.45) and definition of operator $\mathbf{\Delta}$, Eqs. (7.3.98) and (7.3.99), we have

$$\mathbf{Rot}\left(B\left(-\frac{\pi}{2}\right)\right) = \mathbf{\Delta}\mathbf{Rot}\left(B\left(\frac{\pi}{2}\right)\right)\mathbf{\Delta}$$
$$= \mathbf{Rot}(A(\pi))\mathbf{Rot}\left(B\left(\frac{\pi}{2}\right)\right)\mathbf{Rot}(A(\pi)). \tag{8.1.18}$$

With this, relation (8.1.17) can be rewritten as

$$\mathbf{Rot}(Q(\alpha,\beta,\gamma))$$
$$= \mathbf{Rot}\left(A\left(\gamma+\frac{\pi}{2}\right)\right)\mathbf{Rot}\left(B\left(\frac{\pi}{2}\right)\right)\mathbf{Rot}(A(\beta))\mathbf{Rot}\left(B\left(\frac{\pi}{2}\right)\right)\mathbf{Rot}\left(A\left(\frac{\pi}{2}-\alpha\right)\right). \tag{8.1.19}$$

This decomposition involves rotations with three matrices $\mathbf{Rot}(A)$, which are diagonal and fast for computations, and with a dense matrix $\mathbf{Rot}(B(\pi/2))$. Despite $\mathbf{Rot}(B(\pi/2))$ is a constant matrix and can be precomputed and stored once for all this will not reduce the complexity of multiplication, unless some special properties of $\mathbf{Rot}(B(\pi/2))$ are used. Below we consider such a property which enables reduction in the number of operations.

Structure of the constant rotation matrix

From Eq. (3.96) the entries of matrix $\mathbf{Rot}(B(\pi/2))$ can be expressed via real rotation coefficients $H_n^{m'm}(\pi/2)$. Equation (3.3.39) provides an analytical expression for these constants:

$$H_n^{m'm}\left(\frac{\pi}{2}\right) = \varepsilon_m \varepsilon_{m'} (-2)^{-n}\left[\frac{(n+m')!(n-m')!}{(n+m)!(n-m)!}\right]^{1/2} d_n^{m'm},$$

$$n = 0, 1, \ldots, \quad m, m' = -n, \ldots, n, \tag{8.1.20}$$

$$d_n^{m'm} = \sum_{\sigma=\max(0,-(m+m'))}^{\min(n-m,n-m')} (-1)^{\sigma}\binom{n-m}{\sigma}\binom{n+m}{n-m'-\sigma}.$$

The following theorem plays a central role in the decomposition of $\mathbf{Rot}(B(\pi/2))$ into a product of *structured matrices.*

THEOREM 17 *Coefficients $d_n^{m'm}$ defined by Eq. (8.1.20) can be expressed in the form*

$$d_n^{m'm} = \sum_{\tau=\max(m,m')}^{n} (-2)^{n-\tau}\binom{n-m}{n-\tau}\binom{n+\tau}{n+m'}, \tag{8.1.21}$$

$$m, m' = -n, \ldots, n.$$

PROOF. Consider the function

$$f_n^m(x) = (1+x)^{n+m}(1-x)^{n-m}, \qquad m = -n, \ldots, n. \tag{8.1.22}$$

Using the Newton binomial formula for each multiplier, we can expand this function over powers of x as

$$\begin{aligned}
f_n^m(x) &= \sum_{\sigma=0}^{n-m}\sum_{l=0}^{n+m}(-1)^\sigma \binom{n-m}{\sigma}\binom{n+m}{n+m-l} x^{n+m-l+\sigma} \\
&= \sum_{m'=-n}^{n}\ \sum_{\sigma=\max(0,-(m+m'))}^{\min(n-m,n-m')} (-1)^\sigma \binom{n-m}{\sigma}\binom{n+m}{n-m'-\sigma} x^{n-m'} \\
&= \sum_{m'=-n}^{n} d_n^{m'm} x^{n-m'},
\end{aligned} \tag{8.1.23}$$

where we used the definition of $d_n^{m'm}$, Eq. (8.1.20). Function $f_n^m(x)$ can be expanded in power series using the binomial formula in a different way:

$$\begin{aligned}
f_n^m(x) &= (1+x)^{n+m}(1+x-2x)^{n-m} \\
&= (1+x)^{n+m}\sum_{\tau'=0}^{n-m}\binom{n-m}{\tau'}(1+x)^{n-m-\tau'}(-2x)^{\tau'} \\
&= \sum_{\tau=m}^{n}\binom{n-m}{n-\tau}(1+x)^{n+\tau}(-2x)^{n-\tau} \\
&= \sum_{\tau=m}^{n}\sum_{m''=0}^{n+\tau}\binom{n-m}{n-\tau}\binom{n+\tau}{m''} x^{n+\tau-m''}(-2x)^{n-\tau} \\
&= \sum_{\tau=m}^{n}\sum_{m'=-n}^{\tau}(-2)^{n-\tau}\binom{n-m}{n-\tau}\binom{n+\tau}{n+m'} x^{n-m'} \\
&= \sum_{m'=-n}^{n}\left[\sum_{\tau=\max(m,m')}^{n}(-2)^{n-\tau}\binom{n-m}{n-\tau}\binom{n+\tau}{n+m'}\right] x^{n-m'}.
\end{aligned} \tag{8.1.24}$$

Since expansion of $f_n^m(x)$ into a power series is unique, we obtain the statement of the theorem by comparing Eqs. (8.1.23) and (8.1.24). □

Now we can substitute $d_n^{m'm}$ from Eq. (8.1.21) into Eq. (8.1.20). Using the definition of the binomial coefficients (2.1.35), we obtain the following form, symmetric with respect to m and m':

$$H_n^{m'm}\left(\frac{\pi}{2}\right) = \epsilon_m \epsilon_{m'} \left[\frac{(n-m)!(n-m')!}{(n+m)!(n+m')!}\right]^{1/2} \sum_{m''=\max(m,m')}^{n}$$

$$\times \frac{(-2)^{-m''}(n+m'')!}{(n-m'')!(m''-m)!(m''-m')!}. \tag{8.1.25}$$

Equation (8.1.25), yields a proof for the following theorem:

THEOREM 18 *The* $(2n+1)\times(2n+1)$ *matrix of the* nth *invariant subspace of the rotation operator,* $\mathbf{Rot}_n(B(\pi/2))$, *can be decomposed as*

$$\mathbf{Rot}_n\left(B\left(\frac{\pi}{2}\right)\right) = \mathbf{L}_n \mathbf{K}_n^{\mathrm{T}} \mathbf{M}_n \mathbf{K}_n \mathbf{L}_n, \qquad n = 0, 1, \ldots \tag{8.1.26}$$

where $\mathbf{L}_n$ *and* $\mathbf{M}_n$ *are diagonal matrices and* $\mathbf{K}_n$ *is a lower triangular Toeplitz matrix and* $\mathbf{K}_n^{\mathrm{T}}$ *is its transpose. The matrices have the following entries:*

$$L_n^{m'm} = \epsilon_m \left[\frac{(n-m)!}{(n+m)!}\right]^{1/2} \delta_{m'm}, \qquad M_n^{m'm} = \frac{(-2)^{-m}(n+m)!}{(n-m)!}\delta_{m'm},$$

$$m, m' = -n, \ldots, n \qquad K_n^{m'm} = \frac{1}{(m'-m)!}, \tag{8.1.27}$$

$$m' = -n, \ldots, n, \quad m = -n, \ldots, m' \quad K_n^{m'm} = 0, \quad m' < m.$$

PROOF. Indeed, the entries of $\mathbf{Rot}_n(B(\pi/2))$ are $H_n^{m'm}(\pi/2)$, for which Eq. (8.1.25) can be rewritten in the form

$$H_n^{m'm}\left(\frac{\pi}{2}\right) = L_n^{m'm'} L_n^{mm} \sum_{m''=-n}^{n} K_n^{m''m'} M_n^{m''m''} K_n^{m''m}, \tag{8.1.28}$$

$$m, m' = -n, \ldots, n.$$

which is equivalent to Eqs. (8.1.26) and (8.1.27). □

We can also define infinite constant matrices **K**, **L**, and **M** based on the definitions of their subspaces:

$$\mathbf{K} = \sum_{n=0}^{\infty} \oplus \mathbf{K}_n, \qquad \mathbf{L} = \sum_{n=0}^{\infty} \oplus \mathbf{L}_n, \qquad \mathbf{M} = \sum_{n=0}^{\infty} \oplus \mathbf{M}_n, \tag{8.1.29}$$

to have

$$\mathbf{Rot}(B(\pi/2)) = \mathbf{L}\mathbf{K}^{\mathrm{T}}\mathbf{M}\mathbf{K}\mathbf{L}. \tag{8.1.30}$$

This allows us to prove the following theorem.

THEOREM 19 *Let **C** be a p-truncated vector of expansion coefficients over spherical basis functions of length p^2. Then an arbitrary exact rotation transform of this vector can be performed in $O(p^2 \log p)$ operations.*

PROOF. Decomposition (8.1.19) and (1.30) consists of diagonal operators $\mathbf{Rot}(A)$, **L**, and **M**, and block Toeplitz (i.e. Toeplitz for each subspace) operators **K** and $\mathbf{K}^{\mathrm{T}}$. These operators are represented by matrices $p^2 \times p^2$ for an exact rotation transform of vector **C**. Matrix–vector product for diagonal matrices can be performed for $O(p^2)$ operations, while matrix–vector product involving **K** and $\mathbf{K}^{\mathrm{T}}$ can be done for $O(p^2 \log p)$, since the total number of subspaces invariant to the rotation transform is p, and the size of Toeplitz matrices $\mathbf{K}_n$ and $\mathbf{K}_n^{\mathrm{T}}$ for each subspace $n = 0, \ldots, p-1$ does not exceed $(2p+1) \times (2p+1)$. □

REMARK To implement an algorithm based on this decomposition some additional stabilization procedures may be necessary, since the entries of the Toeplitz matrix K (see Eq. (8.1.27) become small for large $(m' - m)$. The decomposition presented in this book differ a bit from the original decompositions presented by Tang [Tan03], where the constant rotation matrix was decomposed into a product of diagonal and Pascal matrices. We refer the interested reader to this work and Ref. [TDG04] where some techniques for stabilization are discussed. These stabilization techniques do not change the overall complexity of the algorithm.

8.1.2 Fast coaxial translation

A general coaxial translation of a p-truncated vector of expansion coefficients can be performed at a cost of $O(p^3)$ operations, as we showed in the earlier chapters. We show in this section multiplication of a truncated coaxial translation matrix by a vector of coefficients can be reduced to a product of diagonal matrices and *Legendre matrices*. The latter matrices are nothing, but matrices describing the *Legendre transform*. Of course, this decomposition does not reduce the complexity if the Legendre transform is performed with $O(p^3)$ complexity (just as decomposition involving Toeplitz matrices has no advantages if the FFT is not available and the Fourier transform is performed with $O(p^2)$ complexity for a vector of length p). However, in the 1990s, exact algorithms for performing the Legendre transform with complexity $O(p^2 \log^2 p)$ were introduced [DH94] (using divide-and-conquer methods and recursions for the associated Legendre functions) and an algorithm for fast evaluation of the Legendre expansions based on the FMM was published by Alpert and Rokhlin [AR91]. A recent publication [ST02] proposes a fast $O(p^2 \log p)$ method that employs the FMM and a divide-and-conquer scheme. The authors of the latter paper found that their method outperforms the $O(p^3)$ direct method for $p \geq 511$, which seems to be too large for the present problem. Indeed for $p \sim 500$ we have $p^3 \sim 10^8$, which is the cost of a single translation (!). The same relates to the algorithm of Ref. [DH94], where some stability problems were reported and it is not clear for which range of p it is more efficient than $O(p^3)$ methods. In any case, here we just wish to state the fact is that the algorithms of this complexity exist and hopefully will be improved in future.

We suggest the interested reader look for details in the original papers and in the reports discussing implementation issues (e.g. [Moh99]), since the description of the exact fast Legendre transform (FLT) is too lengthy to reproduce here. Assuming that the exact $O(p^2 \log^2 p)$ FLT is available, the coaxial translation can be performed for $O(p^2 \log^2 p)$ operations, and therefore, an *exact* (which means machine precision) multiplication of a general truncated translation matrix by a vector of coefficients, decomposed into rotation and coaxial translation transforms, can be performed with complexity $O(p^2 \log^2 p)$.

8.1.2.1 *Decomposition of translation matrix*

Consider the coaxial translation of a p'-truncated vector of expansion coefficients to p-truncated vector using the truncated translation matrix. The truncation numbers p and p' can be equal or different and the size of the matrix for the mth order invariant subspace is $(p - |m| + 1) \times (p' - |m| + 1)$, where maximum $|m|$ is determined by the truncation number p'. The elements of the $S|R$ matrix can be computed using Eq. (7.1.71) and a similar equation can be applied to the $R|R$ matrix with the regular kernel. In component form, we have

$$(S|R)^m_{nn'}(t) = \mathrm{i}^{n-n'} \int_{-1}^{1} \Lambda_s^{(p+p'-1)}(kt;\mu)\underline{P}^m_{n'}(\mu)\underline{P}^m_n(\mu)\mathrm{d}\mu,$$

$$(R|R)^m_{nn'}(t) = \mathrm{i}^{n-n'} \int_{-1}^{1} \Lambda_r^{(p+p'-1)}(kt;\mu)\underline{P}^m_{n'}(\mu)\underline{P}^m_n(\mu)\mathrm{d}\mu, \quad (8.1.31)$$

$$n' = |m|,\ldots,p'-1, \quad n = |m|,\ldots,p-1, \quad m = 0,\pm 1,\ldots,\pm(p'-1),$$

where the singular truncated kernel is specified by Eq. (7.1.69) and the regular kernel has the same expression, with the spherical Bessel function $j_n(kt)$ replaced by $h_n(kt)$. It is noteworthy, that $\Lambda_s^{(p+p'-1)}(kt;\mu)$ and $\Lambda_r^{(p+p'-1)}(kt;\mu)$ are polynomials in μ of degree $p+p'-2$ and the product $\underline{P}^m_{n'}(\mu)\underline{P}^m_n(\mu)$ is a polynomial in μ of degree $n+n' \leq p+p'-2$. Therefore, to compute the elements of coaxial translation matrices one can integrate polynomials of degree that do not exceed $2(p+p'-2)$.

An *exact* integration of polynomials $f(\mu)$ of degree up to $2(p+p'-2)$ can be performed using Gauss–Christoffel quadrature:

$$\int_{-1}^{1} f(\mu)\mathrm{d}\mu = \sum_{l=0}^{N_q-1} w_l f(\mu_l), \quad (8.1.32)$$

where w_l are the weights and μ_l are the abscissas, or nodes, of the quadrature, and N_q is the number of quadrature nodes. The most compact general integration formula is the Gauss formula, where μ_l are zeros of the Legendre polynomial of degree N_q and it is sufficient to set $N_q = p+p'-1$ for the current problem. For $N_q = 2(p+p'-2)+1$ the set of nodes $\{\mu_l\}$ can be arbitrary, since in this case one can determine the weights of the quadrature by solving a corresponding linear system (8.1.32). One can note also that

$$\underline{P}^m_{n'}(\mu)\underline{P}^m_n(\mu) = (1-\mu^2)^{|m|}G^m_{nn'}(\mu), \quad (8.1.33)$$

where $G^m_{nn'}(\mu)$ is a polynomial of degree $n+n'-|m|$ that can be expressed via the product of the *Gegenbauer polynomials*. In this case we can use the Gauss–Christoffel quadrature with weight $(1-\mu^2)^{|m|}$:

$$\int_{-1}^{1} (1-\mu^2)^{|m|} g(\mu)\mathrm{d}\mu = \sum_{l=0}^{N_q^{(m)}-1} w_l^{(m)} g(\mu_l^{(m)}),$$

$$f(\mu) = (1-\mu^2)^{|m|} g(\mu), \quad (8.1.34)$$

where $N_q^{(m)} = p + p' - [(|m| + 3)/2]$ ([ ] here means the integer part) and this is the lowest possible $N_q^{(m)}$ for exact integration of arbitrary polynomials $g(\mu)$ of degree $2(p + p' - 2) - |m|$.

For us it is sufficient to use a general formula (8.1.32), where we can be more specific regarding the weights and nodes if necessary. This quadrature then provides the following representation for elements of translation matrices (8.1.31):

$$(E|F)_{nn'}^{m}(t) = \mathrm{i}^{n-n'} \sum_{l=0}^{N_q-1} w_l \Lambda^{(p+p'-1)}(kt; \mu_l) \underline{P}_{n'}^{m}(\mu_l) \underline{P}_{n}^{m}(\mu_l), \tag{8.1.35}$$

$$n' < p', \quad n < p, \quad E, F = S, R,$$

where Λ is Λ_s or Λ_r depending on the type of translation. The translation operation can be described then as

$$\begin{aligned} \hat{C}_n^m &= \sum_{n'=|m|}^{p'-1} (E|F)_{nn'}^{m}(t) C_{n'}^{m} \\ &= \sum_{l=0}^{N_q-1} \mathrm{i}^{n} \underline{P}_{n}^{m}(\mu_l) w_l \Lambda^{(p+p'-1)}(kt; \mu_l) \left(\sum_{n'=|m|}^{p'-1} \mathrm{i}^{-n'} \underline{P}_{n'}^{m}(\mu_l) C_{n'}^{m} \right), \end{aligned} \tag{8.1.36}$$

or in matrix vector form

$$\hat{\mathbf{C}}^m = \mathbf{i}^{(p)(m)} \underline{\mathbf{L}}^{(p)(m)} \mathbf{W}^{(m)} \mathbf{\Lambda}^{(p+p'-1)(m)}(t) (\underline{\mathbf{L}}^{(p')(m)})^{\mathrm{T}} \overline{\mathbf{i}^{(p')(m)}} \mathbf{C}^m, \tag{8.1.37}$$

where $\underline{\mathbf{L}}^{(p)(m)}$ is a rectangular $(p - |m| + 1) \times N_q^{(m)}$ matrix with entries $\{\underline{P}_n^m(\mu_l)\}$ (it is zero if $p - |m| < 0$), $\mathbf{\Lambda}^{(p+p'-1)(m)}(t)$ is a diagonal $N_q^{(m)} \times N_q^{(m)}$ matrix with entries $\{\Lambda^{(p+p'-1)}(kt; \mu_l)\}$, $\mathbf{W}^{(m)}$ is a diagonal $N_q^{(m)} \times N_q^{(m)}$ matrix with $\{w_l\}$ on the diagonal, and $\mathbf{i}^{(p)(m)}$ is a diagonal matrix of size $(p - |m| + 1) \times (p - |m| + 1)$ with $\{\mathrm{i}^n\}$ on the diagonal. The quadrature weights and nodes can be selected depending on m as in Eq. (8.1.34) or independent (e.g. for Gauss quadrature (Eq. (8.1.32)). In the former case $\mathbf{\Lambda}^{(p+p'-1)(m)}$ and $\mathbf{W}^{(m)}$ will be specific to each subspace, while in the latter case it will be common for all subspaces. This yields the decomposition of the coaxial translation matrix for the mth subspace:

$$(\mathbf{E}|\mathbf{F})_{(\text{coax})}^{(p,p')(m)}(t) = \mathbf{i}^{(p)(m)} \underline{\mathbf{L}}^{(p)(m)} \mathbf{W}^{(m)} \mathbf{\Lambda}^{(p+p'-1)(m)}(t) (\underline{\mathbf{L}}^{(p')(m)})^{\mathrm{T}} \overline{\mathbf{i}^{(p')(m)}}, \tag{8.1.38}$$

where $m = 0, \pm 1, \ldots, \pm(p' - 1)$. A general truncated coaxial translation matrix can be represented in the form

$$(\mathbf{E}|\mathbf{F})^{(p,p')}_{(\mathrm{coax})}(t) = \mathbf{i}^{(p)}\underline{\mathbf{L}}^{(p)}\mathbf{W}\boldsymbol{\Lambda}^{(p+p'-1)}(t)(\underline{\mathbf{L}}^{(p')})^{\mathrm{T}}\overline{\mathbf{i}^{(p')}}, \quad E,F = S,R,$$

$$\mathbf{G} = \sum_{m=-(p'-1)}^{p'-1} \oplus \mathbf{G}^{(m)}, \qquad \mathbf{G} = \mathbf{i}^{(p)}, \underline{\mathbf{L}}^{(p)}, \mathbf{W}, \boldsymbol{\Lambda}^{(p+p'-1)}. \tag{8.1.39}$$

This decomposition is of the same type as that in Eq. (8.1.13). Indeed, we decomposed the dependence of a dense matrix on some parameter (the Euler angle β in the case of rotation matrix, and translation distance t for coaxial translation matrix) to a product of constant matrices and a diagonal matrix, which depends on this parameter. Such decompositions arise from some deep reason related to the fact that translation operators are *normal* and *commute* for different values of the parameter, as do the elementary rotation operators.

8.1.2.2 Legendre transform

While being important from the theoretical point of view, the decomposition (8.1.39) does not bring any savings into the complexity of the matrix–vector multiplication, unless some fast method for computation of the product of a dense constant matrix $\underline{\mathbf{L}}$ by a vector is available. In the case of rotations we were able to decompose the matrix into a product of diagonal and Toeplitz matrices. For coaxial translation it is still unknown whether or not it is possible to decompose $\underline{\mathbf{L}}$ into a product of Toeplitz (Hankel) and diagonal matrices. However, $\underline{\mathbf{L}}$ is a normalized Legendre matrix, related to the *Legendre transform*, which can be performed quickly. Let us define it.

Let $f^{(m)}(\mu) \in L^2[-1,1]$ be some function associated with the mth order subspace and this function can be expanded as

$$f^{(m)}(\mu) = \sum_{n=|m|}^{\infty} C_n^m \underline{P}_n^{|m|}(\mu). \tag{8.1.40}$$

The *Legendre transform of order m* means the transform from the space of functions to the space of expansion coefficients, $f^{(m)}(\mu) \to \{C_n^m\}$, and the *inverse* Legendre transform of order m means transform from the space of coefficients to the space of functions, $\{C_n^m\} \to f^{(m)}(\mu)$. This is similar to the Fourier transform of periodic functions. As in the case of Fourier transform we can define the *discrete Legendre transform of order m*. This is a transform of vector $\mathbf{f}^m = \{f^{(m)}(\mu_l)\}$, which is a vector of samples of $f^m(\mu)$ for some sampling set $\{\mu_l\}$ to vector $\mathbf{C}^m = \{C_n^m\}$. Generally speaking, vectors

$\mathbf{f}^m$ and $\mathbf{A}^m$ are infinite, however, usually, when we speak about the discrete transforms we mean that these vectors are finite.

It is clear now that the inverse discrete Legendre transform of vector $\{C_n^m\}$ of length $p' - |m| + 1$ to the vector of samples $\{f^{(m)}(\mu_l)\}$ of length N_q is performed by multiplication of the transposed Legendre matrix, $(\underline{\mathbf{L}}^{(p')(m)})^{\mathrm{T}}$, defined by Eq. (8.1.37).

Consider now the forward Legendre transform. Due to orthonormality of functions $\{\underline{P}_n^{|m|}(\mu)\}$ (Eq. (7.1.64)) coefficients $\{C_n^m\}$ can be found from Eq. (8.1.40) as

$$C_n^m = \int_{-1}^{1} f^{(m)}(\mu)\underline{P}_n^{|m|}(\mu)\mathrm{d}\mu. \tag{8.1.41}$$

Assuming now that $f^{(m)}(\mu)$ is a *band-limited* function, i.e. the sum (8.1.40) has a finite number of terms. We can see that $f^{(m)}(\mu)\underline{P}_n^{|m|}(\mu)$ is a *polynomial*. The degree of this polynomial does not exceed $2(p' - 2) + p$ and we can use the same type of quadrature, which we used for the decomposition of the coaxial translation operator (8.1.39). This yields

$$C_n^m = \sum_{l=0}^{N_q-1} w_l \underline{P}_n^{|m|}(\mu_l) f^{(m)}(\mu_l), \tag{8.1.42}$$

or in matrix form

$$\mathbf{C}^m = \underline{\mathbf{L}}^{(p)(m)}\mathbf{W}^{(m)}\mathbf{f}^m. \tag{8.1.43}$$

Therefore, the forward Legendre transform of order m can be performed by multiplying the matrix $\underline{\mathbf{L}}^{(p)(m)}\mathbf{W}^{(m)}$ by $\mathbf{f}^m$.

8.1.2.3 Extension and truncation operators

Now let us work out the issue on the length of vectors and the order and type of the quadrature formula in the context, of the FLT as originally introduced by Driscoll and Healy [DH94]. The FLT is developed for functions sampled at zeros of the Chebyshev polynomials, μ_l (or equispaced angles θ_l, $\mu_l = \cos\theta_l$) to relate it with the cosine transform. To perform the forward Legendre transform the maximum degree n of the resulting vector in the mth order subspace should be $n = N_q/2 - 1$ (assume that N_q is even). The same relates to the inverse Legendre transform, the maximum degree n' of the vector, which should be transformed should be $n' = N_q/2 - 1$. As pointed out above, exact quadrature for an arbitrary set of abscissas can be done with $N_q = 2(p + p' - 2) + 1$. This is the case here, since the zeros of the Chebyshev polynomials do not provide any advantage in reduction of this maximum

N_q. We can see then that to perform the inverse FLT we need to take the vector $\mathbf{C}^m$ with maximum degree $n' = N_q/2 - 1 < p + p' - 2$, while for the actual vector we have $n' = p' - 1$. The same relates to the forward FLT, when we produce vector $\hat{\mathbf{C}}^m$ with the maximum $n = N_q/2 - 1 < p + p' - 2$, while we need to have $n' = p - 1$. This shows that for the current problem $N_q/2 = p + p' - 1$.

This inconsistency can be easily resolved by introducing vector *extension* and vector *truncation* operations. The former operations simply extend the length of the vector to the required length by padding it with zeros at the new positions, and the latter operations decrease the length, by neglecting all components, which are above the required length. Therefore, if we have a p-truncated vector $\mathbf{A}^{(p)}$ of length p^2 then q-truncated vector $\mathbf{A}^{(q)}$ of length q^2 can be obtained as

$$\mathbf{A}^{(q)} = \mathbf{Pr}^{(q,p)}\mathbf{A}^{(p)}, \qquad \mathbf{Pr}^{(q,p)} = (\mathbf{Pr}^{(p,q)})^{\mathrm{T}}, \tag{8.1.44}$$

where $\mathbf{Pr}^{(q,p)}$ is a $q^2 \times p^2$ rectangular matrix, which has ones on the main diagonal, and zero everywhere else (at $p = q$ this is the identity operator). For $p < q$ this operator is the extension operator, and for $p > q$ it turns into the truncation, or projection operator.

We also should note that the size of the diagonal matrix $\Lambda^{(p+p'-1)}$ is $[2(N_q/2)]^2 \times [2(N_q/2)]^2 = [2(p + p' - 1)]^2 \times [2(p + p' - 1)]^2$. In other words it is a $2(p + p' - 1)$-truncated matrix. Therefore, the truncation number for this matrix is exactly twice of the truncation number $p + p' - 1$ for the number of terms in expansion (7.1.69), which is indicated as a subscript in $\Lambda^{(p+p'-1)}$. This is an important point about this matrix, which should be kept in mind.

8.1.2.4 Fast coaxial translation algorithm

We note that the original FLT is written in terms of associated Legendre functions $(-1)^m P_n^m(\mu_l)$, while we have normalized functions $\underline{P}_n^{|m|}(\mu_l)$. This also can be easily fixed, since normalization of functions is equivalent to multiplication by corresponding diagonal matrix $N^{(p)(m)}$ of normalization coefficients (see Eq. (7.1.63)):

$$\underline{\mathbf{L}}^{(p)(m)} = \mathbf{N}^{(p)(m)}\mathbf{L}^{(p)(m)}, \qquad \underline{L}_{nl}^{(p)m} = \underline{P}_n^{|m|}(\mu_l),$$
$$L_{nl}^{(p)m} = (-1)^m P_n^m(\mu_l), \qquad N_{nn''}^{(p)m} = \sqrt{\frac{2n+1}{2}\frac{(n-m)!}{(n+m)!}}\,\delta_{nn''}. \tag{8.1.45}$$

Therefore, Eq. (8.1.39) and the above consideration provides the following algorithm for fast coaxial translation transform based on the decomposition

$$(\mathbf{E}|\mathbf{F})^{(p,p')}_{(\text{coax})}(t) = \mathbf{i}^{(p)}\mathbf{N}^{(p)}\mathbf{Pr}^{(p,N_q/2)}(\underline{\mathbf{L}}^{(N_q/2)}\mathbf{W})\mathbf{\Lambda}^{(p+p'-1)}(t)$$

$$\times(\underline{\mathbf{L}}^{(N_q/2)})^{\mathrm{T}}\mathbf{Pr}^{(N_q/2,p')}\mathbf{N}^{(p')}\overline{\mathbf{i}^{(p')}}. \tag{8.1.46}$$

- Multiply diagonal matrices $\overline{\mathbf{i}^{(p')}}$ and $\mathbf{N}^{(p')}$ by a p'-truncated vector $\mathbf{C}^{(p')}$ to obtain $\mathbf{B}^{(p')} = \mathbf{N}^{(p')}\overline{\mathbf{i}^{(p')}}\mathbf{C}^{(p')}$;
- extend the length of vector $\mathbf{B}^{(p')}$ to $\mathbf{B}^{(N_q/2)}$, $N_q = 2(p+p'-1)$, $\mathbf{B}^{(N_q/2)} = \mathbf{Pr}^{(N_q/2,p')}\mathbf{B}^{(p')}$;
- perform inverse FLT of $\mathbf{B}^{(N_q/2)}$, $\mathbf{f}^{(N_q)} = (\mathbf{L}^{(N_q/2)})^{\mathrm{T}}\mathbf{B}^{(N_q/2)}$;
- multiply diagonal matrix of translation kernel $\mathbf{\Lambda}^{(p+p'-1)}$ by $\mathbf{f}^{(N_q)}$, $\mathbf{g}^{(N_q)} = \mathbf{\Lambda}^{(p+p'-1)}(t)\mathbf{f}^{(N_q)}$;
- perform forward FLT of $\mathbf{g}^{(N_q)}$, $\mathbf{A}^{(N_q/2)} = (\mathbf{L}^{(N_q/2)}\mathbf{W})\mathbf{f}^{(N_q)}$;
- p-truncate $\mathbf{A}^{(N_q/2)}$: $\mathbf{A}^{(p)} = \mathbf{Pr}^{(p,N_q/2)}\mathbf{A}^{(N_q/2)}$;
- multiply diagonal matrices $\mathbf{N}^{(p)}$ and $\mathbf{i}^{(p)}$ by $\mathbf{A}^{(p)}$ to obtain the result of translation, $\hat{\mathbf{C}}^{(p)} = \mathbf{i}^{(p)}\mathbf{N}^{(p)}\mathbf{A}^{(p)}$.

Each step of this algorithm has a complexity which does not exceed $O(N_q) = O(p+p')$, except for the forward and inverse Legendre transforms, for which the complexity is $O(N_q \log^2 N_q)$. Hence for p and p' of the same order of magnitude, the total complexity of coaxial translation is $O(p \log^2 p)$.

8.1.2.5 *Precomputation of diagonal matrices*

We should mention that multiplication by diagonal matrix $\mathbf{\Lambda}^{(p+p'-1)}$ requires computation of its elements. If this is performed during the algorithm the algorithm will turn into $O(p^3)$ complexity. To avoid this all matrices $\mathbf{\Lambda}^{(p+p'-1)}(t)$ can be precomputed and stored. This is neither difficult nor an expensive operation, since for the FMM it can be performed only once in the pre-set step for all possible translation distances t, whose number is small. Indeed, for the $R|R$ and $S|S$ translations we have only one distance t for transition between the levels (the distance between the centers of the parent and child boxes is the same for all children). So the total number of values t for which we need to compute the regular kernel matrix $\mathbf{\Lambda}_r^{(p+p'-1)}(t)$ is $l_{\max} - l_{\min}$, where $l_{\max}$ and $l_{\min}$ are the maximum and minimum levels of the MLFMM. The singular kernel matrix $\mathbf{\Lambda}_s^{(p+p'-1)}(t)$ has only 15 different values for each level of space subdivision, l (since in the three-dimensional case there are only 15 different distances from the center of a given box to the centers of all boxes in its E_4-neighborhood, see Chapter 5), and since the $S|R$ translations should be performed for each level from $l_{\min}$ to $l_{\max}$ we have at most $15(l_{\max} - l_{\min} + 1)$ different $\Lambda_s^{(p+p'-1)}(t)$ that can be used in the MLFMM algorithm. Compared to the sizes of the problems to be solved the complexity of precomputation and storage here is low.

8.1.3 Fast general translation

Fast rotation and coaxial translation methods are valuable for decomposition of general translation operators, and for development of other methods. Moreover, the FRT can be used for finding solution of other problems of mathematical physics (e.g. Laplace equation), for asymptotic methods, etc. A remarkable property of the Helmholtz equation is that the general translation performed with truncated matrices can be computed avoiding rotational–coaxial translation decomposition and using the *Fast spherical transform* (FST). The asymptotic complexity of this operation based on the FLT and the FFT is $O(p^2 \log^2 p)$. The choice of the FST or the rotational–coaxial translation decomposition depends on their relative advantages and disadvantages, which we discuss in this section.

8.1.3.1 Decomposition of the translation matrix

We can proceed based on integral representation of the elements of the translation matrices that is similar to the method by which we obtained decomposition of the coaxial translation operator. Equations (7.1.58) or (7.1.59), and their analogs for the $R|R$ expansion coefficients (7.1.33), where we can use the truncated translation kernel, can be rewritten in the form

$$(S|R)_{nn'}^{mm'}(\mathbf{t}) = \mathrm{i}^{n-n'} \int_{S_u} \Lambda_s^{(p+p'-1)}(\mathbf{t}; \mathbf{s}) Y_{n'}^{m'}(\mathbf{s}) Y_n^{-m}(\mathbf{s}) dS(\mathbf{s}),$$

$$(R|R)_{nn'}^{mm'}(\mathbf{t}) = \mathrm{i}^{n-n'} \int_{S_u} \Lambda_r^{(p+p'-1)}(\mathbf{t}; \mathbf{s}) Y_{n'}^{m'}(\mathbf{s}) Y_n^{-m}(\mathbf{s}) dS(\mathbf{s}), \tag{8.1.47}$$

$$n = 0, \ldots, p-1, \quad n' = 0, \ldots, p'-1, \quad m = -n, \ldots, n, \quad m' = -n', \ldots, n'.$$

The point here is that there exist *exact cubature* formulae for integration of band-limited spherical functions of type

$$\Psi^{(N_b)}(\mathbf{s}) = \sum_{n=0}^{N_b-1} \sum_{m=-n}^{n} A_n^m Y_n^m(\mathbf{s}), \tag{8.1.48}$$

where N_b is the *bandwidth* of $\Psi^{(N_b)}(\mathbf{s})$. These cubatures can be written in the form

$$\int_{S_u} \Psi^{(N_b)}(\mathbf{s})dS(\mathbf{s}) = \sum_{j=0}^{N_c-1} w_j \Psi^{(N_b)}(\mathbf{s}_j), \qquad \mathbf{s}_j \in S_u, \tag{8.1.49}$$

where $\mathbf{s}_j$ are the cubature nodes, and w_j are the cubature weights, and $N_c(N_b)$ is the order of cubature. A particular case of this cubature related to the FST will be considered later in this section. Here we can see that the functions under integration (8.1.47) are band-limited and, hence, they can be represented in the following form:

$$(E|F)_{nn'}^{mm'}(\mathbf{t}) = \mathrm{i}^{n-n'} \sum_{j=0}^{N_c-1} w_j \Lambda^{(p+p'-1)}(\mathbf{t}; \mathbf{s}_j) Y_{n'}^{m'}(\mathbf{s}_j) Y_n^{-m}(\mathbf{s}_j), \tag{8.1.50}$$

$$n' < p', \quad n < p,$$

and E, F denote any of S, R. Here Λ is Λ_s or Λ_r depending on the type of translation. The translation operation is a transform of vector $\mathbf{C}$ of length p'^2 into vector $\hat{\mathbf{C}}$ of length p^2 that can be described in component form as

$$\hat{C}_n^m = \sum_{n'=0}^{p'-1} \sum_{m'=-n'}^{n'} (E|F)_{nn'}^{mm'}(\mathbf{t}) C_{n'}^{m'},$$

$$= \sum_{j=0}^{N_c-1} \mathrm{i}^n Y_n^{-m}(\mathbf{s}_j) w_j \Lambda^{(p+p'-1)}(\mathbf{t}; \mathbf{s}_j) \left(\sum_{n'=0}^{p'-1} \sum_{m'=-n'}^{n'} \mathrm{i}^{-n'} Y_{n'}^{m'}(\mathbf{s}_j) C_{n'}^{m} \right) \tag{8.1.51}$$

or in matrix–vector form

$$\hat{\mathbf{C}} = \mathbf{i}^{(p)} \mathbf{Y}^{(p)} \mathbf{W} \mathbf{\Lambda}^{(p+p'-1)}(\mathbf{t}) (\overline{\mathbf{Y}^{(p')}})^{\mathrm{T}} \overline{\mathbf{i}^{(p')}} \mathbf{C}, \tag{8.1.52}$$

where $\mathbf{Y}^{(p)}$ is a rectangular $p^2 \times N_c$ matrix with entries $\{Y_n^{-m}(\mathbf{s}_j)\}$ (assuming that all p^2 components $\{C_n^m\}$ of vector $\hat{\mathbf{C}}$ are ordered and aligned, e.g. see Eq. (7.2.3)), $\mathbf{\Lambda}^{(p+p'-1)}(\mathbf{t})$ is a diagonal $N_c \times N_c$ matrix with entries $\{\mathbf{\Lambda}^{(p+p'-1)}(\mathbf{t}; \mathbf{s}_j)\}$, $\mathbf{W}$ is a diagonal $N_c \times N_c$ matrix with $\{w_j\}$ on the diagonal, and $\mathbf{i}^{(p)}$ is a diagonal matrix of size p^2 with $\{\mathrm{i}^n\}$ on the diagonal according to the vector alignment rule, e.g. Eq. (7.2.3). This yields decomposition of a general translation matrix:

$$(\mathbf{E}|\mathbf{F})^{(p,p')}(\mathbf{t}) = \mathbf{i}^{(p)} \mathbf{Y}^{(p)} \mathbf{W} \mathbf{\Lambda}^{(p+p'-1)}(\mathbf{t}) (\overline{\mathbf{Y}^{(p')}})^{\mathrm{T}} \overline{\mathbf{i}^{(p')}}. \tag{8.1.53}$$

It is rather obvious, that this decomposition is of type (8.1.39), which shows that the dependence of the translation matrix on $\mathbf{t}$ can be decomposed to a product of constant matrices and a diagonal matrix

$\Lambda^{(p+p'-1)}(\mathbf{t})$. While multiplications involving the diagonal matrices are computationally cheap, the major computation complexity comes again from the multiplication of a dense constant matrix $\mathbf{Y}^{(p)}$ (and its transpose) by vectors appearing in the decomposition.

8.1.3.2 *Fast spherical transform*

While the FLT is the most substantial part of the Driscoll–Healy algorithm [DH94], in fact this paper deals with the *spherical transform*. In general, this is a transform from the space of surface functions $\Psi(\mathbf{s}) \in L^2(S_u)$ to the space of expansion coefficients over spherical harmonics. Particular interest is paid to band-limited functions and expansions of type (8.1.48), since in this case the space of coefficients is finite-dimensional and a finite *Discrete Spherical Transform* can be applied. An important issue for the discrete spherical transform is selection of the distribution of sampling nodes over the sphere surface, which provide the most compact cubature. There exist a number of papers dedicated to this issue (e.g. [Tay95]). Perhaps, the simplest case, which, however, enables the use of the FFT and FLT is an *equiangular* grid, for which the following theorem has been proved (we reproduce it in our notation; the reader should pay attention to our definitions of spherical harmonics and the associated Legendre polynomials (see Eqs. (2.1.20) and (2.1.59), and symmetry (2.1.44)), which have normalization factors different from that in the cited paper).

THEOREM 20 (DRISCOLL AND HEALY, 1994) *Let $\Psi^{(N_b)}(\mathbf{s}) \in L^2(S_u)$ have bandwidth N_b. Then for each $|m| \leq n < N_b$ coefficients A_n^m of expansion (8.1.48) can be represented as*

$$\begin{aligned} A_n^m &= \frac{\pi}{N_b} \sum_{l=0}^{2N_b-1} \sum_{l'=0}^{2N_b-1} w_l^{(N_b)} \Psi^{(N_b)}(\theta_l, \varphi_{l'}) Y_n^{-m}(\theta_l, \varphi_{l'}) \\ &= \frac{\pi}{N_b} \sqrt{\frac{2n+1}{4\pi} \frac{(n-m)!}{(n+m)!}} \sum_{l=0}^{2N_b-1} w_l^{(N_b)} \\ &\quad \times \sum_{l'=0}^{2N_b-1} \Psi^{(N_b)}(\theta_l, \varphi_{l'}) e^{-im\varphi_{l'}} (-1)^m P_n^m(\cos\theta_l), \end{aligned} \tag{8.1.54}$$

where the sample points are chosen from the equiangular grid: $\theta_l = \pi(2l+1)/(4N_b)$, $\varphi_{l'} = 2\pi l'/(2N_b)$; and the weights $w_l^{(N_b)}$ play a role analogous to the $\sin\theta$ *factor in the integrals.*

An explicit expression for $w_l^{(N_b)}$ are

$$w_l^{(N_b)} = \frac{2}{N_b}\sin\frac{\pi(2l+1)}{4N_b}\sum_{l''=0}^{N_b-1}\frac{1}{2l''+1}\sin\frac{\pi(2l+1)(2l''+1)}{4N_b}, \quad (8.1.55)$$

and can be found in Ref. [KMHR00].

Note that this theorem follows from a cubature formula of type (8.1.49). Indeed, using the orthonormality of spherical harmonics and cubature (8.1.49) we have

$$\begin{aligned} A_n^m &= \int_{S_u} \Psi^{(N_b)}(\mathbf{s})Y_n^{-m}(\mathbf{s})dS(\mathbf{s}) = \sum_{j=0}^{N_c-1} w_j\Psi^{(N_b)}(\mathbf{s}_j)Y_n^{-m}(\mathbf{s}_j) \\ &= \frac{\pi}{N_b}\sum_{l=0}^{2N_b-1}\sum_{l'=0}^{2N_b-1} w_l^{(N_b)}\Psi^{(N_b)}(\theta_l,\varphi_{l'})Y_n^{-m}(\theta_l,\varphi_{l'}), \quad n < N_b. \end{aligned} \quad (8.1.56)$$

This shows that the maximum bandwidth of function $\Psi^{(N_b)}(\mathbf{s})Y_n^{-m}(\mathbf{s})$ is $2N_b$. Therefore, the cubature with $N_c = 4N_b^2$ and the nodes and weights specified above can be applied for integration of band-limited functions with bandwidth $2N_b$. For integration of functions with bandwidth N_b, we, therefore, have $N_c = N_b^2$ and the following cubature

$$\int_{S_u} \Psi^{(N_b)}(\mathbf{s})dS(\mathbf{s}) = \frac{2\pi}{N_b}\sum_{l=0}^{N_b-1}\sum_{l'=0}^{N_b-1} w_l^{(N_b/2)}\Psi^{(N_b)}(\theta_l,\varphi_{l'}), \quad (8.1.57)$$

$$\theta_l = \pi(2l+1)/(2N_b), \quad \varphi_{l'} = 2\pi l'/N_b.$$

The relation between this equation and Eq. (8.1.49) is rather obvious:

$$\begin{aligned} &\mathbf{s}_j = (\theta_l,\varphi_{l'}), \qquad w_j = \frac{2\pi}{N_b}w_l^{(N_b/2)}, \qquad N_c = N_b^2, \\ &j = 0,\ldots,N_c-1, \quad l,l' = 0,\ldots,N_b-1. \end{aligned} \quad (8.1.58)$$

The first relation here establishes correspondence between index j and the indices l and l'.

Theorem (8.1.54) provides a decomposition of the spherical transform into the Fourier transform with respect to the angle φ and the Legendre Transform with respect to the angle θ. The order to compute them is, first, perform summation over the l' index (FFT) and then multiply by $w_l^{(N_b)}$ and perform the FLT for subspaces of order m.

It is important issue to establish the bandwidth of the functions we are working with and use appropriate extension and truncation operators (8.1.44). As in the case of coaxial translation we note that the bandwidth of the integrands (8.1.47) is $N_b = 2(p + p') - 3$. In fact, we should take $N_b = 2(p + p' - 1)$, since the order of quadrature should be evaluated based on half the bandwidth (see also truncation for coaxial translations). Therefore, the length of the diagonal of matrix of weights $\mathbf{W}$ is $N_c = N_b^2 = 4(p + p' - 1)^2$ and this is the length of the vector to which we apply the forward FST. The result of this operation is a vector which maximum degree n is $N_b/2 - 1 = p + p' - 2$. Therefore, this vector should be truncated to the length of p with the truncation operator $\mathbf{Pr}^{(p,N_b/2)}$. The same relates to extension operation $\mathbf{Pr}^{(N_b/2,p')}$ of the initial vector to which we should apply the forward transform.

This shows that (p, p')-truncated translation matrices can be decomposed as follows:

$$(\mathbf{E}|\mathbf{F})^{(p,p')}(\mathbf{t}) = \mathbf{i}^{(p)}\mathbf{Pr}^{(p,N_b/2)}\mathbf{Fs}^{(N_b/2)}\boldsymbol{\Lambda}^{(p+p'-1)}(\mathbf{t})\mathbf{Is}^{(N_b/2)}\mathbf{Pr}^{(N_b/2,p')}\overline{\mathbf{i}^{(p')}},$$

$$\mathbf{Fs}^{(N_b/2)} = \mathbf{Y}^{(N_b/2)}\mathbf{W}, \qquad \mathbf{Is}^{(N_b/2)} = (\overline{\mathbf{Y}^{(N_b/2)}})^{\mathrm{T}}, \tag{8.1.59}$$

$$N_b = 2(p + p' - 1).$$

Here $\mathbf{Fs}^{(N_b/2)}$ and $\mathbf{Is}^{(N_b/2)}$ are notations for the forward and inverse spherical transforms of vectors of length $(N_b/2)^2$ in the space of the expansion coefficients and N_b^2 in the space of function samples. These translations can be performed with complexity $O(p^2 \log^2 p)$. As in the case of coaxial translations, we note that the size of the diagonal matrix $\boldsymbol{\Lambda}^{(p+p'-1)}(\mathbf{t})$ is N_b^2, or this is a $2(p + p' - 1)$-truncated matrix, while the truncation number for the series for computation of the matrix elements (8.1.45) is half of that. In this decomposition we did not separate the normalization matrices from the matrices for the forward and inverse transforms, since these transforms can be either rewritten in terms of orthonormal spherical harmonics, or the vectors should be multiplied by the corresponding normalization diagonal matrices, similar to the case of the coaxial translation.

8.1.3.3 Precomputation of diagonal matrices

As in the case of coaxial translations the decomposition makes no sense from computational point of view, if $\boldsymbol{\Lambda}^{(p+p'-1)}(\mathbf{t})$ needs to be computed each time, when the translation is performed, since such a computation brings the complexity to $O(p^3)$. Therefore, the values of this matrix should be precomputed and stored. Note, that in contrast to the coaxial translations, where the kernel depends only on $|\mathbf{t}|$, in the general case it also depends on the direction of $\mathbf{t}$. The number of possible $R|R$ translations (and $S|S$ translations) between two neighboring levels is 8, so we need to compute and store $8(l_{\max} - l_{\min})$ diagonals of the regular kernel matrices $\Lambda_r^{(p+p'-1)}(\mathbf{t})$. The number of possible $S|R$ translations for a given level is 216, which results in precomputation and storage of $216(l_{\max} - l_{\min} + 1)$ diagonals of singular kernel matrices $\Lambda_r^{(p+p'-1)}(\mathbf{t})$. Now we note that if we have $N_c^{(l)}$ nodes $\mathbf{s}_j^{(l)}$ for each level l, the cost of precomputations is significant. In any case these computations should be performed and it is much cheaper to do them once and store the results. Some savings (several times in time and memory) can be achieved if we note that the entries of the kernel matrices depend on the translation distance, t, and dot products $(\mathbf{t}\cdot\mathbf{s}_j^{(l)})$. Due to symmetry the number of possible combinations of parameters $(l, t_n^{(l)}, (\mathbf{s}_j^{(l)}\cdot\mathbf{t}_n^{(l)}/t_n^{(l)}))$ is much smaller than the number of combinations $(l, \mathbf{s}_j^{(l)}, \mathbf{t}_n^{(l)})$, where $\mathbf{t}_n^{(l)}$ is the number of possible translation distances for level l. Moreover, for different levels only the modulus of $\mathbf{t}_n^{(l)}$ is different, due to the hierarchical binary data structure $\mathbf{t}_n^{(l)} = 2\mathbf{t}_n^{(l+1)}$. Taking into account that in order to perform an efficient FFT the $N_c^{(l)}$ should be selected as a power of 2, and coarser surface grids are subsets of the finer grids. This means that if $N_c^{(l_1)} = 4N_c^{(l_2)}$, $\{\mathbf{s}_j^{(l_2)}\} \subset \{\mathbf{s}_j^{(l_1)}\}$. This also reduces the number of possible $(\mathbf{t}_n^{(l)}\cdot\mathbf{s}_j^{(l)})$. Our tests show that precomputation and storage of $\Lambda^{(p+p'-1)}(\mathbf{t})$ is an efficient method in the implementation of the translation methods based on diagonalization of translation operators. Other methods for fast computation of the kernel matrices (e.g. interpolation) can be used if they do not increase the asymptotic complexity of the translation.

8.2 LOW- AND HIGH-FREQUENCY ASYMPTOTICS

As translation operators depend on the translation vector $\mathbf{t}$, or more precisely on the vector $k\mathbf{t}$, we consider below some asymptotics when values of the module of this vector are small or large. The case $|k\mathbf{t}| \ll 1$ is identified as the "low-frequency" case and $|k\mathbf{t}| \gg 1$ as the "high-frequency" case. As the asymptotics are considered with respect to the parameter $|k\mathbf{t}|$ it is obvious that the low-frequency case can be applied for arbitrary k, but for small enough $|\mathbf{t}|$ or short translation distances. Similarly, the high-frequency asymptotics are valid for arbitrary k, but for long translation distances. In practical problems the translation distances are also related to the domains of validity of the *R*- and *S*- expansions, so the applicability of these formulae should be checked from the point of view of a particular problem. As shown below, the asymptotic translations can be performed with lower asymptotic complexity than general $O(p^4)$ or $O(p^3)$ methods considered above and here we are interested in $o(p^3)$ algorithms (say of $O(p^2)$ or $O(p^2 \log p)$ type asymptotic complexities). For simplicity we consider below the case of real $k>0$, while the asymptotics can be generalized for complex k.

8.2.1 Low frequencies

8.2.1.1 Exponential sparse matrix decomposition of the R|R matrix

Representation of the $R|R$- and $S|S$ translation operators via the sparse matrix exponential (7.3.77) can be treated as *low-frequency asymptotics*. Equation (7.3.74) can be written in the form

$$(\mathbf{R}|\mathbf{R})(\mathbf{t}) = \sum_{n=0}^{N_t-1} \frac{(kt)^n}{n!} \mathbf{D}_t^n + O\left(\frac{(kt)^{N_t}}{N_t!} \|\mathbf{D}_t\|^{N_t}\right), \tag{8.2.1}$$

where $\|\mathbf{D}_t\|$ is a somehow defined *norm* of the operator $\mathbf{D}_t$. As shown in Chapter 7, $O(N_t p^2)$ operations are sufficient to perform the translation of the p-truncated vector of coefficients of size p^2. The number of terms in the asymptotic series can be selected to provide a necessary accuracy, and for small kt this can be achieved by the summation of very few terms. So this expansion is an example of $O(p^2)$ asymptotic complexity algorithm.

8.2.1.2 Toeplitz/Hankel matrix representations

$R|R$ and $S|S$ translations

Perhaps, Eq. (8.2.1) provides the best practical asymptotic representation for the $R|R$ and $S|S$ translation at low frequencies. However, below we show some other computationally efficient method applicable at very low frequencies. We think that it is important to have it in mind both from a methodological and theoretical point of view. These asymptotics follow from the structure of the translation coefficients given by Eq. (3.2.48). In this equation we note that for low kt the principal term in the expansion of the regular spherical basis functions $R_n^m(\mathbf{t})$ have asymptotics

$$R_n^m(\mathbf{t}) = j_n(kt)Y_n^m\left(\frac{\mathbf{t}}{t}\right) = \frac{(kt)^n}{(2n+1)!!}Y_n^m\left(\frac{\mathbf{t}}{t}\right)[1+O((kt)^2)]. \qquad (8.2.2)$$

So we have

$$(R|R)_{n'n}^{m'm}(\mathbf{t}) = (s|r)_{|n-n'|,n'n}^{m-m',m'm} R_{|n-n'|}^{m-m'}(\mathbf{t})[1+O((kt)^2)]$$

$$= (s|r)_{|n-n'|,n'n}^{m-m',m'm} \frac{(kt)^{|n-n'|}}{(2|n-n'|+1)!!} Y_{|n-n'|}^{m-m'}\left(\frac{\mathbf{t}}{t}\right)[1+O((kt)^2)]. \qquad (8.2.3)$$

Now, using just the principal term of this expansion we can represent the translation operation as follows:

$$\hat{C}_{n'}^{m'} = \sum_{n=0}^{\infty}\sum_{m=-n}^{n}(R|R)_{n'n}^{m'm}(\mathbf{t})C_n^m = \sum_{n=0}^{\infty}\sum_{m=-n}^{n}(s|r)_{|n-n'|,n'n}^{m-m',m'm}R_{|n-n'|}^{m-m'}(\mathbf{t})C_n^m$$

$$= \sum_{n=n'+1}^{\infty}\sum_{m=-n}^{n}(s|r)_{n-n',n'n}^{m-m',m'm}R_{n-n'}^{m-m'}(\mathbf{t})C_n^m + C_{n'}^{m'} +$$

$$\sum_{n=0}^{n'-1}\sum_{m=-n}^{n}(s|r)_{n'-n,n',n}^{m-m',m'm}R_{n'-n}^{m-m'}(\mathbf{t})C_n^m. \qquad (8.2.4)$$

Using relation between the structural $(s|r)_{|n-n'|,n'n}^{m-m',m'm}$ and the Clebsch–Gordan coefficients (3.2.32) and explicit expressions for the latter coefficients (3.2.25) we can represent them as

$$(s|r)_{n-n',n'n}^{m-m',m'm} = (4\pi)^{1/2}\frac{\eta_{n'}^{m'}\eta_{n-n'}^{m-m'}}{\eta_n^m}, \quad n \geq n',$$

$$(s|r)_{n'-n,n'n}^{m-m',m'm} = (4\pi)^{1/2}(-1)^{n-n'}(-1)^{m-m'}\frac{\eta_n^m\eta_{n'-n}^{m-m'}}{\eta_{n'}^{m'}}, \quad n' \geq n, \qquad (8.2.5)$$

where

$$\eta_n^m = \varepsilon_m\frac{(2n)!}{n!}\sqrt{\frac{2n+1}{(n-m)!(n+m)!}}, \qquad m=-n,\ldots,n. \qquad (8.2.6)$$

With this factorization and Eq. (8.2.3) we can rewrite Eq. (8.2.4) in the form

$$\hat{C}_{n'}^{m'} = \hat{C}_{n'}^{(1)m'} + C_{n'}^{m'} + \hat{C}_{n'}^{(2)m'},$$

$$\hat{C}_{n'}^{(1)m'} = (4\pi)^{1/2}\eta_{n'}^{m'} \sum_{n=n'+1}^{\infty} \sum_{m=-n}^{n} \xi_{n-n'}^{m-m'}(t)\left(\frac{1}{\eta_n^m} C_n^m\right), \tag{8.2.7}$$

$$\hat{C}_{n'}^{(2)m'} = (4\pi)^{1/2}\frac{1}{\eta_{n'}^{m'}} \sum_{n=0}^{n'-1} \sum_{m=-n}^{n} \zeta_{n'-n}^{m-m'}(\mathbf{t})(\eta_n^m C_n^m),$$

where $\xi_n^m(\mathbf{t})$ and $\zeta_n^m(\mathbf{t})$ are the following functions:

$$\xi_n^m(\mathbf{t}) = \eta_n^m R_n^m(\mathbf{t}), \qquad \zeta_n^m(\mathbf{t}) = (-1)^{n+m}\xi_n^m(\mathbf{t}). \tag{8.2.8}$$

These relations enable a fast translation of a truncated vector C_n^m to be performed as follows. To obtain $\hat{C}_{n'}^{(1)m'}$ we first multiply C_n^m by $(\eta_n^m)^{-1}$. This is a term-by-term multiplication and requires $O(p^2)$ operations (one can also think about this as a multiplication of a diagonal matrix by vector $\mathbf{C}$). The resulting vector $\tilde{\mathbf{C}}$ should then be multiplied by a dense matrix whose elements are $\xi_{n-n'}^{m-m'}$. The trick here is that such multiplication can be done in $O(p^2 \log p)$ operations since this matrix has a Toeplitz structure. Standard Toeplitz matrices are 2D, while here we have a 4D matrix. However, this case can also be handled by the same idea of multiplication of Toeplitz matrix by a vector and the 2D FFT. In fact, this idea was already implemented and tested for the $S|R$ translation of solutions of the Laplace equation. We will not elaborate here the technical details of this algorithm and refer the interested reader to Ref. [EJB96]. So multiplication using 2D FFT costs $O(p^2 \log p)$ for a matrix with such structure. The obtained vector $\tilde{\tilde{\mathbf{C}}}$ should then be multiplied again by a diagonal matrix of coefficients $(4\pi)^{1/2}\eta_n^m$. The cost of this operation is $O(p^2)$, which totals the cost of obtaining vector $\hat{\mathbf{C}}^{(1)}$ as $O(p^2 \log p)$. A similar procedure with the same computational complexity is required to obtain vector $\hat{\mathbf{C}}^{(2)}$, which brings the cost of computation of the $R|R$ and $S|S$ translation to $O(p^2 \log p)$.

We also note that the availability of the FRT enables construction of a $O(p^2 \log p)$ algorithm based on rotational–coaxial translation decomposition of the translation operator (7.2.22). This method avoids the use of 2D FFT and operations with 4D Toeplitz matrices. The rotational–coaxial translation decomposition includes two rotations and one coaxial translation. The cost of rotations is $O(p^2 \log p)$. The coaxial translation can also be performed with the same complexity. This is not difficult to show using Eqs. (8.2.7), since coaxial translations can be performed for each subspace with respect to order m independently. Setting in this equation $m' = m$ and dropping summations over m we obtain the following equations describing low-frequency coaxial translation of p-truncated vector **C**

$$\hat{C}_{n'}^{m} = \hat{C}_{n'}^{(1)m} + C_{n'}^{m} + \hat{C}_{n'}^{(2)m}, \qquad n' \geq |m|, \qquad m = 0, \pm 1, \ldots \pm (p-1),$$

$$\hat{C}_{n'}^{(1)m} = (4\pi)^{1/2} \eta_{n'}^{m} \sum_{n=n'+1}^{p-1} \xi_{n-n'}^{0}(t\mathbf{i}_z)\left(\frac{1}{\eta_n^m} C_n^m\right), \tag{8.2.9}$$

$$\hat{C}_{n'}^{(2)m} = (4\pi)^{1/2} \frac{1}{\eta_{n'}^{m}} \sum_{n=|m|}^{n'-1} \zeta_{n'-n}^{0}(t\mathbf{i}_z)(\eta_n^m C_n^m).$$

Here $\xi_{n-n'}^{0}$ are elements of an upper-triangular Toeplitz matrix with zeros on the main diagonal and $\zeta_{n'-n}^{0}$ are elements of a lower-triangular Toeplitz matrix with zeros on the main diagonal. The size of the matrix for the mth subspace is $(p - |m|) \times (p - |m|)$ so the cost of translation for this subspace is $O(p \log p)$. The total number of subspaces is $2p - 1$, which yields $O(p^2 \log p)$ algorithm for coaxial translation.

$S|R$ translation

Similar type of asymptotics holds for the $(S|R)$-matrix. The singular spherical basis functions at low frequencies satisfy the asymptotic relation

$$S_n^m(\mathbf{t}) = h_n(kt) Y_n^m\left(\frac{\mathbf{t}}{t}\right)$$

$$= -\frac{\mathrm{i}(2n-1)!!}{(kt)^{n+1}} Y_n^m\left(\frac{\mathbf{t}}{t}\right) \begin{cases} 1 + O((kt)^2), & n \geq 1, \\ 1 + O((kt)), & n = 0. \end{cases} \tag{8.2.10}$$

Equation (3.2.38) then yields

$$(S|R)_{n'n}^{m'm}(\mathbf{t}) = (s|r)_{n+n',n'n}^{m-m',m'm} S_{n+n'}^{m-m'}(\mathbf{t})[1 + O(kt)]. \tag{8.2.11}$$

The structural $(s|r)_{n+n',n'n}^{m-m',m'm}$ and Clebsch–Gordan coefficients are related as Eq. (3.2.32). With the explicit expressions for the latter coefficients (3.2.25) we can represent the structural coefficients as

$$(s|r)_{n+n',n'n}^{m-m',m'm} = (4\pi)^{1/2} (-1)^{n'+m'} \frac{\eta_n^m \eta_{n'}^{m'}}{\eta_{n+n'}^{m-m'}}, \tag{8.2.12}$$

where η_n^m are provided by Eq. (8.2.6).

Therefore, a low-frequency $S|R$ translation of the expansion coefficients can be represented as

$$\hat{C}_{n'}^{m'} = (4\pi)^{1/2} \frac{(-1)^{n'+m'}}{\eta_{n'}^{m'}} \sum_{n=0}^{\infty} \sum_{m=-n}^{n} \varpi_{n+n'}^{m-m'}(\mathbf{t})(\eta_n^m C_n^m), \tag{8.2.13}$$

where

$$\varpi_n^m(\mathbf{t}) = \frac{1}{\eta_n^m} S_n^m(\mathbf{t}).$$

As in the case of the $R|R$ translation we can see that the translation decomposes to the product of a diagonal matrix, a 4D matrix of special structure and another diagonal matrix. The major complexity comes from the multiplication of this matrix of special structure, which is a Toeplitz matrix with respect to indices m and m' and the Hankel matrix with respect to indices n and n'. The product of such a matrix by a vector of size p^2 takes $O(p^2 \log p)$ operations if using the 2D FFT. The details of the technique are described in the paper [EJB96].

We also note that we can use rotational–coaxial translation decomposition to avoid the use of 2D FFT, in the same way as we did above for the $R|R$ translation. Indeed Eq. (8.2.13) shows that coaxial $S|R$ translation for low frequencies can be performed using a Hankel matrix with entries $\varpi_{n+n'}^0$:

$$\hat{C}_{n'}^{m} = (4\pi)^{1/2} \frac{(-1)^{n'+m}}{\eta_{n'}^{m}} \sum_{n=|m|}^{p-1} \varpi_{n+n'}^{0}(t\mathbf{i}_z)(\eta_n^m C_n^m), \tag{8.2.14}$$

$$m = 0, \pm 1, \ldots, \pm(p-1).$$

8.2.1.3 *Renormalization*

Let us compare two low-frequency translation methods based on the expansion of the matrix exponential (8.2.1) and the principal term of asymptotics (8.2.8). A valid question, which one may ask is why should we proceed with a $O(p^2 \log p)$ method if a faster $O(p^2)$ method is available. Moreover, the $O(p^2)$ method seems to be more accurate, ($O((kt)^{N_t})$ vs $O((kt)^2)$)? The answer is that the $O(p^2 \log p)$ method can be more efficient for solution of the Laplace equation or the Helmholtz equation for very low k.

To think a bit more about this issue let us turn to the Helmholtz equation and recall that for $k = 0$ this equation becomes the Laplace equation. The Laplace equation can be solved similarly to the Helmholtz equation using multipole methods and translation theory. However, if we set formally $k = 0$ in expansion (8.2.1) it yields $(\mathbf{R}|\mathbf{R})(\mathbf{t}) = \mathbf{I}$. So our translation operator is an identity operator in this case! We also can check this using Eq. (7.1.36), which for $k = 0$ yields the same result. On the other hand, $(\mathbf{R}|\mathbf{R})(\mathbf{0}) = \mathbf{I}$ is a valid result, since a translation with zero translation vector is the identity operation.

Some insight comes from the consideration of spherical basis functions, which we use for expansion of our solutions. As we can see from Eq. (8.2.2) all the regular basis function of degree $n > 0$ become zero for $k = 0$ and so they do not provide a basis for such k. A similar problem appears for the singular basis functions, which blow up for $k = 0$. Respectively, since the function $\psi(\mathbf{r}; k)$ which we expand over the basis continuously transforms to $\psi(\mathbf{r}; 0)$ regular in the same spatial domain, the R expansion coefficients of this function grow infinitely as $k \to 0$ and the S expansion coefficients tend to zero. To have a proper basis that works well at very low frequencies we need to *renormalize* the basis functions and the expansion coefficients. Renormalization of the basis function is more or less obvious:

$$\underline{R}_n^m(\mathbf{r}) = k^{-n} R_n^m(\mathbf{r}), \qquad \underline{S}_n^m(\mathbf{r}) = k^{n+1} S_n^m(\mathbf{r}), \tag{8.2.15}$$

because this results in finite values

$$\begin{aligned} \lim_{k\to 0} \underline{R}_n^m(\mathbf{r}) &= \frac{r^n}{(2n+1)!!} Y_n^m\left(\frac{\mathbf{r}}{r}\right), \\ \lim_{k\to 0} \underline{S}_n^m(\mathbf{r}) &= -\frac{\mathrm{i}(2n-1)!!}{r^{n+1}} Y_n^m\left(\frac{\mathbf{r}}{r}\right). \end{aligned} \tag{8.2.16}$$

These functions are elementary solutions of the Laplace equation in spherical coordinates. If we use renormalized basis functions (8.2.15) then we should renormalize the expansion coefficients accordingly. So, for the R expansion we have

$$\psi(\mathbf{r}) = \sum_{n=0}^{\infty} \sum_{m=-n}^{n} C_n^m R_n^m(\mathbf{r}) = \sum_{n=0}^{\infty} \sum_{m=-n}^{n} \underline{C}_n^m \underline{R}_n^m(\mathbf{r}), \qquad C_n^m = k^{-n} \underline{C}_n^m, \tag{8.2.17}$$

while for the S expansion we have

$$\psi(\mathbf{r}) = \sum_{n=0}^{\infty} \sum_{m=-n}^{n} C_n^m S_n^m(\mathbf{r}) = \sum_{n=0}^{\infty} \sum_{m=-n}^{n} \underline{C}_n^m \underline{S}_n^m(\mathbf{r}), \qquad C_n^m = k^{n+1} \underline{C}_n^m. \quad (8.2.18)$$

The renormalized expansion coefficients $\underline{C}_n^m$ here should be considered to be $O(1)$.

We need to renormalize the translation matrices as follows:

$$\begin{aligned}
&(\underline{R}|\underline{R})_{n'n}^{m'm}(\mathbf{t}) = k^{n'-n}(R|R)_{n'n}^{m'm}(\mathbf{t}),\\
&\underline{R}_n^m(\mathbf{r}+\mathbf{t}) = \sum_{n'=0}^{\infty} \sum_{m'=-n'}^{n'} (\underline{R}|\underline{R})_{n'n}^{m'm}(\mathbf{t}) \underline{R}_{n'}^{m'}(\mathbf{r}),\\
&(\underline{S}|\underline{S})_{n'n}^{m'm}(\mathbf{t}) = k^{n'-n}(S|S)_{n'n}^{m'm}(\mathbf{t}),\\
&\underline{S}_n^m(\mathbf{r}+\mathbf{t}) = \sum_{n'=0}^{\infty} \sum_{m'=-n'}^{n'} (\underline{S}|\underline{S})_{n'n}^{m'm}(\mathbf{t}) \underline{S}_{n'}^{m'}(\mathbf{r}), \quad |\mathbf{t}| < |\mathbf{r}|,\\
&(\underline{S}|\underline{R})_{n'n}^{m'm}(\mathbf{t}) = k^{n+n'+1}(S|R)_{n'n}^{m'm}(\mathbf{t}),\\
&\underline{S}_n^m(\mathbf{r}+\mathbf{t}) = \sum_{n'=0}^{\infty} \sum_{m'=-n'}^{n'} (\underline{S}|\underline{R})_{n'n}^{m'm}(\mathbf{t}) \underline{R}_{n'}^{m'}(\mathbf{r}), \quad |\mathbf{t}| > |\mathbf{r}|.
\end{aligned} \quad (8.2.19)$$

It is noteworthy that the renormalized matrices $(\underline{\mathbf{R}}|\underline{\mathbf{R}})$ and $(\underline{\mathbf{S}}|\underline{\mathbf{S}})$ are *different* (simply because $(\mathbf{R}|\mathbf{R})$ and $(\mathbf{S}|\mathbf{S})$ are the same). In terms of asymptotic translation as $k \to 0$ this brings an interesting result. Let us consider first the $(\underline{\mathbf{R}}|\underline{\mathbf{R}})$-translation. From Eq. (8.2.4) this type of translation of the renormalized coefficients can be performed as

$$\begin{aligned}
\hat{\underline{C}}_{n'}^{m'} &= k^{n'} \sum_{n=n'+1}^{\infty} \sum_{m=-n}^{n} (s|r)_{n-n',n'n}^{m-m',m'm} k^{n-n'} \underline{R}_{n-n'}^{m-m'}(\mathbf{t}) k^{-n} \underline{C}_n^m + \underline{C}_{n'}^{m'} +\\
&\quad k^{n'} \sum_{n=0}^{n'-1} \sum_{m=-n}^{n} (s|r)_{n'-n,n'n}^{m-m',m'm} k^{n'-n} \underline{R}_{n'-n}^{m-m'}(\mathbf{t}) k^{-n} \underline{C}_n^m\\
&= \sum_{n=n'+1}^{\infty} \sum_{m=-n}^{n} (s|r)_{n-n',n'n}^{m-m',m'm} \underline{R}_{n-n'}^{m-m'}(\mathbf{t}) \underline{C}_n^m + \underline{C}_{n'}^{m'} + O(k^2)\\
&= \hat{\underline{C}}_{n'}^{(1)m'} + \underline{C}_{n'}^{m'} + O(k^2).
\end{aligned} \quad (8.2.20)$$

For the (S|S)-translation we have

$$\begin{aligned}\hat{\underline{C}}_{n'}^{m'} &= k^{-n'} \sum_{n=n'}^{\infty} \sum_{m=-n}^{n} (s|r)_{n-n',n'n}^{m-m',m'm} k^{n-n'} \underline{R}_{n-n'}^{m-m'}(\mathbf{t}) k^{n} \underline{C}_{n}^{m} + \\ &\quad k^{-n'} \sum_{n=0}^{n'-1} \sum_{m=-n}^{n} (s|r)_{n'-n,n'n}^{m-m',m'm} k^{n'-n} \underline{R}_{n'-n}^{m-m'}(\mathbf{t}) k^{n} \underline{C}_{n}^{m} \\ &= \sum_{n=0}^{n'-1} \sum_{m=-n}^{n} (s|r)_{n'-n,n'n}^{m-m',m'm} \underline{R}_{n'-n}^{m-m'}(\mathbf{t}) \underline{C}_{n}^{m} + \underline{C}_{n}^{m} + O(k^2) \\ &= \hat{\underline{C}}_{n'}^{(2)m'} + \underline{C}_{n}^{m} + O(k^2). \end{aligned} \tag{8.2.21}$$

Thus, the (R|R) and (S|S) translations of the renormalized vector $\underline{\mathbf{C}}$ can be performed using Eq. (8.2.7), where we use the corresponding Toeplitz matrices with coefficients

$$\underline{\xi}_{n}^{m}(\mathbf{t}) = \frac{\eta_{n}^{m}}{(2n+1)!!} t^{n} Y_{n}^{m}\left(\frac{\mathbf{t}}{t}\right), \qquad \underline{\zeta}_{n}^{m}(\mathbf{t}) = (-1)^{n+m} \underline{\xi}_{n}^{m}(\mathbf{t}). \tag{8.2.22}$$

Note that these functions are regular elementary solutions of the Laplace equation in spherical coordinates.

Similar consideration can be given to the (S|R) translation. Here we have

$$\begin{aligned}\hat{\underline{C}}_{n'}^{m'} &= k^{n'} \sum_{n=0}^{\infty} \sum_{m=-n}^{n} (s|r)_{n+n',n'n}^{m-m',m'm} k^{-n'-n} \underline{S}_{n+n'}^{m-m'}(\mathbf{t}) k^{-n} \underline{C}_{n}^{m} \\ &= \sum_{n=0}^{\infty} \sum_{m=-n}^{n} (s|r)_{n+n',n'n}^{m-m',m'm} \underline{S}_{n+n'}^{m-m'}(\mathbf{t}) \underline{C}_{n}^{m}. \end{aligned} \tag{8.2.23}$$

Therefore, the (S|R) translation of the renormalized vector $\underline{\mathbf{C}}$ can be performed using Eq. (8.2.13), where the elements of the Hankel–Toepliz matrix are

$$\varpi_{n}^{m}(\mathbf{t}) = -\frac{\mathrm{i}(2n-1)!!}{t^{n+1}} \frac{1}{\eta_{n}^{m}} Y_{n}^{m}\left(\frac{\mathbf{t}}{t}\right). \tag{8.2.24}$$

These functions are singular elementary solutions of the Laplace equation in spherical coordinates. Asymptotic formula (8.2.24) has a residual term $O(k^2)$ for $n \geq 1$ and $O(k)$ for $n = 0$. To have $O(k^2)$ for this n also one need to take one more term in the expansion of the zero-order spherical Hankel function at low argument, which yields (see also Eqs. (8.2.6) and (1.62))

$$\varpi_0^0(\mathbf{t}) = \sqrt{\frac{1}{4\pi}}\left[-\frac{\mathrm{i}}{t} + k + O(k^2)\right]. \tag{8.2.25}$$

In fact, the matrix exponential can be applied in the case of renormalized coefficients. Let us rewrite Eq. (8.3.36) describing action of the differentiation operator $\mathbf{D_t}$ as

$$\hat{\mathbf{C}} = \mathbf{D_t C} = (\mathbf{D_t^{(1)}} + \mathbf{D_t^{(2)}})\mathbf{C} = \hat{\mathbf{C}}^{(1)} + \hat{\mathbf{C}}^{(2)}, \qquad \hat{\mathbf{C}}^{(1)} = \mathbf{D_t^{(1)}C}, \\ \hat{\mathbf{C}}^{(2)} = \mathbf{D_t^{(2)}C}, \tag{8.2.26}$$

or in component form:

$$\hat{C}_n^m = \hat{C}_n^{(1)m} + \hat{C}_n^{(2)m}, \tag{8.2.27}$$

$$\hat{C}_n^{(1)m} = -\frac{1}{2t}[(t_x + \mathrm{i}t_y)b_{n+1}^{-m-1}C_{n+1}^{m+1} + (t_x - \mathrm{i}t_y)b_{n+1}^{m-1}C_{n+1}^{m-1}] + \frac{t_z}{t}a_n^m C_{n+1}^m, \tag{8.2.28}$$

$$\hat{C}_n^{(2)m} = \frac{1}{2t}[(t_x + \mathrm{i}t_y)b_n^m C_{n-1}^{m+1} + (t_x - \mathrm{i}t_y)b_n^{-m}C_{n-1}^{m-1}] - \frac{t_z}{t}a_{n-1}^m C_{n-1}^m. \tag{8.2.29}$$

Now we can see that if $\mathbf{C}$ and $\hat{\mathbf{C}}$ are coefficients of the R expansion, then according to Eq. (8.2.17) we have for the renormalized coefficients:

$$\underline{\hat{C}}_n^{(1)m} = O(k^{-1}), \qquad \underline{\hat{C}}_n^{(2)m} = O(k), \qquad \underline{\hat{\mathbf{C}}} = \underline{\mathbf{D}}_\mathbf{t}^{(1)}\underline{\mathbf{C}}[1 + O(k^2)], \\ \underline{\mathbf{D}}_\mathbf{t}^{(1)} = k\mathbf{D}_\mathbf{t}^{(1)}. \tag{8.2.30}$$

Here action of $\underline{\mathbf{D}}_\mathbf{t}^{(1)}$ is described by Eq. (8.2.28) where one needs to use the renormalized coefficients $\underline{\hat{C}}_n^m$ and $\underline{C}_n^m$ instead of $\hat{C}_n^m$ and C_n^m. Thus, $(\mathbf{R}|\mathbf{R})(\mathbf{t})$ can be represented as the following matrix exponential

$$(\underline{\mathbf{R}}|\underline{\mathbf{R}})(\mathbf{t}) = \sum_{n=0}^{\infty}\frac{t^n}{n!}\underline{\mathbf{D}}_\mathbf{t}^{(1)n} + O(k^2) = e^{t\underline{\mathbf{D}}_\mathbf{t}^{(1)}} + O(k^2), \tag{8.2.31}$$

Similarly we can consider the case when $\mathbf{C}$ and $\hat{\mathbf{C}}$ are coefficients of the S expansion and obtain

$$(\underline{\mathbf{S}}|\underline{\mathbf{S}})(\mathbf{t}) = \sum_{n=0}^{\infty}\frac{t^n}{n!}\underline{\mathbf{D}}_\mathbf{t}^{(2)n} + O(k^2) = e^{t\underline{\mathbf{D}}_\mathbf{t}^{(2)}} + O(k^2), \tag{8.2.32}$$

where action of $\underline{D}_\mathbf{t}^{(2)}$ is described by Eq. (8.2.29), where one should use the renormalized coefficients $\underline{\hat{C}}_n^m$ and $\underline{C}_n^m$ instead of $\hat{C}_n^m$ and C_n^m.

While both the series (8.2.31) and (8.2.32) converge they are written now for renormalized coefficients and $(\underline{\mathbf{R}}|\underline{\mathbf{R}})(\mathbf{0}) = (\underline{\mathbf{S}}|\underline{\mathbf{S}})(\mathbf{0}) = \mathbf{I}$ for $t = 0$, but are finite non-zero matrices for $k = 0$ and $t \neq 0$. Thus translations can be performed for arbitrary t and for zero k. Therefore, for $k = 0$ we obtain expressions for translation of expansion coefficients for solution of the Laplace equation.

If these formulae are used in practice the series will be truncated. The truncation error is determined by t (or more precisely by the ratio of the translation distance to the characteristic scale of the domain a, t/a). For larger t the number of terms to keep can be in the order of $O(p)$, where p^2 is the number of the expansion coefficients to translate. In this case the direct use of the series results in a $O(p^3)$ translation algorithm. On the other hand, translations using Toeplitz matrices require $O(p^2 \log p)$ operations and, therefore, is a faster method for translation of the renormalized coefficients.

We can conclude now that the efficiency of low-frequency translation depends on the size of the domain a, where the initial and translated expansions are applicable. The low-frequency case means $kD_* \ll 1$. If $t/a \ll 1$, use of exponential expansions is more efficient than the utilization of decompositions involving Toeplitz matrices. For $t/a \gg 1$ ($kt \ll 1$) the Toeplitz matrix representation may provide a faster translation method. Similarly for the $S|R$ translation with $t/a \ll 1$, expansion (7.3.83) can be truncated at a relatively small number of terms and is computationally efficient. For $t/a \gg 1$ ($kt \ll 1$) we can use Toeplitz/Hankel matrix decompositions, which have asymptotic complexity $O(p^2 \log p)$. Finally, we refer the reader to Ref. [Tan03] dedicated to the analysis of efficiency of application of translation algorithms using structured matrices for the Laplace equation.

8.2.2 High frequencies

8.2.2.1 Surface delta-function

The asymptotic behavior of translation operators is related to the surface delta function, and here we consider expansion of this function over the spherical harmonics. The mathematical apparatus of the Dirac delta-function as a representative of a class of *generalized functions* is common in mathematical physics and we already used this function in connection with the Green's function, which satisfies the Helmholtz equation everywhere, except for at a singular point. Similarly, one can consider a surface delta-function $\delta(\mathbf{s}; \mathbf{s}_0)$ defined as a function of two unit vectors $\mathbf{s}$ and $\mathbf{s}_0$ which is zero for any $\mathbf{s} \neq \mathbf{s}_0$ and blows up at $\mathbf{s} = \mathbf{s}_0$, while the following relation holds:

$$Y_n^m(\mathbf{s}_0) = \int_{S_u} \delta(\mathbf{s};\mathbf{s}_0) Y_n^m(\mathbf{s}) dS(\mathbf{s}), \quad n = 0, 1, 2, \ldots, \quad m = -n, \ldots, n. \quad (8.2.33)$$

Once we introduce the delta-function, we can interpret the integral as a functional scalar product. Expansion of a surface function over the basis of spherical harmonics is then given by Eq. (2.1.65) and we have

$$\delta(\mathbf{s};\mathbf{s}_0) = \sum_{n=0}^{\infty} \sum_{m=-n}^{n} Y_n^{-m}(\mathbf{s}_0) Y_n^m(\mathbf{s}). \quad (8.2.34)$$

Obviously, these series should diverge at $\mathbf{s} = \mathbf{s}_0$, while they should converge to zero at $\mathbf{s} \neq \mathbf{s}_0$.

We now note that using the addition theorem (2.1.72) we can collapse the inner sum and represent the delta-function as a sum of the Legendre polynomials:

$$\delta(\mathbf{s};\mathbf{s}_0) = \frac{1}{4\pi} \sum_{n=0}^{\infty} (2n+1) P_n(\mathbf{s}_0 \cdot \mathbf{s}). \quad (8.2.35)$$

We can also consider the approximation of the delta-function by finite sums or by a sequence of functions, which in the limit yield the delta function. This can be done in various ways. Here we consider finite surface functions defined as

$$\delta^{(p)}(\mathbf{s};\mathbf{s}_0) = \sum_{n=0}^{p-1} \sum_{m=-n}^{n} Y_n^{-m}(\mathbf{s}_0) Y_n^m(\mathbf{s}) = \frac{1}{4\pi} \sum_{n=0}^{p-1} (2n+1) P_n(\mathbf{s}_0 \cdot \mathbf{s}). \quad (8.2.36)$$

These functions are finite. Moreover, since this is a regular expansion over the spherical basis functions, we can see that the "truncated delta-function" satisfies

$$\int_{S_u} \delta^{(p)}(\mathbf{s};\mathbf{s}_0) Y_n^m(\mathbf{s}) dS(\mathbf{s})$$

$$= \begin{cases} Y_n^m(\mathbf{s}_0), & n = 0, 1, 2, \ldots, p-1, m = -n, \ldots, n, \\ 0, & n = p, p+1, \ldots, m = -n, \ldots, n. \end{cases} \quad (8.2.37)$$

So it is an analog of the delta-function in the space of band-limited functions. This can also be interpreted that $\delta^{(p)}(\mathbf{s};\mathbf{s}_0)$ is the kernal of a *low-pass spherical filter*. Indeed, if we have a surface function $\Psi(\mathbf{s}) \in L^2(S_u)$:

$$\Psi(\mathbf{s}) = \sum_{n=0}^{\infty} \sum_{m=-n}^{n} C_n^m Y_n^m(\mathbf{s})$$

then due to orthonormality of the basis of the spherical harmonics we have

$$\int_{S_u} \delta^{(p)}(\mathbf{s};\mathbf{s}_0)\Psi(\mathbf{s})dS(\mathbf{s}) = \Psi^{(p)}(\mathbf{s}_0) = \sum_{n=0}^{p-1}\sum_{m=-n}^{n} C_n^m Y_n^m(\mathbf{s}_0). \qquad (8.2.38)$$

It is also noticeable that $\delta^{(p)}(\mathbf{s},\mathbf{s}_0)$ can be expressed in a closed explicit form, due to the Christoffel–Darboux summation formula for the Legendre polynomials (e.g. see [AS64]):

$$\sum_{n=0}^{p-1}(2n+1)P_n(\mu)P_n(\mu') = \frac{p}{\mu-\mu'}[P_p(\mu)P_{p-1}(\mu') - P_{p-1}(\mu)P_p(\mu')]. \qquad (8.2.39)$$

If we set here $\mu' = 1$ and use the fact that $P_n(1) = 1$ we obtain

$$\sum_{n=0}^{p-1}(2n+1)P_n(\mu) = p\frac{P_{p-1}(\mu) - P_p(\mu)}{1-\mu}. \qquad (8.2.40)$$

So

$$\delta^{(p)}(\mathbf{s};\mathbf{s}_0) = \frac{p}{4\pi}\frac{P_{p-1}(\mathbf{s}\cdot\mathbf{s}_0) - P_p(\mathbf{s}\cdot\mathbf{s}_0)}{1-(\mathbf{s}\cdot\mathbf{s}_0)}. \qquad (8.2.41)$$

Note that $\delta^{(p)}(\mathbf{s}_0;\mathbf{s}_0)$ is finite, since for $\mathbf{s}\to\mathbf{s}_0$ we have

$$\lim_{\epsilon\to 0}\frac{P_{p-1}(1-\epsilon) - P_p(1-\epsilon)}{\epsilon} = P_p'(1) - P_{p-1}'(1)$$

$$= \frac{p(p+1)}{2} - \frac{p(p-1)}{2} = p, \qquad (8.2.42)$$

where $P_p'(1) = \frac{1}{2}p(p+1)P_p(1) = \frac{1}{2}p(p+1)$ follows from the Legendre differential equation for $P_n(\mu)$. This yields

$$\delta^{(p)}(\mathbf{s}_0;\mathbf{s}_0) = \frac{p^2}{4\pi}. \qquad (8.2.43)$$

As $p\to\infty$ sum (8.2.36) diverges not only at $\mathbf{s} = \mathbf{s}_0$, but for other values of $\mathbf{s}$ too (e.g. for $\mathbf{s} = -\mathbf{s}_0$ we have $\delta^{(p)}(-\mathbf{s}_0;\mathbf{s}_0) = (-1)^p p/4\pi$). Therefore, extension of the class of functions to generalized functions, such as the Dirac delta-function $\delta(\mathbf{s};\mathbf{s}_0)$ cannot be done immediately using $\delta^{(p)}(\mathbf{s};\mathbf{s}_0)$. However, we can consider other approximations of Dirac's delta by using infinitely differentiable functions, e.g.

$$\delta_{(\gamma)}(\mathbf{s};\mathbf{s}_0) = \frac{1-\gamma^2}{4\pi(1-2\gamma\mathbf{s}\cdot\mathbf{s}_0+\gamma^2)^{3/2}}, \qquad \gamma<1, \quad \gamma\to 1. \qquad (8.2.44)$$

In the sense that

$$\lim_{\gamma \to 1} \delta_{(\gamma)}(\mathbf{s}; \mathbf{s}_0) = 0, \qquad \mathbf{s} \cdot \mathbf{s}_0 \neq 1, \tag{8.2.45}$$

and

$$\lim_{\gamma \to 1} \int_{S_u} \delta_{(\gamma)}(\mathbf{s}; \mathbf{s}_0) F(\mathbf{s}) dS = F(\mathbf{s}_0), \tag{8.2.46}$$

where $F(\mathbf{s})$ is a continuous surface function from $L^2(S_u)$. For $|\gamma| < 1$ the generating function for the Legendre polynomials is

$$(1 - 2\gamma \mathbf{s} \cdot \mathbf{s}_0 + \gamma^2)^{-1/2} = \sum_{n=0}^{\infty} P_n(\mathbf{s} \cdot \mathbf{s}_0) \gamma^n. \tag{8.2.47}$$

Since this a convergent power series with the radius of convergence $|\gamma| = 1$ it can be differentiated. Hence,

$$\begin{aligned} \delta_{(\gamma)}(\mathbf{s}; \mathbf{s}_0) &= \frac{1}{4\pi} \left[(1 - 2\gamma \mathbf{s} \cdot \mathbf{s}_0 + \gamma^2)^{-1/2} + 2\gamma \frac{\partial}{\partial \gamma} (1 - 2\gamma \mathbf{s} \cdot \mathbf{s}_0 + \gamma^2)^{-1/2} \right] \\ &= \frac{1}{4\pi} \sum_{n=0}^{\infty} (2n+1) P_n(\mathbf{s} \cdot \mathbf{s}_0) \gamma^n = \sum_{n=0}^{\infty} \sum_{m=-n}^{n} Y_n^m(\mathbf{s}_0) Y_n^{-m}(\mathbf{s}) \gamma^n, \end{aligned} \tag{8.2.48}$$

where the last equality holds due to the addition theorem. So if $F(\mathbf{s})$ in Eq. (8.2.46) is expanded in a series of surface harmonics (we can consider just one harmonic $Y_{n'}^{m'}(\mathbf{s})$) the limiting transition as $\gamma \to 1$ can be naturally performed after integration:

$$\begin{aligned} &\lim_{\gamma \to 1} \int_{S_u} \sum_{n=0}^{\infty} \sum_{m=-n}^{n} Y_n^m(\mathbf{s}_0) Y_n^{-m}(\mathbf{s}) \gamma^n Y_{n'}^{m'}(\mathbf{s}) d\mathbf{s} \\ &= \lim_{\gamma \to 1} \sum_{n=0}^{\infty} \sum_{m=-n}^{n} \gamma^n Y_n^m(\mathbf{s}_0) \int_{S_u} Y_n^{-m}(\mathbf{s}) Y_{n'}^{m'}(\mathbf{s}) d\mathbf{s} = \lim_{\gamma \to 1} \gamma^{n'} Y_{n'}^{m'}(\mathbf{s}_0) \\ &= Y_{n'}^{m'}(\mathbf{s}_0). \end{aligned} \tag{8.2.49}$$

Now we can consider an intermediate p-truncated function

$$\delta_{(\gamma)}^{(p)}(\mathbf{s}; \mathbf{s}_0) = \frac{1}{4\pi} \sum_{n=0}^{p-1} (2n+1) P_n(\mathbf{s}; \mathbf{s}_0) \gamma^n, \quad |\gamma| < 1, \quad p < \infty, \tag{8.2.50}$$

such that both limits with respect to γ and p exist and

$$\delta_{(\gamma)}(\mathbf{s};\mathbf{s}_0) = \delta_{(\gamma)}^{(\infty)}(\mathbf{s};\mathbf{s}_0) = \lim_{p\to\infty} \delta_{(\gamma)}^{(p)}(\mathbf{s};\mathbf{s}_0),$$
$$\delta^{(p)}(\mathbf{s};\mathbf{s}_0) = \delta_{(1)}^{(p)}(\mathbf{s};\mathbf{s}_0) = \lim_{\gamma\to 1} \delta_{(\gamma)}^{(p)}(\mathbf{s};\mathbf{s}_0). \tag{8.2.51}$$

The Dirac delta-function can then be considered as the limit

$$\delta(\mathbf{s};\mathbf{s}_0) = \lim_{\gamma\to 1}\lim_{p\to\infty} \delta_{(\gamma)}^{(p)}(\mathbf{s};\mathbf{s}_0). \tag{8.2.52}$$

8.2.2.2 Principal term of the S|R translation

At high frequencies characterized as $kt \gg 1$ we can derive some useful asymptotics that reduce the cost of translations. First we note that at high frequencies the spherical Hankel functions can be represented as

$$h_n(kt) = \mathrm{i}^{-n} h_0(kt)\left[1 + O\left(\frac{n}{kt}\right)\right]. \tag{8.2.53}$$

Therefore, the truncated singular kernel (8.1.45) has the following behavior for $kt \gg p$:

$$\Lambda_S^{(p)}(\mathbf{t};\mathbf{s}) = h_0(kt)\sum_{n=0}^{p-1}(2n+1)P_n\left(\frac{\mathbf{t}\cdot\mathbf{s}}{t}\right)\left[1 + O\left(\frac{p}{kt}\right)\right]$$
$$= 4\pi h_0(kt)\delta^{(p)}\left(\frac{\mathbf{t}}{t};\mathbf{s}\right)\left[1 + O\left(\frac{p}{kt}\right)\right] \tag{8.2.54}$$

Substituting this expression into Eq. (7.1.58), we can determine the asymptotics of the elements of the (**S**|**R**) translation matrix

$$(S|R)_{nn'}^{mm'}(\mathbf{t}) = 4\pi\mathrm{i}^{n-n'}h_0(kt)\int_{S_u}\delta^{(n+n'+1)}\left(\frac{\mathbf{t}}{t};\mathbf{s}\right)$$
$$\times\left[1 + O\left(\frac{n+n'}{kt}\right)\right]Y_{n'}^{m'}(\mathbf{s})Y_n^{-m}(\mathbf{s})\mathrm{d}S$$
$$= 4\pi\mathrm{i}^{n-n'}h_0(kt)Y_{n'}^{m'}\left(\frac{\mathbf{t}}{t}\right)Y_n^{-m}\left(\frac{\mathbf{t}}{t}\right)\left[1 + O\left(\frac{n+n'}{kt}\right)\right]. \tag{8.2.55}$$

Since this holds for any finite n and n' we can write this in the matrix form using notation for the Kronecker product (8.1.35)

$$(\mathbf{S}|\mathbf{R})^{(p,p')}(\mathbf{t}) = 4\pi h_0(kt)\left[\mathbf{iY}^{(p)}\left(\frac{\mathbf{t}}{t}\right)\otimes\overline{\mathbf{iY}^{(p')}\left(\frac{\mathbf{t}}{t}\right)} + O\left(\frac{p+p'}{kt}\right)\right]. \tag{8.2.56}$$

It is interesting to note that this relation can also be obtained using recurrence relations for the translation coefficients (see Chapter 3). Indeed, all translation coefficients are generated by recurrences and symmetries (Eqs. (3.2.49), (3.2.50), (3.2.66), (3.2.78), and (3.2.79)) from initial values (3.2.5), which for high frequencies, Eq. (8.2.53), turns into

$$(S|R)_{n'0}^{m'0}(\mathbf{t}) = \sqrt{4\pi}(-1)^{n'} S_{n'}^{-m'}(\mathbf{t}) = \sqrt{4\pi} h_0(kt)(-1)^{n'} \mathrm{i}^{-n} Y_{n'}^{-m'}\left(\frac{\mathbf{t}}{t}\right). \quad (8.2.57)$$

Functions

$$Y_{nn'}^{mm'} = \mathrm{i}^{n-n'} Y_{n'}^{m'}\left(\frac{\mathbf{t}}{t}\right) Y_n^{-m}\left(\frac{\mathbf{t}}{t}\right), \quad (8.2.58)$$

satisfy all the recurrences for translation coefficients $(S|R)_{nn'}^{mm'}(\mathbf{t})$ *exactly*. This proves Eq. (8.2.56) in another way, which does not employ the delta-function. We also note the following interesting formal relation that follows from Eqs. (8.2.56) and (7.3.83), expanded using Eq. (8.2.53):

$$\sum_{n=0}^{\infty} (2n+1) P_n(-\mathrm{i}\mathbf{D}_t) = 4\pi \mathrm{i}\mathbf{Y}\left(\frac{\mathbf{t}}{t}\right) \otimes \overline{\mathrm{i}\mathbf{Y}\left(\frac{\mathbf{t}}{t}\right)}. \quad (8.2.59)$$

However, one should be aware that this relation is obtained formally by first applying limit as $k \to \infty$ and then as $p \to \infty$ to the truncated expansions. The change in the order of these limits leads to a different result.

For coaxial translations we have

$$Y_n^m(\mathbf{i}_z) = Y_n^m\left(\frac{\mathbf{t}}{t}\right)\Big|_{\theta_t=0} = \sqrt{\frac{2n+1}{4\pi}} \delta_{m0}. \quad (8.2.60)$$

This shows that the asymptotics of the elements of the coaxial translation matrix are

$$(S|R)_{nn'}^{m}(t) = \mathrm{i}^{n-n'} h_0(kt) \sqrt{(2n+1)(2n'+1)} \left[\delta_{m0} + O\left(\frac{n+n'}{kt}\right)\right]. \quad (8.2.61)$$

We can see that the principal term of the high-frequency asymptotics describes transformation of zonal harmonics only, while transform of other surface modes is due to the higher order terms. If $\mathbf{t} = t\mathbf{i}_z$ the operator $\mathbf{D}_t$ transforms into $\mathbf{D}_z$, so a series similar to Eq. (8.2.59) can be written in this case as well. We can also use notation (7.1.65) to represent the coaxial translation operator as

$$(\mathbf{S}|\mathbf{R})_{(\mathrm{coax})}^{(p,p')}(t) = 2h_0(kt)\left[\mathrm{i}\underline{\mathbf{P}}(1) \otimes \overline{\mathrm{i}\underline{\mathbf{P}}(1)} + O\left(\frac{p+p'}{kt}\right)\right], \quad (8.2.62)$$

$$(\mathbf{S}|\mathbf{R})^{(p,p')(m)}_{(\mathrm{coax})}(t) = 2h_0(kt)\left[\mathbf{i}\underline{\mathbf{P}}^m(1)\otimes\overline{\mathbf{i}\underline{\mathbf{P}}^m(1)} + O\left(\frac{p+p'}{kt}\right)\right]. \quad (8.2.63)$$

These expressions yield the following formal identities

$$\sum_{n=0}^{\infty}(2n+1)P_n(-\mathbf{i}\mathbf{D}_z) = 2\mathbf{i}\underline{\mathbf{P}}(1)\otimes\overline{\mathbf{i}\underline{\mathbf{P}}(1)}, \quad (8.2.64)$$

$$\sum_{n=0}^{\infty}(2n+1)P_n(-\mathbf{i}\mathbf{D}_z^{(m)}) = 2\mathbf{i}\underline{\mathbf{P}}^m(1)\otimes\overline{\mathbf{i}\underline{\mathbf{P}}^m(1)}, \quad (8.2.65)$$

where again we should understand that these identities are the results of $\lim_{p\to\infty}\lim_{k\to\infty}$, which is different from $\lim_{k\to\infty}\lim_{p\to\infty}$. Note that

$$\underline{\mathbf{P}}^m(1) = \mathbf{0}, \qquad m \neq 0. \quad (8.2.66)$$

The fact that the translation matrix can be expressed as the Kronecker product of two vectors yields the fastest possible algorithm of translation, which is an $O(p^2)$ algorithm. Indeed, components of the $S|R$ translated vector of expansion coefficients are

$$\begin{aligned}\hat{C}_n^{(p)m} &= \sum_{n'=0}^{p'-1}\sum_{m'=-n'}^{n'}(S|R)_{nn'}^{mm'}(\mathbf{t})C_{n'}^{m'} \\ &= 4\pi h_0(kt)\mathrm{i}^n Y_n^{-m}\left(\frac{\mathbf{t}}{t}\right)\sum_{n'=0}^{p'-1}\sum_{m'=-n'}^{n'}\mathrm{i}^{-n'}Y_{n'}^{m'}\left(\frac{\mathbf{t}}{t}\right)C_{n'}^{m'} \\ &= 4\pi Q h_0(kt)\mathrm{i}^n Y_n^{-m}\left(\frac{\mathbf{t}}{t}\right). \end{aligned} \quad (8.2.67)$$

Computation of the sum Q here requires $O(p^2)$ operations and can be treated as a dot product of vectors $\mathbf{iY}$ and $\mathbf{C}$. Then we can see that $\hat{\mathbf{C}}$ is proportional to $\mathbf{iY}$ and for computation of this vector we need from $\mathbf{C}$ only the constant Q. Multiplication of a p^2 vector by constant takes $O(p^2)$ operations.

It is amazing that for coaxial translation we need only $O(p)$ operations, since all the mth subspaces of vector $\hat{\mathbf{C}}$ are zero, except of $m = 0$. This also shows that the principal term of the high-frequency asymptotics provides only the axisymmetric field with the axis of symmetry directed along the translation vector. In fact, this is a plane wave from the direction of the multipole source.

So we can see that all the information about $\mathbf{C}$ at asymptotically high frequencies is compressed into the constant Q. This has a clear physical interpretation. Indeed, in the absence of diffraction (which is the case in the high-frequency limit) we can consider a multipole source within the framework of the ray theory. In this case there is no difference between a point multipole source or an "effective" monopole source of certain intensity (e.g. stars in the sky). The intensity of this "effective" monopole source is a function of the spherical polar angles in the reference frame connected with the source. So the point receiver can "see" only the "side" (its intensity at given angles (θ, φ)) of the source which is faced toward the receiver and cannot see the other "side" of the source (the intensity at different angles (θ, φ)). This also follows from the principal term of the far-field expansion of an arbitrary scattered field (2.3.22). If the receiver is not a point then higher order terms might be required to provide sufficient accuracy of the expansion in the receiver domain.

As we could see the translated field is axisymmetric for the coaxial translation, and as soon as the axis $\mathbf{i}_z$ can be selected in the direction of translation vector $\mathbf{t}$ the field near the receiver will be axisymmetric with respect to the axis $\mathbf{t}$. This shows that the Kronecker product of the spherical harmonics in Eq. (8.2.59), in fact, represent forward and inverse rotation transform of the system of coordinates. So, from the rotational–coaxial translation decomposition (8.2.23) and Eqs. (8.2.59) and (8.2.64), we have

$$\mathbf{Rot}^{-1}\left(Q\left(\frac{\mathbf{t}}{t}\right)\right)\left[\mathbf{i}\underline{\mathbf{P}}(1)\otimes\overline{\mathbf{i}\underline{\mathbf{P}}(1)}\right]\mathbf{Rot}\left(Q\left(\frac{\mathbf{t}}{t}\right)\right) = 2\pi\mathbf{i}\mathbf{Y}\left(\frac{\mathbf{t}}{t}\right)\otimes\overline{\mathbf{i}\mathbf{Y}\left(\frac{\mathbf{t}}{t}\right)}, \quad (8.2.68)$$

This relation can also be obtained from Eq. (7.1.77).

8.2.2.3 Non-uniform and uniform asymptotic expansions

The above consideration shows that the principal term of high-frequency expansion provides a *non-uniform* asymptotic expansion, which, first, is valid only in a *small* domain near the center of expansion. Second, the distance from a source point to the center of the initial S expansion is *small* compared to the translation distance, since there is a loss of information about the actual angle at which the true source is located. While the translation formulae in this case are simple, this situation is not acceptable for the application of the FMM, where the sizes of the domains mentioned above are not small, and, in fact, are comparable with the translation distance. In this case some special methods are needed to build a *uniform* expansion with respect to large parameter kt. By uniform expansions we mean that expansion is valid with some accuracy (say $O((kt)^{-1})$) in the domain of expansion comparable with the translation distance.

Understanding of what happens at high frequencies comes from the analysis of the expansion and translation errors due to truncation of infinite series, which is presented in Chapter 9. There we found, that the truncation numbers p' and p for the S and R series should be based on the radius of the sphere enclosing the singularities (sources), a', and the radius of sphere containing the evaluation domain, a (containing the receivers). These truncation numbers are restricted as $p' > ka'$, $p > ka$. Therefore, the residual term of the asymptotic expansion (8.2.56) is small only if $a + a' \ll t$. In case $a + a' \sim t$, we cannot use the asymptotics, simply because the series are infinite (or truncated with large p) and the residual term is comparable with the principal term. This means that a more careful consideration should be given to the asymptotic expansion, which takes into account that for large kt terms of type n/kt for $n \sim p$ cannot be neglected.

We note also that asymptotic series can be convergent or *divergent*. For example, the exponential integral, $E_n(z)$, can be expanded into the following asymptotic series as $|z| \to \infty$ [AS64]:

$$E_n(z) = \int_1^{\infty} \xi^{-n} e^{-\xi z} \, d\xi$$
$$\sim \frac{e^{-z}}{z}\left\{1 - \frac{n}{z} + \frac{n(n+1)}{z^2} - \frac{n(n+1)(n+2)}{z^3} + \cdots\right\}, \qquad (8.2.69)$$
$$\mathrm{Re}(z) > 0, \quad n = 0, 1, \ldots$$

While for $|z| \gg n$ a few first terms of the expansion provide a good approximation for the function, one cannot take an infinite number of terms since the series diverges. Another point is that the series are not applicable for computation of $E_n(z)$ for $n > |z|$, while for say $n \sim |z|/2$ the several first terms behave as a geometric progression and decay exponentially, so again some approximation can be found by taking a few first terms.

A similar thing happens for asymptotic expansion of the coefficients $(S|R)_{nn'}^m(t)$ for large kt. Despite the fact that the series of type (8.2.69) here are finite, the structure of the asymptotics is the same. For these functions several asymptotic regions can be identified, such as $(n + n') \ll kt$, $(n + n') \sim kt$, and $(n + n') \gg kt$. The most complex asymptotic analysis relates to the so-called *transition region*, $(n + n') \sim kt$. We note that the spherical Hankel functions can be considered as a particular case of coefficients $(S|R)_{nn'}^m(t)$, Eq. (3.2.103). Even an analysis of asymptotic behavior of the Hankel functions in the transition region is complex enough (e.g. [Wat58]). Despite our consideration that this type of analysis is useful and should be performed in terms of a search of efficient asymptotic representation of the translation operators, it is obvious that it goes beyond the scope of this book and is a subject of special study. Below we present a simpler case, which provides an expansion of functions $(S|R)_{nn'}^{m}(t)$ into a finite series of type (8.2.69), and, therefore, can be used as an asymptotic formula for computation of $(S|R)_{nn'}^{m}(t)$ for $(n+n') \ll kt$. Of course, the advantage here is not the computation of coefficients $(S|R)_{nn'}^{m}(t)$ which can be computed efficiently using the recurrence relations, but decomposition of a matrix into the sum of matrices, which are Kronecker products of two vectors and which can be multiplied by the vector of coefficients quickly. While we see the obstacles of application of high-frequency asymptotic expansions to represent the entire matrix in the range which includes $(n+n') \sim kt$ we note that asymptotic forms for smaller $(n+n')$ can be used in some other context, e.g. for *block-decomposition* of translation matrices. In this case some part of translation matrix for large kt can be replaced by its simpler asymptotic representation, which may reduce the complexity of the translation operation and be used for design of some other translation schemes.

8.2.2.4 Expansion of coaxial S|R matrix

In this section we are going to prove several theorems and lemmas, which provide *exact* representation of the elements of the coaxial $S|R$ translation matrix. These representations are interesting from a theoretical point of view and can be used to obtain various asymptotics. The main theorem here is

THEOREM 21 *Elements of the coaxial translation matrix defined by Eq. (3.2.86) can be represented in the form*

$$(S|R)_{nn'}^{m}(t) = 2\mathrm{i}^{n-n'} h_0(kt) \sum_{l=|m|}^{p-1} F_{nn'}^{(l)m}(1) \left(\frac{1}{\mathrm{i}kt} \right)^{l}, \qquad \forall p \geq n+n'+1. \quad (8.2.70)$$

where the coefficients $F_{nn'}^{(l)m}(1)$ *are related to the derivatives of the normalized associated Legendre functions* $\underline{P}_n^m(\mu)$ *defined by Eq. (1.63):*

$$F_{nn'}^{(l)m}(1) = \frac{\mathrm{d}^l}{\mathrm{d}\mu^l} [\underline{P}_{n'}^{m}(\mu) \underline{P}_n^m(\mu)] \bigg|_{\mu=1}. \quad (8.2.71)$$

PROOF. The proof of this theorem follows from the integral representation (7.1.68), theorem (8.2.83)–(8.2.85), and the fact that $\underline{P}_{n'}^{m}(\mu)\underline{P}_n^m(\mu)$ is a polynomial of degree $n + n' < p$, whose derivatives for $l = 0, \ldots, |m| - 1$ at $\mu = 1$ are zero. □

Now we have to prove two supporting lemmas and the theorem, which we refer to in the proof.

LEMMA 1 *Let n and q be non-negative integers, and $\gamma_n^{(q)}$ be coefficients defined as*

$$\gamma_n^{(q)} = \frac{2n+1}{2} \int_{-1}^{1} \mu^q P_n(\mu) \mathrm{d}\mu. \tag{8.2.72}$$

Then the following equality holds

$$\sum_{n=n'}^{q} \frac{(n+n')!}{(n-n')!} \gamma_n^{(q)} = \frac{2^{n'}(n')!q!}{(q-n')!}. \tag{8.2.73}$$

PROOF. Coefficients $\gamma_n^{(q)}$ can be treated as the coefficients of the expansion of μ^q over the orthogonal basis of the Legendre polynomials, and therefore, they satisfy relation

$$\mu^q = \sum_{n=0}^{q} \gamma_n^{(q)} P_n(\mu). \tag{8.2.74}$$

(see Eqs. (2.1.39) and (2.1.40)). Differentiating this relation n' times with respect to μ we obtain

$$\frac{q!}{(q-n')!} \mu^{q-n'} = \sum_{n=n'}^{q} \gamma_n^{(q)} P_n^{(n')}(\mu), \qquad P_n^{(n')} = \frac{\mathrm{d}^{n'} P_n}{\mathrm{d}\mu^{n'}}. \tag{8.2.75}$$

Particularly for $\mu = 1$ we have

$$\frac{q!}{(q-n')!} = \sum_{n=n'}^{q} \gamma_n^{(q)} P_n^{(n')}(1). \tag{8.2.76}$$

Derivatives of the Legendre polynomials for $\mu = 1$ can be determined using the representation (2.1.47), which can be n' times differentiated for $m = 0$:

$$P_n^{(n')}(\mu) = \sum_{n''=0}^{n} \frac{(-1)^{n''}(n+n'')!}{2^{n''}(n''!)^2(n-n'')!} (-1)^{n'}(1-\mu)^{n''-n'} \frac{(n'')!}{(n''-n')!}. \tag{8.2.77}$$

For $\mu = 1$ we obtain

$$P_n^{(n')}(1) = \frac{(n+n')!}{2^{n'}(n')!(n-n')!}. \qquad (8.2.78)$$

Substituting this into Eq. (8.2.75) we obtain the statement of the lemma. □

LEMMA 2 *Let l, n', and q be non-negative integers then*

$$\sum_{q=n'}^{l} (-1)^q \binom{l}{q} \frac{q!}{(q-n')!} = (-1)^l l! \delta_{n'l}, \qquad (8.2.79)$$

where the binomial coefficients are defined by Eq. (2.1.35) and $\delta_{n'l}$ is the Kronecker symbol.

PROOF. By definition of the binomial coefficients we have

$$\sum_{q=n'}^{l} (-1)^q \binom{l}{q} \frac{q!}{(q-n')!} = l! \sum_{q=n'}^{l} \frac{(-1)^q}{(l-q)!(q-n')!}. \qquad (8.2.80)$$

and this sum is zero for $n' > l$. For $l = n'$ the sum is $(-1)^l l!$ and for $l > n'$ we can show that it is also zero. For this purpose we denote $m = l - n' > 0$, $m' = q - n'$ to obtain

$$\sum_{q=n'}^{l} \frac{(-1)^q}{(l-q)!(q-n')!} = (-1)^{n'} \sum_{m'=0}^{m} \frac{(-1)^{m'}}{(m-m')!m!}$$

$$= (m')!(-1)^{n'} \sum_{m'=0}^{m} (-1)^{m'} \binom{m}{m'} = 0. \qquad (8.2.81)$$

The latter equality holds due to the definition of the binomial coefficients:

$$\sum_{m'=0}^{m} (-1)^{m'} \binom{m}{m'} = (1-1)^m = 0, \qquad m > 0. \qquad (8.2.82)$$

□

THEOREM 22 *Let $f^{(p)}(kt)$ be a function represented by integral*

$$f^{(p)}(kt) = \int_{-1}^{1} \Lambda_s^{(p)}(kt;\mu) F(\mu) d\mu, \qquad (8.2.83)$$

where $\Lambda_s^{(p)}(kt;\mu)$ is the function defined by Eq. (7.1.69) and $F(\mu)$ is a regular function in $-1 \le \mu \le 1$, expandable into the following Taylor series in the vicinity $\mu = 1$:

$$F(\mu) = \sum_{l=0}^{\infty} \frac{1}{l!} F^{(l)}(1)(\mu-1)^l, \qquad F^{(l)} = \frac{d^l F}{d\mu^l}, \quad l = 0, 1, \ldots, p-1. \quad (8.2.84)$$

Then

$$f^{(p)}(kt) = 2h_0(kt) \sum_{n=0}^{p-1} F^{(n)}(1)\left(\frac{1}{ikt}\right)^n. \qquad (8.2.85)$$

PROOF. Substituting expansion (8.2.84) and expression for $\Lambda_s^{(p)}(kt;\mu)$ Eq. (7.1.69), into definition of $f^{(p)}(kt)$, and taking into account that for regular functions we can exchange the integration and summation signs, we obtain

$$f^{(p)}(kt) = \sum_{l=0}^{\infty} \frac{1}{l!} F^{(l)}(1) \sum_{n=0}^{p-1} (2n+1) i^n h_n(kt) \int_{-1}^{1} P_n(\mu)(\mu-1)^l \, d\mu. \quad (8.2.86)$$

The latter integral can be evaluated using binomial expansion of $(\mu - 1)^l$:

$$\frac{2n+1}{2} \int_{-1}^{1} P_n(\mu)(\mu-1)^l \, d\mu = \sum_{q=0}^{l} (-1)^q \binom{l}{q} \gamma_n^{(q)}, \qquad (8.2.87)$$

where $\gamma_n^{(q)}$ are defined by Eq. (8.2.72). The spherical Hankel functions (2.1.82) can be represented as

$$h_n(kt) = i^{-n} h_0(kt) \sum_{n'=0}^{n} \frac{(n+n')!}{(n')!(n-n')!} \left(-\frac{1}{2ikt}\right)^{n'}. \qquad (8.2.88)$$

We substitute this expression and Eq. (8.2.87) into Eq. (8.2.86) and rearrange the summation as follows:

$$\begin{aligned}
f^{(p)}(kt) &= 2h_0(kt)\sum_{l=0}^{\infty}\frac{1}{l!}F^{(l)}(1)\sum_{n=0}^{p-1}\sum_{n'=0}^{n}\frac{(n+n')!}{(n')!(n-n')!}\left(-\frac{1}{2ikt}\right)^{n'}\\
&\quad\times\sum_{q=0}^{l}(-1)^q\binom{l}{q}\gamma_n^{(q)}\\
&= 2h_0(kt)\sum_{n'=0}^{p-1}\frac{1}{(n')!}\left(-\frac{1}{2ikt}\right)^{n'}\sum_{l=0}^{\infty}\frac{1}{l!}F^{(l)}(1)\sum_{q=n'}^{l}(-1)^q\binom{l}{q}\\
&\quad\times\sum_{n=n'}^{q}\frac{(n+n')!}{(n-n')!}\gamma_n^{(q)}\\
&= 2h_0(kt)\sum_{n'=0}^{p-1}\left(-\frac{1}{ikt}\right)^{n'}\sum_{l=0}^{\infty}\frac{1}{l!}F^{(l)}(1)\sum_{q=n'}^{l}(-1)^q\binom{l}{q}\frac{q!}{(q-n')!}\\
&= 2h_0(kt)\sum_{n'=0}^{p-1}\left(-\frac{1}{ikt}\right)^{n'}\sum_{l=0}^{\infty}(-1)^l F^{(l)}(1)\delta_{n'l}\\
&= 2h_0(kt)\sum_{l=0}^{p-1}F^{(l)}(1)\left(\frac{1}{ikt}\right)^{l}.
\end{aligned} \tag{8.2.89}$$

Here we used the results of the lemmas proved before the theorem (see Eqs. (8.2.73) and (8.2.79)) which enable to collapse three sums in the expansion. □

We note that $\underline{P}_{n'}^{m}(\mu)\underline{P}_{n}^{m}(\mu)$ is a polynomial of μ that have the following structure (see Eqs. (2.1.47) and (7.1.63))

$$\underline{P}_{n'}^{m}(\mu)\underline{P}_{n}^{m}(\mu) = (1-\mu^2)^{|m|}G_n^{|m|}(\mu)G_{n'}^{|m|}(\mu), \tag{8.2.90}$$

where $G_n^m(\mu)$ are polynomials of degree $n-m$, which, in fact, are the normalized *Gegenbauer* polynomials (see [AS64]):

$$G_n^{|m|}(\mu) = N_n^{|m|}\sum_{q=|m|}^{n}(-2)^{-q-|m|}M_{qn}^{|m|}(1-\mu)^q, \tag{8.2.91}$$

where

$$N_n^{|m|} = \sqrt{\frac{2n+1}{2}\frac{(n-|m|)!}{(n+|m|)!}}, \qquad M_{qn}^{|m|} = \frac{(n+q)!}{(q-|m|)!(n-q)!q!}.$$

With these expressions the values of coefficients (8.2.71) can be found explicitly:

$$F_{nn'}^{(l)m}(1) = (-1)^m l! 2^{-l} N_n^{|m|} N_{n'}^{|m|} \sum_{q=|m|}^{n} \sum_{q'=|m|}^{n'} \binom{|m|}{l-q-q'} M_{qn}^{|m|} M_{q'n'}^{|m|}. \quad (8.2.92)$$

Particularly, we have

$$F_{nn'}^{(|m|)m}(1) = \frac{(-1)^{|m|}}{2^{|m|}(|m|)!} \sqrt{\frac{2n+1}{2} \frac{(n+|m|)!}{(n-|m|)!}} \sqrt{\frac{2n'+1}{2} \frac{(n'+|m|)!}{(n'-|m|)!}}. \quad (8.2.93)$$

This provides the principal term of the asymptotic expansion for fixed m, n, and n' as $kt \to \infty$ for the mth order subspace, (see Eq. (8.2.70)):

$$(S|R)_{nn'}^{m}(t) = \mathrm{i}^{n-n'} \frac{h_0(kt)}{(|m|)!} \left(-\frac{1}{2\mathrm{i}kt}\right)^{|m|} \sqrt{\frac{(2n+1)(n+|m|)!}{(n-|m|)!} \frac{(2n'+1)(n'+|m|)!}{(n'-|m|)!}}$$
$$\times [1 + O((kt)^{-1})] \quad (8.2.94)$$

Thus, the principal term of the asymptotic expansions for each subspace can be represented in a form similar to Eq. (8.2.61) via the Kronecker product of two vectors:

$$(\mathbf{S}|\mathbf{R})_{(\mathrm{coax})}^{(p,p')(m)}(t) = 2h_0(kt) \frac{1}{(|m|)!} \left(-\frac{1}{2\mathrm{i}kt}\right)^{|m|} [\mathrm{i}\underline{\mathbf{G}}^{(m)}(1) \otimes \overline{\mathrm{i}\underline{\mathbf{G}}^{(m)}(1)}$$
$$+ O((kt)^{-1})], \quad (8.2.95)$$

where p, $p' = \mathrm{const}$, $kt \to \infty$ and

$$\mathrm{i}\underline{\mathbf{G}}^{(m)}(1) = (\mathrm{i}^{|m|}\underline{G}_{|m|}^{|m|}(1), \mathrm{i}^{|m|+1}\underline{G}_{|m|+1}^{|m|}(1), \ldots)^{\mathrm{T}},$$
$$G_n^{|m|}(1) = \sqrt{\frac{(2n+1)}{2} \frac{(n+|m|)!}{(n-|m|)!}}. \quad (8.2.96)$$

We already discussed the advantages of the representation of functions via the Kronecker product for fast computations.

While more study of the errors in the use of the principal (or a few) terms of asymptotic expansions are needed, we note that as $kt \to \infty$ the truncation numbers are $p + p' - 1 < kt$, and the range of explosive growth of the spherical Hankel functions is outside the range of the truncated matrix dimensions.

8.2.2.5 R|R translation

Principal term

To obtain the principal term for the $R|R$ translation for $kt \gg 1$ we can represent the spherical Bessel functions of the first kind as

$$j_n(kt) = \frac{1}{2}[h_n(kt) + \overline{h_n(kt)}] = \frac{1}{2}[\mathrm{i}^{-n}h_0(kt) + \mathrm{i}^n\overline{h_0(kt)}]\left[1 + O\left(\frac{n}{kt}\right)\right], \tag{8.2.97}$$

where $\overline{h_n(kt)}$ are the spherical Hankel functions of the second kind, which are complex conjugate to the functions of the first kind. If we are interested only in the principal term, (7.3.82) yields

$$(\mathbf{R}|\mathbf{R})(\mathbf{t}) \sim \frac{1}{2}h_0(kt)\sum_{n=0}^{\infty}(2n+1)P_n(-\mathrm{i}\mathbf{D}_\mathbf{t}) + \frac{1}{2}\overline{h_0(kt)} \times \sum_{n=0}^{\infty}(2n+1)(-1)^n P_n(-\mathrm{i}\mathbf{D}_\mathbf{t}). \tag{8.2.98}$$

Now we note that

$$(-1)^n P_n(-\mathrm{i}\mathbf{D}_\mathbf{t}) = P_n(\mathrm{i}\mathbf{D}_\mathbf{t}) = P_n(-\mathrm{i}\mathbf{D}_{-\mathbf{t}}). \tag{8.2.99}$$

The first equality here is a common property of the Legendre polynomials and the second follows from the definition of operator $\mathbf{D}_\mathbf{t}$ (Eq. (7.3.33)). According to Eq. (8.2.59) the sums then can be represented as the Kronecker products of the basis vectors and we obtain

$$(\mathbf{R}|\mathbf{R})(\mathbf{t}) \sim 2\pi\left[h_0(kt)\mathrm{i}\mathbf{Y}\left(\frac{\mathbf{t}}{t}\right)\otimes\overline{\mathrm{i}\mathbf{Y}\left(\frac{\mathbf{t}}{t}\right)} + \overline{h_0(kt)}\mathrm{i}\mathbf{Y}\left(-\frac{\mathbf{t}}{t}\right)\otimes\overline{\mathrm{i}\mathbf{Y}\left(-\frac{\mathbf{t}}{t}\right)}\right]. \tag{8.2.100}$$

Higher order expansions of coaxial operator

A similar consideration can be given to obtain the principal term of asymptotic expansion for the coaxial translation matrix of fixed size for $kt \to \infty$, where we can use results for the $S|R$ matrix. In fact here we have a simpler case, due to the following relation between the coaxial matrix operators:

$$(\mathbf{R}|\mathbf{R})^{(m)}_{(\mathrm{coax})}(t) = \mathrm{Re}\{(\mathbf{S}|\mathbf{R})^{(m)}_{(\mathrm{coax})}(t)\},$$
$$(\mathbf{R}|\mathbf{R})_{(\mathrm{coax})}(t) = \mathrm{Re}\{(\mathbf{S}|\mathbf{R})_{(\mathrm{coax})}(t)\}. \tag{8.2.101}$$

This follows from Eqs. (3.2.99) and (3.2.100), if we note that $j_n(kt) = \mathrm{Re}\{h_n(kt)\}$. The real part of a complex number can be taken by taking half of the sum of this number and its complex conjugate. Hence, we can represent

$$(\mathbf{R}|\mathbf{R})^{(p,p')(m)}_{(\text{coax})}(t) = \tfrac{1}{2}[(\mathbf{S}|\mathbf{R})^{(p,p')(m)}_{(\text{coax})}(t) + \overline{(\mathbf{S}|\mathbf{R})^{(p,p')(m)}_{(\text{coax})}(t)}], \qquad (8.2.102)$$

and use any available decomposition of the coaxial $S|R$ matrix to obtain a decomposition of the $R|R$ matrix. Particularly, we have from Eq. (8.2.95) the principal term for the high-frequency asymptotics

$$(\mathbf{S}|\mathbf{R})^{(p,p')(m)}_{(\text{coax})}(t) = \frac{1}{(|m|)!}\left(-\frac{1}{2ikt}\right)^{|m|}[h_0(kt)\mathbf{i}\underline{\mathbf{G}}^{(m)}(1) \otimes \overline{\mathbf{i}\underline{\mathbf{G}}^{(m)}(1)}$$
$$+ (-1)^{|m|}\overline{h_0(kt)\mathbf{i}\underline{\mathbf{G}}^{(m)}(1)} \otimes \mathbf{i}\underline{\mathbf{G}}^{(m)}(1) + O((kt)^{-1})], \qquad (8.2.103)$$

where $p, p' = \text{const}$, and $kt \to \infty$.

8.3 DIAGONAL FORMS OF TRANSLATION OPERATORS

Finally we come to *another* method for fast function translations, which, in fact, is a $O(p^2)$ method. We emphasize here the word "another", since the method is based on a function representation, different from that we use considering transforms in the space of expansion coefficients over spherical basis functions. The diagonal forms of translation operators for the Helmholtz equation are due to Rokhlin [Rok93] and they are based on the representation of solutions of the Helmholtz equation via *samples* of the signature function (7.1.28).

In fact, there are strong links between the translation method using truncated reexpansion matrices, and we approach the diagonal forms in this chapter from that side, when we introduced the decomposition of truncated translation matrices via constant and diagonal matrices. It appeared that with orthogonal transforms translation matrices can be diagonalized, and if somehow extension and truncation operations could be avoided, we should be able to represent translation matrices as

$$(\mathbf{R}|\mathbf{R})(\mathbf{t}) = (\mathbf{S}|\mathbf{S})(\mathbf{t}) = \mathbf{U}\mathbf{\Lambda}_r(\mathbf{t})\mathbf{U}^{-1}, \qquad (\mathbf{S}|\mathbf{R})(\mathbf{t}) = \mathbf{U}\mathbf{\Lambda}_s(\mathbf{t})\mathbf{U}^{-1}, \quad (8.3.1)$$

which are *similarity* transforms for the matrices. An important observation here is that all translation operators are diagonalized by the *same* constant matrices $\mathbf{U}$ (which is a consequence of the fact that all these operators *commute*, and are *normal*). This observation combined with understanding of what happens in the MLFMM algorithm, where we have a sequence of translations, which decompose the $S|R$ translation operator from the center of any source box n_s to the center of any receiver box n_r according to decomposition of the translation vector $\mathbf{t}_{sr}$:

$$\mathbf{t}_{sr} = \mathbf{t}_{s,s_1} + \mathbf{t}_{s_1,s_2} + \cdots + \mathbf{t}_{s_{l-1},s_l} + \mathbf{t}_{s_l,r_l} + \mathbf{t}_{r_l,r_{l-1}} + \cdots + \mathbf{t}_{r_1,r}, \tag{8.3.2}$$

$$\begin{aligned}(\mathbf{S}|\mathbf{R})(\mathbf{t}_{sr}) = {} & (\mathbf{R}|\mathbf{R})(\mathbf{t}_{r_1,r}) \cdots (\mathbf{R}|\mathbf{R})(\mathbf{t}_{r_l,r_{l-1}})(\mathbf{S}|\mathbf{R})(\mathbf{t}_{s_l,r_l}) \\ & \times (\mathbf{S}|\mathbf{S})(\mathbf{t}_{s_{l-1},s_l}) \cdots (\mathbf{S}|\mathbf{S})(\mathbf{t}_{s_1,s_2})(\mathbf{S}|\mathbf{S})(\mathbf{t}_{s,s_1}),\end{aligned} \tag{8.3.3}$$

results in

$$(\mathbf{S}|\mathbf{R})(\mathbf{t}_{sr}) = \mathbf{U}\boldsymbol{\Lambda}_r(\mathbf{t}_{r_1,r}) \cdots \boldsymbol{\Lambda}_r(\mathbf{t}_{r_l,r_{l-1}})\boldsymbol{\Lambda}_s(\mathbf{t}_{s_l,r_l})\boldsymbol{\Lambda}_r(\mathbf{t}_{s_{l-1},s_l}) \cdots \boldsymbol{\Lambda}_r(\mathbf{t}_{s_1,s_2})\boldsymbol{\Lambda}_r(\mathbf{t}_{s,s_1})\mathbf{U}^{-1}. \tag{8.3.4}$$

Here the entire translation process is represented as a sequence of the $S|S$ translations between boxes of the source hierarchy (upward pass), followed by the $S|R$ translation for some level l, and further by the $R|R$ translations between the boxes of the receiver hierarchy (downward pass).

We can see that the similarity transforms with matrices $\mathbf{U}$ and $\mathbf{U}^{-1}$ here need be performed only once, while all the operations involve only diagonal matrix multiplication. In the method of signature function even these steps can be avoided, since we can immediately build "S expansion" or more precisely the far-field function representation via its signature function *analytically* for the most typical type of sources, such as monopoles, dipoles, or single multipoles, and even more complex functions, such as integrals over a triangular element of the surface, and use the final step of translated function evaluation via direct integration of the signature function, which has the same complexity as the evaluation of the truncated series.

There are some problems in the practical application of the method of the signature function (which is usually called a "method based on diagonal forms"). This is related to problems of error control and variation of the size of the representing vector from level to level (indeed, we can multiply only diagonal matrices of the same size without zeros on the diagonal; otherwise one can select smaller size matrices). Some problems are also known to occur for the application of the method at low frequencies. However, we should emphasize that these problems can be considered from a different point of view and the algorithm can be constructed in a way to avoid any computation of the expansion coefficients over spherical basis functions.

8.3.1 Representations using the far-field signature function

8.3.1.1 Spherical cubatures

Band unlimited functions

As we noted above in the method of signature function we are going to work in a different space, than the space of expansion coefficients. The problem of finite representation of functions holds anyway, and we should represent a function with a vector of finite length P. To do this in a correct way we can start from the final stage of the fast multipole method, when we have a problem of evaluation of function at arbitrary points in the evaluation domain.

Assume now that the signature function $\Psi(\mathbf{s})$ is known for some given target box and represents a regular function in this domain. The problem then is to evaluate integral (7.1.27) to determine potential $\psi(\mathbf{r})$. This integral is an integral over a surface of a unit sphere, and therefore, we can use cubatures of type (8.1.49), which result in the following representation of solution:

$$\psi(\mathbf{r}) = \frac{1}{4\pi} \sum_{j=0}^{N_c-1} w_j \, e^{ik\mathbf{s}_j \cdot \mathbf{r}} \Psi(\mathbf{s}_j) + \epsilon_c, \qquad \mathbf{s}_j \in S_u. \tag{8.3.5}$$

Here ϵ_c is the *error* of cubature. Indeed, for arbitrary function $\Psi(\mathbf{s})$ the integrand in Eq. (7.1.27) is a *band-unlimited* function, and there is no hope to have a finite quadrature with zero error. Let us assume that the quadrature is chosen by some good way, and the error is small enough. In this case we can see that functions

$$r_j(\mathbf{r}) = w_j \, e^{ik\mathbf{s}_j \cdot \mathbf{r}}, \qquad j = 0, \ldots, N_c - 1, \tag{8.3.6}$$

play the role of the regular basis functions $R_n^m(\mathbf{r})$, and the set of $\{\Psi(\mathbf{s}_j)\}$

$$\Psi_j = \Psi(\mathbf{s}_j), \qquad j = 0, \ldots, N_c - 1, \tag{8.3.7}$$

plays the role of the expansion coefficients C_n^m. For a fixed cubature formula ($\{w_j\}$ and $\{\mathbf{s}_j\}$ are specified), computation of functions $r_j(\mathbf{r})$ for some given $\mathbf{r}$ is a procedure of the same complexity as computation of $R_n^m(\mathbf{r})$, if $N_c \sim p^2$. Moreover, the cubature provides the values $\mathbf{s}_j$ at which the function $\Psi(\mathbf{s})$ should be sampled. So the cubature provides a solution of the first problem in the application of the signature function method. That is the problem, how to represent a continuous surface function with a finite vector.

Band limited functions

One of the methods to control the error when using diagonal translations is to limit the bandwidth of functions based on the error estimations. In this case the signature functions are approximated by the band-limited functions and regular translation kernels are replaced by the band-limited (truncated) kernels similar to the singular kernel.

Assume now that the signature function which we need to evaluate at the final stage has a bandwidth p, which is determined by the necessary number of terms in the expansion to have the maximum absolute error in computation of function $\psi(\mathbf{r})$ below ϵ for $r < a$:

$$\Psi^{(p)}(\mathbf{s}) = \sum_{n=0}^{p-1} \mathrm{i}^{-n} C_n^m Y_n^m(\mathbf{s}), \qquad \psi^{(p)}(\mathbf{r}) = \sum_{n=0}^{p-1} C_n^m j_n(kr) Y_n^m(\mathbf{s}), \tag{8.3.8}$$

where

$$|\psi(\mathbf{r}) - \psi^{(p)}(\mathbf{r})| < \epsilon, \qquad r < a.$$

We also assume that the cubature (8.3.5) is exact for functions of bandwidth N_b. Analytical integration of $\Psi^{(p)}(\mathbf{s})$ multiplied by the band-unlimited regular kernel $e^{ik\mathbf{s}\cdot\mathbf{r}}$ results in $\psi^{(p)}(\mathbf{r})$. Owing to the orthogonality of spherical harmonics, the same result can be obtained with truncated kernel $\Lambda_r^{(p)}(\mathbf{r};\mathbf{s})$ (Eq. (8.1.47)). The maximum degree n for product $\Psi^{(p)}(\mathbf{s})\Lambda_r^{(p)}(\mathbf{r};\mathbf{s})$ is $2p-2$, and thus this function has bandwidth $2p-1$. A cubature which is exact for $N_b \geq 2p-1$ will provide zero error of integration. The basis functions (8.3.6) in this case should be modified as

$$r_j^{(p)}(\mathbf{r}) = w_j \Lambda_r^{(p)}(\mathbf{r};\mathbf{s}_j), \qquad j = 0,\ldots,N_c - 1. \tag{8.3.9}$$

Note that while the evaluation of the truncated kernel does not seem to cause any difficulties formally, if each computation of the truncated kernel requires $O(p)$ operations, replacement of $e^{ik\mathbf{s}_j\cdot\mathbf{r}}$ with $\Lambda_r^{(p)}(\mathbf{r};\mathbf{s}_j)$ brings the algorithm to complexity $O(Mp^3)$, where M is the number of evaluation points. While the evaluation stage can be computationally cheaper than the translation stage due to smaller asymptotic constants, for a $O(p^2)$ algorithm this problem should be somehow resolved to claim $O(p^2)$ for the total complexity.

The problem of series summation for evaluation of the truncated kernel function $\Lambda_r^{(p)}(\mathbf{t};\mathbf{s}_j)$ during translations can be resolved, since the set of all possible $\mathbf{t}$ is finite, and this number is usually much smaller than M. Therefore, it can be precomputed and stored as we described above. Once the set of evaluation points is given, the same can be done, of course, with $\Lambda_r^{(p)}(\mathbf{r};\mathbf{s}_j)$ during the preset step. Nevertheless, this requires additional memory in the order of $MN_c = O(Mp^2)$, which can be substantial and limit the size of the problems to be solved.

Another approach is to approximate $\Lambda_r^{(p)}(\mathbf{r};\mathbf{s}_j)$ with some fast computable functions or use interpolation methods. In this way, we should note that the error of approximation should be related to the error of the method and the error of the cubature. Furthermore, it is paradoxical, but $e^{ik\mathbf{s}_j\cdot\mathbf{r}}$, perhaps, is one of the best candidates for the approximation of $\Lambda_r^{(p)}(\mathbf{r};\mathbf{s}_j)$. In this case we will have faster computable functions $r_j(\mathbf{r})$ instead of $r_j^{(p)}(\mathbf{r})$, which should provide estimates for the cubature error ϵ_c. One can also compute $e^{ik\mathbf{s}_j\cdot\mathbf{r}}$ and then apply a fast spherical filter to obtain $\Lambda_r^{(p)}(\mathbf{t};\mathbf{s}_j)$, as described later in this section.

8.3.1.2 Signature functions for multipoles

Consider now the expansion step of the FMM (5.2.14). The goal of this step is to obtain far-field representations for potentials of sources located in the source box. For most typical applications, e.g. solution of the boundary value problems via the BEM, the "sources" are monopoles and dipoles. This can be generalized for arbitrary multipoles. In the most general case the "source" is a sum of multipoles of different degrees (e.g. in multiple scattering problems, when we represent the far field of each scatterer as a sum of multipoles).

Let us derive now analytical expressions for the signature function in these cases.

Monopole

Expansion (3.2.2) provides the following representation for a monopole source located at $\mathbf{r} = \mathbf{r}_s$ near the expansion center $\mathbf{r} = \mathbf{r}_*$:

$$M_0(\mathbf{r}-\mathbf{r}_s) = G(\mathbf{r}-\mathbf{r}_s) = \frac{ik}{\sqrt{4\pi}} S_0^0(\mathbf{r}-\mathbf{r}_s)$$

$$= ik\sum_{n=0}^{\infty}\sum_{m=-n}^{n} R_n^{-m}(\mathbf{r}_s-\mathbf{r}_*)S_n^m(\mathbf{r}-\mathbf{r}_*), \qquad (8.3.10)$$

$$|\mathbf{r}-\mathbf{r}_*| > |\mathbf{r}_s-\mathbf{r}_*|$$

This shows that the coefficients of expansion over the basis S are

$$C_n^m = ikR_n^{-m}(\mathbf{r}_s-\mathbf{r}_*), \qquad n = 0,1,\ldots, \quad m = -n,\ldots,n. \qquad (8.3.11)$$

Equations (7.1.28) and (7.1.40) let us produce the following far-field signature function related to the expansion center $\mathbf{r} = \mathbf{r}_*$:

$$\Psi_{(0)}(\mathbf{s}; \mathbf{r}_s - \mathbf{r}_*) = \sum_{n=0}^{\infty} \sum_{m=-n}^{n} \mathrm{i}^{-n} C_n^m Y_n^m(\mathbf{s})$$

$$= \mathrm{i}k \sum_{n=0}^{\infty} \sum_{m=-n}^{n} \mathrm{i}^n R_n^m(-(\mathbf{r}_s - \mathbf{r}_*)) Y_n^{-m}(\mathbf{s}) = \frac{\mathrm{i}k}{4\pi} \mathrm{e}^{-\mathrm{i}k\mathbf{s}\cdot(\mathbf{r}_s - \mathbf{r}_*)}, \qquad (8.3.12)$$

where one can use plane wave expansion (2.3.6) and the symmetry of the spherical basis functions (3.2.7).

Dipole

The dipole is a multipole of degree 1, and from Eq. (8.2.24) we have

$$M_1(\mathbf{r} - \mathbf{r}_s) = -\mathbf{p}\cdot\nabla_{\mathbf{r}} G(\mathbf{r} - \mathbf{r}_s) = \mathbf{p}\cdot\nabla_{\mathbf{r}_s} G(\mathbf{r} - \mathbf{r}_s). \qquad (8.3.13)$$

where $\mathbf{p}$ is the dipole moment, while $\nabla_{\mathbf{r}}$ and $\nabla_{\mathbf{r}_s}$ are differential operators with respect to the field point $\mathbf{r}$ and the dipole source location $\mathbf{r}_s$. The latter equality enables us to determine the corresponding signature function for the dipole simply by the application of operator $\nabla_{\mathbf{r}_s}$ to the signature function for the monopole (this holds due to the linearity of transform from the space of solutions of the Helmholtz equation to the space of signature functions):

$$\Psi_{(1)}(\mathbf{s}; \mathbf{r}_s - \mathbf{r}_*) = \mathbf{p}\cdot\nabla_{\mathbf{r}_s} \Psi_{(0)}(\mathbf{s}; \mathbf{r}_s - \mathbf{r}_*) = (\mathbf{p}\cdot\mathbf{s}) \frac{k^2}{4\pi} \mathrm{e}^{-\mathrm{i}k\mathbf{s}\cdot(\mathbf{r}_s - \mathbf{r}_*)}. \qquad (8.3.14)$$

Multipoles

This method applies to multipoles of arbitrary degree $l = 1, 2, \ldots$, since $\nabla_{\mathbf{r}}$ in Eq. (2.2.24) can be replaced by $-\nabla_{\mathbf{r}_s}$ as we are concerned with the functions of $\mathbf{r} - \mathbf{r}_s$. So we have the following expression for a multipole of degree l specified by its moments $\mathbf{p}_1, \ldots, \mathbf{p}_l$:

$$\Psi_{(l)}(\mathbf{s}; \mathbf{r}_s - \mathbf{r}_*) = (\mathbf{p}_l\cdot\nabla_{\mathbf{r}_s})\cdots(\mathbf{p}_1\cdot\nabla_{\mathbf{r}_s})\Psi_{(0)}(\mathbf{s}; \mathbf{r}_s - \mathbf{r}_*)$$

$$= \frac{\mathrm{i}k}{4\pi}(-\mathrm{i}k)^l (\mathbf{p}_l\cdot\mathbf{s})\cdots(\mathbf{p}_1\cdot\mathbf{s})\mathrm{e}^{-\mathrm{i}k\mathbf{s}\cdot(\mathbf{r}_s - \mathbf{r}_*)}. \qquad (8.3.15)$$

Usually we are concerned with expansions over spherical basis functions, not multipoles, and in the case it is useful to have an analytical form for multipoles of degree $l = 0, 1, \ldots$ expanded over the spherical basis using Eqs. (2.2.26) and (2.2.23) as

$$M_l(\mathbf{r} - \mathbf{r}_s) = \sum_{n'=0}^{l} \sum_{m'=-n'}^{n'} A_{n'}^{m'} S_{n'}^{m'}(\mathbf{r} - \mathbf{r}_s), \tag{8.3.16}$$

where $A_l^{m'}$ are some given coefficients, which determine the moments of the multipole. By definition of the reexpansion matrices (Eq. (3.2.4))

$$S_{n'}^{m'}(\mathbf{r} - \mathbf{r}_s) = \sum_{n=0}^{\infty} \sum_{m=-n}^{n} (S|S)_{nn'}^{mm'}(-(\mathbf{r}_s - \mathbf{r}_*))S_n^m(\mathbf{r} - \mathbf{r}_*). \tag{8.3.17}$$

Substituting this expression into Eq. (8.3.16) and regrouping terms we obtain

$$\begin{aligned} M_l(\mathbf{r} - \mathbf{r}_s) &= \sum_{n=0}^{\infty} \sum_{m=-n}^{n} \left[\sum_{n'=0}^{l} \sum_{m'=-n'}^{n'} A_{n'}^{m'} (S|S)_{nn'}^{mm'}(-(\mathbf{r}_s - \mathbf{r}_*)) \right] S_n^m(\mathbf{r} - \mathbf{r}_*) \\ &= \sum_{n=0}^{\infty} \sum_{m=-n}^{n} C_{n(l)}^{m} S_n^m(\mathbf{r} - \mathbf{r}_*). \end{aligned} \tag{8.3.18}$$

Thus, the coefficients of the S expansion near the center $\mathbf{r} = \mathbf{r}_*$ are

$$C_{n(l)}^{m} = \sum_{n'=0}^{l} \sum_{m'=-n'}^{n'} A_{n'}^{m'} (S|S)_{nn'}^{mm'}(-(\mathbf{r}_s - \mathbf{r}_*)), \tag{8.3.19}$$

$$n = 0, 1, \ldots, \quad m = -n, \ldots, n.$$

This enables determination of the signature function (7.1.28) that can be found as follows:

$$\begin{aligned} \Psi_{(l)}(\mathbf{s}; \mathbf{r}_s - \mathbf{r}_*) &= \sum_{n'=0}^{l} \sum_{m'=-n'}^{n'} A_{n'}^{m'} \sum_{n=0}^{\infty} \sum_{m=-n}^{n} \mathrm{i}^{-n} (S|S)_{nn'}^{mm'}(-(\mathbf{r}_s - \mathbf{r}_*)) Y_n^m(\mathbf{s}) \\ &= \sum_{n'=0}^{l} \sum_{m'=-n'}^{n'} A_{n'}^{m'} \overline{\sum_{n=0}^{\infty} \sum_{m=-n}^{n} (S|S)_{n'n}^{m'm}(\mathbf{r}_s - \mathbf{r}_*) \mathrm{i}^n Y_n^{-m}(\mathbf{s})} \\ &= \sum_{n'=0}^{l} \sum_{m'=-n'}^{n'} A_{n'}^{m'} \overline{\mathrm{i}^{n'} Y_{n'}^{-m'}(\mathbf{s}) e^{\mathrm{i}k\mathbf{s}\cdot(\mathbf{r}_s - \mathbf{r}_*)}} = e^{-\mathrm{i}k\mathbf{s}\cdot(\mathbf{r}_s - \mathbf{r}_*)} \Psi_{(l)}(\mathbf{s}; 0), \\ \Psi_{(l)}(\mathbf{s}; 0) &= \sum_{n'=0}^{l} \sum_{m'=-n'}^{n'} \mathrm{i}^{-n'} A_{n'}^{m'} Y_{n'}^{m'}(\mathbf{s}). \end{aligned} \tag{8.3.20}$$

Here we used the symmetry property of the $S|S = R|R$ translation coefficients (3.2.62) and the fact that $\{\mathrm{i}^n Y_n^{-m}(\mathbf{s})\}$ is an eigenvector of the

$R|R$ translation matrix with the known eigenvalue (see Eqs. (7.3.56) and (7.3.57)).

It is not difficult to see that this relation can be easily obtained from the fact that $\Psi_{(l)}(\mathbf{s};\mathbf{0})$ is the signature function for expansion of the multipole $M_l(\mathbf{r})$ located at $\mathbf{r}_s = \mathbf{0}$ near the center $\mathbf{r}_* = 0$ (Eqs. (7.1.28), (8.3.16), and (8.3.20)). The shift of the argument in $M_l(\mathbf{r})$ as $M_l(\mathbf{r} - \mathbf{r}_s)$ means translation of this function with translation vector $\mathbf{t} = -\mathbf{r}_s$, and the shift of the expansion center from zero to $\mathbf{r}_*$ means translation with translation vector $\mathbf{t} = \mathbf{r}_*$. Since we speak here only about the S expansions, we can simply apply translation operator (7.1.50) to obtain the required signature function from $\Psi_{(l)}(\mathbf{s};0)$.

Complexity of finite representation

Note now that for multipoles of low degree l (monopoles, dipoles, etc.) sampling of the exact signature function at nodes $\mathbf{s}_j, j = 0, \ldots, N_c - 1 \sim p^2$, is computationally an inexpensive procedure, and the first step of the FMM (5.2.14) has $O(Np^2)$ complexity. However, for "sources" specified by coefficients $\{A_n^m\}$ as a sum of multipoles up to degree p computation of a sum of p^2 terms at p^2 points appears to be an $O(Np^4)$ procedure if performed directly. Obviously this complexity should not be permitted in an $O(Np^2)$ algorithm. The solution to the problem can be of two types. First, we can look at the formulation of the problem, which results in the specification of coefficients $\{A_n^m\}$ and try to reformulate it in terms of the signature function. It may happen that the signature function can be found analytically. Second, if there is no other function representation available, except via coefficients $\{A_n^m\}$ one can apply the *inverse spherical transform* described in the previous sections. Even if the Legendre transform is not fast, this will reduce the problem to $O(p^3)$ complexity for each source due to separation of variables. The FLT combined with the FFT reduces the complexity of the initial representation operation to $O(p^2 \log^2 p)$.

Another issue concerning the complexity of the first step of the FMM (5.2.14) relates to the approximation of the initial signature function by a band-limited function. In this case we need to replace the band-unlimited function $e^{-ik\mathbf{s}\cdot(\mathbf{r}_s - \mathbf{r}_*)}$ which figures in all signature functions and is associated with relative location of the source to the center of expansion, by its truncated analog. The issue one should take care about is the same as mentioned above for final evaluation step. The algorithm for computation of the truncated regular kernel should not be of complexity $O(p)$, since in this case the entire algorithm will turn into $O(Np^3)$ method. The methods to reduce this complexity can be similar to that used for the evaluation step.

8.3.2 Translation procedures

8.3.2.1 Algorithm using band-unlimited functions

The simplest and the fastest version of the signature function method in the FMM can be described by the following prescription:

- The FMM algorithm is performed for representing vectors of constant length.
- Based on error analysis and cubature, determine the truncation number for singular kernel, N_t, the length of the representing vectors N_c. The representing vectors are the vectors of samples of the signature functions at the nodes of the cubature $\mathbf{s}_j$, $j = 0, \ldots, N_c - 1$.
- Determine the number of levels of the FMM, precompute and store all diagonals of the diagonalized $S|R$ translation matrices for all possible translation vectors in the process $\mathbf{t}_n$, $\mathbf{\Lambda}_s^{(N_t)}(\mathbf{t}_n)$, which are diagonal matrices of size $N_c \times N_c$ (diagonals of $\mathbf{\Lambda}_r(\mathbf{t}_n)$ can also be precomputed and stored to speed up the process).
- For the first step of the FMM generate initial representations. In case of monopoles, dipoles, etc. use analytical expressions for the signature function.
- Apply the FMM procedure until the final summation step. In this procedure the $S|S$ and $R|R$ translations with vector $\mathbf{t}_n$ are multiplication of the diagonal matrices $\mathbf{\Lambda}_r(\mathbf{t}_n)$ by the representing vector, and the $S|R$ translations with vector $\mathbf{t}_n$ are multiplications of matrices $\mathbf{\Lambda}_s^{(N_t)}(\mathbf{t}_n)$ by the representing vectors.
- At the final summation step (5.2.21) use basis (8.3.6) to evaluate representations at the receiver points.

In this prescription $N_t = O(p)$, $N_c = O(p^2)$ and therefore, the complexity of the algorithm for N sources and $M \sim N$ receivers is $O(p^2N)$ or $O((p^2 + \alpha \log N)N)$ or $O(p^2N \log N)$, depending on the design of the FMM (implementation of data structures, optimization, and effective dimensionality of source and receivers distribution). Moreover, this perfectly fits the model of translation, which we used to study the complexity of the FMM procedures with $\nu = 1$, when the size of the representing vector is constant.

8.3.2.2 *Numerical tests and discussion*

We implemented this algorithm and compared its performance with the algorithm which uses $O(p^3)$ translation method based on rotational–coaxial translation decomposition of the reexpansion matrices (see Chapter 7). Some results of this test are shown in Figs. 8.1 and 8.2.

First, following a general methodology of the multilevel FMM implementation we determined (experimentally) truncation numbers p which we can use for the method operating in the space of expansion coefficients, the truncation number for the singular kernel N_t, and the length of the representing vectors N_c for the signature function methods. In the test implementation we used the spherical grid, which is equispaced with respect to angle φ and corresponds to the Gauss quadrature for integration with respect to angle θ, so the cubature

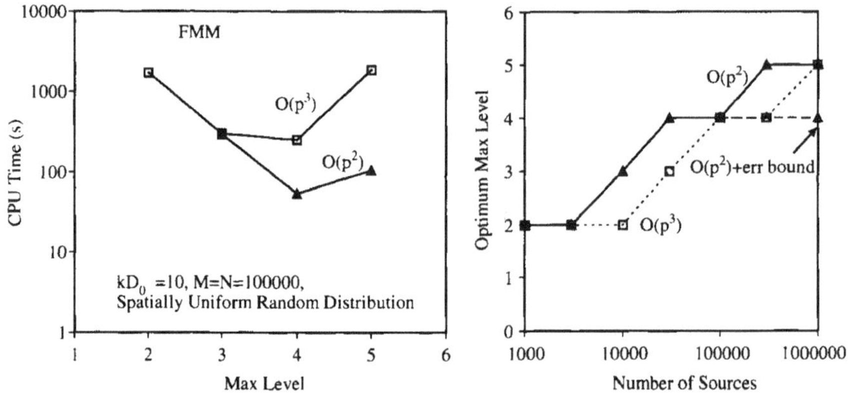

FIGURE 8.1 The plot on the left illustrates dependences of the CPU time of the FMM on the maximum level of space subdivision for the method operating in the space of expansion coefficients (the $O(p^3)$ method) for $p = 8$ and the method using band-unlimited signature function (the $O(p^2)$ method) for $N_c = 200$, $N_t = 9$. The maximum absolute error in the solution of the Helmholtz equation at 1000 representative points, ϵ, for both the methods was below 4×10^{-4}. The CPU time was measured for a PC with dual 3.2 GHz Xeon processor and 3.5 GB RAM, where only 25% of the CPU resources (time and memory) were activated. The plot on the right shows the optimum maximum level of space subdivision as a function of the number of sources ($N = M$). Here $p = 7$ for the $O(p^3)$ method, while parameters of the other methods are the same as on the left. The curves marked by the filled triangles correspond to the unconstrained optimization. For the curves marked by the triangles and squares optimization is performed based on the error constraint, $\epsilon < 10^{-3}$.

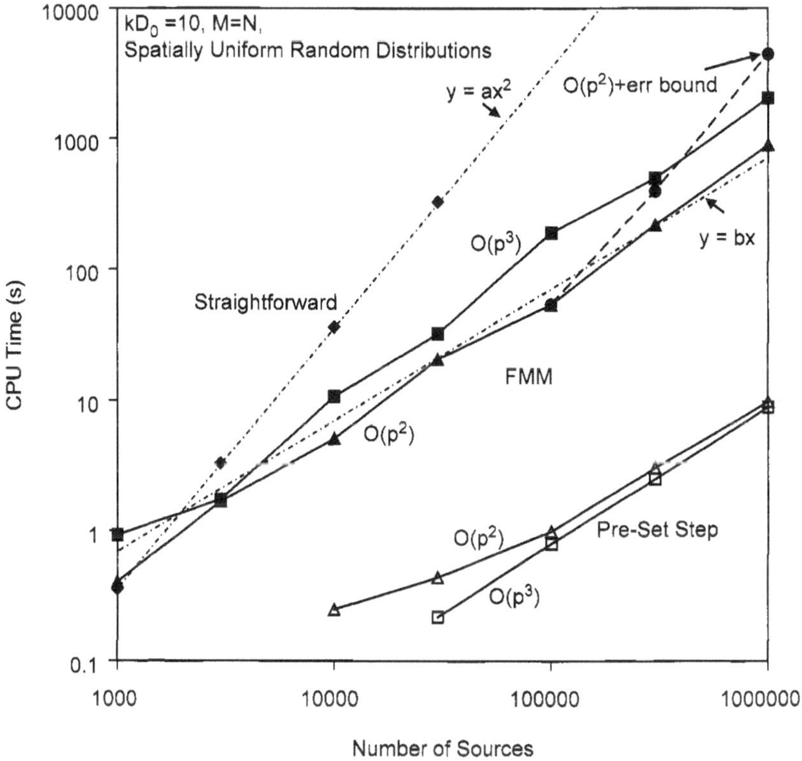

FIGURE 8.2 The CPU time vs the size of the problem for the straightforward method (the diamonds), the optimized $O(p^3)$ method (the filled squares, $p = 7$), the optimized $O(p^2)$ method without error constraints (the filled triangles, $N_c = 200$, $N_t = 9$), and the optimized $O(p^2)$ method with constraints (the filled circles, the same N_c and N_t). The curves marked by the squares and triangles show the CPU time dependencies for the pre-set step of the methods. The methods and the computational resources are the same as for Fig. 8.1.

formula, exact for functions of bandwidth N_b, was

$$\int_{S_u} \Psi^{(N_b)}(\mathbf{s})dS(\mathbf{s}) = \sum_{j=0}^{N_c-1} w_j \Psi^{(N_b)}(\mathbf{s}_j)$$

$$= \frac{2\pi}{N_b} \sum_{l=0}^{N_b/2-1} \sum_{l'=0}^{N_b-1} w_l^{(N_b/2)} \Psi^{(N_b)}(\theta_l, \varphi_{l'}), \tag{8.3.21}$$

$$\cos\theta_l = \mu_l^{(N_b/2)}, \quad \varphi_{l'} = \frac{2\pi l'}{N_b},$$

and $\mu_l^{(N_b/2)}$ and $w_l^{(N_b/2)}$ are the abscissas and weights of the Gaussian quadrature of order $N_b/2$ ($P_{N_b/2}(\mu_l^{(N_b/2)}) = 0$, $l = 0, \ldots, N_b/2$, see e.g. Ref. [AS64]). In the example for $kD_0 = 10$ illustrated below and which can be computed with level independent truncation numbers, we set $N_b = 20$, which corresponds to $N_c = 200$, which is, therefore, the length of the representing vector. The kernel truncation number was $N_t = 9$, that is slightly below $N_b/2$. For problems with $N \geq 10{,}000$ monopole sources of unit intensity with a uniform random spatial distribution over the computational domain (box of size $D_0 \times D_0 \times D_0$) resulted in the absolute maximum error in the solution of the Helmholtz equation $10^{-4} < \epsilon < 10^{-3}$. This error was determined by comparisons of the solution with straightforward summation of the sources and evaluation of the potential at 1000 randomly distributed points in the domain. The reason why we measured the error on the set of size smaller than M, which we used to have $M = N$, is because the computation of the "exact" solution using straightforward method at larger N was unaffordably large ($O(MN)$ method). Of course, this provides only a probabilistic measure of the error. We found that to obtain the same order of accuracy for the method using the reexpansion matrices the truncation number should be $p \geq 7$.

The second step is to determine the optimum maximum subdivision level, for each N, subject to the constraint that the error should not exceed 10^{-3}. This is an important step, since the speed of the algorithm substantially depends on this optimization. The left chart in Fig. 8.1 illustrates this point. Computations for the $O(p^3)$ method using reexpansion matrices are performed for $p = 8$. The optimum $l_{\max}$ value for both the methods compared in the present case is 4, and if the maximum level is not selected properly, then the CPU time for the solution of the same problem increases several times (e.g. for the $O(p^3)$ computations with $l_{\max} = 2$ or 5 are about seven times slower than the computations with $l_{\max} = 4$). The plot on the right of Fig. 8.1 shows the results of optimization for a range of N. It is seen that $l_{\max}$ for the faster algorithm is the same or larger than the same parameter for the slower method. However, we found some interesting effect for larger N (cases $N = 3 \times 10^5$ and 10^6 are displayed on the chart), that while the optimum level in terms of speed should be five for the signature function method, the error jumps drastically ($\epsilon \approx 10^{-2}$), while computation with $l_{\max} = 4$ for the same size problem provided in these cases acceptable errors ($\epsilon < 4 \times 10^{-4}$). As a result to stay within the specified error in this case we are forced to make computations in non-optimal regime, which slows down the algorithm as illustrated in Fig. 8.2.

Figure 8.2 presents a comparative performance test of the two methods considered. We can see immediately that both the methods provide faster solution of the problem than the straightforward method for problems of size $N \gtrsim 2000$. This "break-even" point, of course, depends on the algorithm implementation, but provides an idea of what size of problems can be solved faster using the FMM opposed to the direct summation. We can also see that the straightforward method is almost perfectly scaled as $O(N^2)$, while the FMM algorithm with different types of translation and function representation is scaled almost as $O(N)$ or $O(N(\log N + A))$. The computations made here are optimized in a sense that for each N we used a specific maximum level, shown in Fig. 8.2 (the chart on the right). We can also see that the use of the $O(p^2)$ translation method can reduce the computational time by several times in some range of N. However, for larger N the algorithm functions in a non-optimal regime to stay within the specified error, as discussed above. This leads to the effect, which we can see for $N = 10^6$ that the $O(p^2)$ method is slower than the $O(p^3)$ method, which performs well in terms of the error at the optimum level of space subdivision. In fact, this means that the method using band-unlimited functions is less *stable* than the method operating in the space of expansion coefficients to the *length of the chain* of translations performed by the FMM, which is proportional to $l_{\max}$.

Finally, we should note that the CPU time plotted as the FMM execution time is measured for the part of the algorithm, which can be performed many times for the same sets of sources and receivers (which is the case for iterative solution of a system of linear equations). The overhead, which need be performed only once for given data sets, in fact, is cheap (a few percent) for the way which we selected for the management of the data structure based on the minimum memory requirements (we did not use such things as precomputation of the lists of children and E_2- and E_4-neighbors for each box, but the children and neighbors were determined "on-the-fly" by bit-interleaving and set intersection routines for each box in the data hierarchies during the run time as needed). There exists an opportunity for trading memory for speed, when the lists of children and neighbors are precomputed and stored, which will decrease the run time, increase the pre-set (setting up the data structure) time, and increase the storage requirements. We also note that there is also some relatively small speed and memory overhead in the pre-set step for the signature function method, which is related to precomputation and storage of the singular kernel $\Lambda_s^{(N_t)}(\mathbf{t}_n)$, as discussed above.

8.3.2.3 Deficiencies of the signature function method

While we found that the method using band-unlimited functions that achieves the theoretical minimum for asymptotic complexities works in a certain range of kD_0 and outperforms the method using the translation matrices, we found that its capabilities are limited. The problems here are related to the error of the method. First, we found that while the increase of p for a given kD_0 increases the accuracy of computations for the $O(p^3)$ method (e.g. we could compute the solution with accuracy 10^{-7}), we could not achieve the same accuracy using the $O(p^2)$ method for any settings (variations of N_t and N_c). Second, for low kD_0 and high kD_0 the fastest method was simply inaccurate for any settings, and so cannot be used as is for computations neither for high, nor for low frequency range. Third, as discussed above, we found that we cannot increase the levels in the FMM to achieve the optimum performance in the case of use of the $O(p^2)$ method. The method looses its advantages compared to the slower one.

There are several possible reasons for the limited applicability of the method of signature functions. One reason is related to some intrinsic feature of the signature function method. The reason is related to computation of the singular kernel and the loss of the accuracy due to summation of functions of different magnitudes. Indeed, the modulus of the spherical Hankel functions $h_n(kt)$ for fixed kt grows with n (for details see Chapter 9). For $n < kt$ this growth is very slow, and there are no problems to sum up these functions with coefficients (7.1.45) and compute the singular kernel. However, for $n > kt$ the modulus $|h_n(kt)|$ grows *exponentially* with n (see Fig. 8.3).

This causes problems for low frequencies, $kt \lesssim 1$ since in this case the magnitude of spherical Hankel functions grow with n starting with $n = 1$. If for $kt = 1$ we need, say 10 terms to achieve some accuracy, ($N_t = 10$), then in the sum of modes from $n = 0$ to 9 we unavoidably lose the information about the zero mode if computations are performed with double precision arithmetic (128 bits). This may be fixed using some special methods (the simplest, but not the most efficient, is to increase the number of bits in float number representation, but this increases the length of function representation, slows down all multiplication operations, and slows down the algorithm), which we are not going to consider here, since the accuracy can be achieved by using, e.g. reexpansion matrices, and low frequency renormalization, as we discussed earlier in this chapter.

An analysis of error bounds performed in Chapter 9 shows that at higher frequencies we should not have this type of accuracy limitation.

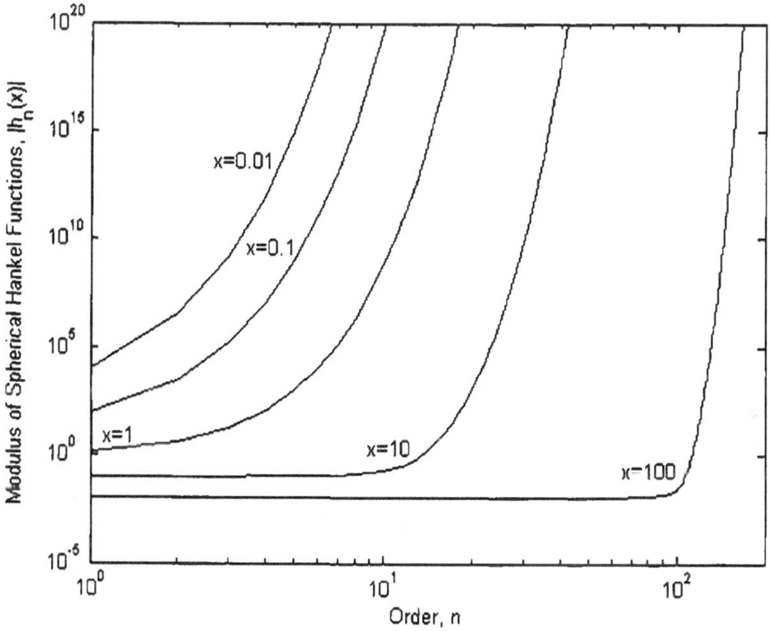

FIGURE 8.3 The modulus of spherical Hankel functions $|h_n(x)|$ as a function of order n at fixed x.

Indeed, the $S|R$ translations occur between the boxes of the same level, which can be enclosed by spheres of the same radius a, and here $t = a + b$, $b/a = \sigma > 1$. In terms of truncated vectors of coefficients and matrices, this means that for the $S|R$ translations both the initial and translated vectors can be truncated with the same truncation number p. The bandwidth of the singular kernel will then be $N_t = 2p - 1$. The truncation number for the high-frequency region is given by Eq. (9.1.41), and this yields

$$N_t = 2p - 1 \sim 2ka + \left(3\ln\frac{1}{\epsilon}\right)^{2/3}(ka)^{1/3}, \tag{8.3.22}$$

where ϵ is the maximum absolute error of approximation. Comparing this number with $kt = ka(1 + \sigma)$, we can see that

$$kt - N_t \sim ka\sigma - \left(3\ln\frac{1}{\epsilon}\right)^{2/3}(ka)^{1/3} = ka\sigma - \left[\left(\frac{3}{ka}\ln\frac{1}{\epsilon}\right)^{2/3}\right]. \tag{8.3.23}$$

This shows that for large ka we have $kt > N_t$, and even $kt - N_t \to \infty$ as $ka \to \infty$ and means that computations of the truncated singular kernel is a numerically stable procedure.

Therefore, the error observed in the FMM computations with the method of signature functions at higher frequencies has a different nature. Even a first look at the above prescription for the signature function method shows that it is different from the model, which we considered for the FMM in Chapters 5 and 6. In the prescription the size of the representing vectors is constant, but according to the model it should change, since longer series are required for coarser levels, if kD_0 is above the breakdown level l_{b*} (see Eq. (8.1.5)). This more or less is consistent with kD_0 for which numerical tests show a catastrophic loss in the accuracy of computations. The reason here is that this algorithm is written in terms of *band-unlimited* functions.

There is a strong correlation between the error of integration, the bandwidth of the integrand, and the number of nodes. Obviously, if we use band-unlimited signature functions $\Psi(\mathbf{s})$, while the errors and truncation numbers are derived for functions $\psi^{(p)}(\mathbf{r})$ approximating solution of $\psi(\mathbf{r})$ there is no guarantee that errors in the computation of $\Psi(\mathbf{s})$ due to the use of finite representations should not cause larger errors for the evaluation of integrals. Moreover, it is clear that the multiplication of a band-limited function $\Psi^{(p)}(\mathbf{s})$, by the truncated singular kernel $\Lambda_s^{(N_t)}(\mathbf{t};\mathbf{s})$ results in a function of bandwidth $p + N_t - 1$ for which only the first $N_t - p$ modes are correct. In the chain of translations the $S|R$ translation is followed by the multiplication of the signature function by the regular kernel $\Lambda_r(\mathbf{t};\mathbf{s})$. If we take it as it is, i.e. the band unlimited function, then the incorrect modes immediately affect all the spectrum, since multiplication of two spherical harmonics of degree n and n' produces a function with modes from $|n - n'|$ to $n + n'$. If we truncate the regular kernel with some N_t' then only the first $N_t - p - N_t' + 2$ modes are correct, etc. So the error of truncation propagates towards the low modes for each translation. It is obvious, that if not immediately, but after one or two $R|R$ translations the error will affect all the modes. This may not be catastrophic if the error is small (and we observed in the above computational example that there is a case when the algorithm produces some acceptable error), or may make the procedure completely incorrect for larger errors.

To avoid such situations we can introduce some *error control* operations, which leave us only with the modes that guarantee proper computation of necessary number of modes for $\Psi(\mathbf{s})$ and provide those modes to obtain $\psi^{(p)}(\mathbf{r})$ (which is guaranteed by some theory to be close to $\psi(\mathbf{r})$).

There are a number of publications related to this issue (e.g. see [CMSJ01]). The error control operations, which are known as *interpolation* and *anterpolation*, increase or decrease the size of the representing vector to keep the information in a band-limited function during the translation process. We combine these operations under one term *spherical filtering*, or *filtering of spherical harmonics*, which can also be found in the literature [JCA97]. We consider below how to apply and perform fast filtering to be consistent with the $O(p^2)$ translation complexity of the signature method. In the context of the comparison of the two approaches to translation we should mention, that the translation with filtering is a property and advantage of the method operating in the functional space of expansion coefficients (indeed we always have only a limited number of modes due to the truncation of the vector of coefficients, and below we will show this formally).

8.3.2.4 *Algorithms using band-limited functions*

To proceed further let us be more specific with the cubature formula used. While more compact cubatures can be used (e.g. Eq. (8.3.21)), let us accept Eq. (8.1.57), which is a good model for study. Furthermore, we have already obtained some results with this cubature, it can be used for the FST, and the equiangular grid is convenient for up- or down-sampling of surface functions.

Relation to translations in the space of coefficients

Let us now consider how translations in the space of coefficients can be mapped to translations in the space of signature functions. For this purpose we turn our attention to decomposition (8.1.59) of the general $p^2 \times p'^2$ translation matrix, which is exact. It provides a model and relations between the kernel truncation number N_t, the length of representing vector N_c, and the bandwidth of the signature function. This formula applies to the p'-truncated vector of coefficients $\mathbf{C}^{(p')}$ and produces a p-truncated vector of coefficients $\hat{\mathbf{C}}^{(p)}$. In the signature function method vector $\boldsymbol{\Psi}^{(N'_c)} = \{\Psi^{(p')}(\mathbf{s}'_{j'})\}$ of size N'_c is used to represent function $\Psi^{(p')}(\mathbf{s})$ instead of $\mathbf{C}^{(p')}$ and vector $\boldsymbol{\Psi}^{(N_c)} = \{\Psi^{(p)}(\mathbf{s}_j)\}$ is used to represent function $\hat{\Psi}^{(p)}(\mathbf{s})$ instead of $\hat{\mathbf{C}}^{(p)}$. While formally one may relate N'_c to p' and N_c to p, which seems natural, in the sequence of translations we may skip some unnecessary operations, if we have *a priori* knowledge on what we are going to do with the result of translation. We have such knowledge in the FMM algorithm, since we know (from the error analysis) the bandwidth of all functions in the algorithm. As soon as we know this, we can relate N'_c to p' and p. Now we note that $\mathbf{\Psi}^{(N'_c)}$ can be treated as the result of the operation:

$$\mathbf{\Psi}^{(N'_c)} = \mathbf{Is}^{(N'_b/2)}\mathbf{Pr}^{(N'_b/2,p')}\overline{\mathbf{i}^{(p')}}\mathbf{C}^{(p')},$$
$$N'_b = 2(p+p'-1), \quad N'_c = (N'_b)^2. \tag{8.3.24}$$

which is a result of the spherical transform of $\mathbf{C}^{(p')}$. Equation (8.1.59) then provides

$$\hat{\mathbf{C}}^{(p)} = (\mathbf{E}|\mathbf{F})^{(p,p')}(\mathbf{t})\mathbf{C}^{(p')} = \mathbf{i}^{(p)}\mathbf{Pr}^{(p,N'_b/2)}\mathbf{Fs}^{(N'_b/2)}\mathbf{\Lambda}^{(p+p'-1)}(\mathbf{t})\mathbf{\Psi}^{(N'_c)}. \tag{8.3.25}$$

This is not that we want to compute, and we convert this to $\mathbf{\Psi}^{(N_c)}$, which can be found similarly to $\mathbf{\Psi}^{(N'_c)}$ (Eq. (8.3.24)). In Eq. (8.3.24) we should reset parameters to take into account that $\mathbf{\Psi}^{(N_c)}$ will further be translated to obtain samples of the function $\tilde{\Psi}^{(p'')}(\mathbf{s})$ which is a function with bandwidth p''. Combining these results we obtain the following representation of translation operation in the space of samples of signature functions:

$$\hat{\mathbf{\Psi}}^{(N_c)} = \mathbf{Is}^{(N_b/2)}\mathbf{Pd}_{(p)}^{(N_b/2,N'_b/2)}\mathbf{Fs}^{(N'_b/2)}\mathbf{\Lambda}^{(p+p'-1)}(\mathbf{t})\mathbf{\Psi}^{(N'_c)},$$
$$N_b = 2(p+p''-1), \qquad N_c = N_b^2, \tag{8.3.26}$$

where we took into account that $\overline{\mathbf{i}^{(p)}}\mathbf{i}^{(p)}$ is the identity matrix, and denoted

$$\mathbf{Pr}^{(N_b/2,p)}\mathbf{Pr}^{(p,N'_b/2)} = \mathbf{Pd}_{(p)}^{(N_b/2,N'_b/2)}. \tag{8.3.27}$$

Operator $\mathbf{Pd}_{(p)}^{(q,q')}$ can be called the *zero-padding* operator. From the definition of extension and truncation operators (see Eq. (8.1.44) and around) this operator is a matrix of size $q^2 \times q'^2$ ($q \geq p$, $q' \geq p'$) whose elements are zero but for p^2 ones, which are elements of a $p^2 \times p^2$ identity matrix block at the upper left corner of the matrix. This can also be seen as an $p^2 \times p^2$ identity matrix *extended* to size $q^2 \times q'^2$ by putting zeros into the new positions. This operator acts on a vector of length q'^2 by taking its first p^2 components, and putting them into the first p^2 positions of the resulting vector of size q^2, all other components of which are zeros.

With this notation we can represent translation operator $(\mathrm{E}|\mathrm{F})^{(p,p')}(\mathbf{t})$ which acts in the space of samples of signature functions (so this is a matrix of size $N_c \times N'_c$)

$$\hat{\mathbf{\Psi}}^{(N_c)} = (\mathrm{E}|\mathrm{F})^{(p,p')}(\mathbf{t})\mathbf{\Psi}^{(N'_c)}, \qquad (\mathrm{E}|\mathrm{F})^{(p,p')}(\mathbf{t}) = \mathbf{Ft}^{(p,2p'+p-2)}\mathbf{\Lambda}^{(p+p'-1)}(\mathbf{t}), \tag{8.3.28}$$

where

$$\mathbf{Ft}^{(p,2p'+p-2)} = \mathbf{Is}^{(N_b/2)}\mathbf{Pd}_{(p)}^{(N_b/2,N_b'/2)}\mathbf{Fs}^{(N_b'/2)} \qquad (8.3.29)$$

is a matrix of size $N_c \times N_c'$ which performs filtering of function of bandwidth $2p'+p-2$ sampled on a spherical grid of size N_c' to produce a function of bandwidth p sampled on a spherical grid of size N_c. Indeed, as follows from Eq. (8.3.24) $\boldsymbol{\Psi}^{(N_c')}$ is a vector of samples of a function of bandwidth p'. Being multiplied by the diagonal matrix $\boldsymbol{\Lambda}^{(p+p'-1)}$, which can be thought of as samples of function of bandwidth $p'+p-1$ on the same grid it produces a vector of samples of a function of bandwidth $2p'+p-2$ (the rule here is to sum the bandwidths and subtract 1). The obtained vector of samples should be resampled on a new grid (or on the same grid if it does not change) to represent a function of bandwidth p.

Trading speed for memory

A natural question which may appear is, whether translation in the space of coefficients is optimal in terms of minimum size of the grid at which we should keep information about the signature function of bandwidth p?. This is an important question from the point of view of optimization of memory and speed of computations. Equation (7.1.41) shows that to recover function $\psi^{(p)}(\mathbf{r})$ it is sufficient to keep information about its signature function $\Psi^{(p)}(\mathbf{s})$ that enables exact integration of functions of bandwidth $N_b - 2p - 1$. In case we use the equiangular grid it results in a grid of size $N_c = N_b^2 = (2p-1)^2$. Comparing this with the sizes of grids that are needed for translations in the space of coefficients, we can see that the grid can be smaller, and so the representing vector can be of smaller length.

Consider now the translation process, e.g. for the $S|S$ translation of $\Psi^{(p')}(\mathbf{s})$, which is the signature function for $\psi^{(p')}(\mathbf{r})$ in the S basis centered at some point $\mathbf{r} = -\mathbf{t}$. Function $\psi^{(p')}(\mathbf{r})$ is an approximation of the true solution $\psi(\mathbf{r})$ in some domain. Assume that the bandwidth of the translation kernel is N_t. When we multiply this kernel by the function $\Psi^{(p')}(\mathbf{s})$ we generate a new function $\hat{\Psi}^{(N_t+p'-1)}(\mathbf{s}) = \Lambda_r^{(N_t)}(\mathbf{t};\mathbf{s})\Psi^{(p')}(\mathbf{s})$ of bandwidth $N_t + p' - 1$. If it is sufficient to approximate the function $\psi(\mathbf{r})$ with $\hat{\psi}^{(p)}(\mathbf{r})$ in the new domain, then we have

$$\hat{\psi}^{(p)}(\mathbf{r}) = \frac{1}{4\pi}\int_{S_u} \Lambda_s^{(p)}(\mathbf{r};\mathbf{s})\Lambda_r^{(N_t)}(\mathbf{t};\mathbf{s})\Psi^{(p')}(\mathbf{s})dS(\mathbf{s}). \qquad (8.3.30)$$

The bandwidth of the function $\Lambda_s^{(p)}(\mathbf{r};\mathbf{s})\Psi^{(p')}(\mathbf{s})$ is $p+p'-1$, and, therefore, to compute this integral exactly with minimum possible N_t we need to set

$N_t = p + p' - 1$. The cubature should then be exact for functions of bandwidth $N_b = 2p + 2p' - 3$. It is sufficient to have the equiangular cubature with the number of nodes $N_c'' = N_b^2 = 4(p + p' - 1)^2$, which we choose.

So the process of translation of the signature function, sampled on minimal grids can be described as a composition of the following operators (matrices):

$$\hat{\boldsymbol{\Psi}}^{(N_c)} = (\mathrm{E}|\mathrm{F})^{(p,p')}(\mathbf{t})\boldsymbol{\Psi}^{(N_c')},$$
$$(\mathrm{E}|\mathrm{F})^{(p,p')}(\mathbf{t}) = (\mathbf{Dn})^{(N_c,N_c'')}\boldsymbol{\Lambda}^{(p+p'-1)}(\mathbf{t})(\mathbf{Up})^{(N_c'',N_c')}. \tag{8.3.31}$$

Here $(\mathbf{Up})^{(N_c'',N_c')}$ is the operator of *upsampling*, or *interpolation*, which is a matrix of size $N_c'' \times N_c'$, $\boldsymbol{\Lambda}^{(p+p'-1)}$ is the diagonal translation matrix of size $N_c'' \times N_c''$ with the $(p + p' - 1)$-truncated kernel function, and $(\mathbf{Dn})^{(N_c,N_c'')}$ is the operator of *downsampling*, or *anterpolation*, that is a matrix of size $N_c' \times N_c''$. For the minimal equiangular grid we have $N_c = (2p - 1)^2$, $N_c' = (2p' - 1)^2$, $N_c'' = (2p' + 2p - 2)^2$.

Now we can see that the price to pay for smaller size of the representing vectors is two dense matrix multiplications, instead of one multiplication using the filtering matrix (8.3.29). It is a trade-off issue between the speed and memory, whether to use this decomposition with smaller length of the representing vector and perform two dense matrix multiplications for each translation, or to use the fact that translations are used in a sequence, and specify the representing vector as a function sampled on finer grid of size N_c''.

Since translation operations are performed in a sequence, we can return to larger size representing vectors and Eq. (8.3.28), where the filtering matrix is

$$\mathbf{Ft}^{(p,2p'+p-2)} = (\mathbf{Up})^{(N_c,N_c'')}(\mathbf{Dn})^{(N_c'',N_c')} \tag{8.3.32}$$

and we changed notations to the notations of equations (8.3.28) and (8.3.29). For the equiangular grid we have $N_c' = (2p + p' - 2)^2$, $N_c = (2p + p'' - 2)^2$, and $N_c'' = (2p - 1)^2$, where p'' is the bandwidth of the signature function for the next step.

Algorithm with filtering

We note that in this example we selected some fixed cubature, which can be changed, and the sizes of the grid, which were uniquely defined by the prescribed truncation numbers. The requirement for the size of grid can be relaxed; so the grids can be of larger size, sufficient for an exact integration of a function of bandwidth $2p' + 2p - 2$ for translation of a function of bandwidth p' to function of bandwidth p. This relaxation of the requirement enables, first, to use grids with some convenient number of nodes (e.g. some power of 2) for application of fast filters, and, second, to determine some acceptable size, such that an increase or decrease in the truncation number leads to a convenient refining or coarsening of the grid. Indeed, while in the model of the FMM we assumed that p is either constant (for levels $l \leq l_{b*}$) or doubles (for levels $l > l_{b*}$), the actual behavior of p for high frequencies is a bit more complex (see Chapter 9). Nevertheless, with the relaxed requirement it is not difficult to find some grid sizes when an increase in p results in the increase in N_c not more than four times, since the dependence of p on the size of the domain a can be majorated in the high-frequency region by a linear dependence, $p = \alpha ka$ (see Chapter 9).

Now let us return to the FMM. Dropping some details, such as how one should compute and store diagonal matrices, we can formulate the prescription as follows:

- Based on the error analysis, determine the bandwidth, p_l, for each level of subdivision $l = l_{\min,\ldots,}l_{\max}$.
- Based on the cubature formula and filtering methods, determine the lengths of the representing vectors $N_c^{(l)}$. These lengths should be sufficient to integrate a function of bandwidth $2p_l + 2p_{l-1} - 2$ at levels $l = l_{\min} + 1, \ldots, l_{\max}$, and functions of bandwidth $4p_{l_{\min}} - 2$ for level $l_{\min}$.
- Generate initial representations based on analytical formulae, where the band unlimited regular kernel is replaced by its truncated analog on the grid associated with level $l_{\max}$.
- Perform the FMM procedure until the final summation step. In this procedure each translation from level l' to l $(l, l' = l_{\min}, \ldots, l_{\max})$ is a multiplication of a diagonal band-limited matrix with $(p_l + p_{l'} - 1)$-truncated kernel, followed by the filtering operation, which enables filtering of function of bandwidth $2p_{l'} + p_l - 2$ sampled on the grid of size $N_c^{(l)}$ to function of bandwidth p_l sampled on the grid of size $N_c^{(l')}$. This is consistent with Eq. (8.3.28).
- At the final summation step use basis (8.3.9) with the truncation number appropriate for the maximum level to evaluate representations at the receiver points.

This algorithm shows that the complexity of translation operation will be determined by the complexity of the filtering procedure, since this involves multiplication of a dense $N_c^{(l)} \times N_c^{(l')}$ matrix by the vector. While some fast methods are available, the filtering operation is much more expensive than multiplication with the diagonal matrix. A couple of remarks can also be made concerning the length of the vectors. Even assuming that it is sufficient to perform a procedure with vectors of a fixed size, the length N_c for equiangular grid appears to be $(4p-2)^2 \sim 16p^2$, which is 16 times larger than for translations using matrix representations. This size can be reduced by the selection of some other cubature (e.g. if the Legendre nodes are used for integration over angle θ, we have size $\sim 8p^2$), but in any case this cannot be smaller than $\sim 4p^2$. As discussed earlier one can trade speed for memory by using a decomposition of the filtering matrix, (8.3.29) or (8.3.32).

Finally, we note that in the context of application of filters the efficiency of the matrix-based translation methods in the space of expansion coefficients should be revised. Indeed, there is no advantage to use the method of signature function (which in addition requires larger memory) if the filtering is performed with the same complexity as translation in the space of coefficients (say using the forward and inverse FST). So, while the algorithm with band-unlimited functions is really fast, the complexity of the method increases when the error control functions are introduced. In any case, the translation in the signature function method can be done with complexity $O(p^2 \log^2 p)$ (in the case of the use of the FST and maybe faster with other filtering methods), and despite some problems which are to be resolved concerning the initial representation and the final evaluation kernels (again by trading memory for complexity), this method remains to be among the fastest translation methods.

8.3.3 Fast spherical filtering

Understanding that at least one fast spherical filtering method is available via decomposition (8.3.29) and the FST, let us consider some other methods with asymptotic complexity $O(p^2 \log^2 p)$ or better, which may bring an advantage for the signature function method. It is obvious, that to do this we should not use decomposition (8.3.29), but use the advantage of Eq. (8.3.28) which presents filtering as one operator.

8.3.3.1 Integral representation of spherical filter

Let us return to representation of translation operators in the space of continuous functions. As we found, it is sufficient to perform translation with the regular or singular translation kernel of bandwidth $p + p' - 1$, where p' and p are the bandwidths of the initial and translated functions determined by the error bounds. Unfiltered translation of the signature function $\Psi^{(p')}(\mathbf{s})$ can then be represented as

$$\hat{\Psi}^{(2p'+p-2)}(\mathbf{s}) = \Lambda^{(p+p'-1)}(\mathbf{t};\mathbf{s}')\Psi^{(p')}(\mathbf{s}'). \tag{8.3.33}$$

Using integral representations of translation coefficients (e.g. Eq. (7.1.58)) and signature function (7.1.28) we can show that the translation with truncated reexpansion matrix is equivalent to the following operation in the space of signature functions:

$$\begin{aligned}
\hat{\Psi}^{(p)}(\mathbf{s}) &= \sum_{n=0}^{p-1}\sum_{m=-n}^{n} \mathrm{i}^{-n} \sum_{n'=0}^{p'-1}\sum_{m'=-n'}^{n'} (E|F)_{nn'}^{mm'}(\mathbf{t}) C_{n'}^{m'} Y_n^m(\mathbf{s}) \\
&= \int_{S_u} \Lambda^{(p+p'-1)}(\mathbf{t};\mathbf{s}') \left[\sum_{n'=0}^{p'-1}\sum_{m'=-n'}^{n'} \mathrm{i}^{-n'} C_{n'}^{m'} Y_{n'}^{m'}(\mathbf{s}')\right] \\
&\quad \times \left[\sum_{n=0}^{p-1}\sum_{m=-n}^{n} Y_n^{-m}(\mathbf{s}')Y_n^m(\mathbf{s})\right] \mathrm{d}S(\mathbf{s}') \\
&= \int_{S_u} \Lambda^{(p+p'-1)}(\mathbf{t};\mathbf{s}')\Psi^{(p')}(\mathbf{s}')\delta^{(p)}(\mathbf{s}';\mathbf{s})\mathrm{d}S(\mathbf{s}') \\
&= \int_{S_u} \hat{\Psi}^{(2p'+p-2)}(\mathbf{s}')\delta^{(p)}(\mathbf{s}';\mathbf{s})\mathrm{d}S(\mathbf{s}'),
\end{aligned} \tag{8.3.34}$$

where we used definitions (8.2.36) and (8.3.33). In fact, we could obtain the final result immediately, if we recall Eq. (8.2.37), and interpretation of the band-limited function $\delta^{(p)}(\mathbf{s}';\mathbf{s})$ as a kernel of a low-pass spherical filter. Nevertheless, it is useful to have this derivation which proves the fact that we obtained by some discrete procedures from a different viewpoint and provides a different representation of the filtering matrix, via kernel $\delta^{(p)}(\mathbf{s}';\mathbf{s})$.

Indeed, using a cubature exact for functions of bandwidth $2p'+2p-3$ or, to simplify, for $N_b' = 2(p'+p-1)$, with N_c' nodes, we can represent the latter equation in the form

$$\hat{\Psi}^{(p)}(\mathbf{s}_j) = \sum_{j'=0}^{N_c'-1} w_{j'} \hat{\Psi}^{(2p'+p-2)}(\mathbf{s}_{j'}')\delta^{(p)}(\mathbf{s}_{j'}';\mathbf{s}_j), \qquad j = 0,\ldots,N_c-1, \tag{8.3.35}$$

where $\mathbf{s}_j$ are the nodes of the grid at which function $\hat{\Psi}^{(p)}(\mathbf{s})$ should be sampled.

Function $\delta^{(p)}(\mathbf{s}';\mathbf{s})$ has different representations. We proceed with the following decomposition based on the Christoffel–Darboux formula for the normalized associated Legendre functions (7.1.63):

$$(\mu' - \mu) \sum_{n=|m|}^{p-1} \underline{P}_n^m(\mu')\underline{P}_n^m(\mu) = a_{p-1}^m[\underline{P}_p^m(\mu')\underline{P}_{p-1}^m(\mu) - \underline{P}_{p-1}^m(\mu')\underline{P}_p^m(\mu)], \tag{8.3.36}$$

where a_{p-1}^m are the differentiation coefficients (2.2.8). With this equation the truncated surface delta-function (8.2.36) can be represented as

$$\begin{aligned}\delta^{(p)}(\mathbf{s}';\mathbf{s}) &= \sum_{m=-(p-1)}^{p-1} \sum_{n=|m|}^{p-1} Y_n^{-m}(\mathbf{s}')Y_n^m(\mathbf{s}) \\ &= \frac{1}{2\pi} \sum_{m=-(p-1)}^{p-1} e^{im(\varphi-\varphi')} a_{p-1}^m \left[\frac{\underline{P}_p^m(\mu')\underline{P}_{p-1}^m(\mu)}{\mu' - \mu} - \frac{\underline{P}_{p-1}^m(\mu')\underline{P}_p^m(\mu)}{\mu' - \mu} \right],\end{aligned} \tag{8.3.37}$$

where (θ, φ) and (θ', φ') are the spherical polar angles of surface points $\mathbf{s}$ and $\mathbf{s}'$, respectively, and $\mu = \cos\theta$, $\mu' = \cos\theta'$.

8.3.3.2 Separation of variables

This representation for fast multiplication of the filtering matrix by a vector is valid for an arbitrary grid. However, in order to take an advantage for the use of the FFT with respect to the angular variable φ it is almost obligatory to use equiangular grid with respect to this angle. The first step of filtering algorithm then is to perform the FFT with respect to angle φ. The weights of the cubature in the case of equiangular grid with respect to φ depend only on the θ coordinates of the cubature nodes θ_j' (e.g. see dependence (8.1.55) of the weights for the equiangular grid). Furthermore, we will consider below only grids which are *Cartesian products* of 1D grids with respect to φ and θ, so $\mathbf{s}_j = (\theta_l, \varphi_q)$ and Eq. (8.1.58) provides an example of such a type of surface grid. Substituting Eq. (8.3.37) into Eq. (8.3.35) we obtain the following formula for filtering on grids of the described type:

$$
\begin{aligned}
\hat{\Psi}^{(p)}(\mu_l, \varphi_q) = & \sum_{m=-(p-1)}^{p-1} \frac{e^{im\varphi_q}}{2\pi} a_{p-1}^m \\
& \times \sum_{l'=0}^{N_l'-1} w_{l'} \left[\frac{\underline{P}_p^m(\mu_{l'}')\underline{P}_{p-1}^m(\mu_l)}{\mu_{l'}' - \mu_l} - \frac{\underline{P}_{p-1}^m(\mu_{l'}')\underline{P}_p^m(\mu_l)}{\mu_{l'}' - \mu_l} \right] \\
& \times \sum_{q'=0}^{N_q'-1} e^{-im\varphi_{q'}'} \hat{\Psi}^{(2p'+p-2)}(\mu_{l'}', \varphi_{q'}'), \\
& \mu_{l'}' = \cos \theta_{l'}', \quad \mu_l' = \cos \theta_l, \quad l = 0, \ldots, N_l - 1, \quad q = 0, \ldots, N_q - 1, \\
& N_l N_q = N_c, \quad N_l' N_q' = N_c',
\end{aligned}
\tag{8.3.38}
$$

and the correspondence between $w_{j'}$ and $w_{l'}$ follows from the mapping of indices $j' \to (l', q')$.

Let us assume now that we have the equiangular grid with respect to both the angles θ and φ which is convenient for analysis. As we found from the consideration of sizes of the grid related to the bandwidths of the integrand, the equiangular grid which enables exact integration of functions of bandwidth $N_b = 2(p + p' - 1)$, which is the case for integral (8.3.34), has parameters

$$
N_l' = N_q' = N_b = 2(p + p' - 1). \tag{8.3.39}
$$

Denote now

$$
\begin{aligned}
\hat{\Psi}^m(\mu_{l'}') &= \frac{2\pi}{N_q'} \sum_{q'=0}^{N_q'-1} e^{-im'\varphi_{q'}'} \hat{\Psi}^{(2p'+p-2)}(\mu_{l'}', \varphi_{q'}') \\
&= \frac{2\pi}{N_q'} \sum_{q'=0}^{N_q'-1} e^{-im'\varphi_{q'}'} [e^{i(p-1)\varphi_{q'}'} \hat{\Psi}^{(2p'+p-2)}(\mu_{l'}', \varphi_{q'}')],
\end{aligned}
\tag{8.3.40}
$$

$m' = 0, \ldots, N_q' - 1, \quad m = m' - (p - 1),$

If the forward and inverse *Discrete Fourier Transforms* are defined as

$$g_m = \frac{2\pi}{N}\sum_{q=0}^{N-1} e^{-im\varphi_q} f_q, \qquad f_q = \sum_{q=0}^{N-1} e^{im\varphi_q} g_m, \qquad \varphi_q = \frac{2\pi q}{N},$$
$$m, q = 0, \ldots, N-1, \qquad \{g_m\} = (\mathrm{DFT})_{(N)}\{f_q\}, \qquad \{f_q\} = (\mathrm{DFT})^{-1}_{(N)}\{g_m\}, \tag{8.3.41}$$

then Eq. (8.3.40) can be represented in the form

$$\{\hat{\Psi}^m(\mu'_{l'})\} = \{\hat{\Psi}^{(2p'+p-2)}_{m'}(\mu'_{l'})\} = (\mathrm{DFT})_{(N'_q)}\{e^{i(p-1)\varphi'_{q'}}\hat{\Psi}^{(2p'+p-2)}(\mu'_{l'}, \varphi'_{q'})\}, \tag{8.3.42}$$

where $m = m' - (p-1)$.

We now note that we need to transform only the components $m = -(p-1), \ldots, (p-1)$, or $m' = 0, \ldots, 2p-2$, while the Fourier transform provides an excessive number of points. The components, which are not needed can then be excluded from consideration, or, in other words the obtained result can be truncated. We can see that N'_l Fourier transforms should be performed to obtain all necessary $\hat{\Psi}^{(2p'+p-2)(m)}(\mu'_{l'})$ for further operations. Since $N'_l = O(p)$ and the length of each vector to which we apply the Fourier transform is $N'_q = O(p)$, then this operation if performed with the FFT will have complexity $O(p^2 \log p)$. Further we apply operation which is the second sum in Eq. (8.3.38). The result of this operation can be denoted as

$$\hat{\Phi}^m(\mu_l) = a^m_{p-1}\sum_{l'=0}^{N'_l-1} w_{l'}\left[\frac{\underline{P}^m_p(\mu'_{l'})\underline{P}^m_{p-1}(\mu_l)}{\mu'_{l'} - \mu_l} - \frac{\underline{P}^m_{p-1}(\mu'_{l'})\underline{P}^m_p(\mu_l)}{\mu'_{l'} - \mu_l}\right]\hat{\Psi}^m(\mu'_{l'}),$$
$$l = 0, \ldots, N_l - 1, \quad m = -(p-1), \ldots, p-1, \tag{8.3.43}$$

or

$$\{\hat{\Phi}^m(\mu_l)\} = (\mathrm{LF})^{(m)}_{(N_l N'_l)}\{\hat{\Psi}^m(\mu'_{l'})\}, \tag{8.3.44}$$

where via $(\mathrm{LF})^{(m)}_{(N_l N'_l)}$ we denoted the transform that we can call the *Legendre filter* (to be consistent with the terminology "Spherical filter", "Spherical transform" and "Legendre transform" used earlier), which applies to N'_l samples of some function at $\mu'_{l'} \in (-1, 1)$, $l' = 0, \ldots, N'_l - 1$ and produces N_l samples of a filtered function at some $\mu_l \in (-1, 1)$, $l = 0, \ldots, N_l - 1$. We will postpone consideration of this operation and its complexity to Section 8.3.3.3, but note here that this function is specified at all nodes of the grid where we need to have the filtered function. Now we need to perform the inverse Fourier transform to obtain $\hat{\Psi}^{(p)}(\mu_l, \varphi_q)$. We can do this by the application $N_l = 2(p + p'' - 1)$ inverse DFTs to $\{\hat{\Phi}^m(\mu_l)\}$ considered as vectors at fixed μ_l. As the size of these vectors, $2p - 1$, is smaller than $N_q = N_l = 2(p + p'' - 1)$ they should be extended to the required size by putting zeros into the new positions, since the extension operation in the space of coefficients does not change the function bandwidth. Technically, we can replace m with $m' = m + p - 1$ to have the index changing from 0 to $N_q - 1$, with the notice that

$$\hat{\Psi}^{(p)}(\mu_l, \varphi_q) = \frac{N_{q'}}{4\pi^2} \sum_{m=-(p-1)}^{p-1} e^{im\varphi_q} \hat{\Phi}^m(\mu_l)$$

$$= \frac{N_{q'}}{4\pi^2} e^{-i(p-1)\varphi_q} \sum_{m'=0}^{2p-2} e^{im'\varphi_q} \hat{\Phi}^m(\mu_l), \qquad (8.3.45)$$

Here the coefficient came from the normalization factor in Eq. (8.3.40) and can be incorporated into the DFT or the quadrature weights $w_{l'}$. We can also see that our definition of the DFT leads to the multiplication of the initial and resulting vector by some constant diagonal matrices $\{e^{i(p-1)\varphi'_{q'}}\}$ and $\{e^{-i(p-1)\varphi_q}\}$. The set of the inverse Fourier transforms can be performed with complexity $O(p^2 \log p)$, and, thus, the total complexity of the filter is $O(p^2 \log p)$ if the Legendre filtering can be performed for $O(p^2 \log p)$ operations or faster. Otherwise, the complexity will be determined by the number of operation needed to perform $2p - 1$ operations specified by Eq. (8.3.44).

8.3.3.3 Legendre filter

Decomposition of filtering matrix

Let us now consider in detail the filtering operation (Eq. (8.3.44)). As defined $(LF)^{(m)}_{(N_l N'_l)}$ is a rectangular matrix of size $N_l \times N'_l$ for any m. To simplify notations we will temporarily denote it as $\mathbf{L}^{(m)}$. The elements of this matrix are finite and there is no singularity at $\mu'_{l'} = \mu_l$, while Eq. (8.3.43) provides some uncertainty of type 0/0 or $\infty \cdot \infty$. To avoid this type of uncertainty let us, first, consider elements $L^m_{ll'}$ of this matrix corresponding to $\mu'_{l'} = \mu_l$.

This can be done easily using Eq. (8.3.36), which shows that

$$L^m_{ll'} = w_{l'} \sum_{n=|m|}^{p-1} \underline{P}^m_n(\mu'_{l'}) \underline{P}^m_n(\mu_l) = w_{l'} \sum_{n=|m|}^{p-1} [\underline{P}^m_n(\mu_l)]^2, \qquad \mu_l = \mu'_{l'}. \quad (8.3.46)$$

Since we have N_l different μ_l, and N'_l different $\mu'_{l'}$, the number of cases $\mu_l = \mu'_{l'}$ cannot exceed $\min(N_l, N_{l'})$. Therefore, we can represent $\mathbf{L}^{(m)}$ as a sum of two matrices:

$$\mathbf{L}^{(m)} = \mathbf{K}^{(m)} + \mathbf{M}^{(m)}, \tag{8.3.47}$$

where $\mathbf{M}^{(m)}$ is a sparse matrix populated with no more than $\min(N_l, N_{l'})$ non-zero elements $L_{ll'}^m$ at positions of matrix $\mathbf{L}^{(m)}$ corresponding $\mu_l = \mu'_{l'}$, and $\mathbf{K}^{(m)}$ is a dense matrix, which consists of elements of matrix $\mathbf{L}^{(m)}$ with zeros at positions where $\mu_l = \mu'_{l'}$. In case there are no situations with $\mu_l = \mu'_{l'}$ we have $\mathbf{M}^{(m)} = \mathbf{0}$, $\mathbf{K}^{(m)} = \mathbf{L}^{(m)}$. Multiplication of matrix $\mathbf{M}^{(m)}$ by an arbitrary vector is a computationally cheap operation of complexity $O(\min(N_l, N_{l'})) = O(p)$ and complexity of multiplication for $2p-1$ subspaces (all ms) and input vectors is $O(p^2)$.

Consider now the matrix $\mathbf{K}^{(m)}$. Since it has zeros in place of elements $\mu_l = \mu'_{l'}$ it can be decomposed into a sum (or difference) of two non-singular matrices $\mathbf{K}^{(m)(1)}$ and $\mathbf{K}^{(m)(2)}$:

$$\mathbf{K}^{(m)} = a_{p-1}^m (\mathbf{K}^{(m)(1)} - \mathbf{K}^{(m)(2)}), \tag{8.3.48}$$

where elements of the introduced matrices are

$$K_{ll'}^{(m)(1)} = \frac{\underline{P}_p^m(\mu_l) w_{l'} \underline{P}_{p-1}^m(\mu'_{l'})}{\mu_l - \mu'_{l'}}, \qquad K_{ll'}^{(m)(2)} = \frac{\underline{P}_{p-1}^m(\mu_l) w_{l'} \underline{P}_p^m(\mu'_{l'})}{\mu_l - \mu'_{l'}}, \tag{8.3.49}$$

$$l = 0, \ldots, N_l - 1, \quad l' = 0, \ldots, N'_l - 1.$$

It is not difficult to see that these matrices can be decomposed further as products of diagonal matrices and rectangular *Cauchy matrix*, **Ca**:

$$\mathbf{K}^{(m)(1)} = \underline{\mathbf{P}}_p^{(m)} \mathbf{Ca} \mathbf{W}' \underline{\mathbf{P}}_{p-1}^{(m)'}, \qquad \mathbf{K}^{(m)(2)} = \underline{\mathbf{P}}_{p-1}^{(m)} \mathbf{Ca} \mathbf{W}' \underline{\mathbf{P}}_p^{(m)'}, \tag{8.3.50}$$

where $\mathbf{W}'$, $\underline{\mathbf{P}}_{p-1}^{(m)'}$, and $\underline{\mathbf{P}}_p^{(m)'}$ are $N'_l \times N'_l$ diagonal matrices, with diagonal elements $w_{l'}$, $\underline{P}_{p-1}^m(\mu'_{l'})$, and $\underline{P}_p^m(\mu'_{l'})$, respectively, and $\underline{\mathbf{P}}_p^{(m)}$ and $\underline{\mathbf{P}}_{p-1}^{(m)}$ are $N_l \times N_l$ diagonal matrices with entries $\underline{P}_p^m(\mu_l)$, and $\underline{P}_{p-1}^m(\mu_l)$. The Cauchy matrix here is defined as $N'_l \times N_l$ matrix with entries

$$Ca_{ll'} = \begin{cases} (\mu_l - \mu'_{l'})^{-1}, & \mu_l \neq \mu'_{l'}, \\ 0, & \mu_l = \mu'_{l'}, \end{cases} \tag{8.3.51}$$

$$l = 0, \ldots, N_l - 1, \qquad l' = 0, \ldots, N'_l - 1.$$

It is clear now that the complexity of the Legendre filtering is related to the complexity of multiplication of the dense Cauchy matrix by an input vector, since the complexity of multiplication of diagonal matrices is $O(p)$ for a given subspace.

Fast Cauchy matrix multiplication

The decomposition also shows that filtering in any subspace m is reduced to multiplication by the same Cauchy matrix of the same size, which depends only on the grids used for sampling and integration. There are several fast algorithms for matrix–vector multiplication involving the Cauchy matrix. It is remarkable that these matrices belongs to the class of structured matrices, for which divide-and-conquer type algorithms are applicable. An *exact* multiplication of the Cauchy matrix by a vector of size $O(p)$ can be done with complexity $O(p \log^2 p)$ [VS01]. This results in $O(p^2 \log^2 p)$ total complexity of the spherical filtering for vectors of length $O(p^2)$, via the decomposition as shown above. While the algorithm for fast multiplication here is different from that used in the FLT or FST, it is interesting that the complexity is the same. For more accurate comparisons and practical values of these algorithm additional research is required.

Another algorithm which shows in practice complexity $O(p)$ for multiplication of the Cauchy matrix by a vector of size of order p is based on utilization of 1D FMM [DGR96], [YR98]. We also implemented and studied the accuracy and performance of the FMM for the Cauchy matrix multiplication and found it quite fast and accurate [GDB03]. Some issue, which may puzzle the reader in this context is that the FMM is an approximate method, and so the error of this method should somehow be related to the errors of the larger scale FMM, which we use for solution of the 3D Helmholtz equation. In fact, the error of 1D FMM in this case reaches machine precision (say double precision computations) for low truncation numbers and in this case there is *no difference* in terms of accuracy of computations between the "approximate FMM" and "exact" methods, since the error for both methods exists and is determined by the *roundoff* errors. We note that this issue is important from the point of theory of computation, since the FMM should be classified in this case as an $O(p \log p)$ method (due to increase of the truncation number with the increase in the size of the problem to stay within the error bound), but if computations are performed with finite machine precision the method is $O(p)$ as, in this case, the error is not determined by the truncation size, but by the finite number of bits which represent float numbers.

For the details of the FMM algorithm (errors, truncation numbers, optimizations, and performance) we refer the reader to the cited literature. Here we just repeat our earlier remark that in practice the speed of the method is determined not only by the asymptotic complexity, but also by the asymptotic constant. For example, it is known (and we confirm this based on our experience with the FMM) that $O(p)$ 1D FMM is slower than $O(p \log p)$ FFT, and an algorithm, which uses, say 10 FFTs can be faster than the algorithm that uses 1 FMM for computation of the same problem. To the best of our knowledge, currently no comparisons are available between exact divide-and-conquer $O(p \log^2 p)$ methods for multiplication with the Cauchy matrix and the FMM, while such a study could be interesting.

8.4 SUMMARY

In this chapter we considered algorithms which can be used for reduction of complexity of translation. First we showed some exact decompositions of the rotation, truncated coaxial translation, and truncated general translation matrices. Fast rotation algorithm has complexity $O(p^2 \log p)$ and can be used in various methods of translations. The other two algorithms relied on the translation matrix diagonalization via fast Legendre and fast spherical transforms, which can be performed with complexity $O(p^2 \log^2 p)$, and this, to the best of our knowledge, is the fastest complexity which can be currently achieved for a single translation with truncated translation matrices. Further, we considered low- and high-frequency asymptotic expansions. We showed that for very low frequencies an arbitrary translation can be performed with complexity $O(p^2 \log p)$. For higher frequencies we obtained complete asymptotic representations that enable any order asymptotic expansions for coaxial translation operators. These expansions, however, are derived for fixed sizes of the truncated translation matrices as $kt \to \infty$ and further research is needed to obtain uniform asymptotic representation of translation operators. Finally, we came to the diagonal forms of the translation operators interpreted as a method of signature function. This method achieves complexity $O(p^2)$ for translations when implemented for the class of band-unlimited functions. Theoretically this is the lowest complexity of translations. While our tests show that the method works and can be practical, its accuracy is limited at low frequencies and the algorithm is unstable for a sequence of translations. We showed that this instability can be damped by the application of the spherical filters, which control the function bandwidths. These filters, however, have $O(p^2 \log^2 p)$ or $O(p^2 \times \log p)$ complexity (the latter complexity can be achieved by using 1D FMM used for matrix–vector products involving Cauchy matrices). A comparative analysis of different filtering methods with complexities close to $O(p^2)$ should be performed in the future to show their practical value.

CHAPTER 9

Error Bounds

The numerical implementation of multipole methods relies on the availability of good error bounds. Indeed, whether one uses truncated series, truncated matrices or integral representations, that are treated numerically, errors appear, and the goal of this chapter is to provide error analysis to ensure that multipole methods are practical and that the error can be controlled. In the previous chapters we assumed that for each level of the FMM there exists some truncation number that ensures that the truncated expansions and translations are accurate. In this chapter we will provide a theoretical justification for this and establish some error bounds (though usually they overestimate actual errors), which can be used for proper design of the FMM.

The subject of error bounds for the FMM continues to be an area of active research, and there are several previous works that discuss this issue, including Refs. [Dar00, KSC99, Rah96]. While they sometimes contain some contradictory results, they are dedicated to establishing error bounds for single source expansions, and the interested reader should refer to these papers directly. Here, we establish error bounds for both this case and for expansions of multipoles, and analyze the error caused by the use of truncated translation matrices. All results obtained are illustrated by comparing the derived bounds with actual computations. Throughout this chapter we assume that $k \in \mathbb{R}$ and $k > 0$.

9.1 TRUNCATION ERRORS FOR EXPANSIONS OF MONOPOLES

Let us consider the truncation error of a monopole source expansion for the two settings shown in Fig. 9.1. In the first case the source whose

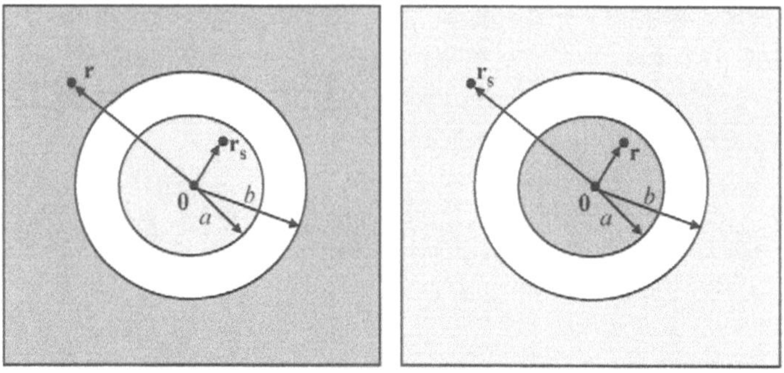

FIGURE 9.1 Schematic for evaluation of the expansion of a source located in the light gray domain ($\mathbf{r}_s$). The evaluation point ($\mathbf{r}$) is located in the dark gray area.

position is characterized by coordinates $\mathbf{r}_s$ is located inside a sphere of radius a, $|\mathbf{r}_s| < a$, and the problem here is to evaluate the error, ϵ_p, of the truncated expansion (9.2.3):

$$h_0(k|\mathbf{r}-\mathbf{r}_s|) = \sqrt{4\pi}S_0^0(\mathbf{r}-\mathbf{r}_s) = 4\pi\sum_{n=0}^{p-1}\sum_{m=-n}^{n} R_n^{-m}(\mathbf{r}_s)S_n^m(\mathbf{r}) + \epsilon_p, \tag{9.1.1}$$

$$r_s = |\mathbf{r}_s| < a, \quad r = |\mathbf{r}| > b, \quad a < b$$

for an arbitrary point $\mathbf{r}$ located outside the sphere of radius $b > a$. In the second case shown in Fig. 9.1 the problem is to evaluate the error for an arbitrary point $|\mathbf{r}| < a$ when $|\mathbf{r}_s| > b$. However, in this case the mathematical problem will be same as described by Eq. (9.1.1) since we can simply exchange symbols $\mathbf{r}_s \rightleftarrows \mathbf{r}$ and the error of series truncation will be same as in the first case. This also follows from the well-known reciprocity principle for the Helmholtz equation. Note that $h_0(k|\mathbf{r}-\mathbf{r}_s|) = [4\pi/(ik)]G(\mathbf{r}-\mathbf{r}_s)$, so while this does not influence the relative error, the absolute errors should be corrected accordingly if one considers monopoles specified in the form $G(\mathbf{r}-\mathbf{r}_s)$. The reason why we use h_0 instead of G is that $h_0(kr)$ is a function of the dimensionless parameter kr only, while respective G depends on the dimensional quantity k and dimensionless kr.

The truncation error in Eq. (9.1.1) can be expressed in the following form using the addition theorem for spherical harmonics (2.1.70):

$$\begin{aligned}\epsilon_p &= 4\pi \sum_{n=p}^{\infty} \sum_{m=-n}^{n} R_n^{-m}(\mathbf{r}_s) S_n^m(\mathbf{r}) \\ &= 4\pi \sum_{n=p}^{\infty} j_n(kr_s) h_n(kr) \sum_{m=-n}^{n} Y_n^{-m}\left(\frac{\mathbf{r}_s}{r_s}\right) Y_n^m\left(\frac{\mathbf{r}}{r}\right) \\ &= \sum_{n=p}^{\infty} (2n+1) j_n(kr_s) h_n(kr) P_n\left(\frac{\mathbf{r}_s \cdot \mathbf{r}}{r_s r}\right). \end{aligned} \tag{9.1.2}$$

The absolute value of the Legendre polynomials for an arbitrary angle between $\mathbf{r}_s$ and $\mathbf{r}$ does not exceed 1. Therefore, this error is bounded as

$$\begin{aligned}|\epsilon_p| &= \left| \sum_{n=p}^{\infty} (2n+1) j_n(kr_s) h_n(kr) P_n\left(\frac{\mathbf{r}_s \cdot \mathbf{r}}{r_s r}\right) \right| \\ &\leq \sum_{n=p}^{\infty} (2n+1) |j_n(kr_s)| |h_n(kr)| \left| P_n\left(\frac{\mathbf{r}_s \cdot \mathbf{r}}{r_s r}\right) \right| \\ &\leq \sum_{n=p}^{\infty} (2n+1) |j_n(kr_s)| |h_n(kr)|. \end{aligned} \tag{9.1.3}$$

Below we only consider the case when the wavenumbers are real and positive. In this case the modulus of the spherical Hankel function of an arbitrary order is a monotonic decaying function of kr. Indeed, we have (e.g. see Ref. [AS64])

$$|h_n(kr)|^2 = (kr)^{-2} \sum_{l=0}^{n} \frac{(n+l)!(2l)!}{(n-l)!(l!)^2} (2kr)^{-2l}, \tag{9.1.4}$$

which shows that $|h_n(kr)|^2$ is a sum of negative degrees of kr with positive coefficients, so that its derivative is strictly negative.

This fact brings us to conclusion that the maximum of $|h_n(kr)|$ is reaching at the lowest possible kr, which is kb. Thus, we have from Eq. (9.1.3)

$$\begin{aligned}|\epsilon_p| &\leq \sum_{n=p}^{\infty} (2n+1) |j_n(kr_s)| |h_n(kr)| \leq \sum_{n=p}^{\infty} (2n+1) |j_n(kr_s)| |h_n(kb)| \\ &= |\epsilon_p^{(b)}|. \end{aligned} \tag{9.1.5}$$

We emphasize the importance of this result, which shows that $\epsilon_p(\mathbf{r}_s, \mathbf{r})$ is bounded by some value on the sphere of radius b for an arbitrary $\mathbf{r}$, $|\mathbf{r}| > b$. So the p truncated series stays within the error bound $|\epsilon_p^{(b)}|$ for any $\mathbf{r}$ in the region considered.

Consider now the upper bound for $|j_n(kr_s)|$. Combining the expression for the spherical Bessel functions via Bessel functions (9.1.25) with the estimate for the upper bound (e.g. see Ref. [AS64]) of the latter functions:

$$|J_\nu(kr_s)| \leq \frac{(kr_s)^\nu}{2^\nu \Gamma(\nu+1)}, \tag{9.1.6}$$

we obtain

$$|j_n(kr_s)| \leq \frac{(kr_s)^n}{1 \cdot 3 \cdots (2n+1)} = \frac{(kr_s)^n}{(2n+1)!!} \leq \frac{(ka)^n}{(2n+1)!!}. \tag{9.1.7}$$

Substituting this bound into Eq. (9.1.5) we obtain

$$|\epsilon_p(\mathbf{r}_s, \mathbf{r})| \leq \sum_{n=p}^{\infty} \frac{(ka)^n}{(2n-1)!!} |h_n(kb)| = |\epsilon_p^{(a,b)}|. \tag{9.1.8}$$

Again we can see that the error can be bounded by some value on the sphere of radius a for an arbitrary $\mathbf{r}_s$, $|\mathbf{r}_s| < a$. So the p truncated series stays within the error bound $|\epsilon_p^{(a,b)}|$ for any $\mathbf{r}_s$ belonging to the smaller sphere.

Figure 9.2 shows a map of actual truncation error $|\epsilon_p|$ obtained using Eq. (9.1.2), where we set $r_s = a$, $r = b$, when vectors $\mathbf{r}_s$ and $\mathbf{r}$ are colinear, $\mathbf{r}_s \cdot \mathbf{r} = ab$ and the ratio b/a is fixed. Here we can indicate three asymptotic regions: the low frequency region, $kb \lesssim 1$, where p for given $|\epsilon_p|$ varies slowly and is determined by lower values of parameter kb; the high frequency region, $kb \gg 1$, where for fixed $|\epsilon_p|$ the truncation number grows with ka; and the transition region between these two extreme cases. Below we will discuss the boundaries of these regions which are plotted as lines on this graph.

9.1.1 Behavior of spherical Hankel functions

We start with analysis of dependences of spherical Hankel functions $h_n(kr)$ on n which enter the expansion of source about a new center. First, we note the following inequality for factorials, which is helpful for

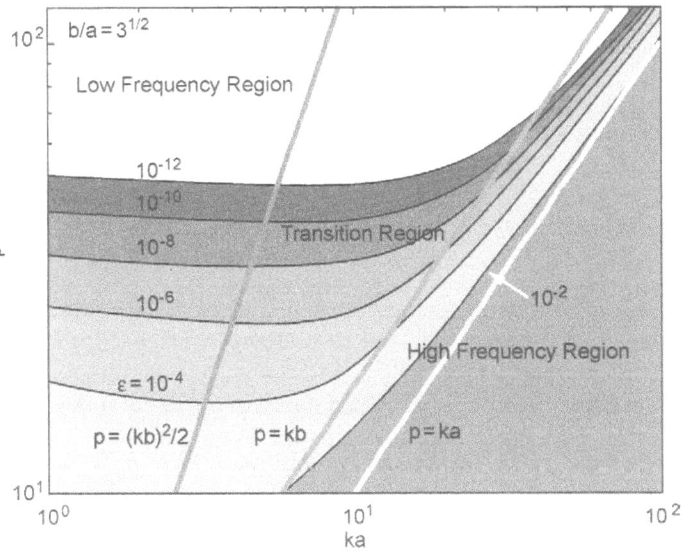

FIGURE 9.2 Computed regions for the maximum absolute error of the expansion of the function $h_0(k|\mathbf{r} - \mathbf{r}_s|)$ in the (ka, p) plane. The error inside the shaded regions varies between the values indicated on the contour boundaries separating the domains. The computations are performed for the "worst" location of the source, expansion center and evaluation point.

simplification of estimates:

$$\frac{(2n)!}{(n!)^2} \leq 2^{2n}, \qquad n = 0, 1, 2, \ldots. \tag{9.1.9}$$

Indeed, this holds for $n = 0$ and we have for $n \geq 1$

$$(2n)! = (2n - 1)!!(2n)!! < (2n)!!(2n)!! = 2^{2n}(n!)^2, \tag{9.1.10}$$

where $(2n - 1)!! = 1 \cdot 3 \cdots (2n - 1)$ and $(2n)!! = 2 \cdot 4 \cdots (2n) = 2^n n!$.

Substituting this into Eq. (9.1.4) we obtain the following bound for the modulus of the spherical Hankel functions:

$$|h_n(kr)|^2 \leq (kr)^{-2} \sum_{l=0}^{n} g_l^{(n)}(kr), \qquad g_l^{(n)}(kr) = \frac{(n + l)!}{(n - l)!}(kr)^{-2l}, \tag{9.1.11}$$

$$l = 0, \ldots, n, \quad n = 0, 1, \ldots.$$

Let us compare the functions $g_l^{(n)}(kr)$ and $g_{l+1}^{(n)}(kr)$. We have

$$f_l^{(n)}(kr) = \frac{g_{l+1}^{(n)}(kr)}{g_l^{(n)}(kr)} = \frac{(n+l+1)!(n-l)!}{(n-l-1)!(n+l)!}(kr)^{-2}$$
$$= (n+l+1)(n-l)(kr)^{-2}, \qquad (9.1.12)$$
$$l = 0, \ldots, n-1.$$

For fixed n and kr this ratio monotonically decays as a function of l for non-negative l. Therefore, if $f_0^{(n)}(kr) \leq 1$, then $f_l^{(n)}(kr) < 1$ for $l = 1, \ldots, n - 1$. This means that $g_{l+1}^{(n)}(kr) < g_l^{(n)}(kr)$ and the largest term in sum (9.1.11) is $g_0^{(n)}(kr)$. As follows from Eq. (9.1.12), condition $f_0^{(n)}(kr) \leq 1$ is equivalent to

$$n(n+1) \leq (kr)^2, \qquad (9.1.13)$$

which holds for

$$n \leq n_*(kr) = \tfrac{1}{2}\{[4(kr)^2 + 1]^{1/2} - 1\}. \qquad (9.1.14)$$

If this condition holds, we have the following inequality:

$$|h_n(kr)| \leq \left[\frac{1}{(kr)^2}\sum_{l=0}^{n} g_0^{(n)}(kr)\right]^{1/2} = \frac{1}{kr}[(n+1)g_0^{(n)}(kr)]^{1/2}$$
$$= \frac{(n+1)^{1/2}}{kr}, \qquad (9.1.15)$$
$$n \leq n_*(kr).$$

The other limiting case is achieved when $f_{n-1}^{(n)}(kr) \geq 1$. In this case the minimum of $f_l^{(n)}(kr)$ over all $l = 0, \ldots, n-1$ is greater or equal to 1 and, therefore, $g_{l+1}^{(n)}(kr) \geq g_l^{(n)}(kr)$ for any l and the largest term in sum (9.1.11) is $g_n^{(n)}(kr)$. From Eq. (9.1.12) we can see that condition $f_{n-1}^{(n)}(kr) \geq 1$ is equivalent to

$$2n \geq (kr)^2, \qquad (9.1.16)$$

which holds for

$$n \geq n_{**}(kr) = \tfrac{1}{2}(kr)^2. \qquad (9.1.17)$$

Hence, in this case we have for larger n

$$|h_n(kr)| \leq \left[\frac{1}{(kr)^2}\sum_{l=0}^{n} g_n^{(n)}(kr)\right]^{1/2} = \frac{(n+1)^{1/2}[(2n)!]^{1/2}}{(kr)^{n+1}}, \qquad (9.1.18)$$

$$n \geq n_{**}(kr).$$

Note that for $n_*(kr) > 1$ we always have $n_{**}(kr) > n_*(kr)$. Therefore, the values of n from the region $n_*(kr) < n < n_{**}(kr)$ correspond to the case where $f_l^{(n)}(kr) > 1$ for $l < l_*$ and $f_l^{(n)}(kr) < 1$ for $l > l_*$ where $0 < l_* < n-1$. Thus, the terms in sum (9.1.11) grow for $l = 0, \ldots, l_* - 1$ and decay for $l = l_* + 1, \ldots, n-1$. The maximum value is reached for $l = l_*$ and in this case we have

$$|h_n(kr)| \leq \left[\frac{1}{(kr)^2}\sum_{l=0}^{n} g_{l_*}^{(n)}(kr)\right]^{1/2} = \frac{1}{(kr)^{l_*+1}}\left[(n+1)\frac{(n+l_*)!}{(n-l_*)!}\right]^{1/2}, \qquad (9.1.19)$$

$$n_*(kr) < n < n_{**}(kr).$$

The value of l_* can be determined from condition $f_{l_*}^{(n)}(kr) = 1$, which yields

$$(n + l_* + 1)(n - l_*) = (kr)^2. \qquad (9.1.20)$$

The only positive root of this quadratic equation is

$$l_*(n; kr) = \tfrac{1}{2}\left[\sqrt{4[n^2 + n - (kr)^2] + 1} - 1\right] \qquad (9.1.21)$$

and, in fact, the closest integer to this value should be assigned as l_*.

This consideration of the modulus of the spherical Hankel function indicates three regions for n where the behavior of $h_n(kr)$ is different. The region of smaller n, $n \leq n_*(kr)$, where all $|h_n(kr)|$ for fixed n decay as $(kr)^{-1}$ and the dependence on n for fixed kr is relatively weak, the region of larger n, $n \geq n_*(kr)$, where $|h_n(kr)|$ exponentially grows with n, while decays as $(kr)^{-n-1}$ for fixed n, and the transition region, $n_*(kr) < n < n_{**}(kr)$, which matches these two extreme cases and where the function behavior is more complex.

9.1.2 Low frequency error bounds and series convergence

In this section we consider the case when the truncation number p is located in the region $p \geq n_{**}(kb)$. For such p we can substitute inequality

(9.1.18) into Eq. (9.1.8) to obtain

$$
\begin{aligned}
|\epsilon_p^{(a,b)}| &\le \frac{1}{kb}\sum_{n=p}^{\infty}\frac{(n+1)^{1/2}[(2n)!]^{1/2}}{(2n-1)!!}\left(\frac{a}{b}\right)^n \\
&= \frac{1}{kb}\sum_{n=p}^{\infty}\left[\frac{(n+1)(2n)!!}{(2n-1)!!}\right]^{1/2}\left(\frac{a}{b}\right)^n \\
&= \frac{1}{kb}\sum_{n=p}^{\infty}\left[\frac{2n(n+1)(2n-2)!!}{(2n-1)!!}\right]^{1/2}\left(\frac{a}{b}\right)^n \\
&< \frac{1}{kb}\sum_{n=p}^{\infty}[2n(n+1)]^{1/2}\left(\frac{a}{b}\right)^n. \qquad (9.1.22)
\end{aligned}
$$

To evaluate the latter sum we note that

$$
\begin{aligned}
&[n(n+1)]^{1/2} < n+1 = m+p+1 \le (p+1)(m+1), \\
&m = n-p, \quad n \ge p, \quad p \ge 2.
\end{aligned} \qquad (9.1.23)
$$

This yields

$$
\begin{aligned}
&|\epsilon_p^{(a,b)}| < \frac{\sqrt{2}}{kb}(p+1)\left(\frac{a}{b}\right)^p\sum_{m=0}^{\infty}(m+1)\left(\frac{a}{b}\right)^m = \frac{\sqrt{2}}{ka}\frac{(p+1)\sigma^{-p-1}}{(1-\sigma^{-1})^2}, \\
&\sigma = \frac{b}{a},
\end{aligned} \qquad (9.1.24)
$$

since

$$
\frac{1}{(1-\sigma^{-1})^2} = \sum_{m=0}^{\infty}(m+1)\sigma^{-m}, \qquad \sigma > 1 \qquad (9.1.25)
$$

is a derivative of geometric progression.

First, this result proves that series given by Eq. (9.1.8) converge for $b/a > 1$, or, in other words, series (9.2.3) converges *absolutely* in the regions indicated (these series also converge *uniformly*, since we are able to bound the error in the entire region by the maximum taken on the region boundary). Second, it shows that for $p \ge n_{**}(kb)$ the error decays exponentially with p and the exponent is determined by the ratio $\sigma = b/a$. Finally, it provides us with an analytical formula which establishes the error bound.

While this error bound can be improved, we checked the result numerically by comparisons of the obtained analytical result with straightforward computations for the low frequency region shown in Fig. 9.2. This region can be interpreted as a high truncation number region, $p \geq n_{**}(kb)$ (so the points in the region are located above the line $p = n_{**}(kb) = \frac{1}{2}(kb)^2$) or as a low frequency region, $kb \leq (2p)^{1/2}$ (so the points in the region are located on the left-hand side from the line $p = n_{**}(kb) = \frac{1}{2}(kb)^2$). The comparisons are shown in Fig. 9.3.

It is seen that qualitatively the theoretical curves behave as the curves obtained in computations and so the low frequency behavior is explained by the theory. Quantitatively, error bound (9.1.24) overestimates the actual p required to achieve the prescribed error approximately by 10. We consider this as a satisfactory theoretical result, since in practical computations the graph of type shown in Fig. 9.2 can be precomputed to have tighter bounds for p. We also found that approximation of the error bound:

$$|\epsilon_p^{(a,b)}| < \frac{1}{ka} \frac{\sigma^{-p-1}}{(1-\sigma^{-1})^{3/2}}, \tag{9.1.26}$$

is close to the actual curves in the region considered. According to this approximation p can be determined as a function of prescribed

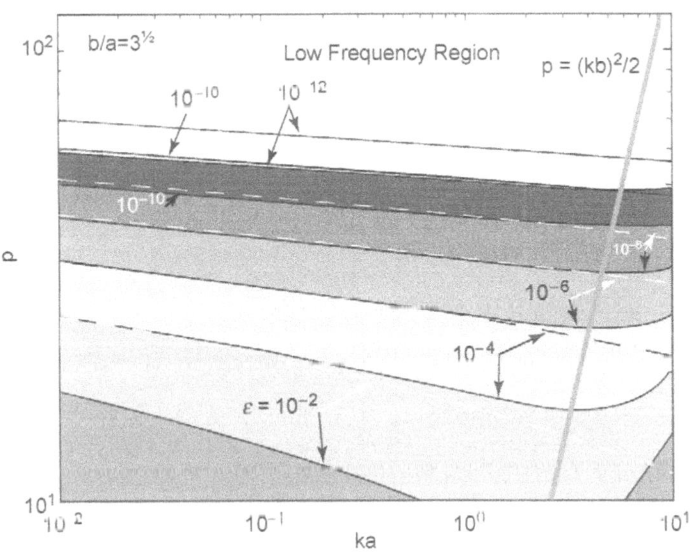

FIGURE 9.3 Comparison of theoretical errors (the dashed lines) and computed bounds for the low frequency expansions of $h_0(k|\mathbf{r} - \mathbf{r}_s|)$ in the worst case. Notations are same as in Fig. 9.2.

absolute error, ϵ, as

$$p = -\frac{\ln[\epsilon ka(1-\sigma^{-1})^{3/2}]}{\ln \sigma} - 1. \tag{9.1.27}$$

These curves are displayed in Fig. 9.4 which also shows that the actual values of p required to achieve the prescribed error are a bit smaller than provided by Eq. (9.1.27). This ensures that it can be used for computation with the parameters in the figure. We also checked this formula for a range of σ and found that in the range $1.1 < \sigma < 5$ the approximation is quantitatively close to the actual truncation numbers needed for a specified error, while for larger σ it may underestimate the actual p.

We also note that sometimes it is more convenient and practical to operate with relative error rather than an absolute error. The relative error in the present case can be easily derived from the absolute error taking into account that $|h_0(kr)| = (kr)^{-1}$. In this case we can rewrite approximation (9.1.26) and Eq. (9.1.27) in the form:

$$|\epsilon_{p,\mathrm{rel}}^{(a,b)}| < \frac{\sigma^{-p}}{(1-\sigma^{-1})^{3/2}}, \qquad p = -\frac{\ln[\epsilon_{\mathrm{rel}}(1-\sigma^{-1})^{3/2}]}{\ln \sigma}. \tag{9.1.28}$$

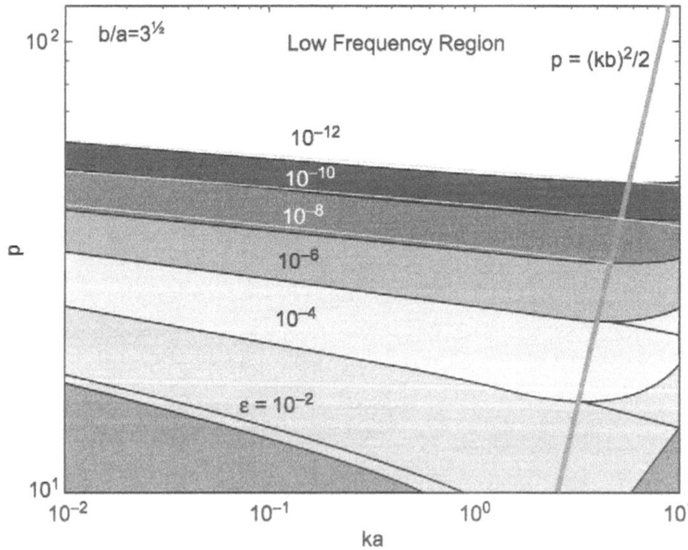

FIGURE 9.4 Comparison of theoretical errors (the dashed lines) and computed bounds for low frequency approximation (9.1.28) with straightforward computations. See Figs. 9.2 and 9.3 for further descriptions.

As we can see for a prescribed relative error ϵ_{rel} the truncation number p is *constant* as a function of ka in the low frequency region. This constant depends only on ϵ_{rel} and σ.

9.1.3 High frequency asymptotics

First, we note that expression (9.1.14) for $n_*(kb)$ can be simplified due to the following inequality:

$$kb > n_*(kb) = \tfrac{1}{2}\{[4(kb)^2 + 1]^{1/2} - 1\} > kb - \tfrac{1}{2}. \qquad (9.1.29)$$

Indeed, since we deal with integers, the approximation

$$n_*(kb) \approx kb \qquad (9.1.30)$$

for the lower bound which determines the behavior of the spherical Hankel functions is as good as the exact formula.

Figure 9.2 shows that to achieve a specified error the truncation numbers p can be much smaller than $n_{**}(kb)$. Moreover, at higher frequencies, $kb \gg 1$, the situation $p < n_*(kb)$ occurs. As follows from Eq. (9.1.15), in this case, the dependence of spherical Hankel functions on n is rather weak. This means that in the high frequency limit, it is not the growth of $h_n(kb)$, but the decay of $j_n(ka)$ for large n that determines the actual error. We can also state this important fact as that it is a, and not b, that determines the error bound at high frequencies.

Figure 9.5 shows typical dependencies of the spherical Bessel functions on the order n for fixed ka. It is clearly seen that there are two regions of n separated by value

$$n_*(ka) \approx ka. \qquad (9.1.31)$$

For $n < ka$ functions oscillate and there is no any decay in the amplitude for growing n. This shows that at high frequencies the truncation numbers should stay within the region

$$p > n_*(ka). \qquad (9.1.32)$$

In contrast to this behavior functions for $n > ka$ monotonically decay with growing n, and this decay is very fast. So the region $n \sim ka$ is an asymptotic region where the decay of $j_n(ka)$ supposedly determines the error.

The width of this asymptotic region can be determined from the following expansion of the Bessel functions for large orders (e.g. see Ref. [AS64]):

$$J_\nu(\nu + z\nu^{1/3}) = 2^{1/3}\nu^{-1/3}\,\mathrm{Ai}(-2^{1/3}z) + O(\nu^{-1}), \qquad (9.1.33)$$

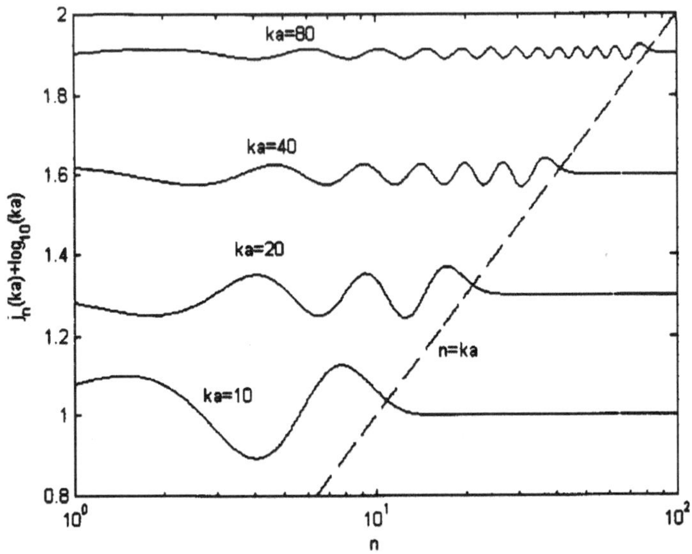

FIGURE 9.5 Dependence of the spherical Bessel functions $j_n(ka)$ in the order n at different fixed ka shown near the curves. These curves are shifted vertically according to ka.

where Ai is the *Airy function* of the first kind. This shows that for large fixed ka the Bessel functions should be asymptotically similar to each other as functions of the variable:

$$\eta = \frac{n - ka}{(ka)^{1/3}}. \tag{9.1.34}$$

Figure 9.6 shows results of direct computations. The values of the normalized functions are close to each other for a wide range of ka which proves that η defined by Eq. (9.1.34) is a proper choice of scale. We also see that the error as a function of η decays fast and, therefore, the truncation number can be determined by

$$p = ka + \eta_\epsilon (ka)^{1/3}, \tag{9.1.35}$$

where η_ϵ is the value of variable η_ϵ, for which the error is equal to the acceptable error ϵ.

To determine η_ϵ we can use the relation (9.1.25) between the spherical and regular Bessel functions and Eqs. (9.1.33) and (9.1.35) that yield

$$j_p(ka) = \sqrt{\frac{\pi}{2ka}} J_{p+1/2}(ka) \sim \sqrt{\frac{\pi}{2ka}} 2^{1/3} p^{-1/3} \operatorname{Ai}(2^{1/3}\eta). \tag{9.1.36}$$

Furthermore, we note that the values of η to reach errors of order 10^{-4} and smaller are rather large, $\eta \gg 1$ (see Fig. 9.6). This means that we can use

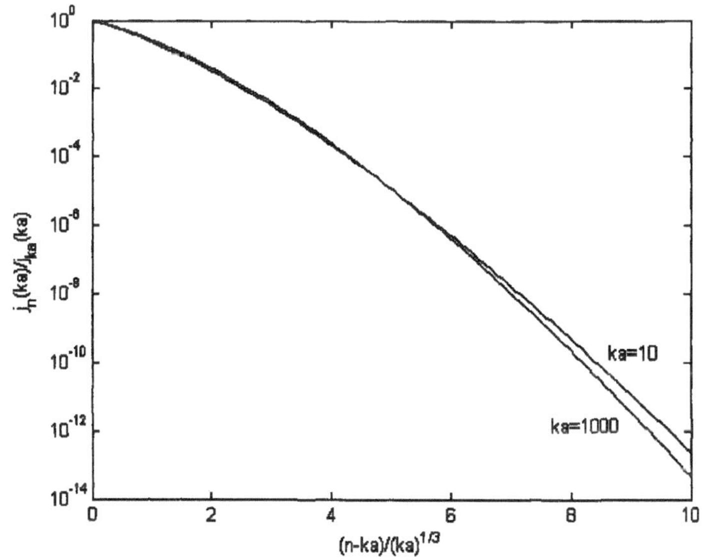

FIGURE 9.6 Dependence of the ratio $j_n(ka)/j_{ka}(ka)$ on the stretched variable in the region $n \geq ka$.

large argument asymptotics to estimate the decay rate of the Airy functions (e.g. see Ref. [AS64]):

$$\mathrm{Ai}(2^{1/3}\eta) \sim \tfrac{1}{2}\pi^{-1/2}(2^{1/3}\eta)^{-1/4}\exp\left(-\tfrac{1}{3}(2\eta)^{3/2}\right). \tag{9.1.37}$$

Now, since there is an exponential decay of terms for increasing n in the series (9.1.3) in the region considered, the error is determined by the first term in the truncated "tail", so using Eqs. (9.1.36), (9.1.37), and the bound for the modulus of the Hankel functions, Eq. (9.1.15), we obtain

$$|\epsilon_p| \lesssim (2p+1)|j_p(ka)||h_p(kb)| \sim \sigma^{-1}p^{7/6}(2\eta)^{-1/4}(ka)^{-3/2}\exp\left(-\frac{1}{3}(2\eta)^{3/2}\right),$$
$$\sigma = \frac{b}{a}. \tag{9.1.38}$$

The principal term of asymptotics of p for large ka is the first term in the right-hand side of Eq. (9.1.35). Substituting this value into the latter formula and using $(2\eta)^{-1/4} < 1$ we obtain

$$|\epsilon_p| \sim \sigma^{-1}(ka)^{-1/3}\exp\left(-\frac{1}{3}(2\eta)^{3/2}\right). \tag{9.1.39}$$

We can solve this equation with respect to η and denote $\epsilon = |\epsilon_p|$ to determine

$$\eta_\epsilon \sim \frac{1}{2}\left[3\ln\frac{1}{\epsilon\sigma} - \ln(ka)\right]^{2/3}. \tag{9.1.40}$$

This formula shows that η_ϵ depends on σ and ka. Usually, we have $\sigma \sim 1$, $\epsilon^{-1} \gg ka$, and by this reason the dependence on ka can be neglected for simplicity. So, combining Eqs. (9.1.35) and (9.1.40) we obtain

$$p = ka + \frac{1}{2}\left(3\ln\frac{1}{\epsilon\sigma}\right)^{2/3}(ka)^{1/3}. \tag{9.1.41}$$

This type of high frequency behavior was also obtained by Rokhlin [Rok93] and Chew et al. [CMSJ01].

Figure 9.7 plots high frequency dependences (Eq. (9.1.41)) of the truncation number on ka for different ϵ. As we can see, the approximation is good in the range $p < kb$ and deviates from actual bounds in the transition region. Note that the latter observation contradicts the statement, which can be found in the book [CMSJ01], that Eq. (9.1.41) is applicable for any ka.

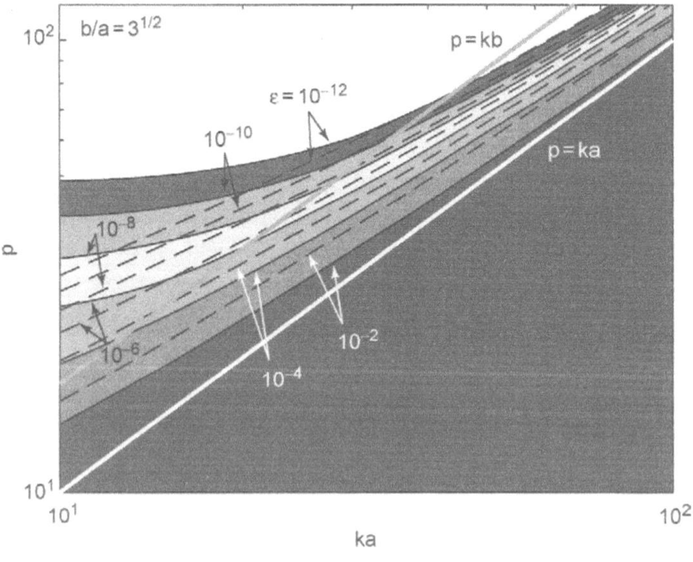

FIGURE 9.7 Comparison of the high frequency asymptotics (9.1.41) shown by the dashed lines with results of straightforward computation of the expansion error. See the caption of Fig. 9.2 for further information.

9.1.4 Transition region and combined approximation

Some approximations for the dependence of the truncation number on *ka* and other parameters can be obtained in the transition region. Here we have $p > kb$. In this region the modulus of the spherical Hankel function grows exponentially, while the spherical Bessel functions decay exponentially with n. So the error is determined by the product of these two exponents, and expansions using the Airy functions of the first and the second kind can be considered.

However, it is easier to obtain some satisfactory approximation of function behavior in the transition region by combining the two limiting cases considered. For example, we found that approximation:

$$p = \left\{ \left[\frac{1}{\ln \sigma} \ln \frac{1}{\epsilon ka(1-\sigma^{-1})^{3/2}} + 1 \right]^4 + \left[ka + \frac{1}{2} \left(3 \ln \frac{1}{\epsilon \sigma} \right)^{2/3} (ka)^{1/3} \right]^4 \right\}^{1/4} \tag{9.1.42}$$

is good enough to approximate the actual truncation error for the region of parameters shown in Fig. 9.8. It is easy to see how approximation (9.1.42) is constructed. It combines the low frequency and high frequency approximations (9.1.27) and (9.1.41). So, for low and high *ka* it coincides

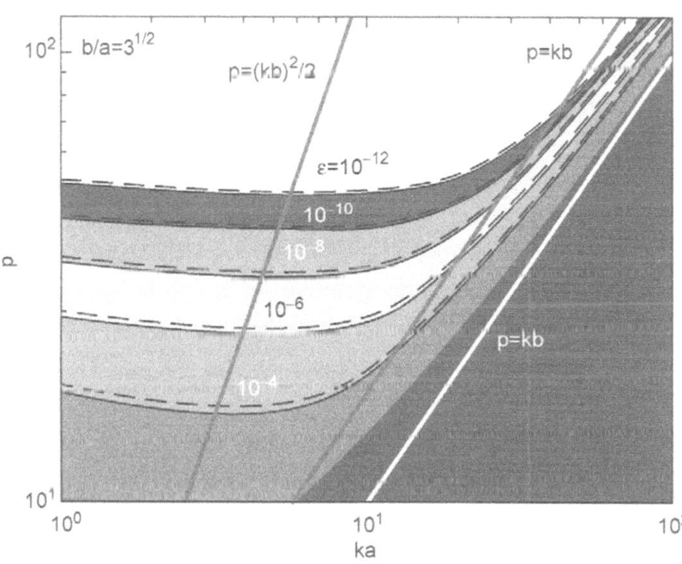

FIGURE 9.8 Comparison of the approximation (9.1.42) with straightforward computations, for the same case as in Fig. 9.2.

asymptotically with the limiting cases, while in the transition region it is also acceptable for estimation. In this context, we note that the computation of the actual error map is relatively easy, so while the low frequency and high frequency approximations are important for preliminary evaluations, the truncation numbers in the transition region can be evaluated directly.

The error map also shows that simpler type of dependence of the truncation number on *ka* can be used. In this simplified approximation we assume that the high frequency formula (9.1.33) can be applied for $p < kb$, while for the low frequency region we can use $p = p_* = \text{const}$ which overestimates the actual truncation number in a wide range of low frequencies. To determine this constant, we can first find the intersection point of the curve (9.1.33) and line $p = kb$ $(= \sigma ka)$:

$$(ka)_* + \frac{1}{2}\left(3\ln\frac{1}{\epsilon\sigma}\right)^{2/3}(ka)_*^{1/3} = \sigma(ka)_*. \tag{9.1.43}$$

This yields

$$(ka)_* = \frac{3}{2^{3/2}(\sigma-1)^{3/2}}\ln\frac{1}{\epsilon\sigma}. \tag{9.1.44}$$

So the constant for the low frequency region can be found as $p_* = \sigma(ka)_*$, which yields

$$p_* = \frac{3\sigma}{2^{3/2}(\sigma-1)^{3/2}}\ln\frac{1}{\epsilon\sigma}. \tag{9.1.45}$$

This results in the following combined approximation for the truncation number:

$$p = \begin{cases} p_*, & ka \leq (ka)_*, \\ ka + \frac{1}{2}\left(3\ln\frac{1}{\epsilon\sigma}\right)^{2/3}(ka)^{1/3}, & ka \geq (ka)_*, \end{cases} \tag{9.1.46}$$

where the values p_* and $(ka)_*$ depend only on ϵ and σ and are given by Eqs. (9.1.44) and (9.1.45).

9.2 TRUNCATION ERRORS FOR EXPANSIONS OF MULTIPOLES

In a more general setting the truncation error can be estimated for expansion of multipoles expressed as sums of singular spherical basis functions. The illustrations of the problems considered remain the same (see Fig. 9.1). Generally, a multipole of degree n centered at $\mathbf{r} = \mathbf{r}_s$ can be expressed in forms (2.2.23) and (2.2.26), while here we consider the error associated with the highest mode of the multipole:

$$\psi_n(\mathbf{r} - \mathbf{r}_s) = \sum_{m=-n}^{n} C_n^m S_n^m(\mathbf{r} - \mathbf{r}_s), \tag{9.2.1}$$

where C_n^m are some coefficients related to the moments of the multipole. The spherical basis functions can be translated to a new expansion center according to Eq. (9.2.4). If the series are truncated with the first p degrees we obtain

$$\psi_n(\mathbf{r} - \mathbf{r}_s) = \sum_{m=-n}^{n} \sum_{n'=0}^{p-1} \sum_{m'=-n'}^{n'} C_n^m (S|S)_{n'n}^{m'm}(-\mathbf{r}_s) S_{n'}^{m'}(\mathbf{r}) + \epsilon_p, \tag{9.2.2}$$

$$r_s = |\mathbf{r}_s| < a, \quad r = |\mathbf{r}| > b, \quad a < b.$$

where $(S|S)_{n'n}^{m'm} = (R|R)_{n'n}^{m'm}$ are the reexpansion coefficients and we are looking for the evaluation of $|\epsilon_p|$.

Using integral representation of the reexpansion coefficients (7.1.33), definition of spherical basis functions (2.1.102), and addition theorem for spherical harmonics (2.1.72), we can express the residual term in Eq. (9.2.2) as follows:

$$\begin{aligned}
\epsilon_p &= \sum_{n'=p}^{\infty} \sum_{m'=-n'}^{n'} \sum_{m=-n}^{n} C_n^m \mathrm{i}^{n'-n} \int_{S_u} \mathrm{e}^{-ik\mathbf{s}\cdot\mathbf{r}_s} Y_n^m(\mathbf{s}) Y_{n'}^{-m'}(\mathbf{s}) \mathrm{d}S(\mathbf{s}) S_{n'}^{m'}(\mathbf{r}) \\
&= \sum_{n'=p}^{\infty} h_{n'}(kr) \sum_{m=-n}^{n} C_n^m \mathrm{i}^{n'-n} \int_{S_u} \mathrm{e}^{-ik\mathbf{s}\cdot\mathbf{r}_s} Y_n^m(\mathbf{s}) \\
&\quad \times \left[\sum_{m'=-n'}^{n'} Y_{n'}^{-m'}(\mathbf{s}) Y_{n'}^{m'}\left(\frac{\mathbf{r}}{r}\right) \right] \mathrm{d}S(\mathbf{s}) \\
&= \frac{1}{4\pi} \sum_{n'=p}^{\infty} (2n'+1) h_{n'}(kr) \sum_{m=-n}^{n} C_n^m \mathrm{i}^{n'-n} \\
&\quad \times \int_{S_u} \mathrm{e}^{-ik\mathbf{s}\cdot\mathbf{r}_s} Y_n^m(\mathbf{s}) P_{n'}\left(\frac{\mathbf{s}\cdot\mathbf{r}}{r}\right) \mathrm{d}S(\mathbf{s}).
\end{aligned} \tag{9.2.3}$$

Further, we can use the plane wave expansion (2.3.5) to obtain:

$$\epsilon_p = \sum_{n'=p}^{\infty} \frac{(2n'+1)}{4\pi} h_{n'}(kr) \sum_{m=-n}^{n} C_n^m i^{n'-n} \sum_{n''=0}^{\infty} (2n''+1) i^{n''} j_{n''}(kr_s)$$
$$\times \int_{S_u} P_{n''}\left(\frac{-\mathbf{s}\cdot\mathbf{r}_s}{r_s}\right) Y_n^m(\mathbf{s}) P_{n'}\left(\frac{\mathbf{s}\cdot\mathbf{r}}{r}\right) dS(\mathbf{s}). \quad (9.2.4)$$

Note now that the limits of summation over n'' can be taken from $n'' = |n' - n|$ to $n'' = n' + n$. Indeed, the integral in Eq. (9.2.4) can be written as

$$\int_{S_u} P_{n''}\left(-\mathbf{s}\cdot\frac{\mathbf{r}_s}{r_s}\right) Y_n^m(\mathbf{s}) P_{n'}\left(\frac{\mathbf{s}\cdot\mathbf{r}}{r}\right) dS(\mathbf{s})$$
$$= \frac{(4\pi)^2}{(2n'+1)(2n''+1)} \sum_{m'=-n'}^{n'} \sum_{m''=-n''}^{n''} Y_{n'}^{m'}\left(\frac{\mathbf{r}}{r}\right) Y_{n''}^{m''}\left(-\frac{\mathbf{r}_s}{r_s}\right)$$
$$\times \int_{S_u} Y_n^m(\mathbf{s}) Y_{n'}^{-m'}(\mathbf{s}) Y_{n''}^{-m''}(\mathbf{s}) dS(\mathbf{s})$$
$$= \frac{4\pi i^{n-n'-n''}}{(2n'+1)(2n''+1)} \sum_{m'=-n'}^{n'} \sum_{m''=-n''}^{n''} Y_{n'}^{m'}\left(\frac{\mathbf{r}}{r}\right) Y_{n''}^{m''}\left(-\frac{\mathbf{r}_s}{r_s}\right) (s|r)_{n''n'n}^{m''m'm},$$

where the numerical coefficients $(s|r)_{n''n'n}^{m''m'm}$ are given by Eq. (3.2.31). Since

$$(s|r)_{n''n'n}^{m''m'm} = 0, \qquad \text{for } n'' < |n - n'|, \qquad n'' > n + n', \quad (9.2.5)$$

which is a result of the relation between $(s|r)_{n''n'n}^{m''m'm}$ and the Clebsch–Gordan coefficients (3.2.32), we can see that the limits of summation can be changed as stated.

Let us now define the following surface function:

$$\Psi_n(\mathbf{s}) = \sum_{m=-n}^{n} i^{-n} C_n^m Y_n^m(\mathbf{s}), \quad (9.2.6)$$

which characterizes the magnitude and multipole moments of ψ_n specified by Eq. (9.2.1). With this definition we can rewrite Eq. (9.2.4) in the form:

$$\epsilon_p = \frac{1}{4\pi} \sum_{n'=p}^{\infty} (2n'+1) i^{n'} h_{n'}(kr) \sum_{n''=|n'-n|}^{n+n'} (2n''+1) i^{n''} j_{n''}(kr_s)$$
$$\times \int_{S_u} \Psi_n(\mathbf{s}) P_{n''}\left(-\mathbf{s}\cdot\frac{\mathbf{r}_s}{r_s}\right) P_{n'}\left(\frac{\mathbf{s}\cdot\mathbf{r}}{r}\right) dS(\mathbf{s}). \quad (9.2.7)$$

Now we can estimate the modulus of the residual term taking into account that $P_n \leq 1$:

$$|\epsilon_p| \leq \frac{1}{4\pi}\sum_{n'=p}^{\infty}(2n'+1)|h_{n'}(kr)|\sum_{n''=|n-n'|}^{n+n'}(2n''+1)|j_{n''}(kr_s)|$$
$$\times\int_{S_u}|\Psi_n(\mathbf{s})|\left|P_{n''}\left(-\mathbf{s}\cdot\frac{\mathbf{r}_s}{r_s}\right)\right|\left|P_{n'}\left(\frac{\mathbf{s}\cdot\mathbf{r}}{r}\right)\right|dS(\mathbf{s})$$
$$\leq \max_{\mathbf{s}}|\Psi_n(\mathbf{s})|\sum_{n'=p}^{\infty}(2n'+1)|h_{n'}(kr)|\sum_{n''=|n'-n|}^{n+n'}(2n''+1)|j_{n''}(kr_s)|. \qquad (9.2.8)$$

The latest inequality shows that

$$|\epsilon_p| = O((2p+1)|h_p(kr)|\max_{\mathbf{s}}|\Psi_n(\mathbf{s})|), \qquad \text{if } p \leq n. \qquad (9.2.9)$$

Indeed, in this case there exists $n'' = 0$ in the latter sum and coefficients near the spherical Hankel functions will not decay. Therefore, to provide better estimate, we need to have $n' \geq p > n$. Moreover, if $n'' \leq kr_s$ we have $|j_{n''}(kr_s)| \sim |j_0(kr_s)|$ and, again the best estimate which can be obtained with Eq. (9.2.8) is given by Eq. (9.2.9). Thus, to have something better than Eq. (9.2.9) we need to have

$$p > n + ka. \qquad (9.2.10)$$

In this case for arbitrary location of the multipole source inside a sphere of radius a function $|j_{n''}(kr_s)|$ decays when n' grows. Taking into account that in this case we have $|j_{n''+1}(ka)| < |j_{n''}(ka)|$ and $|j_{n''}(kr_s)| \leq |j_{n''}(ka)|$ for $n'' \geq n' - n$, we determine

$$\sum_{n''=n'-n}^{n+n'}(2n''+1)|j_{n''}(kr_s)| < |j_{n'-n}(ka)|\sum_{n''=n'-n}^{n+n'}(2n''+1)$$
$$= (2n+1)(2n'+1)|j_{n'-n}(ka)|. \qquad (9.2.11)$$

We can also use inequality $|h_{n'}(kr)| \leq |h_{n'}(kb)|$ (see Eq. (9.1.5)) to obtain the following bound:

$$|\epsilon_p| < (2n+1)\max_{\mathbf{s}}|\Psi_n(\mathbf{s})|\sum_{n'=p}^{\infty}(2n'+1)^2|h_{n'}(kb)||j_{n'-n}(ka)|, \qquad (9.2.12)$$

$p > n + ka.$

We can compare this formula with that for a monopole (9.1.5), which corresponds to case $n = 0$. The comparison shows that exponentially decaying terms $|h_{n'}(kb)||j_{n'}(ka)|$ are same for both estimates, but we obtained bigger coefficient, $(2n'+1)^2$, instead of $(2n'+1)$, which is a result of rough evaluation of the surface integral (see Eq. (9.2.8)). While this evaluation can be improved, we should understand that the exponent of decay of $|h_{n'}(kb)||j_{n'}(ka)|$ is important for the rate of series convergence and the polynomial factors in the series play rather minor role. So we can stay with Eq. (9.2.12) and consider different frequency regions as for a monopole to establish convergence and truncation error bounds. As this result is obtained, we can consider sums of ψ_l, $l = 0, \ldots, n$, which represents an arbitrary sum of multipoles of degree below n. The expansion error for this sum is determined by its highest mode (assuming that $|\Psi_l(\mathbf{s})|$ are of the same order), since $p > n + ka$.

9.2.1 Low frequency error bounds and series convergence

First, we want to show absolute and uniform convergence of the series. For sufficiently low kb or high $n' \geq p \geq n_{**}(kb) = \frac{1}{2}(kb)^2$ we can use inequalities (9.1.7) and (9.1.18), which after substitution into Eq. (9.2.12) yield

$$\begin{aligned}|\epsilon_p| &< (2n+1)\max_{\mathbf{s}}|\Psi_n(\mathbf{s})| \sum_{n'=p}^{\infty}(2n'+1)^2 \frac{(n'+1)^{1/2}[(2n')!]^{1/2}}{(2n'-2n+1)!!}\frac{(ka)^{n'-n}}{(kb)^{n'+1}} \\ &< \frac{(2n+1)}{(ka)^{n+1}}\max_{\mathbf{s}}|\Psi_n(\mathbf{s})|\sigma^{-p} \\ &\quad\times \sum_{m=0}^{\infty}\frac{(2m+2p+1)^{3/2}(m+p+1)^{1/2}(2m+2p+1)!!}{(2m+2p-2n+1)!!}\sigma^{-m-1}, \\ \sigma &= \frac{b}{a}. \end{aligned} \tag{9.2.13}$$

Then we can use the following inequality:

$$\frac{(2m+2p+1)!!}{(2m+2p-2n+1)!!} \leq \frac{(2p+1)!!(m+n)!}{(2p-2n+1)!!m!n!}, \tag{9.2.14}$$

$$m = 0, 1, \ldots, \quad p > n.$$

One can prove this inequality by consideration of the following sequence

$$c_m = \frac{(2p-2n+1)!!(2m+2p+1)!!m!n!}{(2p+1)!!(2m+2p-2n+1)!!(m+n)!}, \tag{9.2.15}$$

$$c_0 = 1, \quad m = 0, 1, \ldots,$$

where it is not difficult to show that $c_{m+1}/c_m \leq 1$, $m = 0, 1, \ldots$.

Then, since for truncation numbers $p - n \geq 3$ and $m = 0, 1, \ldots$ we have the following inequality:

$$(2m+2p+1)^{3/2}(m+p+1)^{1/2} < 2^{3/2}(m+p+1)^2$$
$$\leq 2^{3/2}(p-n)^2(m+n+1)(m+n+2). \tag{9.2.16}$$

We can obtain from Eqs. (9.2.13), (9.2.14), and (9.2.16):

$$|\epsilon_p| < \frac{2^{3/2}(p-n)^2(2n+1)}{(ka)^{n+1}n!} \frac{(2p+1)!!}{(2p-2n+1)!!} \max_{\mathbf{s}} |\Psi_n(\mathbf{s})| \sigma^{-p}$$
$$\times \sum_{m=0}^{\infty} \frac{(m+n+2)!}{m!} \sigma^{-m-1}. \tag{9.2.17}$$

The latter sum can be expressed via elementary functions, as follows:

$$\sum_{m=0}^{\infty} \frac{(m+n+2)!}{m!} \sigma^{-m-1} = (n+2)!(\sigma-1)^{-n-3}. \tag{9.2.18}$$

This yields

$$|\epsilon_p| < \frac{2^{3/2}(p-n)^2(2n+1)(n+1)(n+2)}{(kb)^{n+1}} \frac{(2p+1)!!}{(2p-2n+1)!!}$$
$$\times \max_{\mathbf{s}} |\Psi_n(\mathbf{s})| (1-\sigma^{-1})^{-n-3} \sigma^{-(p+2)}. \tag{9.2.19}$$

By this way we showed the absolute convergence of the series, which is also uniform, since the bounds for $r_s = a$ and $r = b$ are applicable for arbitrary $r_s \leq a$ and $r_s \geq b$.

Note that

$$\frac{(2p+1)!!}{(2p-2n+1)!!} < (2p+1)^n. \tag{9.2.20}$$

Hence, according to Eq. (9.2.19), for fixed n the error decays exponentially with p as σ^{-p} which is the same exponent as for the monopole source (see Eq. (9.1.24)). Polynomial factors of type p^n do not affect the exponential decay.

We can also notice the presence of factor $(kb)^{-n-1}$ in Eq. (9.2.19) which is due to the degree of multipole n. While for low $kb < 1$ this affects the absolute error, the bounds for the relative error are approximately same as for expansion of the monopole source. Indeed, for $kb < 1$ and $n \gtrsim 1$ we have $n > n_{**}(kb) = \frac{1}{2}(kb)^2$, so we can take for a point on the sphere of radius b (see Eq. (9.1.27)):

$$|\psi_n| \sim \left|\Psi_n\left(\frac{\mathbf{r}}{r}\right)\right| h_n(kr) \sim |\Psi_n(\mathbf{s})| h_n(kb) \sim \max_{\mathbf{s}} |\Psi_n(\mathbf{s})| \frac{(2n-1)!!}{(kb)^{n+1}}. \tag{9.2.21}$$

The relative expansion error can then be estimated as

$$|\epsilon_{p.\mathrm{rel}}| = \frac{|\epsilon_p|}{|\psi_n|} \lesssim \frac{(p-n)^2(2n+\frac{5}{2})^3(2p+1)!!}{(2p-2n+1)!!(2n-1)!!}(1-\sigma^{-1})^{-n-3}\sigma^{-(p+2)}, \tag{9.2.22}$$

$$p > n \gtrsim 1, \quad kb \lesssim 1.$$

This does not depend on k, b and a, but on the ratio $\sigma = b/a$ and the error decays as the ratio increases.

9.2.2 High frequency asymptotics

As in the case of a monopole source, we can consider the case of high frequencies characterized by conditions:

$$kb \gtrsim p \gtrsim n + ka. \tag{9.2.23}$$

In this case the error is determined by the rate of decay of Bessel functions for large orders. The truncation number for a multipole of degree n generalizes dependence (9.1.35) as

$$p = n + ka + \eta_\epsilon (ka)^{1/3}. \tag{9.2.24}$$

This follows from Eq. (9.2.12) if we assume that $\max_{\mathbf{s}} |\Psi_n(\mathbf{s})| \sim 1$ and take the principal term of asymptotics as we did in the case of monopoles. For $n = 0$ the formula then turns into relation for monopoles (Eq. (9.1.35)) with η_ϵ given by Eq. (9.1.40), where its dependence on ka can be neglected for small enough ϵ. This results in

$$p = n + ka + \frac{1}{2}\left(3 \ln \frac{1}{\epsilon\sigma}\right)^{2/3} (ka)^{1/3}. \tag{9.2.25}$$

For $ka \gg 1$ and $n \sim 1$ the truncation number is close to that of monopole (9.1.41). This shows, for example, that truncation number for dipole expansion is larger than the truncation number of monopole expansion just by 1, which is a small addition to p. The differencc in two-term asymptotics occurs for large n comparable with the first correction to the formula, $n \sim \eta_\epsilon (ka)^{1/3} \gg 1$. Note also that condition (9.2.23) holds for the range of frequencies and n, such that

$$k(b - a) \geq n. \qquad (9.2.26)$$

If this condition is violated, expansions in transition regions should be considered.

9.3 TRANSLATION ERRORS

9.3.1 $S|S$ translations

9.3.1.1 Problem

To understand the errors in the FMM due to translations, we start with the picture shown in Fig. 9.9 which is related to the multipole-to-multipole $S|S$ translation. Here a monopole source is located at $\mathbf{r} = \mathbf{r}_s$ inside a sphere of radius a and we are looking for the evaluation of the truncation errors at some point $\mathbf{r}$ located outside a sphere of radius b. The previously evaluated truncation numbers provide a specified error of expansion. However, a typical problem in the FMM is that the source potential, $h_0(k|\mathbf{r} - \mathbf{r}_s|)$, is approximated by an S expansion about $\mathbf{r} = -\mathbf{t}$ in the region

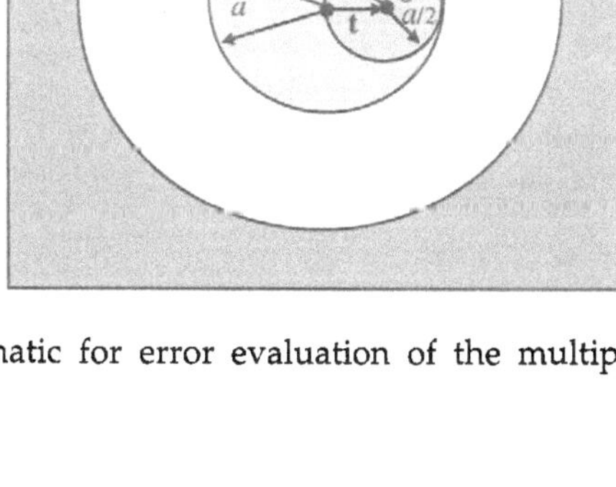

FIGURE 9.9 Schematic for error evaluation of the multipole-to-multipole $S|S$ translation.

outer to the sphere of radius $b/2$ centered at $\mathbf{r} = -\mathbf{t}$, $|\mathbf{t}| = a/2$:

$$\psi^{(p')}(\mathbf{r}) = \sum_{n'=0}^{p'-1} \psi_{n'}(\mathbf{r}+\mathbf{t}) = \sum_{n'=0}^{p'-1} \sum_{m'=-n'}^{n'} C_{n'}^{m'} S_n^m(\mathbf{r}+\mathbf{t}),$$

$$h_0(k|\mathbf{r} - \mathbf{r}_s|) - \psi^{(p')}(\mathbf{r}) = \epsilon_{p'}, \qquad r'_s = |\mathbf{r}_s + \mathbf{t}| < \frac{1}{2}a, \tag{9.3.1}$$

$$r' = |\mathbf{r} + \mathbf{t}| > \frac{1}{2}b, \qquad a < b,$$

where the functions $\psi_n(\mathbf{r})$ are defined by Eq. (9.2.1) and p' is the truncation number. By an approximate (truncated) translation of the expansion coefficients $C_n^m \to \hat{C}_n^m$ we produce a new series centered at $\mathbf{r} = 0$ and valid for any $\mathbf{r}$ located outside the largest sphere:

$$\hat{\psi}^{(p)}(\mathbf{r}) = \sum_{n=0}^{p-1} \hat{\psi}_n(\mathbf{r}) = \sum_{n=0}^{p-1} \sum_{m=-n}^{n} \hat{C}_n^m S_n^m(\mathbf{r}),$$

$$h_0(k|\mathbf{r} - \mathbf{r}_s|) - \hat{\psi}^{(p)}(\mathbf{r}) = \epsilon_p, \qquad r_s = |\mathbf{r}_s| < a, \qquad r = |\mathbf{r}| > b, \tag{9.3.2}$$

$$a < b.$$

Our goal is to specify the finite transform $C_{n'}^{m'} \to \hat{C}_n^m$ $(n' < p', n < p)$ and evaluate ϵ_p which combines the error of approximation $\epsilon_{p'}$ and the error of truncated translation operation.

We assume that the maximum error is determined by the error at the boundaries of the domain at $r'_s = |\mathbf{r}_s + \mathbf{t}| = a/2$ and $r = |\mathbf{r}| = b$. So, further we can consider only this case and apply the error bound for other locations of the source and evaluation points inside the domains.

9.3.1.2 Solution

First, consider translation using rectangular truncated translation matrix. In this case we define the translated coefficients as

$$\hat{C}_n^m = \sum_{n'=0}^{p'-1} \sum_{m'=-n'}^{n'} (S|S)_{nn'}^{mm'}(\mathbf{t}) C_{n'}^{m'}, \qquad n = 0, \ldots, p-1, \;\; m = -n, \ldots, n. \tag{9.3.3}$$

Since function $\psi^{(p')}(\mathbf{r})$ from Eq. (9.3.1) can be represented in the form:

$$\psi^{(p')}(\mathbf{r}) = \sum_{n=0}^{\infty} \sum_{m=-n}^{n} \sum_{n'=0}^{p'-1} \sum_{m'=-n'}^{n'} (S|S)_{nn'}^{mm'}(\mathbf{t}) C_{n'}^{m'} S_n^m(\mathbf{r}), \qquad (9.3.4)$$

then the difference between functions $\hat{\psi}^{(p)}(\mathbf{r})$, defined by Eqs. (9.3.2) and (9.3.3) and $\psi^{(p')}(\mathbf{r})$ is

$$\delta = \psi^{(p')}(\mathbf{r}) - \hat{\psi}^{(p)}(\mathbf{r}) = \sum_{n'=0}^{p'-1} \delta_{n'},$$
$$\delta_{n'} = \sum_{n=p}^{\infty} \sum_{m=-n}^{n} \sum_{m'=-n'}^{n'} (S|S)_{nn'}^{mm'}(\mathbf{t}) C_{n'}^{m'} S_n^m(\mathbf{r}). \qquad (9.3.5)$$

The absolute value of $\delta_{n'}$ can be evaluated using Eq. (9.2.8) which is applicable for translation of the n'th degree functions (Eq. (9.2.2)). Taking into account current notations and $|\mathbf{t}| = a/2$ we determine

$$|\delta_{n'}| < \max_{\mathbf{s}} |\Psi_{n'}(\mathbf{s})| \sum_{n=p}^{\infty} (2n+1)|h_n(kb)| \sum_{n''=|n-n'|}^{n+n'} (2n''+1)\left|j_{n''}\left(\frac{ka}{2}\right)\right|, \qquad (9.3.6)$$

where

$$\Psi_{n'}(\mathbf{s}) = \sum_{m'=-n'}^{n'} i^{-n'} C_{n'}^{m'} Y_{n'}^{*m'}(\mathbf{s}), \qquad n' = 0, \ldots, p'-1 \qquad (9.3.7)$$

Assume now that $\max_{\mathbf{s}} |\Psi_{n'}(\mathbf{s})|$ is bounded as

$$\max_{\mathbf{s}} |\Psi_{n'}(\mathbf{s})| < c_{n'}\left|j_{n'}\left(\frac{ka}{2}\right)\right|, \qquad n' = 0, \ldots, p'-1, \qquad (9.3.8)$$

where $c_{n'}$ are bounded by some polynomials of n'. The idea of such type of bound comes from expansion of a single monopole located at $\mathbf{r}_s$, where $|\mathbf{r}_s + \mathbf{t}| = a/2$ (see Eq. (9.1.1)). This holds for direct expansion and we are going to show further that the same type of bound where $a/2$ should be replaced by the appropriate translation distance holds for translated functions as well (see Eq. (9.3.27)). Indeed, if we prove that Eq. (9.3.27) follows from Eq. (9.3.8), then by induction we can see that Eq. (9.3.27) holds for any S expansion in the multilevel FMM due to the hierarchical doubling of the domain size. The basis for this induction is provided by the initial S expansion of the monopole source.

Substituting Eq. (9.3.8) into inequality (9.3.6) we obtain for $p \gtrsim p'$ $(n \gtrsim n')$

$$|\delta_{n'}| < c_{n'} \sum_{n=p}^{\infty} (2n+1)|h_n(kb)| \sum_{n''=n-n'}^{n+n'} (2n''+1)\left|j_{n'}\left(\frac{ka}{2}\right)\right|\left|j_{n''}\left(\frac{ka}{2}\right)\right|. \quad (9.3.9)$$

To estimate this error, we prove the following two lemmas.

LEMMA 3 *Let $n' + n'' > ka$, $ka \geq 0$. Then the following inequality holds:*

$$\left|j_{n'}\left(\frac{ka}{2}\right)\right|\left|j_{n''}\left(\frac{ka}{2}\right)\right| < \frac{\pi}{ka} J_{n'+n''+1}(ka), \quad (9.3.10)$$

where j_n are the spherical and J_n are the ordinary Bessel functions of the first kind.

PROOF. The following integral representation of the Bessel function product for $n' + n'' > -2$ and so for $n' + n'' > ka$ takes place [Wat58]

$$J_{n'+1/2}\left(\frac{ka}{2}\right)J_{n''+1/2}\left(\frac{ka}{2}\right) = \frac{2}{\pi}\int_0^{\pi/2} J_{n'+n''+1}(ka\cos\theta)\cos[(n''-n')\theta]\mathrm{d}\theta. \quad (9.3.11)$$

For $n' + n'' + 1 > ka$ function $J_{n'+n''+1}(ka\cos\theta)$ is a strictly positive monotonically growing function of its argument, which is not negative within the limits of integration. Therefore,

$$\left|J_{n'+1/2}\left(\frac{ka}{2}\right)J_{n''+1/2}\left(\frac{ka}{2}\right)\right| \leq \frac{2}{\pi}\int_0^{\pi/2} |J_{n'+n''+1}(ka\cos\theta)||\cos[(n-n')\theta]|\mathrm{d}\theta$$
$$\leq |J_{n'+n''+1}(ka)|\frac{2}{\pi}\int_0^{\pi/2}|\cos[(n-n')\theta]|\mathrm{d}\theta \leq |J_{n'+n''+1}(ka)|. \quad (9.3.12)$$

Using the relation between the spherical and usual Bessel functions (2.1.25) we obtain

$$\left|j_{n'}\left(\frac{ka}{2}\right)\right|\left|j_{n''}\left(\frac{ka}{2}\right)\right| = \frac{\pi}{ka}\left|J_{n'+1/2}\left(\frac{ka}{2}\right)J_{n''+1/2}\left(\frac{ka}{2}\right)\right| < \frac{\pi}{ka}|J_{n'+n''+1}(ka)|. \quad \square$$

LEMMA 4 *Let $n > ka - \frac{1}{2}$, $ka \geq 0$. Then the following inequality holds:*

$$\frac{\pi}{ka} J_{n+1}(ka) \leq 2\sqrt{\frac{\pi e}{2n+1}} j_n(ka). \tag{9.3.13}$$

PROOF. Consider the following integral representation of Bessel functions [Wat58]:

$$J_\nu(\nu x) = \frac{1}{\pi} \int_0^\pi e^{-\nu F(\theta, x)} d\theta,$$
$$F(\theta, x) = \ln \frac{\theta + \sqrt{\theta^2 - x^2 \sin^2\theta}}{x \sin\theta} - \cot\theta \sqrt{\theta^2 - x^2 \sin^2\theta}. \tag{9.3.14}$$

Setting $\nu x = ka$ we obtain

$$J_{\nu+1}(ka) = \frac{1}{\pi} \int_0^\pi e^{-(\nu+1)F(\theta, ka/(\nu+1))} d\theta,$$
$$J_{\nu+1/2}(ka) = \frac{1}{\pi} \int_0^\pi e^{-(\nu+1/2)F(\theta, ka/(\nu+1/2))} d\theta. \tag{9.3.15}$$

$F(\theta, x)$ is monotonic function of x and θ [Wat58], and so

$$F\left(\theta, \frac{ka}{\nu+1}\right) \geq F\left(\theta, \frac{ka}{\nu+1/2}\right),$$
$$F\left(\theta, \frac{ka}{\nu+1/2}\right) \geq F\left(0, \frac{ka}{\nu+1/2}\right). \tag{9.3.16}$$

Then we have

$$\begin{aligned} J_{\nu+1}(ka) &= \frac{1}{\pi} \int_0^\pi e^{-(\nu+1)F(\theta, ka/(\nu+1))} d\theta \leq \frac{1}{\pi} \int_0^\pi e^{-(\nu+1)F(\theta, ka/(\nu+1/2))} d\theta \\ &= \frac{1}{\pi} \int_0^\pi e^{-(\nu+1/2)F(\theta, ka/(\nu+1/2))} e^{-\frac{1}{2}F(\theta, ka/(\nu+1/2))} d\theta \\ &\leq e^{-\frac{1}{2}F(0, ka/(\nu+1/2))} \frac{1}{\pi} \int_0^\pi e^{-(\nu+1/2)F(\theta, ka/(\nu+1/2))} d\theta \\ &= e^{-\frac{1}{2}F(0, ka/(\nu+1/2))} J_{\nu+1/2}(ka). \end{aligned} \tag{9.3.17}$$

For $ka < \nu$ we have

$$F\left(0, \frac{ka}{\nu+1/2}\right) > \ln \frac{\nu+\frac{1}{2}+\sqrt{(\nu+\frac{1}{2})^2-(ka)^2}}{ka} - 1 > \ln \frac{\nu+\frac{1}{2}}{ka} - 1 = \ln \frac{2\nu+1}{2eka}. \tag{9.3.18}$$

Substituting this inequality into Eq. (9.3.17) we get

$$J_{\nu+1}(ka) \leq \sqrt{\frac{2eka}{2\nu+1}} J_{\nu+1/2}(ka). \tag{9.3.19}$$

Using the relation between the spherical and usual Bessel functions (2.1.25), we obtain a statement of the lemma:

$$\frac{\pi}{ka} J_{n+1}(ka) \leq \frac{\pi}{ka}\sqrt{\frac{2eka}{2n+1}} J_{n+1/2}(ka) = \frac{\pi}{ka}\sqrt{\frac{2eka}{2n+1}}\sqrt{\frac{2ka}{\pi}} j_n(ka) = 2\sqrt{\frac{\pi e}{2n+1}} j_n(ka). \tag{9.3.20}$$

□

COROLLARY 9 *The results of the above lemmas can be combined in the following inequality:*

$$\left| j_{n'}\left(\frac{ka}{2}\right)\right|\left| j_{n''}\left(\frac{ka}{2}\right)\right| < 2\sqrt{\frac{\pi e}{2n'+2n''+1}} j_{n'+n''}(ka), \tag{9.3.21}$$
$$n' + n'' > ka, \quad ka \geq 0.$$

Let us apply the result of the first lemma to the inequality (9.3.9). We assume that $n \geq p > ka$, in which case $J_{n+m}(ka) < J_{n+1}(ka)$ for $m > 1$. We then have

$$\begin{aligned} |\delta_{n'}| &< c_{n'} \sum_{n=p}^{\infty} \frac{\pi}{ka}(2n+1)|h_n(kb)| \sum_{n''=n-n'}^{n+n'} (2n''+1) J_{n'+n''+1}(ka) \\ &< c_{n'} \sum_{n=p}^{\infty} \frac{\pi}{ka}(2n+1)|h_n(kb)| J_{n+1}(ka) \sum_{n''=n-n'}^{n+n'} (2n''+1) \\ &= (2n'+1)c_{n'} \sum_{n=p}^{\infty} \frac{\pi}{ka}(2n+1)^2 |h_n(kb)| J_{n+1}(ka). \end{aligned} \tag{9.3.22}$$

Using inequality (9.3.13) we obtain

$$|\delta_{n'}| < 2\sqrt{\pi e}(2n'+1)c_{n'} \sum_{n=p}^{\infty} (2n+1)^{3/2} |h_n(kb)| j_n(ka). \tag{9.3.23}$$

Note that this estimate can be improved both for low and large ka taking into account that the Bessel functions decay exponentially with n for $n > ka$ and the principal term (9.3.22) makes the major contribution into the sum. In any case, since functions $|h_n(kb)| j_n(ka)$ decay exponentially with n, the polynomial and constant factors play a minor role in the estimate of large p and we can rewrite Eq. (9.3.23) as

$$|\delta_{n'}| \lesssim (2n'+1)c_{n'} |\epsilon_p^{(s)}|, \qquad |\epsilon_p^{(s)}| \sim \sum_{n=p}^{\infty} |h_n(kb)| j_n(ka), \tag{9.3.24}$$

where $|\epsilon_p^{(s)}|$ is the error of expansion of single source near the center $\mathbf{r} = 0$ (see Eq. (9.1.3)). Since $c_{n'}$ depend on n' as polynomials (in fact, low order polynomials), the overall error $|\delta|$ (Eq. (9.3.5)) is bounded as

$$|\delta| \leq \sum_{n'=0}^{p'-1} |\delta_{n'}| \lesssim d_{p'} |\epsilon_p^{(s)}|, \tag{9.3.25}$$

where $d_{p'}$ is bounded by a polynomial of p'. Since $p \gtrsim p'$ and $\epsilon_p^{(s)}$ asymptotes to the exponential function of p for large p we have

$$|\delta| \sim |\epsilon_p^{(s)}|, \qquad p \gg 1. \tag{9.3.26}$$

Therefore, we showed that in case when the expansion coefficients are bounded as Eq. (9.3.8) the error of translation is of the same order of magnitude as the error of expansion of a source about a new center of expansion, $\mathbf{r} = 0$. We can also state that the expansion of source about $\mathbf{r} = 0$ can be represented by series (9.1.3) which are absolutely and uniformly bounded. Thus, for p selected according to the estimate of $|h_n(kb)| j_n(ka)$, the principal terms of expansion coefficients are bounded as

$$|\hat{\Psi}_n(\mathbf{s})| < c_n |j_n(ka)|, \qquad \hat{\Psi}_n(\mathbf{s}) = \sum_{m=-n}^{n} \mathrm{i}^{-n} \hat{C}_n^m Y_n^m(\mathbf{s}), \tag{9.3.27}$$

$$n = 0, \ldots, p-1.$$

This provides the following simple rule for selection of p for the S|S translations for the monopole source: The truncation number p for the translated expansion should be equal to the truncation number for a direct expansion. In this case the translation error is of the same order of magnitude as the error of the direct expansion.

For example, consider the $S|S$ translation of a monopole source when $ka = 20$, $ka/2 = 10$ with prescribed error $\epsilon = 10^{-6}$ and $\sigma = \sqrt{3}$. Approximation (9.1.42) shows that, in this case, we can take $p' = p(\epsilon, ka/2, \sigma) \approx 26$, $p = p(\epsilon, ka, \sigma) \approx 35$. So translation of the truncated expansion with $p'^2 = 676$ terms can be performed using a truncated translation matrix of size $p^2 \times p'^2$ which provides $p^2 = 1225$ terms for representation of the translated function. We estimate that the error, in this case, will be of order 10^{-6}.

9.3.2 Multipole-to-local $S|R$ translations

9.3.2.1 Problem

Figure 9.10 illustrates the multipole-to-local $S|R$ type of translation. A monopole source is located at $\mathbf{r} = \mathbf{r}_s$ inside a sphere of radius a and we are looking for the evaluation of the truncated translation errors at some point $\mathbf{r}$ located inside the other sphere of the same radius. The distance between a sphere center and the closest point of the other sphere is b and therefore the translation distance is $t = |\mathbf{t}| = a + b$. We place the origin of the reference frame at the center of a sphere which includes the evaluation point $\mathbf{r}$.

As in the previous case, we assume that the source field can be approximated outside the sphere of radius b shown in Fig. 9.10 by truncated series:

$$\psi^{(p')}(\mathbf{r}) = \sum_{n'=0}^{p'-1} \sum_{m'=-n'}^{n'} C_{n'}^{m'} S_n^m(\mathbf{r} + \mathbf{t}), \qquad h_0(k|\mathbf{r} - \mathbf{r}_s|) - \psi^{(p')}(\mathbf{r}) = \epsilon_{p'}, \tag{9.3.28}$$

$$r'_s = |\mathbf{r}_s + \mathbf{t}| < a, \qquad r' = |\mathbf{r} + \mathbf{t}| > b, \qquad a < b.$$

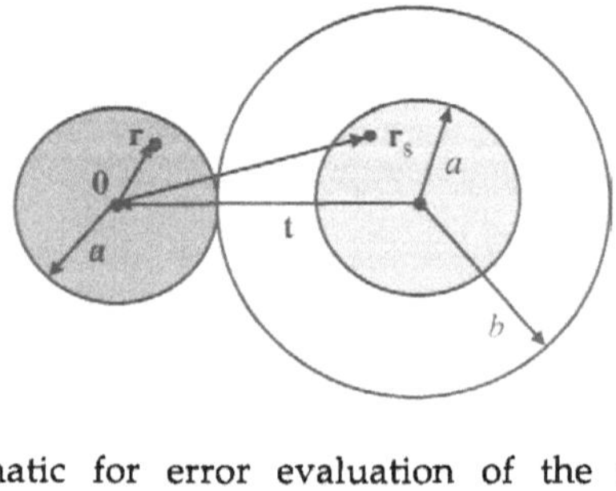

FIGURE 9.10 Schematic for error evaluation of the multipole-to-local $S|R$ translation.

The same field can be approximated inside the darker sphere shown in Fig. 9.10 by the following series:

$$\hat{\psi}^{(p)}(\mathbf{r}) = \sum_{n=0}^{p-1} \sum_{m=-n}^{n} \hat{C}_n^m R_n^m(\mathbf{r}), \qquad h_0(k|\mathbf{r} - \mathbf{r}_s|) - \hat{\psi}^{(p)}(\mathbf{r}) = \epsilon_p, \tag{9.3.29}$$

$$r_s = |\mathbf{r}_s| > a + b, \qquad r = |\mathbf{r}| < a, \qquad a < b.$$

As in the case of the $S|S$ translation, we will try to evaluate the error of expansion (9.3.29) when $\hat{C}_n^m$ are computed using a finite translation scheme. We note that the error can be estimated when the source and evaluation points are located on the boundaries of the domains, $r \times '_s = |\mathbf{r}_s + \mathbf{t}| = a$ and $r = |\mathbf{r}| = a$. We consider this case here.

9.3.2.2 *Solution*

Translation can be performed using rectangular truncated translation matrix. The translated coefficients are then defined as

$$\hat{C}_n^m = \sum_{n'=0}^{p'-1} \sum_{m'=-n'}^{n'} (S|R)_{nn'}^{mm'}(\mathbf{t}) C_{n'}^{m'}, \qquad n = 0, \ldots, p-1, \quad m = -n, \ldots, n. \tag{9.3.30}$$

Since function $\psi^{(p')}(\mathbf{r})$ from Eq. (9.3.28) can be represented in the form:

$$\psi^{(p')}(\mathbf{r}) = \sum_{n=0}^{\infty} \sum_{m=-n}^{n} \sum_{n'=0}^{p'-1} \sum_{m'=-n'}^{n'} (S|R)_{nn'}^{mm'}(\mathbf{t}) C_{n'}^{m'} R_n^m(\mathbf{r}), \tag{9.3.31}$$

the difference between functions $\hat{\psi}^{(p)}(\mathbf{r})$, defined by Eqs. (9.3.29) and (9.3.30) and $\psi^{(p')}(\mathbf{r})$ is

$$\delta = \psi^{(p')}(\mathbf{r}) - \hat{\psi}^{(p)}(\mathbf{r}) = \sum_{n'=0}^{p'-1} \delta_{n'},$$

$$\delta_{n'} = \sum_{n=p}^{\infty} \sum_{m=-n}^{n} \sum_{m'=-n'}^{n'} (S|R)_{nn'}^{mm'}(\mathbf{t}) C_{n'}^{m'} R_n^m(\mathbf{r}). \tag{9.3.32}$$

Using integral representation of the elements of the $S|R$ translation matrix (7.1.58) and definition of surface functions $\Psi_{n'}(\mathbf{s})$ (Eq. (9.3.7)), we obtain the following expression for $\delta_{n'}$:

$$\delta_{n'} = \sum_{n=p}^{\infty} j_n(kr) \sum_{m=-n}^{n} \mathrm{i}^n \int_{S_u} \Lambda_S^{(n+n')}(\mathbf{t};\mathbf{s})\Psi_{n'}(\mathbf{s})Y_n^{-m}(\mathbf{s})Y_n^m\left(\frac{\mathbf{r}}{r}\right)\mathrm{d}S$$

$$= \frac{1}{4\pi}\sum_{n=p}^{\infty}(2n+1)\mathrm{i}^n j_n(kr)\int_{S_u}\Lambda_S^{(n+n')}(\mathbf{t};\mathbf{s})\Psi_{n'}(\mathbf{s})P_n\left(\frac{\mathbf{s}\cdot\mathbf{r}}{r}\right)\mathrm{d}S$$

$$= \frac{1}{4\pi}\sum_{n=p}^{\infty}(2n+1)\mathrm{i}^n j_n(kr)\sum_{n''=|n-n'|}^{n+n'}(2n''+1)\mathrm{i}^{n''}h_{n''}(kt)$$

$$\times \int_{S_u} P_{n''}\left(\frac{\mathbf{s}\cdot\mathbf{t}}{t}\right)\Psi_{n'}(\mathbf{s})P_n\left(\frac{\mathbf{s}\cdot\mathbf{r}}{r}\right)\mathrm{d}S. \quad (9.3.33)$$

In the latter equality, we put the lower limit of summation $n'' = |n - n'|$ instead of $n'' = 0$, since the integral of the products of spherical harmonics is zero for $n'' < |n - n'|$ (see discussion near Eqs. (9.2.4)–(9.2.7)). The absolute value of $\delta_{n'}$ can then be evaluated as

$$|\delta_{n'}| < \max_{\mathbf{s}} |\Psi_{n'}(\mathbf{s})| \sum_{n=p}^{\infty}(2n+1)|j_n(ka)| \sum_{n''=|n-n'|}^{n+n'}(2n''+1)|h_{n''}(kt)|. \quad (9.3.34)$$

From the consideration of the $S|S$ translation of a monopole source, we found that $\max_{\mathbf{s}} |\Psi_{n'}(\mathbf{s})|$ should be bounded according to the size of the expansion domain (see Eqs. (9.3.8) and (9.3.27)). We also note that $|h_{n''}(kt)|$ is a monotonically growing function of n'' and therefore in the latter sum all terms can be majorated by the maximum term. This yields

$$|\delta_{n'}| < c_{n'}|j_{n'}(ka)| \sum_{n=p}^{\infty}(2n+1)|j_n(ka)||h_{n+n'}(kt)| \sum_{n''=|n-n'|}^{n+n'}(2n''+1)$$

$$= c_{n'}(2n'+1)\sum_{n=p}^{\infty}(2n+1)^2|j_n(ka)||j_{n'}(ka)||h_{n+n'}(ka+kb)|. \quad (9.3.35)$$

In the last expression we replaced t with $a + b$.

Now let us prove the following lemma.

LEMMA 5 *The following inequality for Bessel and spherical Hankel functions of real argument holds:*

$$|J_{n'}(ka)||h_{n+n'}(ka+kb)| < |h_n(kb)|, \qquad ka, kb > 0, \quad n, n' = 0, 1, 2, \ldots. \quad (9.3.36)$$

PROOF. The modulus of the Bessel functions of real argument can be represented as [Wat58, p. 444]

$$J_\nu^2(x) + Y_\nu^2(x) = \frac{8}{\pi^2}\int_0^\infty K_0(2x \sinh \eta)\mathrm{e}^{-2\nu\eta}\,\mathrm{d}\eta, \tag{9.3.37}$$

where K_0 is the modified Bessel function of the second kind, which also has the following integral representation [Wat58, p. 446]:

$$K_0(y) = \int_0^\infty \mathrm{e}^{-y\cosh\xi}\,\mathrm{d}\xi. \tag{9.3.38}$$

These two formulae and relation between the spherical and ordinary Bessel functions (2.1.25)–(2.1.27) yield the following integral representation of the modulus of the spherical Hankel function:

$$|h_n(x)|^2 = \frac{4}{\pi x}\int_0^\infty\int_0^\infty \mathrm{e}^{-2x\sinh\eta\cosh\xi}\cosh[(2n+1)\eta]\mathrm{d}\xi\,\mathrm{d}\eta. \tag{9.3.39}$$

The hyperbolic cosine $\cosh x \geq 1$. It also satisfies inequality:

$$\begin{aligned}\cosh(x+y) &= \tfrac{1}{2}\mathrm{e}^{x+y} + \tfrac{1}{2}\mathrm{e}^{-x-y} = \mathrm{e}^{y}(\tfrac{1}{2}\mathrm{e}^{x} + \tfrac{1}{2}\mathrm{e}^{-x-2y})\\ &\leq \mathrm{e}^{y}(\tfrac{1}{2}\mathrm{e}^{x} + \tfrac{1}{2}\mathrm{e}^{-x}) = \mathrm{e}^{y}\cosh x, \quad y > 0.\end{aligned} \tag{9.3.40}$$

Combining this with Eq. (9.3.39), where the integrand is strictly positive function, we obtain

$$\begin{aligned}|h_{n+n'}(kb+ka)|^2 &= \frac{4}{\pi k(b+a)}\int_0^\infty\int_0^\infty \mathrm{e}^{-2k(b+a)\sinh\eta\cosh\xi}\\ &\quad\times\cosh[(2n+2n'+1)\eta]\mathrm{d}\xi\,\mathrm{d}\eta\\ &\leq \frac{4}{\pi k(b+a)}\int_0^\infty\int_0^\infty \mathrm{e}^{2n'\eta-2ka\sinh\eta\cosh\xi}\mathrm{e}^{-2kb\sinh\eta\cosh\xi}\\ &\quad\times\cosh[(2n+1)\eta]\mathrm{d}\xi\,\mathrm{d}\eta\\ &\leq \frac{4}{\pi k(b+a)}\int_0^\infty\int_0^\infty \mathrm{e}^{2n'\eta-2ka\sinh\eta}\mathrm{e}^{-2kb\sinh\eta\cosh\xi}\\ &\quad\times\cosh[(2n+1)\eta]\mathrm{d}\xi\,\mathrm{d}\eta\\ &\leq \frac{4}{\pi k(b+a)}\max_{0\leq\eta<\infty}\mathrm{e}^{2n'\eta-2ka\sinh\eta}\int_0^\infty\int_0^\infty \mathrm{e}^{-2kb\sinh\eta\cosh\xi}\\ &\quad\times\cosh[(2n+1)\eta]\mathrm{d}\xi\,\mathrm{d}\eta\\ &= \frac{b}{b+a}|h_n(kb)|^2\max_{0\leq\eta<\infty}\mathrm{e}^{2n'\eta-2ka\sinh\eta}\\ &< |h_n(kb)|^2\max_{0\leq\eta<\infty}\mathrm{e}^{2n'\eta-2ka\sinh\eta}.\end{aligned} \tag{9.3.41}$$

We note that for $n' \le ka$ the maximum of the exponential function appeared in the above estimate is 1 and reached at $\eta = 0$, since $\sinh \eta \ge \eta$. Since $|J_{n'}(ka)| \le 1$ for any n' and real ka we can see that

$$|J_{n'}(ka)||h_{n+n'}(ka+kb)| \le |h_{n+n'}(ka+kb)| < |h_n(kb)|, \qquad n' \le ka. \tag{9.3.42}$$

Now we need to prove the statement of the lemma for $n' > ka$. In this case the maximum of the exponential function is reached at finite $\eta = \eta_* > 0$, which can be found by taking derivative of that exponent and equalizing it to zero. This yields

$$\cosh \eta_* = \frac{n'}{ka}, \qquad \sinh \eta_* = \sqrt{\frac{n'^2 - (ka)^2}{ka}},$$
$$\eta_* = \ln(\cosh \eta_* + \sinh \eta_*) = \ln \frac{n' + \sqrt{n'^2 - (ka)^2}}{ka}. \tag{9.3.43}$$

Substituting this result into Eq. (9.3.41) we obtain

$$|h_{n+n'}(kb+ka)| < |h_n(kb)|\, e^{n'\eta_* - ka \sinh \eta_*}$$
$$= |h_n(kb)| \exp\left[n' \ln \frac{n' + \sqrt{n'^2 - (ka)^2}}{ka} - \sqrt{n'^2 - (ka)^2} \right], \tag{9.3.44}$$

where $n' > ka$. The last step in the proof is inequality for $|J_{n'}(ka)|$ for $n' > ka$ [AS64]:

$$|J_{n'}(ka)| \le \frac{(ka)^{n'}\, e^{\sqrt{n'^2-(ka)^2}}}{[n' + \sqrt{n'^2 - (ka)^2}]^{n'}}$$
$$= \exp\left[\sqrt{n'^2 - (ka)^2} - n' \ln \frac{n' + \sqrt{n'^2 - (ka)^2}}{ka} \right], \tag{9.3.45}$$

where $n' > ka$. Combining Eqs. (9.3.44) and (9.3.45) we obtain the statement of the lemma for $n' > ka$. □

COROLLARY 10 *The following inequalities hold:*

$$|j_{n'}(ka)||h_{n+n'}(ka+kb)| < |h_n(kb)|,$$
$$ka, kb > 0, \quad n = 0, 1, 2, \ldots, \quad n' = 0, 1, 2, \ldots, [ka], \tag{9.3.46}$$

$$|j_{n'}(ka)||h_{n+n'}(ka+kb)| < \left(\frac{e\pi}{2n'-1}\right)^{1/2} |h_n(kb)|, \qquad ka, kb > 0, \tag{9.3.47}$$
$$n = 0, 1, 2, \ldots, \qquad n' > ka > 0.$$

PROOF. The first inequality follows from the proof of the above lemma, where for $n' \leq ka$ it was sufficient to use the estimate for Bessel functions $|J_{n'}(ka)| \leq 1$. The spherical Bessel functions satisfy the same inequality, $|j_{n'}(ka)| \leq 1$, and so inequality (9.3.46) holds. The second inequality follows from Eqs. (9.3.19) and (9.3.36) where we can set $\nu = n' - \frac{1}{2}$ and use the relation between the spherical and ordinary Bessel functions (2.1.25). □

Inequalities (9.3.46) and (9.3.47) show that the sum in Eq. (9.3.35) converges absolutely and uniformly with the same exponential convergence rate as we observed for p truncated expansions of the source. This yields estimate (9.3.26) for $|\delta|$ specified by Eq. (9.3.32), since the same speculations as we did for the $S|S$ translations can be repeated for the evaluation of the error of the $S|R$ translation (see Eqs. (9.3.24) and (9.3.25)). A peculiarity of the $S|R$ translation is that we can select $p = p'$ as soon as the radius of the sphere for the use of the translated R expansion, a, is same as the maximum distance from the source to the point $\mathbf{r} = -\mathbf{t}$ (the light and the dark gray spheres in Fig. 9.10 are of the same size).

Finally, we note that the $S|R$ translation produces coefficients $\hat{C}_n^m$ that define a surface function, whose modes are bounded as

$$\max_{\mathbf{s}} |\hat{\Psi}_n(\mathbf{s})| < c_n |h_n(kb)|, \qquad \hat{\Psi}_n(\mathbf{s}) = \sum_{m=-n}^{n} \mathrm{i}^{-n} \hat{C}_n^m Y_n^m(\mathbf{s}), \tag{9.3.48}$$
$$n = 0, \ldots, p-1,$$

where c_n are some coefficients that can grow with n not faster than some polynomial of n. This follows from the fact that the direct R expansion of the monopole source near $\mathbf{r} = \mathbf{0}$ has modes bounded as Eq. (9.3.48) (see Fig. 9.1 and discussion after Eq. (9.1.1)). Since the truncated direct and translated expansions have the same error evaluated by the absolutely and uniformly convergent majorating series, the orders of the expansion coefficients should be same.

9.3.3 Local-to-local $R|R$ translations

9.3.3.1 Problem

Figure 9.11 illustrates the local-to-local $R|R$ translation in the multilevel FMM. Here a monopole source is located at $\mathbf{r} = \mathbf{r}_s$ outside a sphere of radius b and we are looking for the evaluation of the truncation errors at some point $\mathbf{r}$ located inside a sphere of radius $a/2$ centered at $\mathbf{r} = \mathbf{0}$. This sphere is located completely inside larger sphere of radius a concentric with the sphere of radius b.

Assume that we have a local expansion centered at $\mathbf{r} = -\mathbf{t}$, $|\mathbf{t}| = a/2$, which approximates the potential of the source inside the sphere of radius a:

$$\psi^{(p')}(\mathbf{r}) = \sum_{n'=0}^{p'-1} \sum_{m'=-n'}^{n'} C_{n'}^{m'} R_n^m(\mathbf{r}+\mathbf{t}), \qquad h_0(k|\mathbf{r}-\mathbf{r}_s|) - \psi^{(p')}(\mathbf{r}) = \epsilon_{p'}, \tag{9.3.49}$$

$$r'_s = |\mathbf{r}_s + \mathbf{t}| > b, \qquad r' = |\mathbf{r}+\mathbf{t}| < a, \qquad a < b.$$

By an approximate (truncated) translation of the expansion coefficients $C_n^m \to \hat{C}_n^m$, we produce a new series centered at $\mathbf{r} = \mathbf{0}$ and valid for any $\mathbf{r}$ located inside the smallest sphere:

$$\hat{\psi}^{(p)}(\mathbf{r}) = \sum_{n=0}^{p-1} \sum_{m=-n}^{n} \hat{C}_n^m R_n^m(\mathbf{r}), \qquad h_0(k|\mathbf{r}-\mathbf{r}_s|) - \hat{\psi}^{(p)}(\mathbf{r}) = \epsilon_p, \tag{9.3.50}$$

$$r_s = |\mathbf{r}_s| > b - \frac{1}{2}a, \qquad r = |\mathbf{r}| < \frac{1}{2}a, \qquad a < b.$$

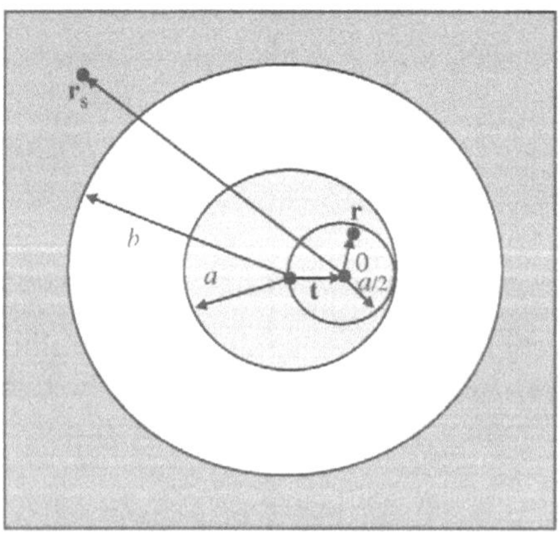

FIGURE 9.11 Schematic for error evaluation of the local-to-local $R|R$ translation.

As before, our goal is to specify finite transform $C_{n'}^{m'} \to \hat{C}_n^m$ $(n' < p', n < p)$ and evaluate ϵ_p which combines the error of approximation $\epsilon_{p'}$ and the error of application of truncated translation. Again we assume that the maximum error is determined by the error at the boundaries of the domain at $r'_s = |\mathbf{r}_s + \mathbf{t}| = b$ and $r = |\mathbf{r}| = a/2$. So, further we can consider only this case and apply the error bound for other locations of the source and evaluation points inside the domains.

9.3.3.2 *Solution*

If translation is performed using the rectangular truncated translation matrix we have

$$\hat{C}_n^m = \sum_{n'=0}^{p'-1} \sum_{m'=-n'}^{n'} (R|R)_{nn'}^{mm'}(\mathbf{t}) C_{n'}^{m'}, \qquad n = 0, \ldots, p-1, \quad m = -n, \ldots, n. \tag{9.3.51}$$

Since function $\psi^{(p')}(\mathbf{r})$ from Eq. (9.3.49) can be represented in the form:

$$\psi^{(p')}(\mathbf{r}) = \sum_{n=0}^{\infty} \sum_{m=-n}^{n} \sum_{n'=0}^{p'-1} \sum_{m'=-n'}^{n'} (R|R)_{nn'}^{mm'}(\mathbf{t}) C_{n'}^{m'} R_n^m(\mathbf{r}), \tag{9.3.52}$$

the difference between functions $\hat{\psi}^{(p)}(\mathbf{r})$ defined by Eq. (9.3.50) and $\psi^{(p')}(\mathbf{r})$ is

$$\delta = \psi^{(p')}(\mathbf{r}) - \hat{\psi}^{(p)}(\mathbf{r}) = \sum_{n'=0}^{p'-1} \delta_{n'},$$
$$\delta_{n'} = \sum_{n=p}^{\infty} \sum_{m=-n}^{n} \sum_{m'=-n'}^{n'} (R|R)_{nn'}^{mm'}(\mathbf{t}) C_{n'}^{m'} R_n^m(\mathbf{r}). \tag{9.3.53}$$

Evaluation of the modulus of $\delta_{n'}$ can be performed in the same way as the one used for the derivation of estimates (9.3.6) and (9.3.34). This yields

$$|\delta_{n'}| < \max_{\mathbf{s}} |\Psi_{n'}(\mathbf{s})| \sum_{n=p}^{\infty} (2n+1) \left| j_n\left(\frac{ka}{2}\right)\right| \sum_{n''=|n-n'|}^{n+n'} (2n''+1) \left| j_{n''}\left(\frac{ka}{2}\right)\right|. \tag{9.3.54}$$

For R expansion of a monopole source $\max_{\mathbf{s}} |\Psi_{n'}(\mathbf{s})|$ is bounded as (9.3.48).

Therefore, $|\delta_{n'}|$ are bounded as

$$|\delta_{n'}| < c_{n'} \sum_{n=p}^{\infty} (2n+1)\left|j_n\left(\frac{ka}{2}\right)\right| \sum_{n''=|n-n'|}^{n+n'} (2n''+1)\left|j_{n''}\left(\frac{ka}{2}\right)\right| |h_{n'}(kb)|. \tag{9.3.55}$$

LEMMA 6 *The following inequality for Bessel and spherical Hankel functions of real argument holds:*

$$\begin{aligned} &\left|J_{|n-n'|+n''}\left(\frac{ka}{2}\right)\right| |h_{n'}(kb)| < \left|h_n\left(kb - \frac{ka}{2}\right)\right|, \\ &kb > \frac{ka}{2} > 0, \quad n, n', n'' = 0, 1, 2, \ldots, \end{aligned} \tag{9.3.56}$$

PROOF. For $n' < n$ due to the fact that $|h_\nu(x)|$ monotonically grow with ν and decay with x and the Bessel functions are bounded, $|J_\nu(x)| \leq 1$, we have

$$\left|J_{|n-n'|+n''}\left(\frac{ka}{2}\right)\right| |h_{n'}(kb)| \leq |h_{n'}(kb)| < \left|h_n\left(kb - \frac{ka}{2}\right)\right|. \tag{9.3.57}$$

So now the case $n' \geq n$ should be considered. If $n' \geq |n - n'| + n''$ we can use inequality (9.3.36) which can be rewritten as

$$\left|J_{|n-n'|+n''}\left(\frac{ka}{2}\right)\right| \left|h_{n'}\left(kb - \frac{ka}{2} + \frac{ka}{2}\right)\right| < \left|h_{n'-n''-|n-n'|}\left(kb - \frac{ka}{2}\right)\right|. \tag{9.3.58}$$

For $n' \geq n$ condition $n' \geq |n - n'| + n''$ results in $n \geq n''$ and $n' - n'' - |n - n'| = n - n''$. Due to the fact that $|h_n(x)|$ monotonically grow with n we obtain

$$\left|J_{|n-n'|+n''}\left(\frac{ka}{2}\right)\right| |h_{n'}(kb)| < \left|h_{n-n''}\left(kb - \frac{ka}{2}\right)\right| \leq \left|h_n\left(kb - \frac{ka}{2}\right)\right|. \tag{9.3.59}$$

The last case remains to be considered is $n' < |n - n'| + n''$. In this case we can again use the property of monotonic decay of $|h_\nu(x)|$ for growing ν and inequality (9.3.36) (considered for $n = 0$):

$$\begin{aligned} \left|J_{|n-n'|+n''}\left(\frac{ka}{2}\right)\right| |h_{n'}(kb)| &< \left|J_{|n-n'|+n''}\left(\frac{ka}{2}\right)\right| \left|h_{|n-n'|+n''}\left(kb - \frac{ka}{2} + \frac{ka}{2}\right)\right| \\ &< \left|h_0\left(kb - \frac{ka}{2}\right)\right| \leq \left|h_n\left(kb - \frac{ka}{2}\right)\right|. \end{aligned} \tag{9.3.60}$$

□

COROLLARY 11 *The following inequalities hold for $kb > ka/2 > 0$:*

$$\left|j_{|n-n'|+n''}\left(\frac{ka}{2}\right)\right||h_{n'}(kb)| < \left|h_n\left(kb - \frac{ka}{2}\right)\right|, \quad n, n', n'' = 0, 1, 2, \ldots, \quad n' \leq \max\left(n, \frac{ka}{2}\right), \tag{9.3.61}$$

$$\left|j_{|n-n'|+n''}\left(\frac{ka}{2}\right)\right||h_{n'}(kb)| < \left(\frac{e\pi}{2n' + 2n'' - 2n - 1}\right)^{1/2}\left|h_n\left(kb - \frac{ka}{2}\right)\right|, \quad n, n', n'' = 0, 1, \ldots, \quad n' > \max\left(n, \frac{ka}{2}\right). \tag{9.3.62}$$

PROOF. The first inequality follows from the proof of the above lemma, where for $n' < n$ it was sufficient to use the estimate for Bessel functions $|J_{|n-n'|+n''}(ka/2)| \leq 1$. The spherical Bessel functions are bounded by the same constant, $|j_{|n-n'|+n''}(ka/2)| \leq 1$, and so Eq. (9.3.61) holds for $n' < n$. In the case $ka/2 > n' \geq n$ the first inequality holds because inequality (9.3.46) can be used in this case instead of inequality (9.3.36) (so J can be replaced with j in the proof of the lemma). The second inequality follows from Eq. (9.3.47) which in this case replaces inequality (9.3.36) used in the proof of the lemma. □

Note that due to the accuracy that is accepted in our proof we can simplify inequalities (9.3.61) and (9.3.62) by combining them into one inequality that holds for $kb > ka/2 > 0$:

$$\left|j_{n''}\left(\frac{ka}{2}\right)\right||h_{n'}(kb)| < (e\pi)^{1/2}\left|h_n\left(kb - \frac{ka}{2}\right)\right|, \quad n, n' = 0, 1, \ldots, \quad n'' = |n - n'|, |n - n'| + 1, \ldots. \tag{9.3.63}$$

This estimate enables us to rewrite inequality (9.3.55) in the form:

$$|\delta_{n'}| < (e\pi)^{1/2} c_{n'} \sum_{n=p}^{\infty} (2n+1)\left|j_n\left(\frac{ka}{2}\right)\right|\left|h_n\left(kb - \frac{ka}{2}\right)\right| \sum_{n''=|n-n'|}^{n+n'} (2n''+1)$$
$$= c_{n'}(e\pi)^{1/2}(2n'+1) \sum_{n=p}^{\infty} (2n+1)^2\left|j_n\left(\frac{ka}{2}\right)\right|\left|h_n\left(kb - \frac{ka}{2}\right)\right|. \tag{9.3.64}$$

This result is similar to that obtained for the $S|R$ translations. Sum (9.3.64) converges absolutely and uniformly with the same exponential convergence rate as for p truncated expansions of the source. As in the case of the $S|R$ and $S|S$ translations, estimate (9.3.26) for $|\delta|$ specified now by Eq. (9.3.53) holds by the same reasons as we described for the $S|S$ translations, since the same speculations as we did for the $S|S$ translations can be repeated for the evaluation of the error of the $S|R$ translation (see Eqs. (9.3.24) and (9.3.25)).

Finally, we do the same type of remark as for the other types of translations that the $R|R$ translation produces coefficients $\hat{C}_n^m$ that define a surface function, whose modes are bounded as

$$\max_{\mathbf{s}} |\hat{\Psi}_n(\mathbf{s})| < c_n \left| h_n\left(kb - \frac{ka}{2}\right)\right|, \qquad \hat{\Psi}_n(\mathbf{s}) = \sum_{m=-n}^{n} \mathrm{i}^{-n} \hat{C}_n^m Y_n^m(\mathbf{s}), \tag{9.3.65}$$

$$n = 0, \ldots, p-1.$$

where c_n are some coefficients which can grow with n not faster than some polynomial of n.

9.3.4 Some remarks

Here, we considered error bounds associated with translation of a single monopole using truncated translation matrices in a simplified setting. This allowed us to use some theorems from the theory of Bessel and Hankel functions. Here, several issues can be clarified concerning practical use of the obtained results.

First, based on this approach it is not difficult to evaluate how $\Psi_{n'}(\mathbf{s})$ contributes to $\hat{\Psi}_n(\mathbf{s})$ to see whether or not some elements of the truncated matrices can be set to zero to achieve the same accuracy. We performed such a study, and the reader can also do something similar, which shows that in the sense of uniform error contribution the truncation of matrices performed by the rectangular way ($p^2 \times p'^2$) is optimal (other opportunities include triangular, trapezoidal, etc. truncations). We found that the $p^2 \times p'^2$ dense rectangular matrix is the *minimal* translation matrix which can ensure the error bounds determined by the accuracy of expansions in the respective domains (numbers p and p') in the case of the $S|R$ and the $R|R$ translations. In the case of the $S|S$ translation a trapezoidal matrix (the $p^2 \times p'^2$ $S|S$ matrix, whose elements $(S|S)_{nn'}^{mm'}$ set to zero for $2n' - n \geq p$) provides the minimal truncation. Even though it exists this truncation does not bring too much computational savings.

Second, we assumed some special setting (like the source and evaluation points are located on the boundary of the domains), where the error is presumably the largest. One may complain that the Bessel functions are oscillating and it may happen that, e.g. $j_n(ka)$ (or $j_n(ka/2)$) is zero on the boundary for some n or ka, and that a particular truncation number obtained cannot be used, since the evaluation is not correct in this case. This is a good observation, if we speak about exact p, and, the actual error of truncation is some oscillating function on p. There are several ways to resolve this problem. One way is to note that if, say, $j_n(ka)$ is zero, then $j_{n+1}(ka)$ is not zero for this value of ka and its module is close to the maximum, so the truncation number can be shifted by 1 to obtain a correct result. Another way, which one may select for more accurate proofs, is to say that, in fact, we consider the domain of *characteristic* size a, which means that we do not specify a exactly, may vary it in some range and consider some a from this range, which provides the maximum error.

Third, the above considerations can be easily modified for the case of arbitrary multipole sources, and we have all necessary material for this in the chapter. In the case when the initial function for expansion is not a monopole, but say a multipole of degree n with the magnitude of the maximum mode of order 1 for some $n \neq 0$, then, as we showed above, this leads basically to the shift of the truncation number p to $p + n$. In the practically important case of a dipole (e.g. for the boundary element method) this shifts p by 1 and this type of corrections are within the errors of approximation of p as a function of ka, b/a, and the prescribed accuracy, ϵ.

Fourth, we note that truncations are performed with p when the truncated "tail" of the series converges absolutely and uniformly and is majorated by an *exponential* convergent series. The meaning of this is that a change of p in this region by one leads to one order of magnitude for the error change (of course, this depends on the exponent, but any way this is a small number of order 1). It is not difficult in the program to change p by 1 or by 5, and this almost does not affect the speed and memory at larger p. So variations of the truncation number are usually much smaller than the truncation number itself.

Finally, we note that the errors are estimated in a strong *norm*. We selected it as the absolute value of the function maximum over the entire evaluation domain. Solution of the Helmholtz equation may not require such a strong norm, since in many applications this is an equation in the *frequency domain*, which should be somehow convolved with other functions, etc. So other norms such as L_2 can be considered and this should relax requirements for truncation numbers.

9.3.5 FMM errors

To check the relation of the actual errors observed in the FMM to theoretical error bounds, we performed several error tests by running the FMM code for spatially uniform random distributions of monopole sources and receivers using $O(p^3)$ translation methods based on reexpansion matrix decompositions. Our tests showed that fast methods described in Chapter 7, namely the rotational–coaxial translation decomposition and sparse matrix decomposition with $N_t = p + p' - 1$, have the same accuracy as the $O(p^4)$ method using the truncated reexpansion matrix without any decomposition. So we do not make any difference between the error of these methods and below we discuss two typical results.

9.3.5.1 *Low and moderate frequencies*

Figure 9.12 shows dependencies of the maximum absolute error:

$$\epsilon = \max_{m=0,\ldots,M-1} \left| \psi_{\mathrm{FMM}}(\mathbf{y}_m) - \sum_{n=0}^{N-1} G(\mathbf{x}_n - \mathbf{y}_m) \right| \tag{9.3.66}$$

in the FMM computation of the sum of Green's functions centered at $\mathbf{x}_n$ and evaluated at $\mathbf{y}_m$ for $N = M = 1000$ and random uniform distribution of sources and receivers inside a cube of size $D_0 \times D_0 \times D_0$ for two values of parameter kD_0, which can be classified as a "low frequency" case. It is seen that the FMM error decays exponentially in this case, and relatively low truncation numbers provide a good approximation.

To compare these errors with the theoretical error, we can rewrite Eq. (9.1.26) obtained for the error of the zero-order spherical Hankel function in terms of the error of computation of Green's function:

$$\epsilon_p < \frac{k}{4\pi ka} \frac{\sigma^{-p-1}}{(1 - \sigma^{-1})^{3/2}} = \frac{1}{4\pi a} \frac{\sigma^{-p-1}}{(1 - \sigma^{-1})^{3/2}}. \tag{9.3.67}$$

This shows that the low frequency bound does not depend on k and for a domain of fixed size D_0 the same bound can be applied for all computations, as soon as they are classified as being in the "low frequency case". Therefore, for a fixed domain and the maximum level of space subdivision in the FMM the dependence of ϵ_p on p is controlled only by a single parameter $\sigma = b/a$.

To understand what should be the value of this parameter for the FMM, let us consider the domains of expansion validity in the FMM. These domains are shown in Fig. 9.13. The picture on the left applied to the S expansions, where the sources are located in the light gray box,

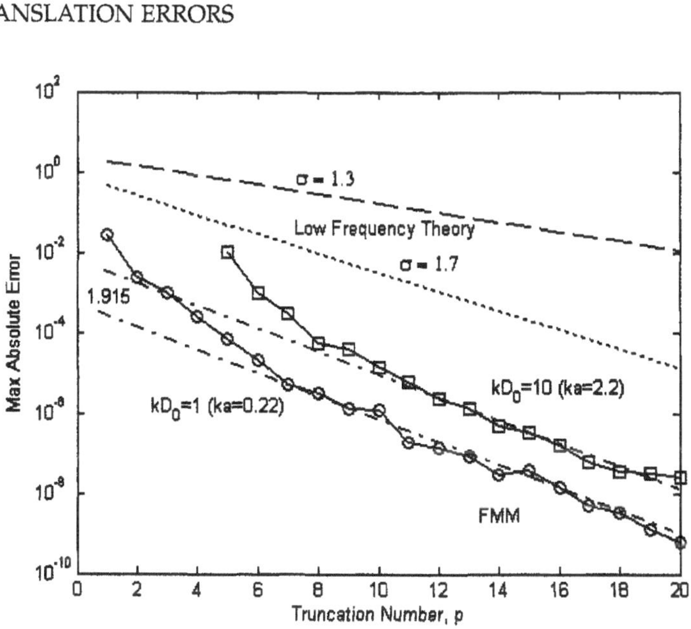

FIGURE 9.12 The dependencies of the maximum absolute error on the truncation number in the FMM computations (the curves marked by the circles and squares). The field was generated by a random spatially uniform distribution of 1000 monopole sources (Green's functions) and evaluated at 1000 spatially uniform random locations. *ka* indicated near the curves is based on the radius of the sphere, *a*, circumferencing the box at level 2 of space subdivision. Theoretical bounds obtained for a single Green's function in low frequency approximation are shown by the dashed and the dotted lines for two different σ. The dash-dotted lines show dependence $\epsilon = A\sigma^{-p}$ with some constant A and $\sigma = 1.915$.

while the evaluation points are located in the darker gray boxes, shows that in the three-dimensional case, we have $\sigma = b/a = \sqrt{3} \approx 1.7$. A similar situation takes place for the R expansions (exchange the source and receiver domains) and for the evaluation of the truncation, one can use the same σ. For the $S|R$ translation, however, we have a different situation. Here for a box of a unit size $a = \sqrt{3}/2$, while $b = 2 - \sqrt{3}/2$. This results in lower $\sigma = b/a = 4/\sqrt{3} - 1 \approx 1.3$. Since for smaller σ the error is larger, it seems that $\sigma \approx 1.3$ should be used for the theoretical error bound.

The bounds corresponding to Eq. (9.3.67) with these two ratios are shown in Fig. 9.12. This figure demonstrates, however, an interesting and rather strange behavior of the error at the first look. First, notice that the actual error is many times smaller than the theoretical error for a *single* source. If we have N sources one can bound the error rather with $N\epsilon_p$, not ϵ_p, since we can create a situation, when the sources are located somehow coherently that the error should sum up. For random distributions this

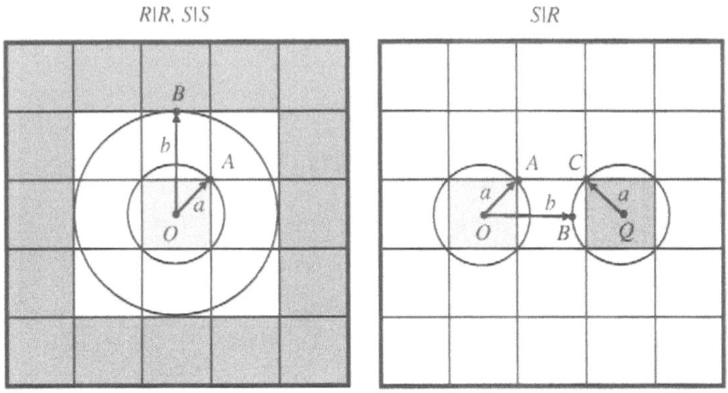

FIGURE 9.13 Illustration for determination of parameters a and b required for theoretical error bounds for the space partitioning by cubic boxes. The chart on the left shows a and b for the $S|S$ and $R|R$ translations. The chart on the right shows the $S|R$ translation scheme.

does not happen, which may be because of some cancellation of oscillating complex fields for different sources at a given receiver location. We are also certain that the sources are not located in the worst position on the boundary, which was used for the error bound establishment. In any case, while some speculations can be done concerning explanation of a quantitative observation of so low an actual error, more important from the point of view of understanding, in our opinion, is the decay exponent in the dependence $\epsilon(p)$.

The actual decay exponent seems different from what the theory predicts. We checked that the dependence $\epsilon(p)$ can be approximated well with some function of type $\epsilon(p) \sim 2^{-p}$, while the theory predicts $\epsilon(p) \sim \sigma^{-p}$ with $\sigma \approx 1.3$. The reason for this may be the following. The error bound is estimated for the worst location of the source and receiver points, which is the case when the source, receiver, and the expansion centers are located on the *same* line and the distance between the source and receiver is $b - a$. As we can see from illustration of the expansion domains shown in Fig. 9.13, this situation never happens. Indeed, in the case when the source and evaluation points are located on the boundaries of the domain (points A and B on the left in Fig. 9.13 and points A and C on the right in the figure). So even in the case when the source and evaluation points are located on corresponding spheres, there is a non-zero (or not π) angle θ between the radius vectors of the source and evaluation points in the reference frames related to the expansion centers. This means that the modulus of the Legendre polynomials in the series, $P_n(\cos\theta)$ are not 1, but smaller, and, in fact, the rate of the series convergence should be faster, than predicted by the "worst" case analysis. While some accurate considerations can be performed to determine what is the effect of this angular dependence on the error, some heuristic speculations can provide hints. For example, one can note that the error bound is determined by the radii that in their turn are determined by points A and C. On the other hand, if we have a source located at point A, then its R expansion near the center Q evaluated at point C has $a = \sqrt{3}/2$, while the distance from A to Q, which is b for source expansion is $|AQ| = \sqrt{11}/2$. Therefore, the effective parameter σ, in this case, is $\sigma = \sqrt{11/3} \approx 1.915$. As before we did not achieve the "worst" case, since vectors QA and QC are not colinear, but we came closer to it. Functions $\epsilon(p) \sim \sigma^{-p}$ for $\sigma = 1.915$ are plotted for illustration in Fig. 9.13 and show that this σ is close to that observed in the numerical experiments.

9.3.5.2 *Higher frequencies*

The same problem was solved with the FMM for higher frequency range. Figure 9.14 shows the error plots for dependence (9.3.66) on the truncation number. In this case the theoretical error was determined according to

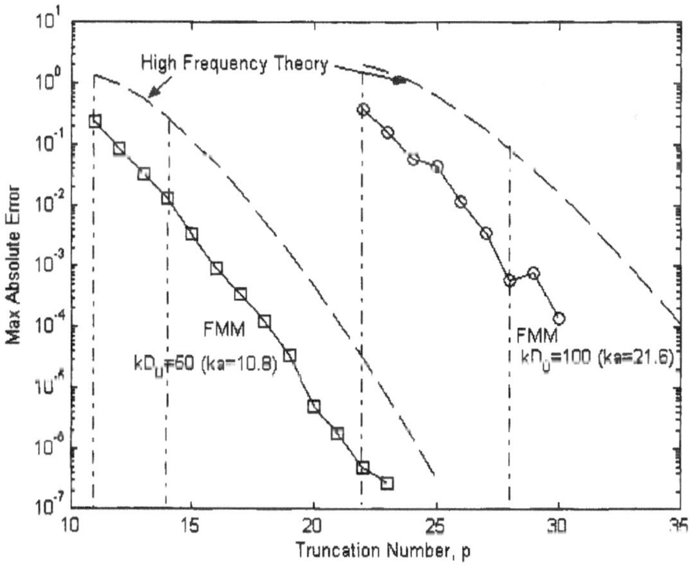

FIGURE 9.14 Comparisons of the error obtained in the FMM computations (curves marked by the squares and circles) with the predictions of the high frequency theory (dashed curves). The vertical dash-dotted lines show formal bounds for the high frequency theory applicability. The FMM settings and notations are same as for Fig. 9.12.

Eqs. (9.1.35) and (9.1.39) which was rewritten in terms of dependence of ϵ on p for Green's function error (so scaling factor was introduced):

$$\epsilon_p = \frac{k}{4\pi\sigma(ka)^{1/3}} \exp\left\{-\frac{1}{3}\left[\frac{2(p-ka)}{(ka)^{1/3}}\right]^{3/2}\right\}. \tag{9.3.68}$$

This formula is valid for the range of truncation numbers $ka < p < kb$. This range, for each case, is shown by the dash-dotted vertical lines in the figure for $\sigma = b/a = 4/\sqrt{3} - 1$. While formally the error bound is valid for $p < kb$ we can see that the equation is applicable in a wider range, which again can be explained by the effect of enlarging of the efficient σ. In the case of higher frequencies this practically does not affect the error bound (9.3.68), but extends the range of applicability of this estimate.

The figure shows that qualitatively the behavior of the error is described well. Again we can observe that quantitatively the actual error is much smaller than it is predicted by the theory for a given p even though we did not multiply the theoretical error bound by N. Possible explanation of this effect is same as in the low frequency case. We can notice that in terms of the values of truncation number p the theoretical predictions are pretty good, since they overestimate the actual p which can be used in computations by a low number, like 3 or 5, which is much smaller than the value of p itself. Finally, we note that in terms of high frequency computations, some care should be taken on roundoff errors and recursive computations of the matrix entries. The errors in these computations can affect the actual error, which otherwise should stay within the theoretical predictions.

9.4 SUMMARY

Here, we considered the errors associated with expansion and translation of monopoles and multipoles. We identified three regions where the behavior of the expansion error is qualitatively different: the low frequency region, the high frequency region, and the transition region. We showed absolute and uniform convergence of the series, and found that in any case $p > ka$, where a is the distance from the source or receiver to the expansion center and provided some approximations and illustrations using the error map. We then analyzed the errors of translation based on a simplified example of monopole translation. This consideration can be extended to translation of multipole sources. By proving several theorems concerning bound of the Bessel and Hankel functions, we showed that any type of translation can be performed using

$p^2 \times p'^2$ rectangularly truncated translation matrices, where p' is the truncation number, sufficient for approximation of the initial expansion in the domain Ω', from which translation is performed, and p is the truncation number, sufficient for approximation of the initial expansion in the subdomain $\Omega \subset \Omega'$, to which translation is performed.

CHAPTER 10

Fast Solution of Multiple Scattering Problems

The statement of the problem of multiple scattering from N objects was described in Chapter 4, where we first showed how multipole methods can be applied to solve it for the case of many spherical scatterers, and then formulated it in more general terms involving T-matrices. Once the T-matrices of the scatterers are known, a larger scattering matrix can be formed, where each block is responsible for the pairwise interaction between the objects. In general, in the three-dimensional case with arbitrary spatial location of scatterers, and when the infinite vectors of expansion coefficients can be truncated to the size $P = p^2$, the computational complexity of the problem is $O(N^3 p^6)$. This arises from a need to solve a linear system of size $Np^2 \times Np^2$. Of course, if the frequencies are sufficiently low, then p can be of the order 10 or so and we have just a few (say, 10) scatterers that one needs to solve a 1000×1000 system, which is not a large problem and can be handled by an ordinary PC in reasonable time. However, if we speak about higher frequencies, when p can be, e.g. $10-10^2$, and larger scale problems, e.g. $N \sim 10^2-10^5$, it is obvious that direct solvers are simply not applicable, since in this case we are faced with the inversion of matrices of sizes from $10^4 \times 10^4$ to $10^9 \times 10^9$. A user then needs to wait for hours for solution of the smallest matrix from this range on his or her PC, while the larger problem is definitely out of memory and speed resources even for today's supercomputers.

Problems of this type appear in practice, e.g. for modeling of disperse systems, light scattering from nanoparticles, inverse scattering problems, design of optical and acoustic bandgap materials, etc. These problems are of a physical nature, and normally one is required to solve not just one single problem, but rather conduct some parametric study and design, which may require thousands and even millions of different solutions of the problem in different settings (e.g. variation of sizes, distributions, etc., plus the study usually must be done over a frequency range).

This motivates the development of faster methods, which should speed up the solution by factors of millions and billions to make the solution of such problems commonplace by physicists, engineers, and applied researchers.

Below we consider several steps which lead to fast algorithms with low memory requirements, and enable such type of reductions in computational complexity. We also refer the reader to other literature on computation of multiple scattering problems (e.g. [KC98]), where the FMM is applied in a different way and some opportunities for algorithm speed up with the FFT are discussed.

10.1 ITERATIVE METHODS

The first substantial reduction in complexity can be achieved by using iterative methods for linear system solution. Assuming that each iteration involves one direct large dense matrix–vector multiplication of complexity $O(N^2p^4)$ and that the iterative process converges to the required accuracy in N_{iter} steps, we immediately obtain complexity $O(N_{\text{iter}}N^2p^4)$ instead of $O(N^3p^6)$. For example, if $N \sim 100$, $p \sim 10$, and $N_{\text{iter}} \sim 10$ the savings could be of order 10^3 times. Of, course, convergence can be slower, but anyway, if $N_{\text{iter}} \ll Np^2$ the iterative procedure brings substantial savings. We note that in this context there exist methods (e.g. based on Krylov subspaces), which guarantee that for $N_{\text{iter}} \sim Np^2$ exact solution will be achieved in exact arithmetic, while in practice this occurs at much smaller N_{iter}.

The scattering problem can be considered as an abstract mathematical problem, in which case a solid arsenal of different iterative methods starting from simple techniques, to various versions of conjugate gradient, and Krylov subspace methods (e.g. GMRES) are available and can be tried. Convergence of these methods can be improved by using well-designed *preconditioners*, block-decompositions, etc. Description and analysis of the entire area of iterative techniques are certainly beyond the scope of this book, and we refer the interested reader to appropriate textbooks [Kel95, Saa92, Saa03].

10.1.1 Reflection method

Another way one can proceed is based on a physical insight into the problem. In fact, a good iterative method for a physical problem, in our opinion, should rather combine ideas and advantages appearing from different viewpoints, than put them into competing positions.

Some physics-based speculations are very useful for the design of a proper iterative methods. Indeed, even the highly popular subspace-based iterative techniques such as the Krylov methods can be understood from the physics of the problem.

For example, one can understand that a passive scatterer does not generate energy. All the energy in the system of scatterers comes from the incident wave and the primary interaction is the interaction of the scatterer with the incident wave. A part of the scattered energy travels to infinity (or is partially absorbed by the boundaries of the domain). This results in *energy losses*. Some scattered waves reach other scatterers, reflect from them, and comes back. While this effect can be strong, only a part of energy come back, and this generally makes this secondary "effective" incident wave weaker than the primary. The process can be repeated many times, and the number of reflections can be substantial. However, if we assume that at each reflection only part of the energy comes back, anyway this should be less than 100% and the process based on consideration of subsequent reflections should converge *exponentially*.

For example, assume that 80% of the scattered energy comes back to some scatterer, and this number is repeated for each reflection. Then the 50th reflection will have about 10^{-5} of the primary energy, and it may happen that this accuracy is sufficient and the iteration process may stop. The exponent of decay (ln(0.8) in the above example) depends on the relative area of a sphere surrounding the given scatterer covered by the other scatterers, absorption properties of the scatterers, and other parameters. We note that the energy, e.g. the acoustic energy of the field can be expressed via a sum of moduli of the expansion coefficients over different modes (since the modes are independent). So the type of convergence we can expect in the case of the above method, which can be called the *reflection method,* is a convergence in the space of coefficients, which is exactly what we are looking for, since our goal is to solve the system of linear equations for the expansion coefficients.

In reality, a more complex situation may occur, e.g. for substantially dense packings of a large number of scatterers. In this case, while some losses of energy can be expected for scatterers close to the cluster boundary, the clusters closer to the center may emit approximately the same amount of energy, as they receive. Indeed, imagine that all the scatterers are approximately in the same situation, and so almost all the energy which goes out from a given scatterer comes to some other scatterers. In turn if admittance is zero and reflection is perfect they scatter some energy, whose sum at the location of the given scatterer may be approximately the same as the energy radiated by that scatterer. This is due to the relative *symmetry* of the scatterers and the fact that the energy can disappear from the system only through the boundaries of the cluster, which presumably are far from the center. In such a situation the convergence of the reflection method can be very slow, and some other techniques need to be considered. Particularly, the reflection method itself can be modified, by performing additional iterations for some part of the domain occupied by scatterers, or by treating several scatterers as a single object with a respective T-matrix for the system (see Chapter 4). In spirit the reflection method is close to the Neumann iteration used for the solution of integral equations of the second kind, and the Börn approximation [CK98].

From a mathematical viewpoint the reflection method is a method of simple iterations. We do not claim that this particular method is the best, but because it offers a good illustration of a fast iterative method, it is worth studying further. Indeed, any iterative method which requires a matrix–vector multiplication at each step can be used with the fast multipole method, as long as it converges and is *stable* to the errors in this multiplication.

Let us now consider some details of the reflection method. Equation (4.4.16) provides basic relations which we need to solve to determine vectors of coefficients $\mathbf{A}^{(q)}$ for each scatterer, $q = 1, \ldots, N$. These equations also involve vectors $\mathbf{B}^{(q)}$ that are vectors describing the "effective" incident field near the qth scatterer due to all other scatterers. According to Eq. (4.4.16) they can be found by the application of translation operators to all other vectors of coefficients and summing up the result. In the iteration process we can build sequences of vectors $\mathbf{A}^{(q)}$ and $\mathbf{B}^{(q)}$, which, hopefully, converges to the true solution:

$$\mathbf{A}_0^{(q)}, \mathbf{A}_1^{(q)}, \ldots, \qquad \mathbf{B}_0^{(q)}, \mathbf{B}_1^{(q)}, \ldots, \qquad \lim_{j\to\infty} \mathbf{A}_j^{(q)} = \mathbf{A}^{(q)},$$
$$\lim_{j\to\infty} \mathbf{B}_j^{(q)} = \mathbf{B}^{(q)}, \qquad q = 1, \ldots, N. \tag{10.1.1}$$

Equation (4.4.16) can then be replaced by the following system of iterative approximations:

$$\mathbf{B}_j^{(q)} = \sum_{q' \neq q} (\mathbf{S}|\mathbf{R})(\mathbf{r}'_{q'q})\mathbf{A}_j^{(q')}, \qquad \mathbf{A}_j^{(q)} = \mathbf{T}^{(q)}(\mathbf{B}_{j-1}^{(q)} + \mathbf{E}^{(q)}),$$
$$q = 1, \ldots, N, \; j = 0, 1, \ldots. \tag{10.1.2}$$

The meaning of these equations is rather obvious. Indeed, if we have some approximation $\mathbf{A}_j^{(q)}$, $q = 1, \ldots, N$, which is a result of taking into account j reflections, $j = 0, 1, \ldots$, then $(\mathbf{B}_j^{(q)} + \mathbf{E}^{(q)})$ are coefficients of the "effective" incident field near the qth scatterer. We can then determine new coefficients $\mathbf{A}_{j+1}^{(q)}$ for the scatterer in this incident field and repeat the process.

The process starts with some initial guess, and we choose

$$\mathbf{A}_0^{(q)} = \mathbf{T}^{(q)}\mathbf{E}^{(q)}, \qquad (\mathbf{B}_{-1}^{(q)} = \mathbf{0}), \qquad q = 1, \ldots, N, \tag{10.1.3}$$

that are the coefficients for the scattered field, which do not take into account interaction between the scatterers. The initial guess can be taken in a different way, if there is some guess or theory providing an approximate value for the expansion coefficients. As a condition for termination of the process, we can use a condition such as

$$\max_q \|\mathbf{A}_{j+1}^{(q)} - \mathbf{A}_j^{(q)}\| < \epsilon, \tag{10.1.4}$$

where $\| \ \|$ is some norm to measure the distance between two vectors (e.g. $\|\mathbf{A}\| = \max_{n,m} |A_n^m|$).

Figure 10.1 illustrates the reflection method applied for solution of the scattering problem for three spheres in the setting described in Chapter 4. We took the solution of the truncated linear system (4.4.16) with an appropriate p as an "exact" solution and plotted the absolute maximum error in the coefficients

$$\epsilon_j = \max_q \max_{n<p} \max_m |A_{(j)n}^{(q)m} - A_n^{(q)m}|, \qquad j = 1, 2, \ldots \tag{10.1.5}$$

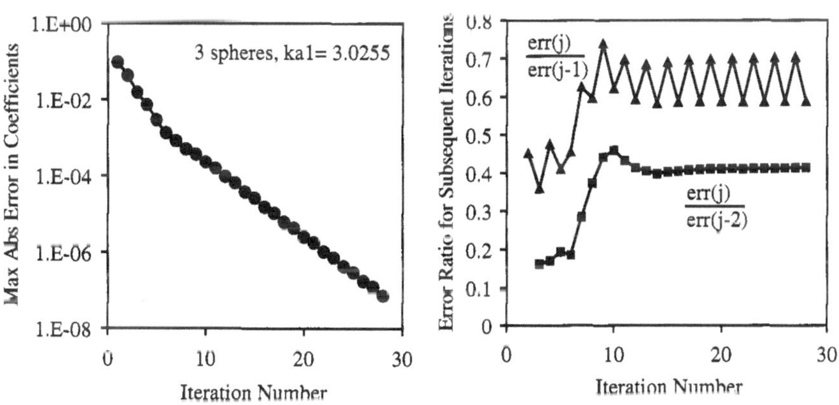

FIGURE 10.1 Illustration of convergence of the iterative reflection method in the case of scattering from a cluster of three spheres (geometry shown in Fig. 4.13). The left chart shows the error measured between the jth iteration and exact solution of the truncated system. The chart on the right shows the error ratio of the jth to the $(j-1)$th iteration (the curve marked by the triangles), and the error ratio of the jth to the $(j-2)$th iteration (squares).

as a function of j. We can see that after several initial iterations the convergence has an exponential character (the plot on the left). To be more precise we plotted the ratios $\epsilon_j/\epsilon_{j-1}$ and $\epsilon_j/\epsilon_{j-2}$ on the right as a function of j. While the decay exponent oscillates for subsequent iterations, these oscillations have some regular character, and in fact the ratio $\epsilon_j/\epsilon_{j-2}$ stays almost constant for $j > 15$. The convergence obtained numerically is in a good agreement with speculations related to the energy losses, which predicts the exponential decay. In this context we also note that an additional "loss" of the energy occurs due to its redistribution between the modes at each scattering act. As soon as we operate with a truncated system, the energy loss also includes the loss of the energy in the higher modes of the field, which are truncated.

10.1.2 Generalized minimal residual and other iterative methods

To solve the linear system,

$$\mathbf{LA} = \mathbf{E} \tag{10.1.6}$$

approximately, any iteration method builds a sequence of approximations to solution

$$\mathbf{A}_0^{(q)}, \mathbf{A}_1^{(q)}, \ldots; \qquad \lim_{j\to\infty} \mathbf{A}_j^{(q)} = \mathbf{A}^{(q)}, \qquad q = 1, \ldots, N, \ \ \mathbf{A} = \{\mathbf{A}^{(q)}\}, \tag{10.1.7}$$

where $\mathbf{A}_0^{(q)}$ is some initial guess, and stops when a convergence criterion is achieved. While understanding of the ideas behind the iterative methods is important (and we suggest reading the literature mentioned above), from the point of view of practical use, one can fortunately usually treat them as "black box" solvers. Several high-quality user-ready software in various versions and implementations are available. All these "black boxes" including the GMRES require from the user to provide a routine, which takes as input an arbitrary vector $\mathbf{F}$ and returns the matrix–vector product $\mathbf{LF}$.

If we turn to the equation to be solved, we can see that the matrix $\mathbf{L}$ can be decomposed as

$$\mathbf{L} = \mathbf{T}^{-1} - (\mathbf{S}|\mathbf{R}), \tag{10.1.8}$$

where $\mathbf{T}^{-1}$ is a diagonal matrix for spherical scatterers and block-diagonal otherwise, and $(\mathbf{S}|\mathbf{R})$ is the matrix consisting of the $(\mathbf{S}|\mathbf{R})(\mathbf{r}'_{qq'})$ blocks. Assuming that the product $\mathbf{T}^{-1}\mathbf{A}$ can be computed rapidly, we can use the

FMM to accelerate computation of $\mathbf{B} = (\mathbf{S}|\mathbf{R})\mathbf{A}$, or

$$\mathbf{B}_j^{(q)} = \sum_{q' \neq q} (\mathbf{S}|\mathbf{R})(\mathbf{r}'_{q'q})\mathbf{A}_j^{(q')}, \qquad q = 1, \ldots, N, \; j = 0, 1, \ldots \tag{10.1.9}$$

10.1.2.1 Preconditioners

Our computations show that the unpreconditioned GMRES slows down substantially as N increases and relatively high volume fractions of the spheres (in this case, the number of iterations increases proportionally to N). If $N_{\text{iter}} = O(N)$ then the method that requires $O(N \log N)$ per matrix–vector multiplication will require $O(N^2 \log N)$ operations to solve the problem. We found that this drawback can be fixed in many cases by the use of properly designed preconditioners. The preconditioners are some matrices, and one can use schemes with the right, left, or both right and left preconditioners. For example, the GMRES with the right preconditioner $\mathbf{M}^{-1}$ solves the problem:

$$\mathbf{L}\mathbf{M}^{-1}(\mathbf{M}\mathbf{A}) = \mathbf{E}, \tag{10.1.10}$$

which is formally equivalent to the original problem for arbitrary invertible matrix $\mathbf{M}$. If solution of the system $\mathbf{MA} = \mathbf{F}$ for an arbitrary input $\mathbf{F}$ can be obtained faster than solution of the original system $\mathbf{LA} = \mathbf{F}$, and $\mathbf{M}$ is a good approximation of $\mathbf{L}(\mathbf{M} \approx \mathbf{L})$, then one can expect substantial reduction in the number of iterations. One problem with the use of preconditioners with the FMM is that often in the literatures preconditioners are devised for algorithms that have $O(N^2)$ matrix–vector multiplies, or for sparse matrix problems. If these preconditioners are used with the FMM they can destroy the overall complexity of the algorithm.

Here we show that for the present problem, efficient preconditioners can be constructed, *that also use the FMM*. This prescription maintains the overall complexity of the algorithm and we describe this below.

10.1.2.2 Flexible GMRES

The original preconditioned GMRES method assumes that matrix–vector product $\mathbf{G} = \mathbf{M}^{-1}\mathbf{F}$ can be computed exactly (which provides an orthogonal basis for Krylov's subspace). In other words, the linearity of operator that produces $\mathbf{G}$ from given $\mathbf{F}$ is substantial. However, computation of matrix $\mathbf{M}^{-1}$ or solution of the system $\mathbf{MG} = \mathbf{F}$ with accuracy ϵ_{pre} not less than the accuracy prescribed for convergence of the main iteration loop, $\epsilon_{\text{pre}} < \epsilon$, may require not less computational work than solution of the system using unpreconditioned GMRES. Taking into account that there is a substantial freedom for selection of the preconditioner and its approximate nature, we can try to reduce the number of iterations by computing it with the aid of unpreconditioned GMRES (internal or preconditioning loop), with $\epsilon_{\text{pre}} \gg \epsilon$. However, this produces rather poor results, due to the non-linear dependence of **G** from **F** in this case.

A method that is free of this drawback is known as flexible GMRES [Saa93]. From the user's point of view, we can treat it again as a "black box" and use available implementations of the FGMRES. This "black box" requires two user-specified routines, which provide the matrix–vector product **LA** for the external iteration loop and **MG** for the internal loop for arbitrary input vectors **A** and **G**. The termination condition for the external loop is specified by accuracy ϵ, while the same type of criterion for the internal loop is specified by accuracy ϵ_{pre}. Of course, this creates an optimization problem, on how to select ϵ_{pre}, and one should solve it somehow in the process of tuning of the algorithm. Our experience shows that this is affected by the particular preconditioner used and on the parameters that specify the problem (e.g. density and number of scatterers, surface impedances, etc.).

10.2 FAST MULTIPOLE METHOD

The second substantial reduction in computational complexity comes from the use of the FMM for matrix–vector multiplication. The use of the FMM also provides a substantial memory reduction from $O(N^2p^4)$ (to store the system matrix) to about $O(Np^2)$. In fact, any iterative solver requires a multiplication of the system matrix by some input vector at each step, and if this is performed directly the cost is $O(N^2p^4)$. The FMM is able to reduce this cost to $O(Np^3)$ if $O(p^3)$ translation methods are used, or to $O(Np^2 \log^\alpha p)$, $\alpha = 0, 1, 2$, when using the theoretically fastest methods. Therefore, a combination of iterative methods and the FMM is able to reduce the complexity from $O(N^3p^6)$ to $O(N_{\text{iter}}Np^3)$ or $O(N_{\text{iter}}Np^2 \log^\alpha p)$. For example, if we assume that $N_{\text{iter}} \sim p \sim 10$, $N \sim 10^3$, and we use a $O(p^3)$ matrix-based translation method (as we did to obtain the results shown below) this reduces the complexity in the order of $O(N^2p^2)$ or 10^8 times! Such a reduction makes the use of simulation in design possible. The scattering problem of this size can be solved on a PC (this is what we actually did to illustrate this chapter).

While the FMM can be treated as a method for fast computation of matrix–vector products, the way how this product is computed in the case of multiple scattering problems has some specifics and can be formulated in terms of far-field and near-field expansions of the scattered field. We discuss these details and approach in the next subsections.

10.2.1 Data structures

For application of the multilevel FMM we use hierarchical space partitioning with the octree, as described in Chapter 4. Since we deal with objects of finite dimension in the case of scatterers, not source points, we should modify some definitions and impose certain constraints, which are important for algorithm optimization.

We say that a scatterer q belongs to box (n, l) if the center $\mathbf{r}'_q$ of the smallest sphere enclosing the scatterer is located inside that box (for spherical scatters this sphere is simply the surface of the scatterer). Any box may intersect a scatterer, which does not belong to it (see Fig. 10.2).

However, we require that the size of the smallest box (at the maximum level of space partitioning) is determined as below to ensure both correctness of the FMM procedure and requirements of validity of multipole reexpansions on the surfaces of neighbor scatterers. If two scatterers q and q' belong to two different boxes at the finest level $(n_q, l_{\max})$

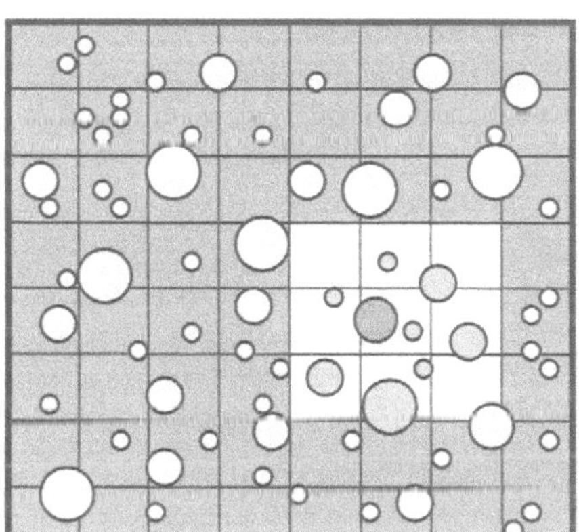

FIGURE 10.2 Illustration of space partitioning and data structure for multiple scattering problem. The minimal spheres circumferencing each scatterer are shown (so an actual scatterer of an arbitrary shape is located inside the sphere). The dark gray sphere belongs to the light gray box. The spheres, which are a bit lighter, show the neighbor spheres. These are the spheres which belong (in terms of their centers) to the E_2 neighborhood of the light gray box. The light gray spheres are the far spheres; they belong to the E_3 neighborhood of the light gray box (the centers are located outside the E_2 neighborhood; see also Fig. 5.14).

and $(n_{q'}, l_{\max})$, which are not neighbors, then the field scattered by scatterer q' can be described by the local expansion of type (4.3.5) near the center of box $(n_q, l_{\max})$ in the entire domain occupied by the sphere surrounding the scatterer q. The maximum distance from the center of box $(n_{q'}, l_{\max})$ to the point on the q'th scatterer surface is $D_{l_{\max}}\sqrt{3}/2 + a_{\max}$, where $a_{\max}$ is the maximum possible size of the minimal sphere enclosing the scatterer, $a_{\max} = \max(a_1, \ldots, a_N)$, and $D_{l_{\max}}$ is the size of the box at the finest level. The minimum possible distance from the center of box $(n_{q'}, l_{\max})$ to a point on the surface of the qth scatterer is then $3D_{l_{\max}}/2 - a_{\max}$. The above requirement will be satisfied if $3D_{l_{\max}}/2 - a_{\max} > D_{l_{\max}}\sqrt{3}/2 + a_{\max}$, which shows that we should have

$$D_{l_{\max}} > \frac{4}{3-\sqrt{3}} a_{\max}, \qquad l_{\max} < \log_2\left(\frac{3-\sqrt{3}}{4}\frac{D_0}{a_{\max}}\right), \tag{10.2.1}$$

where D_0 is the size of the computational domain. Note that this condition also provides that the far-field reexpansion about the center of box $(n_q, l_{\max})$ of the field scattered by scatterer q is valid at any point on the surface of the scatterer q'.

A stronger limitation on $l_{\max}$ follows from formal requirements of the translation theory. Again, we can consider two closest boxes $(n_{q'}, l_{\max})$ and $(n_q, l_{\max})$, which are not neighbors, and try to evaluate the effect of scatterer q' on q. The multipole series representing the potential of the scatterer q', whose center in the worst case is located at distance $D_{l_{\max}}\sqrt{3}/2$ from the center of box $(n_{q'}, l_{\max})$ should be translated first to the center of this box, then to the center of box $(n_q, l_{\max})$ and finally to the center of scatterer q with a guarantee that the final local expansion is valid within the sphere of radius $a_{\max}$. This can be provided if $2D_{l_{\max}} > D_{l_{\max}}\sqrt{3}/2 + D_{l_{\max}}\sqrt{3}/2 + a_{\max}$, which is a condition that spheres of radii $D_{l_{\max}}\sqrt{3}/2$ and $D_{l_{\max}}\sqrt{3}/2 + a_{\max}$ co-centered with boxes $(n_{q'}, l_{\max})$ and $(n_q, l_{\max})$, respectively, do not intersect ($2D_{l_{\max}}$ is the distance between the box centers). This condition can also be rewritten as

$$D_{l_{\max}} > \frac{1}{2-\sqrt{3}} a_{\max} \approx 3.73 a_{\max}, \qquad l_{\max} < \log_2\left[\left(2-\sqrt{3}\right)\frac{D_0}{a_{\max}}\right].$$

Further limitation on $l_{\max}$ should be imposed, since we use truncated expansions. The truncation error is a function of the truncation number, which in turn depends on the distance between the center of the domain of local expansion and the closest point on the sphere surrounding the box containing the multipole, $b_{l_{\max}} = 2D_{l_{\max}} - D_{l_{\max}}\sqrt{3}/2$. The dimensionless parameter

$$\sigma_{\min} = b_{l_{\max}}/(D_{l_{\max}}\sqrt{3}/2 + a_{\max}) > 1$$

then plays an important role for selection of the truncation number (see the section below) ($p \to \infty$ for $\sigma_{\min} \to 1$). If this parameter is specified, then we have the following modification for the condition for $D_{l_{\max}}$:

$$D_{l_{\max}} \geq \frac{2\sigma_{\min}}{4-(1+\sigma_{\min})\sqrt{3}} a_{\max}, \quad l_{\max} \leq \log_2\left[\frac{4-(1+\sigma_{\min})\sqrt{3}}{2\sigma_{\min}} \frac{D_0}{a_{\max}}\right],$$

$$\sigma_{\min} > 1.$$

It can be seen that in any case $\sigma_{\min} < 4/\sqrt{3} - 1 \approx 1.31$.

This condition, however, provides a constraint for the maximum level and it may happen that it is strong enough to force the method to function in a non-optimal regime (in terms of point sources). Note that in some test cases we found numerically that the algorithm may continue to perform satisfactorily when $l_{\max}$ exceeds the value prescribed by the above conditions.

10.2.2 Decomposition of the field

The field near the qth scatterer can be represented using Eq. (4.4.5) as

$$\psi(\mathbf{r}) = \psi_{\text{in}}(\mathbf{r}) + \psi_q(\mathbf{r}) + \psi_{\text{nei}}^{(q)}(\mathbf{r}) + \psi_{\text{far}}^{(q)}(\mathbf{r}),$$

$$\psi_{\text{nei}}^{(q)}(\mathbf{r}) = \sum_{\mathbf{r}_{q'} \in E_2(n_q, l_{\max}), q' \neq q} \overline{\mathbf{A}^{(q')}} \cdot \mathbf{S}(\mathbf{r}_{q'}), \tag{10.2.2}$$

$$\psi_{\text{far}}^{(q)}(\mathbf{r}) = \sum_{\mathbf{r}_{q'} \in E_3(n_q, l_{\max})} \overline{\mathbf{A}^{(q')}} \cdot \mathbf{S}(\mathbf{r}_{q'}),$$

where the sum of scattered potentials entering $\psi_{\text{in}}^{(\text{eff})(q)}(\mathbf{r})$ (see Eq. (4.4.5)) is separated into the sum of potentials of scatterers belonging to the neighborhood of the qth scatterer and the sum of potentials of scatterers located in the far-field domain of the qth scatterer. For compactness of notation we use the dot product of vector of coefficients and the singular spherical basis functions:

$$\bar{\mathbf{A}} \cdot \mathbf{S}(\mathbf{r}) = \sum_{n=0}^{\infty} \sum_{m=-n}^{n} A_n^m S_n^m(\mathbf{r}). \tag{10.2.3}$$

Potentials $\psi_{\text{in}}(\mathbf{r})$, $\psi_{\text{nei}}^{(q)}(\mathbf{r})$, and $\psi_{\text{far}}^{(q)}(\mathbf{r})$ can be expanded in series near $\mathbf{r}'_q$ with respect to local basis $\mathbf{R}(\mathbf{r}_q)$, similar to Eq. (4.4.6):

$$\psi_{\text{nei}}^{(q)}(\mathbf{r}) = \overline{\mathbf{B}_{\text{nei}}^{(q)}} \cdot \mathbf{R}(\mathbf{r}_q), \qquad \psi_{\text{far}}^{(q)}(\mathbf{r}) = \overline{\mathbf{B}_{\text{far}}^{(q)}} \cdot \mathbf{R}(\mathbf{r}_q). \tag{10.2.4}$$

Therefore, the potential of the effective incident field for the qth scatterer (Eq. (4.4.6)) can be represented as

$$\psi_{\text{in}}^{(\text{eff})(q)}(\mathbf{r}) = \overline{\mathbf{E}^{(\text{eff})(q)}} \cdot \mathbf{R}(\mathbf{r}_q), \qquad \mathbf{E}^{(\text{eff})(q)} = \mathbf{E}^{(q)} + \mathbf{B}_{\text{nei}}^{(q)} + \mathbf{B}_{\text{far}}^{(q)}. \tag{10.2.5}$$

The solution that satisfies the boundary conditions on this sphere can be written using the second equation in Eq. (4.4.16):

$$\mathbf{A}^{(q)} = \mathbf{T}^{(q)}(\mathbf{E}^{(q)} + \mathbf{B}_{\text{nei}}^{(q)} + \mathbf{B}_{\text{far}}^{(q)}). \tag{10.2.6}$$

The first equation in Eq. (4.4.16) and definition of the neighbors field result in

$$\mathbf{B}_{\text{nei}}^{(q)} = \sum_{\mathbf{r}_{q'} \in E_2(n_q, l_{\max}), q' \neq q} (\mathbf{S}|\mathbf{R})(\mathbf{r}'_{q'q})\mathbf{A}^{(q')}, \qquad q = 1, \ldots, N. \tag{10.2.7}$$

Now let us turn to the computations of $\mathbf{B}_{\text{far}}^{(q)}$, which are the coefficients of the local expansion of $\psi_{\text{far}}^{(q)}(\mathbf{r})$ (see Eq. (10.2.4)). The potential $\psi_{\text{far}}^{(q)}(\mathbf{r})$ can be considered as a sum of multipole *point* sources located outside the neighborhood of the box containing the qth scatterer. In the iterative process the coefficients $\mathbf{A}^{(q')}$, which determine these multipoles, are given. Therefore, for fast summation of the sources in terms of their potential at the receiver location we can use the FMM. In fact, in the current problem we do not need to perform the final summation step, since all we need from the FMM are the coefficients $\mathbf{D}^{(n,l_{\max})}$ for each target box, i.e. box containing scatterers. As soon as we are able to obtain $\mathbf{D}^{(n_q,l_{\max})}$ for the box to which the qth scatterer belongs, we can represent $\psi_{\text{far}}^{(q)}(\mathbf{r})$ in the form:

$$\psi_{\text{far}}^{(q)}(\mathbf{r}) = \overline{\mathbf{D}^{(n_q,l_{\max})}} \cdot \mathbf{R}(\mathbf{r} - \mathbf{r}_*^{(n_q,l_{\max})}), \qquad q = 1, \ldots, N, \tag{10.2.8}$$

where $\mathbf{r}_*^{(n_q,l_{\max})}$ is the center of box $(n_q, l_{\max})$. This is not what we are looking for, since we need to determine the coefficients of the local expansion near $\mathbf{r}'_q$. The last step can be performed by using the $(\mathbf{R}|\mathbf{R})$-translation matrix, which relates the coefficients of R-expansions for two expansion centers located in the same target box:

$$\mathbf{B}_{\text{far}}^{(q)} = (\mathbf{R}|\mathbf{R})(\mathbf{r}'_q - \mathbf{r}_*^{(n_q,l_{\max})})\mathbf{D}^{(n_q,l_{\max})}, \qquad q = 1, \ldots, N. \tag{10.2.9}$$

This completes the solution of the problem, since it enables us to express $\mathbf{B}^{(q)} = \mathbf{B}_{\text{far}}^{(q)} + \mathbf{B}_{\text{nei}}^{(q)}$ as a function of $\{\mathbf{A}^{(q')}\}$, that can be computed with complexity $O(N\ \text{CostTrans}(p^2))$.

Note that the derivation of the expression for $\mathbf{B}_{\text{far}}^{(q)}$ (Eq. (10.2.8)) can be done in a more formal way. Indeed, by definition we have

$$\mathbf{B}_{\text{far}}^{(q)} = \sum_{\mathbf{r}_{q'} \in E_3(n_q, l_{\max})} (\mathbf{S}|\mathbf{R})(\mathbf{r}'_{q'q})\mathbf{A}^{(q')}, \qquad q = 1, \ldots, N. \tag{10.2.10}$$

The translation matrix can be decomposed as

$$(\mathbf{S}|\mathbf{R})(\mathbf{r}'_{q'q}) = (\mathbf{R}|\mathbf{R})(\mathbf{r}'_q - \mathbf{r}_*^{(n_q, l_{\max})}) \cdots (\mathbf{R}|\mathbf{R})(\mathbf{r}_*^{(m', l+1)} - \mathbf{r}_*^{(m,l)}) \circ (\mathbf{S}|\mathbf{R})(\mathbf{r}_*^{(m,l)} - \mathbf{r}_*^{(n,l)})(\mathbf{S}|\mathbf{S})(\mathbf{r}_*^{(n,l)} - \mathbf{r}_*^{(n', l+1)}) \cdots (\mathbf{S}|\mathbf{S})(\mathbf{r}_*^{(n_{q'}, l_{\max})} - \mathbf{r}'_{q'}), \tag{10.2.11}$$

which is nothing but a representation of the translation process for a single multipole source with coefficients $\mathbf{A}^{(q')}$ in the multilevel FMM. The hierarchical data structure used in the method ensures that all steps of this chain of operators, except the first and the last one, are independent of locations of the source and receiver points, $\mathbf{r}'_{q'}$ and $\mathbf{r}'_q$, respectively, a big advantage. In this case summation (10.2.10) can be performed for all receivers simultaneously with low complexity, as was analyzed in Chapters 5 and 6.

10.2.3 Algorithm for matrix–vector multiplication

We return to the first equation in Eq. (4.4.16), which incorporates the major computational complexity of matrix–vector multiplication. Indeed, we can see from Eq. (4.4.19) that the system matrix $\mathbf{L}$ can be decomposed to diagonal blocks of the T-matrices and the rest, whose multiplication by vector $\mathbf{A}$ is nothing but reformulation of the first equation in Eq. (4.4.16). This multiplication should be performed in the reflection method, or any other iterative method (10.1.2), where the input vector can be different from $\mathbf{A}$. Assuming that $\mathbf{A}$ is given, we can then summarize what we said in the sections above in the following algorithm to obtain $\mathbf{B}$, which is the result of matrix–vector product.

- Based on the parameters of the problem, determine $a_{\max}$ and D_0 and generate the hierarchical data structure, subject to the constraint (10.2.1), where the source and receiver sets are same ($\mathbb{Y} = \mathbb{X}, M = N$) and the set $\mathbb{X}$ consists of the coordinates of the centers of minimal enclosing sphere for each scatterer.

- For each scatterer determine the coefficients of the S expansion near the center of the box at the finest level, to which this scatterer belongs using the multipole-to-multipole $S|S$ translation of the coefficients $\mathbf{A}^{(q)}$ to the center of this box:

$$\tilde{\mathbf{B}}^{(q)}(\mathbf{r}_*^{(n_q,l_{\max})}) = (\mathbf{S}|\mathbf{S})(\mathbf{r}_*^{(n_q,l_{\max})} - \mathbf{r}'_q)\mathbf{A}^{(q)}, \qquad q = 1,\ldots,N. \quad (10.2.12)$$

- Treat the set of these vectors as a result of the first step of the upward pass of the multilevel FMM, Eq. (5.2.14), and apply all other steps of the FMM procedure (5.2.15)–(5.2.19) until the final summation step, to obtain coefficients $\mathbf{D}^{(n_q,l_{\max})}$ for all target boxes (which is the same as the set of source boxes).
- Replace the final summation step with the following procedure:
  - For each target box $R|R$ translate $\mathbf{D}^{(n_q,l_{\max})}$ to the location of all receivers $(\forall \mathbf{r}'_q \in E_1(n_q, l_{\max}))$ in the box. According to Eq. (10.2.9) this results in the determination of the coefficients $\mathbf{B}^{(q)}_{\text{far}}$. This replaces the evaluation of the expansion in the final summation (5.2.21).
  - For all sources q' different from q and located in the neighborhood of the box $(\forall \mathbf{r}'_q \in E_1(n_q, l_{\max}), \forall \mathbf{r}'_{q'} \in E_2(n_q, l_{\max}), q' \neq q)$ compute $\mathbf{B}^{(q)}_{\text{nei}}$ using the $S|R$ translation of coefficients $\mathbf{A}^{(q')}$ followed by summation, according to Eq. (10.2.7). This replaces the straightforward summation in the final summation step (Eq. (5.2.21)).

We note that as the algorithm enables computation of vector $\mathbf{B}$ of total length Np^2 by multiplication of dense matrix $Np^2 \times Np^2$ by vector of length Np^2 (note that this Np^2 can be much larger than N) for expense of only $O(Np^3)$ or even $O(Np^2 \log^\alpha p)$ operations. It is interesting that this asymptotic complexity is *lower* than any known complexity of dense matrix multiplication and comparable only with multiplication involving very sparse matrices. Indeed, the fastest method for multiplication of structured matrices of size $Np^2 \times Np^2$ (e.g. Toeplitz matrices) is based on the FFT, which provides complexity $O(Np^2 \log(Np^2))$. This asymptotic complexity is larger than, say, $O(Np^2 \log p)$ if $\log N \gg \log p$. If $p \sim \log N$, then $O(Np^3)$ FMM-based algorithm has the same complexity as the FFT.

A final remark for this subsection is that the T-matrix for each scatterer is a $p^2 \times p^2$ matrix and we need to multiply N diagonal blocks by respective vectors of size p^2. In the case of spherical scatterers the T-matrix is diagonal and this multiplication is an $O(Np^3)$ procedure. However, in the general case, the T-matrix can be dense, and if there are no fast methods available to speed up the multiplication, then the computation of diagonal blocks of the system matrix **L** (Eq. (4.4.19)) will take $O(Np^4)$ operations, which is larger than $O(Np^3)$, when we use fast matrix-based translations (see Chapter 7). In this case the dominant complexity moves from computation of a larger dense part of the matrix, to the smaller part of block-diagonal matrix.

10.2.4 Complexity of the FMM

The ratio of the size of the scatterers to the wavelength or ka is one of the main parameters of the problem. In the computations we were interested mostly with diffraction effects, which are strong at low and moderate ka, and we varied it in the range $ka \sim 10^{-1}-10$.

It is noteworthy that for the current problem, the parameter kD_0, based on the size of computational domain, can be large enough even for small or moderate ka since it depends on the number of scatterers and the density of the scatterer cluster (or intersphere distances). An important quantity for the FMM is the parameter $\beta = a/D_0$. Indeed, consider a simplified situation, when all the scatterers are of the same size. In this case the volume occupied by the scatterers is $V = \frac{4}{3}\pi a^3 N$ and so the volume fraction of the scatterers in the computational domain is

$$\alpha = \frac{V}{D_0^3} = \frac{4}{3}\pi N \beta^3. \qquad (10.2.13)$$

The range of α can vary from a very small value (diluted disperse systems) to some value, which is of order 1. The problem of finding the densest possible sphere packing is known since 1611, when Kepler proposed that for the densest packing $\alpha_{\max} \approx 0.74$ (*Kepler's Conjecture*). It is amazing, but this problem has not yet been solved completely, while some bounds are proven (e.g. $\alpha_{\max} < 0.78$). We note only that for the densest random packing $\alpha_{\max} \approx 0.64$ [JN92].

In any case Eq. (10.2.13) shows that

$$\beta = \left(\frac{3\alpha}{4\pi N}\right)^{1/3}, \qquad (10.2.14)$$

and, therefore, this quantity is bounded by $O(N^{-1/3})$. Hence, if ka and N are specified as parameters, then we have

$$kD_0 = ka\beta^{-1} = N^{1/3}ka\left(\frac{4\pi}{3\alpha}\right)^{1/3} > N^{1/3}ka\left(\frac{4\pi}{3\alpha_{\max}}\right)^{1/3} > 1.75N^{1/3}ka, \qquad (10.2.15)$$

where we took into account that $\alpha_{\max} < 0.78$. This inequality shows that when the number of scatterers is large (e.g. $N \sim 10^6$), even for relatively small scatterer sizes compared to the wavelength (e.g. $ka = 1$), we have large kD_0 ($kD_0 > 175$) for this example. In practice we usually deal with non-dense packings and so D_0 can be several times larger.

10.2.4.1 Complexity and translation methods for large problems

If we consider the FMM computations, then kD_0, not ka, determines the truncation numbers to use. For example, the radius of the sphere enclosing a box at the minimum level of space subdivision $l_{\min} = 2$ is $(\sqrt{3}/8)D_0$. This determines $p = p_{\max}$ at this level and, respectively, p at finer levels $l > l_{\min}$ until the breakdown level l_{b*} (see Chapter 6). For large kD_0, according to Eqs. (9.1.35) and (10.2.15), we then have

$$p_{\max} \sim \frac{\sqrt{3}}{8} kD_0 + O((kD_0)^{1/3}) = O(\alpha^{-1/3} N^{1/3} ka). \tag{10.2.16}$$

This shows that the use of the FMM with $O(p^3)$ translation methods and constant p result in an algorithm of complexity $O(Np^3) = O(N^2)$. Fortunately, these methods can be used for solution of the present problem, if the truncation number is made level dependent and reduces by a factor of two when we proceed to the next finer level. This is based on the model of the FMM described in Chapter 6. As the analysis shows, for spatial distributions of data points, the complexity of the FMM in this case will be $O(p_{\min}^3 N \log N)$, where for $ka = O(1)$

$$p_{\min} \sim 2^{-l_{b*}} p_{\max} \sim \frac{\sqrt{3}}{8} \kappa_b = O(1), \tag{10.2.17}$$

and expression for l_{b*} is provided by Eq. (6.1.5).

According to the complexity theory of Chapter 6, $O(p^4)$ translation methods result in $O(N^{4/3})$ overall complexity of the FMM and $o(p^3)$ translation methods yield an $O(N)$ algorithm. We note here that despite the method of signature functions with filtering that can be realized in $O(p^2 \log p)$ operations, the current problem is formulated in terms of expansion coefficients, and so it is more natural to use translation methods operating in this space, which are of $O(p^3)$ and $O(p^2 \log^2 p)$ complexity. Indeed, the operation of conversion of all expansion coefficients to the samples of the signature function and back takes $O(Np^2 \log^2 p)$ operations using the FST, which keeps the total complexity of the FMM using this approach at the $O(Np^2 \log^2 p)$ level.

10.2.4.2 *Smaller problems or low frequencies*

There exist practically important low frequency scattering problems and problems with relatively small N that can be efficiently solved with the FMM, while the direct linear system solution or iterations based on straightforward matrix–vector multiplication are computationally expensive. From the viewpoint of the FMM we can characterize them as problems with the breakdown level $l_{b*} \leq l_{\min} = 2$. Using Eqs. (6.1.5) and (10.2.15) this condition can be written in the form:

$$N^{1/3} ka \left(\frac{4\pi}{3\alpha} \right)^{1/3} \leq 4\kappa_{\mathrm{b}}. \quad (10.2.18)$$

If this condition holds, then the problem can be solved with constant p, which is determined by the error bounds of the FMM. This p can be the same as the truncation number for coefficients $\{A_n^m\}$ or different, which depends on several factors, e.g. the minimum distance between two scatterers. Indeed, closer scatterers, larger series are needed to represent their interaction. However, this is a different type of truncation, since the interaction of the neighbor scatterers is computed directly, not via additional reexpansion and translation procedures.

While Eq. (10.2.18) may seem to be restrictive, it determines some range of N and ka that is realizable in problems of high interest for physics and engineering. Indeed, at $\kappa_{\mathrm{b}} \sim 10$ and $\alpha \sim 0.1$ the restriction is $N^{1/3} ka \lesssim 10$. For $ka = 1$ this condition holds for $N \lesssim 10^3$, and for $ka = 0.1$ it holds for $N \lesssim 10^6$. Of course, for $ka \gtrsim 10$ we should exclude this case from consideration.

As shown in Chapter 6, for $l_{b*} \leq l_{\min}$ the FMM outperforms the straightforward method at any complexity of translation of type $O(p^{2\nu})$, since p is constant. We also checked this numerically and found that even the $O(p^4)$ translation method based on computation of the truncated reexpansion matrix without any decomposition is practical. In the present case computations with low truncation numbers, e.g. $p \sim 3$–5, can provide maximum absolute errors in solutions of order $\epsilon \sim 10^{-2}$–10^{-3}, which may be sufficient for applied problems. For such low p the asymptotic complexity of the method plays the role not more important than the asymptotic constants (e.g. for $p = 3$ we have $p^4 = 3p^3$). Certainly, there are no reasons to use slower methods if a faster method with the same accuracy and stability is available, and at low p there is a different type of competition between the translation methods, and we found that for low p the $O(p^4)$ method is the fastest.

10.2.5 Truncation numbers

Our tests support the above theoretical finding that to achieve stable and fast computations with a prescribed accuracy, the truncation numbers for the multipole expansion of the scattered field for single spheres should be treated differently from the truncation numbers used in the FMM. To avoid mixing notations, we will mark the latter numbers as p_{FMM} or level-dependent $p_2, \ldots, p_{l_{\max}}$, while keep p for the former. Selection of the truncation numbers depends on the acceptable error of computations, which is determined by the rate of convergence of the series being truncated. Consider first the selection of p.

The series under consideration converge absolutely and uniformly only for $p \gtrsim ka$. For relatively low ka ($ka \lesssim 10$) one can select p as

$$p = [ka] + p_0(ka, \epsilon, \sigma), \qquad p_0 \geq 1, \ \ \sigma = b/a \geq 1, \ \ \epsilon > 0, \tag{10.2.19}$$

where p_0 depends on the prescribed accuracy ϵ and the distance b between the center of the scatterer and the closest point on the surface of the closest scatter. We suggest evaluation of p_0 based on error maps (or their approximations) obtained as a solution of the scattering problem for two spheres placed in the field of plane incident wave, $\psi_{\mathrm{in}}(\mathbf{r}) = e^{i\mathbf{k}\cdot\mathbf{r}}$. This problem can be solved very fast based on direct linear system solution for coaxial spheres as described in Chapter 4, where the parameters p_0, ka and σ can be varied within the range under consideration, while the error of the solution ϵ can be determined a posteriori, as the error in boundary conditions.

Indeed, we note that since any radiating basis function satisfies the Helmholtz equation, the expansion truncated with an arbitrary p will satisfy the Helmholtz equation and the radiation condition. So the only equation that is not satisfied exactly by the approximate solution is the boundary condition (4.4.1). Once an approximate solution is computed, i.e. the expansion coefficients $\{\mathbf{A}^{(q)}\}$ are found, we can compute the normal derivative on the surface for any scatterer using the differentiation theorem (2.2.12) for the spherical basis functions in an arbitrary direction specified by the unit vector $\mathbf{n} = (n_x, n_y, n_z)$:

$$\begin{aligned}\frac{1}{k}\mathbf{n}\cdot\nabla S_n^m(\mathbf{r}) &= \frac{1}{2}(n_x - in_y)[b_{n+1}^{-m-1} S_{n+1}^{m+1}(\mathbf{r}) - b_n^m S_{n-1}^{m+1}(\mathbf{r})] \\ &+ \frac{1}{2}(n_x + in_y)[b_{n+1}^{m-1} S_{n+1}^{m-1}(\mathbf{r}) - b_n^{-m} S_{n-1}^{m-1}(\mathbf{r})] \\ &+ n_z[a_{n-1}^m S_{n-1}^m(\mathbf{r}) - a_n^m S_{n+1}^m(\mathbf{r})].\end{aligned} \tag{10.2.20}$$

Since solution of the problem is obtained in the space of coefficients, operation of taking the normal derivative can be performed fast using the differential operator $\mathbf{D_n}$ (Eq. (7.3.33)) acting in this space. Therefore, as the truncated solution is computed, the left-hand side of Eq. (4.4.1) can be determined, and there are no additional errors for derivative computation or approximation. If we sample the surface with M points, then we can compute the following errors in boundary conditions:

$$\epsilon_m^{(bc)} = \frac{M|\Delta_{bc}(\mathbf{y}_m)|}{\left[\sum_m \left|\psi|_{S_q}(\mathbf{y}_m)\right|^2\right]^{1/2}}, \qquad \epsilon^{(bc)} = \max_m \epsilon_m^{(bc)},$$

$$\Delta_{bc}(\mathbf{y}_m) = a_q\left(\frac{\partial \psi}{\partial n}(\mathbf{y}_m) + i\sigma_q \psi(\mathbf{y}_m)\bigg|_{S_q}\right), \quad m = 1, \ldots, M. \tag{10.2.21}$$

Figure 10.3 illustrates dependences $p_0(ka, \epsilon^{(bc)}, \sigma)$ for two fixed values of σ and a range of ka. The spheres are sound-hard, $\sigma_q = 0$, $q = 1, 2$.

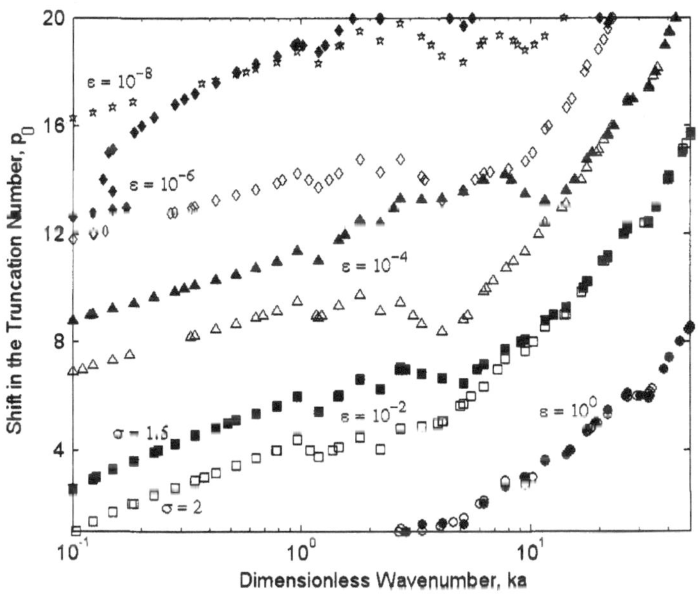

FIGURE 10.3 Dependences of the shift in the truncation number, p_0, defined by Eq. (10.2.19) on the dimensionless wavenumber ka for two sound-hard spheres of equal size. The curves are computed to provide the error $\epsilon^{(bc)}$ in boundary conditions shown near the curves. Two sets of curves marked by the filled and empty symbols correspond to different distances between the spheres ($\sigma = 1.5$ and 2, respectively).

Here for each curve $\epsilon^{(bc)}$ is fixed (p_0 is not integer due to interpolation). $\epsilon^{(bc)}$ was computed according to Eq. (10.2.21) where each sphere was sampled with 181 equispaced points with respect to the spherical polar angle θ. For the low frequency region, $ka \lesssim 1$ and $\epsilon^{(bc)} \gtrsim 10^{-4}$, we have $p_0 \sim p_{00}(\epsilon, \sigma) + \log(ka)$, which in the semilogarithmic coordinates, is a linear dependence. For higher frequencies, $ka \gg 1$, we have $p_0 \sim (ka)^{\nu}$, where theoretically $\nu = 1/3$, which is quite close to our computations. The dependence on σ in this range is rather weak (when σ is not very close to unity), while for lower frequencies the truncation number substantially depends on this parameter. Figure 10.4 illustrates dependencies $p_0(ka, \epsilon^{(bc)}, \sigma)$ for fixed $ka = 1$ and varying σ. It is seen that if the spheres are very close to each other ($\sigma \sim 1$) the truncation number should be high to provide small errors. However, if it is sufficient to compute a solution with, say $\sim 1\%$ of relative error, these computations can be achieved with $p_0 \sim 10$ even when the spheres touch each other.

For larger ka the truncation number can be selected based on the high frequency asymptotics of the spherical Bessel functions (Eq. (9.1.41)). In this equation the truncation number weakly depends on σ, while such a dependence can be introduced. Relation (9.1.41) is qualitatively consistent with the behavior of the error observed in computations where σ is not very close to 1. To relate it to the actual error observed for

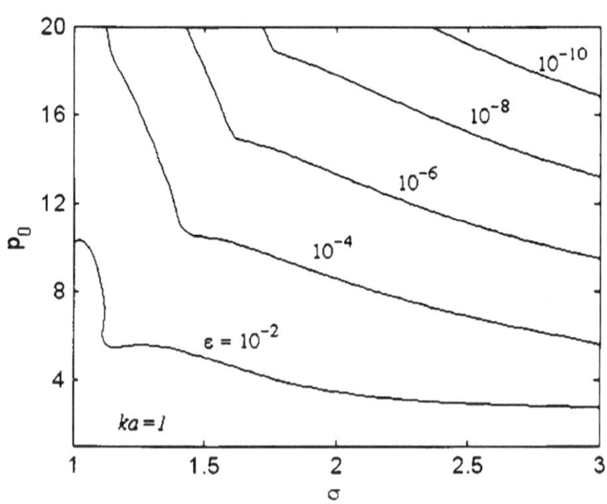

FIGURE 10.4 Dependences of the shift in the truncation number, p_0, defined by Eq. (10.2.19) on the dimensionless parameter σ characterizing the distance between the spheres for two sound-hard spheres of equal size. The curves are computed to provide the error $\epsilon^{(bc)}$ in boundary conditions shown near the curves.

solution of two-sphere scattering problem, one should relate ϵ to $\epsilon^{(bc)}$, and we found that for the range plotted in Fig. 10.3 one can set $\epsilon \sim 10^{-2}\epsilon^{(bc)}$ to fit results for $ka > 1$.

Similar formulae can be applied for the computation of the truncation numbers used in the FMM, in which case we need to specify what we mean by a and σ here. The parameter a can be selected as the radius of the smallest sphere that encloses the box of level l plus $a_{\max}$ and σ is formally limited by the value $\sigma < 1.31$ (see discussion on $l_{\max}$ above, while more accurate analysis is required here, since the "worst case" analysis usually substantially overestimates actual errors). Since the size of this box is $D_l = 2^{-l}D_0$, one can set

$$a_l = 2^{-l-1}3^{1/2}D_0 + a_{\max}, \qquad p_l = P(ka_l, \epsilon, \sigma). \tag{10.2.22}$$

In computations we used this formula for automatically setting the level-dependent truncation numbers. In computations we also performed an a posteriori actual error check using Eq. (10.2.21), which is valid for an arbitrary number of spheres.

We note that for the present problem, when the number of scatterers is large, for fixed ka and volume fraction of scatterers we have $D_0 \sim N^{1/3}a$, the truncation numbers p_l required in the FMM can be substantially larger than p. In other words, if the wavelength is comparable with the size of the scatterer, then the size of the computational domain is much larger than the wavelength for large N.

We can also note that in the case when the spheres are close or may touch each other, p can be larger than p_l to provide the required accuracy. To avoid unnecessary computations in this case, translations for the far-field expansions at the finest level can be made to obtain only $p_{l_{\max}}^2$ terms, since terms of degree n corresponding to $p_{l_{\max}} \leq n < p$ are due to the influence of close neighbors. Moreover, in computations of the near field we can subdivide the set of neighbor spheres into two sets: those that are closer than some prescribed b_* (or σ_*), for which the interaction should be computed using all p terms, and the other neighbors, whose interactions with the given sphere can be efficiently computed using only $p_{l_{\max}}$ terms. We found numerically that these tricks may bring substantial savings in computational time.

10.2.6 Use of the FMM for preconditioning in the GMRES

The FMM provides a good tool for building a preconditioner. The preconditioners that can be constructed are based on decompositions of the dense $(\mathbf{S}|\mathbf{R})$ matrix in Eq. (10.1.8):

$$(\mathbf{S}|\mathbf{R}) = (\mathbf{S}|\mathbf{R})_{\text{nei}} + (\mathbf{S}|\mathbf{R})_{\text{far}}, \qquad \mathbf{M}_{\text{sparse}} = \mathbf{T}^{-1} - (\mathbf{S}|\mathbf{R})_{\text{nei}},$$

$$\mathbf{M}_{\text{dense}} = \mathbf{T}^{-1} - (\mathbf{S}|\mathbf{R})_{\text{far}}, \qquad (10.2.23)$$

where $(\mathbf{S}|\mathbf{R})_{\text{nei}}$ is a sparse matrix, which includes only interactions between the scatterers located in the same neighborhood (say, at the maximum level of space subdivision), and $(\mathbf{S}|\mathbf{R})_{\text{far}}$ is the matrix, which includes far-field interactions. This decomposition, in fact, is decomposition $\mathbf{B} = \mathbf{B}_{\text{nei}} + \mathbf{B}_{\text{far}}$ in terms of the resulting vector $\mathbf{B}$. The matrix–vector products $\mathbf{M}_{\text{sparse}}\mathbf{A}$ or $\mathbf{M}_{\text{dense}}\mathbf{A}$ involving preconditioners $\mathbf{M}_{\text{sparse}}$ and $\mathbf{M}_{\text{dense}}$ can then be computed faster than $\mathbf{LA}$, since only part of the FMM operations is needed to perform the computations. At first look, it seems more reasonable to use $\mathbf{M}_{\text{sparse}}$ as the preconditioner, since presumably matrix–vector multiplication in this case should be cheaper. However, there are two reasons, both working in the same direction, why the use of $\mathbf{M}_{\text{dense}}$ may make more sense.

The first reason is based on the efficiency of the preconditioner in terms of the process convergence. Despite the fact that the distance between the neighbor scatterers is smaller and interaction between any pair of neighbors is stronger than between the remote particles, the number of neighbors is also substantially smaller. Since this interaction decays as r^{-1}, while the number of particles in the far field increases as r^3 (for a uniform distribution), the effect of the far field on the given scatterer dominates for larger domain sizes and larger number of scatterers. The most computational work is then to iterate the solution for proper determination of the far field, and that is what the unpreconditioned method does.

The second reason is a feature of the FMM, which appears to have been used for the first time here. It is based on the efficiency of the preconditioner in terms of the speed of matrix–vector multiplication. It may seem paradoxical, but in the FMM as described above, computation of $(\mathbf{S}|\mathbf{R})_{\text{far}}\mathbf{A}$ can be much faster than the computation of $(\mathbf{S}|\mathbf{R})_{\text{nei}}\mathbf{A}$. This happens due to limitations of type (10.2.1) and the following equations on the maximum level of space subdivision and takes place for relatively high volume fraction of scatterers (for low volume fractions $(\mathbf{S}|\mathbf{R})_{\text{nei}}\mathbf{A}$ can be computed faster). Indeed, if $l_{\max}$ is small enough, then the most computational work is spent for direct computations of matrix–vector products $\mathbf{B}_{\text{nei}} = (\mathbf{S}|\mathbf{R})_{\text{nei}}\mathbf{A}$ (for $l_{\max} = 1$ we have $(\mathbf{S}|\mathbf{R})_{\text{nei}} = (\mathbf{S}|\mathbf{R})$ and need to multiply matrix by vector directly for cost $O(N^2)$). Increase of $l_{\max}$ reduces the number of direct matrix–vector products, while increasing the total number of translations. There exist some optimum $l_{\max}^{(\mathrm{opt})}$ where the costs of translations and direct computations are balanced. It appears that for high volume fractions of scatterers $l_{\max}^{(\mathrm{opt})} > l_{\max}$, where $l_{\max}$ is limited by the condition derived earlier (we checked this numerically for some spatially uniform random distributions of the scatterers), and so the use of $\mathbf{M}_{\mathrm{dense}}$ is preferable.

10.3 RESULTS OF COMPUTATIONS

The algorithm described above had been implemented on Fortran language with some Matlab interface, which enables data input and postprocessing. All computations were performed in double precision complex arithmetic. To speed up the FMM we used precomputation and storage of the data structure and all translation and rotation matrices. To test the algorithm we focused on a simplified case, when all the scatterers are spheres, while the sizes and impedances could be selected arbitrarily. Even this simplified, but practically important case is multiparametric and a complete investigation of the multidimensional parameter space seems unrealistic. For example, in the tests we varied the number of scatterers, their spatial and size distributions, frequencies, incident fields (sources and plane waves), and the parameters controlling the FMM performance (maximum level of space subdivision and truncation numbers).

10.3.1 Typical pictures and settings

Two typical cases, showing random distributions of spheres inside a box (computational domain), are shown in Fig. 10.5. The sphere radii here were distributed uniformly and randomly in the range $[0.5, 1.5]$, where the mean radius, a, is selected as a spatial scale, and all dimensional parameters depending on the length scale can be measured in terms of this radius (e.g. ka). Small arrows on the pictures show the direction of the incident plane wave, and the dots show the imaging plane, which is not physically present but serves for the imaging of the cross-sections of the three-dimensional field. The advantage of the multipole methods is that, as soon as all expansion coefficients for the scattered field are computed, one can determine the value of the total field either on the surface or at any spatial point without any grid (meshless method). Accuracy of imaging of spatial and surface distributions is, therefore, *separated*, from the accuracy of computations.

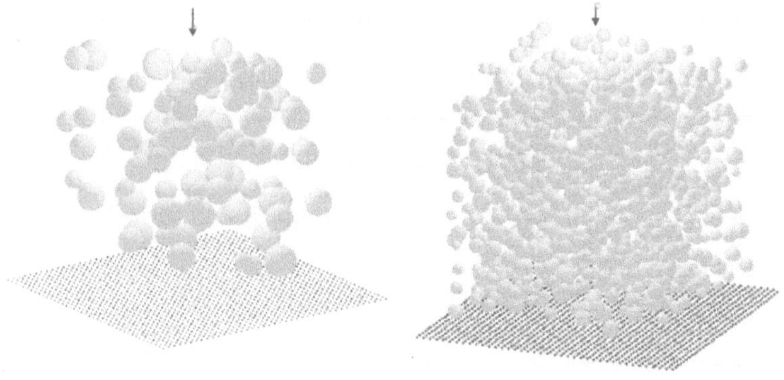

FIGURE 10.5 Typical multiple scattering problem settings for spheres. The scatterers are distributed randomly and uniformly inside the computational domain. The arrows show direction of the incident plane wave. The incident field can also be modeled by distribution of point sources and a set of plane waves. The dots show the imaging plane.

10.3.1.1 FMM for spatial imaging/field calculation

The three-dimensional field imaging problem can also be considered as a large computational task. Assume that we have N scatterers and p^2 coefficients for each scatterer, which are computed by the method described above, say, for expense of $O(N_{\text{iter}}Np^3)$ operations. Then, we want to "image" the field at M points, located at the imaging plane or elsewhere. This means that we need to compute the complex potential of the Helmholtz equation at these M points. If we do this directly, then we need to spend $O(Np^2M)$ operations, since the field of each scatterer should be evaluated at any given point. In the case $M \sim pN_{\text{iter}}$ this problem appears to be of the same complexity as solution of the problem itself! In fact, for a good quality image one should use thousands of imaging points, and usually $M \gg pN_{\text{iter}}$, in which case the imaging problem or field calculation is a problem of larger complexity than solution of the three-dimensional Helmholtz equation.

A natural way to handle this problem is to use the FMM, as soon as the code is available for solution of the basic problem. The code can be designed by combining ideas for point receivers and finite size scattering objects. Indeed, we now set the general FMM algorithm by such a way, when the set of sources is the set of N scatterer centers, and the set of receivers is M field imaging points. The octree-based space partitioning is applied to the new data structure. As soon as the expansion coefficients

$\mathbf{A}^{(q)}$, $q = 1, \ldots, N$ are computed, we apply the FMM procedure described above until the final summation step, at which we know $\mathbf{D}^{(n,l_{\max})}$ for each target box. Then, we compute the potential for all receivers $\mathbf{y}_m$ located inside this box as provided by Eqs. (10.2.8), (4.4.3), and (4.4.4):

$$\psi(\mathbf{y}_m) = \psi_{\mathrm{in}}(\mathbf{y}_m) + \overline{\mathbf{D}^{(n,l_{\max})}} \cdot \mathbf{R}(\mathbf{y}_m - \mathbf{r}_*^{(n,l_{\max})}) + \sum_{\mathbf{r}_q \in E_2(n,l_{\max})} \overline{\mathbf{A}^{(q)}} \cdot \mathbf{S}(\mathbf{y}_m - \mathbf{r}'_q), \qquad \forall \mathbf{y}_m \in E_1(n, l_{\max}). \quad (10.3.1)$$

Here the first term in the right-hand side is the known potential of the incident field at the receiver location, the second term is the potential of the field scattered by all scatterers located outside the neighborhood of the receiver box, and the third term is a field scattered by all scatterers located in the neighborhood of the receiver and found by direct evaluation.

Therefore, the complexity of the imaging problem reduces from $O(Np^2M)$ to $O(Np^3 + Mp^2)$ for a translation complexity $O(Np^3)$. Optimization of the maximum level of space subdivision can be performed as described in Chapter 6, which reduces the complexity. In this context we can note that, in the case when the imaging points are located on the surface (e.g. an imaging plane), which is a manyfold of dimensionality lower than 3, the use of the adaptive fast multipole method described in Chapter 6 might be efficient.

Figure 10.6 illustrates the speckle patterns on the image plane obtained using the FMM both for solution of the problem and for imaging of the field. The grayscale images are computed for monochromatic fields of different frequencies and then put in the RGB channels to produce a color image corresponding to polychromatic wave scattering. These pictures are frames of a movie that animates the wave propagation in time, by tracking the lines of constant phase (this movie can be viewed on the authors' web site).

10.3.1.2 Surface imaging

As soon as the solution of the problem (multipole expansion coefficients) is obtained, surface distributions of the complex potential and/or its normal derivative can be obtained using Eq. (4.3.18), where the index q can take any value from 1 to N. This is a low complexity computational problem, since the expansion coefficients include all interactions between the scatterers, and there is no need to use additional FMM runs for the determination of the surface distributions. Figure 10.7 illustrates distribution of the amplitude of the complex potential of the Helmholtz equation at $ka = 1$ over surfaces of 100 spheres.

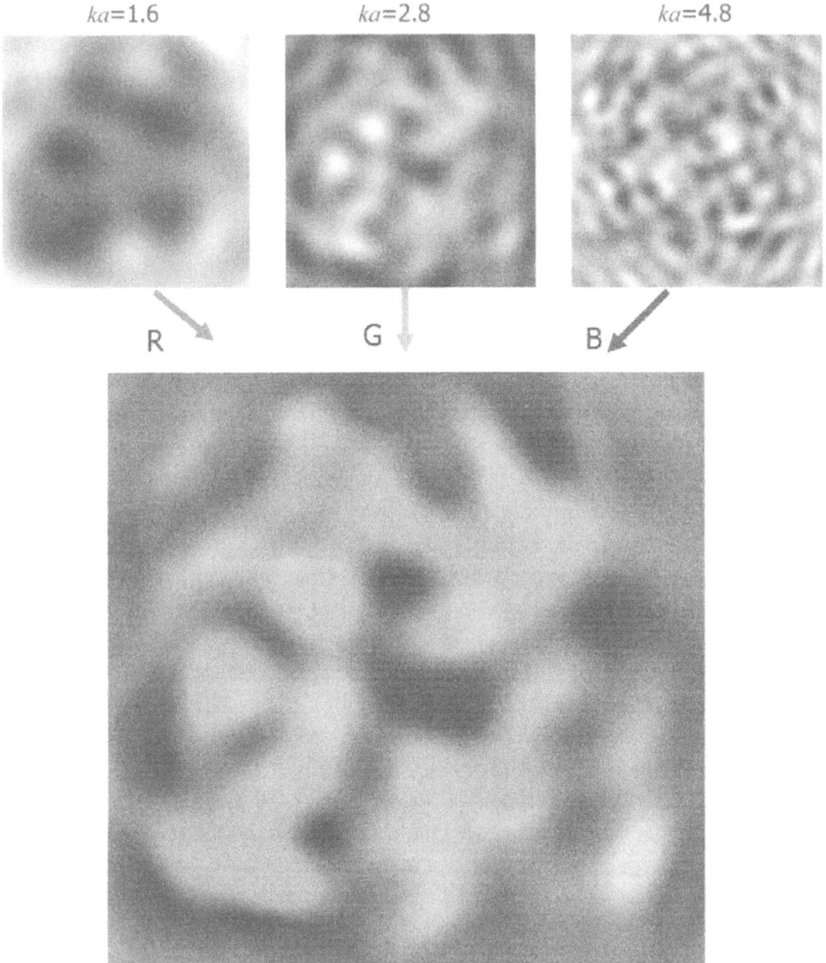

FIGURE 10.6 Speckle patterns on the imaging plane computed using the FMM for scattering of a plane wave from 100 scatterers (the geometry is shown in Fig. 10.5 on the left). Computations are made with three different *ka* to produce grayscale images, which further put into the RGB channels to produce a polychromatic image. For color version, see Plate F after Chapter 10.

10.3.2 A posteriori error evaluation

For these tests we used Eq. (10.2.19) to obtain the truncation number where we varied p_0, and relations (9.1.41) and (10.2.22) for the FMM

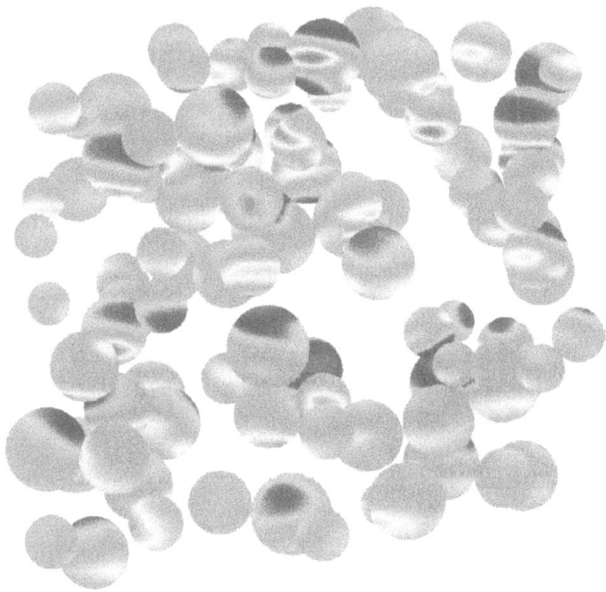

FIGURE 10.7 An example of surface imaging for scattering from a set of 100 spheres ($ka = 1$). The numerical solution is obtained with the FMM. A colormap used to represent the amplitude of the surface potential. For color version, see Plate G after Chapter 10.

procedure, where ϵ was specified to be the same as for the convergence of the iterative process. After the iterative process (GMRES) was terminated, we computed the potential at M sampling points and the relative errors in boundary conditions using Eq. (10.2.21). The spheres were distributed randomly inside a cubic box with the restriction that the distance between the sphere centers exceeds some prescribed minimum distance between the sphere centers, $d_{\min}$. Tests were made for the incident wave $\psi_{\text{in}}(\mathbf{r}) = e^{i\mathbf{k}\cdot\mathbf{r}}$.

An example of the computed surface error is shown in Fig. 10.8. This figure illustrates the behavior of $|\psi||_{S_q}(\mathbf{y}_m)$ and $\Delta_{\text{bc}}(\mathbf{y}_m)$ for $M = 3200$ sampling points allocated on the surfaces of 100 equal size sound-hard spheres for $p = 11$ ($ka = 1$, $kD_0 \approx 30$, $\sigma_q = 0$, $\alpha = 0.042$, $d_{\min}/a = 3$, $l_{\max} = 3$, $\epsilon = 10^{-10}$). It is seen that the maximum absolute error of computations in this case is less than 10^{-4}, while the average error of computations is of the order of 10^{-6}.

The parameter $d_{\min}/a$ has a substantial effect on the accuracy. Figure 10.9 illustrates this effect for the same settings as for 10.8, where

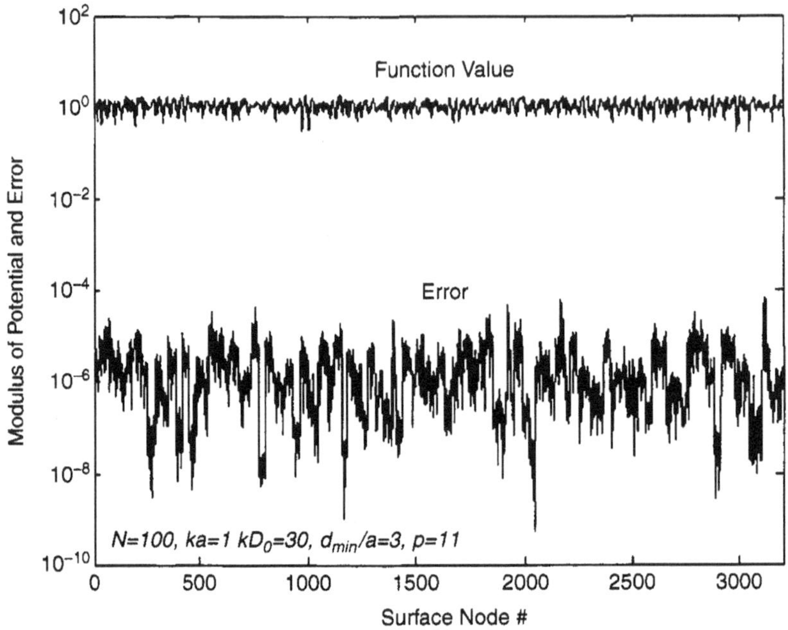

FIGURE 10.8 The modulus of the potential and the absolute error in boundary conditions obtained a posteriori as functions on the surface node index. The surface of each of 100 spheres was sampled by 32 points.

the truncation number and $d_{\min}/a$ were varied. The relative error $\epsilon^{(bc)}$ defined by Eq. (10.2.21) that can also be assigned as the relative error in the "infinity" norm L^{∞} is much larger (about one order of magnitude) than the average, or L^2-norm error, $\epsilon_2^{(bc)}$, defined as

$$\epsilon_2^{(bc)} = \frac{1}{M}\left\{\sum_{m=1}^{M} [\epsilon_m^{(bc)}]^2\right\}^{1/2}, \tag{10.3.2}$$

with $\epsilon_m^{(bc)}$ specified in Eq. (10.2.21). It is seen that even for the case when the spheres may touch ($d_{\min}/a = 2$) computations can be performed with relatively small truncation numbers if an accuracy $\epsilon^{(bc)}$ of the order of a few percent is acceptable (that might be sufficient for comparisons with experiments, field imaging, etc.). For computations with high precision the truncation numbers should be increased, which may substantially slow down the computation process. One of the methods to treat this drawback is to introduce truncation numbers p_q for each scatterer, $q = 1, \ldots, N$, which should depend on the distance

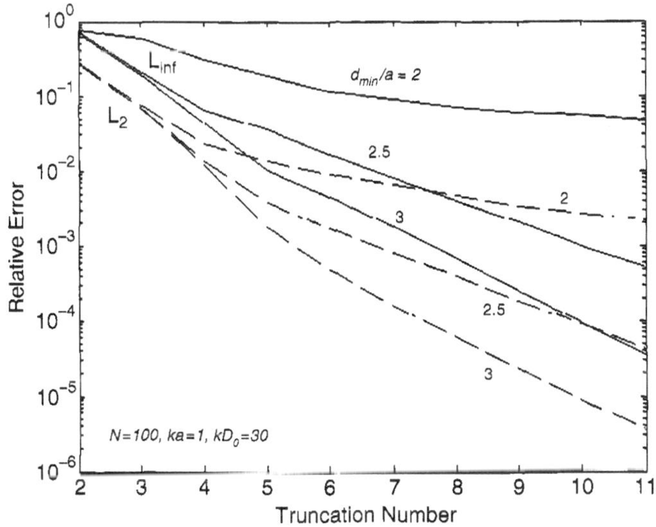

FIGURE 10.9 Dependences of the relative errors in the infinite (the solid lines) and quadratic (the dashed lines) norms on the truncation number p for the same parameters as shown in Fig. 10.8. The errors are computed for random distributions with different ratios $d_{\min}/a$ shown near the curves.

from the qth sphere to its closest neighbor sphere. In the present study, however, we did not perform modifications, since our goal was to develop and test the basic algorithm.

We note that a posteriori error evaluation as we did in the tests is an expensive procedure. So, for solution of multiple problems the errors can be studied for smaller size problems before running large tasks, or some methods for fast a posteriori error control should be developed and employed. We leave the development of such methods to interested readers.

10.3.3 Convergence

Several factors may affect the rate of convergence of the iterative method and more or less complete study involves investigation of the multiparametric space. In the present tests we usually fixed all parameters for the case and varied one or two parameters to see their effect on the convergence rate and accuracy of computations. Figure 10.10 shows the effect of the frequency on the convergence of unpreconditioned GMRES. For the case illustrated, 200 equal size sound-hard spheres with a

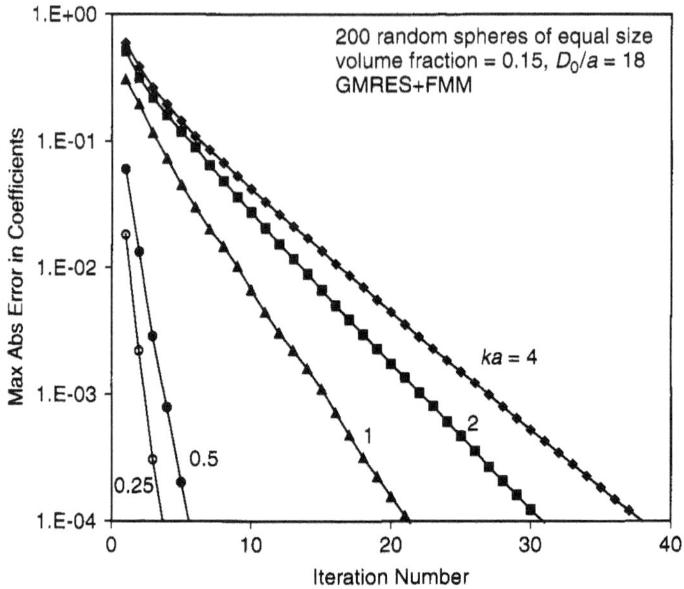

FIGURE 10.10 Convergence of the unpreconditioned GMRES at different dimensionless wavenumbers *ka* shown near the curves.

random distribution of their centers were generated as described above ($d_{\min}/a = 2.3$). The truncation numbers were selected according to Eq. (10.2.19) with a constant $p_0 = 10$ for all cases, while the FMM truncation numbers were determined according to Eq. (9.1.41) with $\epsilon = 10^{-4}$. It is seen that in this case the number of iterations to achieve specified accuracy increases with *ka*. An a posteriori error check for the boundary conditions at 6400 points shows that, for this case, the error $\epsilon^{(bc)}$ defined by relation (10.2.21) does not exceed 5×10^{-3} for all cases (while $\epsilon_2^{(bc)}$ defined by Eq. (10.3.2) was about one order of magnitude smaller). We found that the iteration process for higher frequencies ($ka \gtrsim 5$, $kD_0 \gtrsim 100$) diverges, and special methods for the treatment of higher frequencies are required. Several methods can be tried in this case, such as using preconditioners based on high frequency asymptotic expansions of the system matrix, and some relaxation methods, where, e.g. the impedance of the spheres, σ_q, can be gradually adjusted during the iteration process. Indeed, we did some preliminary computations with fixed $ka = 4$ and varying σ_q and found that, say for $\sigma_q/k \sim 0.1$, the number of iterations to achieve the same accuracy drops by a factor of two.

Properly designed preconditioners can substantially reduce the computation time to obtain a solution of specified accuracy. Some results of the use of GMRES and FGMRES with the right dense preconditioner $\mathbf{M}_{\text{dense}}$ are shown in the table below.

Preconditioner	External iterations	Internal iterations	CPU time (s)
No preconditioner	58	0	8340
Right dense, $\epsilon_{\text{pre}} = 0.1$	11	87	3460
Right dense, $\epsilon_{\text{pre}} = 0.01$	10	165	4525

This table shows computational results for 10,000 sound-hard spheres of equal size with a randomly periodic distribution (blocks of 80 spheres with random distribution of centers, $d_{\min}/a = 2$, $\alpha = 0.2$, were repeated 125 times to occupy a larger cube). Here $D_0/a = 60$ and $ka = 0.5$, $l_{\max} = 4$, $\epsilon = 10^{-4}$, $p = 7$ and the CPU time was measured for 3.2 MHz dual Xeon processor (3.5 GB RAM, 25% CPU resource allocation for a task). The table shows that the number of iterations for the external loop can be drastically reduced due to the use of preconditioner. The price to pay for this is the internal iteration loop (the table shows the total number of internal iterations), which also uses GMRES to obtain solution of equation $\mathbf{M}_{\text{dense}}\mathbf{A} = \mathbf{F}$. However, for the present case matrix–vector multiplication $\mathbf{M}_{\text{dense}}\mathbf{A}$ was seven times faster than performing multiplication $\mathbf{LA}$ (both using the FMM). This resulted in a decrease of the overall computational time. Solution of equation $\mathbf{M}_{\text{dense}}\mathbf{A} = \mathbf{F}$ can be performed with much less accuracy, ϵ_{pre}, (shown in the first column of the table), than the accuracy required for the external iterative process. As we mentioned above, this can be a subject for optimization, since the increase of the accuracy for internal iterations reduces the number of external iterations and increases the number of internal iterations.

10.3.4 Performance study

To determine the dependence of the CPU time required for solution of the problem, and check the consistency of the actual results with the theoretical predictions, we conducted a performance study. As a test distribution, we used randomly periodic distributions of equal size sound-hard spheres (blocks of 80 spheres with random distribution of centers, $d_{\min}/a = 2$ (so the spheres may touch), were repeated m^3 times to occupy a larger cube, $m = 1, \ldots, 5$). Since the volume fraction and the wavenumber in the tests were fixed ($\alpha = 0.2$, $ka = 0.5$), the increase in the number of spheres also means the increase of the size of the computational domain.

We found that to reduce the CPU time for the present case, one should select the computational domain size larger than the size of the box, which encloses all the spheres. The size of the computational domain can be determined based on $l_{\max}$ as

$$D_0 = \frac{2\sigma_{\min}}{4 - (1 + \sigma_{\min})\sqrt{3}} 2^{l_{\max}} a_{\max}. \tag{10.3.3}$$

Setting here $l_{\max} = 2, 3, \ldots$ we can determine a discrete set of D_0. For computations we selected the minimum of D_0 from this set, which encloses all the spheres. Although $\sigma_{\min}$ here should be larger than 1, we found with the a posteriori error check that $\sigma_{\min} = 1$ provides stable and accurate results in this case. Note that for performance tests we accepted relative errors $\epsilon^{(bc)}$ of about 2% (the errors $\epsilon_2^{(bc)}$ were one order of magnitude smaller). To provide this order of errors, we set $\epsilon = 10^{-4}$ for the global iteration convergence, while the error for preconditioner was set to $\epsilon_{\text{pre}} = 0.2$. In all computations we used a truncation number of $p = 15$. This substantially exceeds $p_{l_{\max}} \sim 8$ and the choice of p was dictated by a desire to compute accurately the case of touching spheres. Since the excessive value $p = 15$ is only due to the influence of the neighbor spheres, for each sphere we took the contribution of the neighbors with $d/a < 1.3$ into harmonics of degree $n = 8, \ldots, 15$, while neglecting the effect of other scatterers on these harmonics. A posteriori error check showed that the relative errors specified above were achieved.

Figure 10.11 shows the measurements of the CPU time for various operations of the algorithm based on the FGMRES with right preconditioner. It is seen that the computation time scales almost linearly for a single matrix–vector multiplication in the internal or external loop using the FMM. This agrees well with the theoretical predictions of complexity $O(N \log N)$. Due to the growth of the number of iterations both in the internal and external loops with N, the total CPU time scales as $O(N^{1.25})$. This shows that some average number of internal–external iterations, N_{iter}, is proportional to $N^{0.25}$ or so in the present case. As it was discussed above, an optimum can be sought by varying ϵ_{pre} and, possibly, ϵ to perform better, while we kept these parameters for the illustration case fixed. Note that for large N and unpreconditioned GMRES we found that the number of iterations N_{iter}

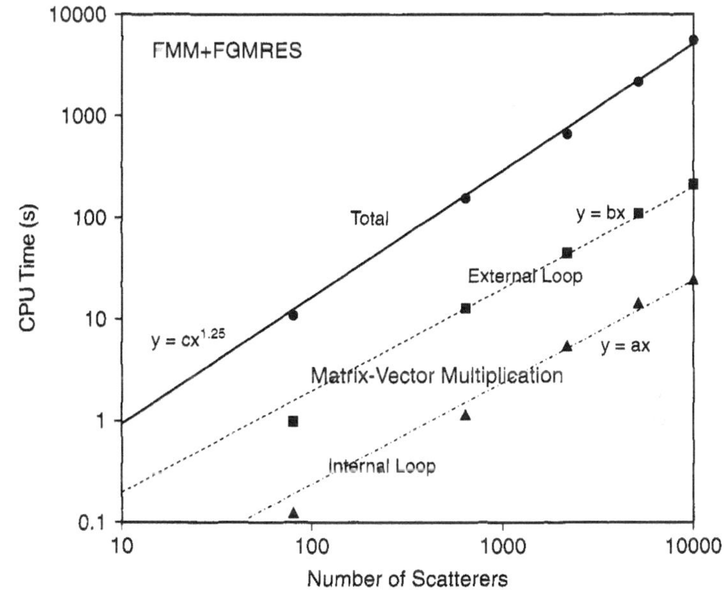

FIGURE 10.11 Dependences of the CPU time (Xeon 3.2 GHz, 3.5 GB RAM) on the number of scatterers ($ka = 0.5$, $p = 15$, $\epsilon = 10^{-4}$) for randomly periodic distribution of sound-hard spheres of equal size. The filled triangles and squares show the CPU time required for matrix–vector multiplication using the FMM for the preconditioner and full matrix, respectively. The filled circles show the total CPU time for the execution of the algorithm. The lines show interpolation of data.

grows proportionally to N for $N \gtrsim 2000$, which substantially reduces the efficiency of the algorithm.

It is also noteworthy that even for $N = 80$ the size of the complex system matrix **L** is $Np^2 \times Np^2 = 18{,}000 \times 18{,}000$ which would require several hours to solve the problem directly, while the present method requires only 10 s for this operation. We also found that solution of this problem using GMRES without preconditioners and the FMM for matrix–vector multiplication takes about 75 s of CPU time for the same machine. Solution of a larger problem ($N = 10{,}000$) requires inversion of a complex matrix of size $2.25 \times 10^6 \times 2.25 \times 10^6$, which is unsolvable problem for the computational resources indicated above, while the problem can be computed with the method described above for about 1 h. We should also note that the actual speed of the algorithm substantially depends on the programming skills of various procedures and experienced developers can drop the CPU time several times just using smart implementation.

10.4 SUMMARY

In this chapter, as an example, we considered application of the FMM to the solution of multiple scattering problem. We showed that the use of iterative methods substantially reduces the complexity of the straightforward multipole reexpansion method. We provided a novel formulation of the problem which can be efficiently solved with the FMM and enables drastic reduction in the complexity of the algorithm. We also pointed out that some related problems, such as spatial field imaging, can also be solved with the FMM. Finally, we discussed some particular issues concerning the relation of truncation numbers to the size of the problem, convergence of iterations, optimization of the FMM in this case, and provided some numerical examples and illustrations.

Color Plates

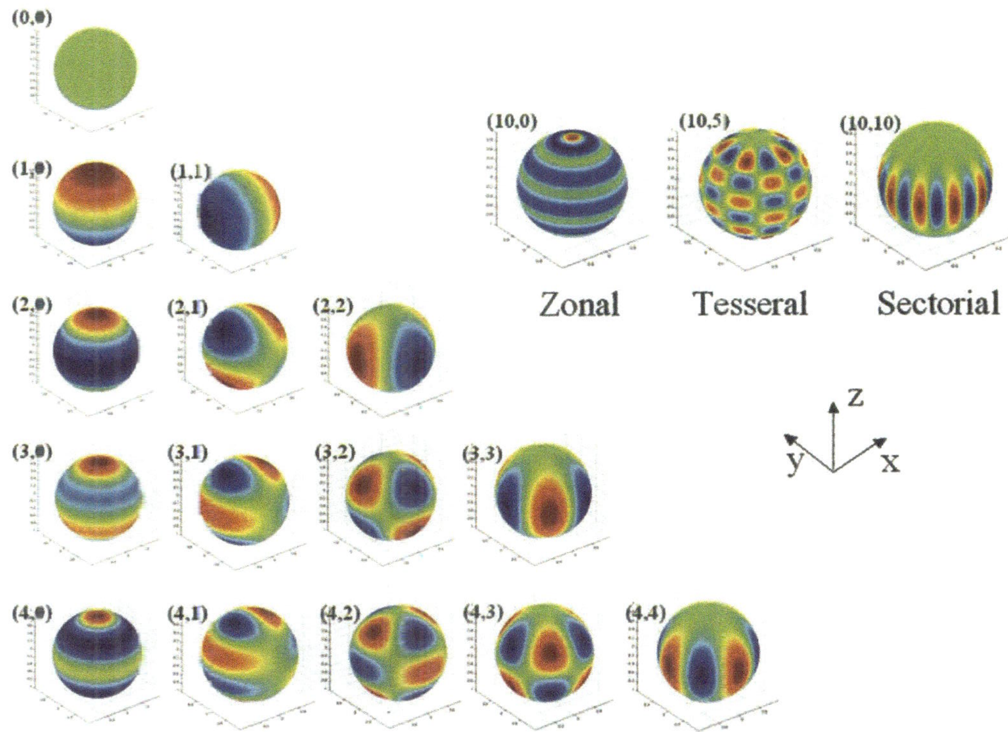

Plate A: Spherical harmonics. Illustration is made by plotting the real parts of spherical harmonics $Y_n^m(\theta, \varphi)$ on a sphere. The color corresponds to the value of $\mathrm{Re}\{Y_n^m(\theta, \varphi)\}$. The first number near the spheres shows the degree (n) and the second number shows the order (m) of the harmonics.

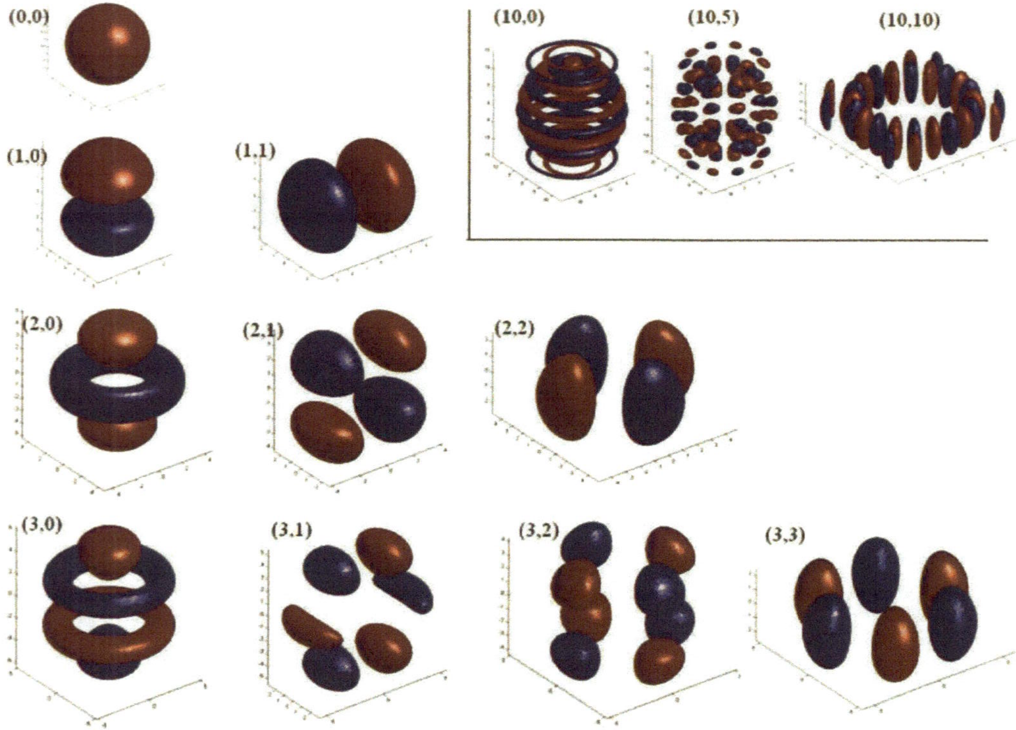

Plate B: Isosurfaces of spherical basis functions $\text{Re}\{R_n^m(\mathbf{r})\} = \text{const}$ for various degree n and order m indicated in parenthesis as (n, m) near each mode.

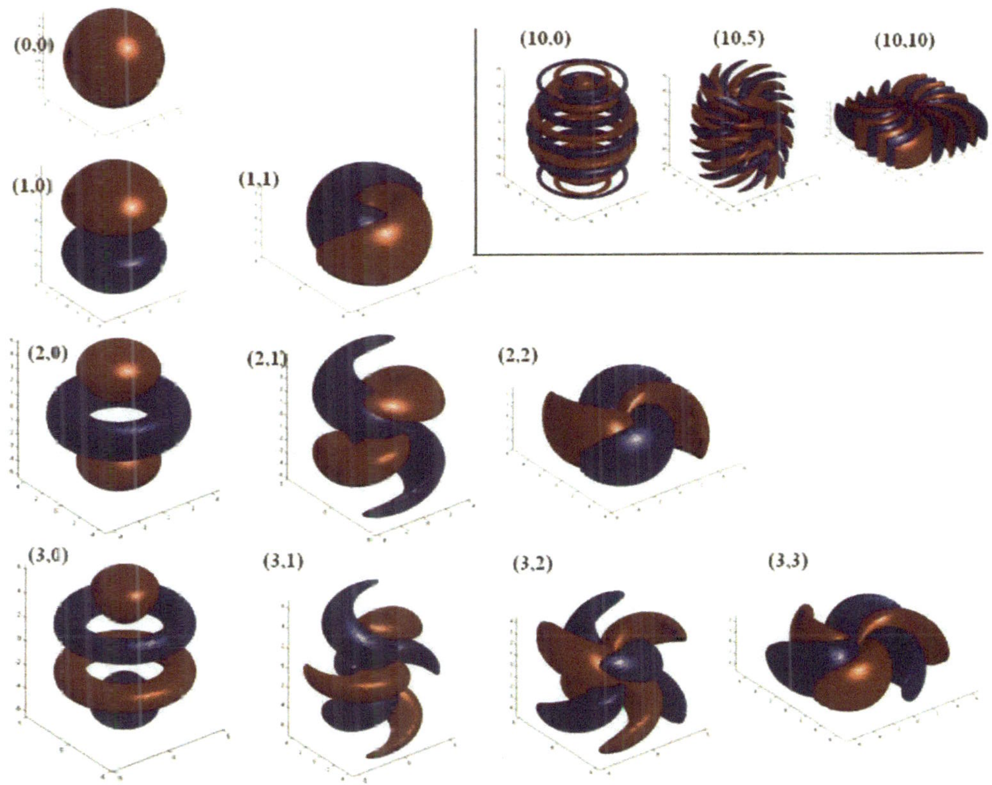

Plate C: Isosurfaces of spherical basis functions $\mathrm{Re}\{S_n^m(\mathbf{r})\} = \mathrm{const}$ for various degree n and order m indicated in parenthesis as (n, m) near each mode.

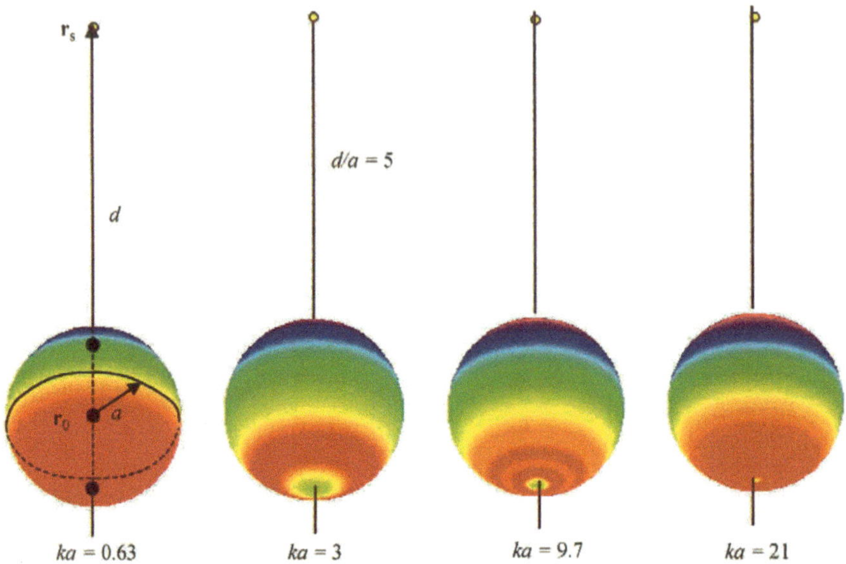

Plate D: Source above a sphere. The surface potential $|\psi|_S|$ is plotted for various ka and fixed $d/a = 5$. Because of the axisymmetric nature of the problem only a dependence on the incidence angle θ is observed.

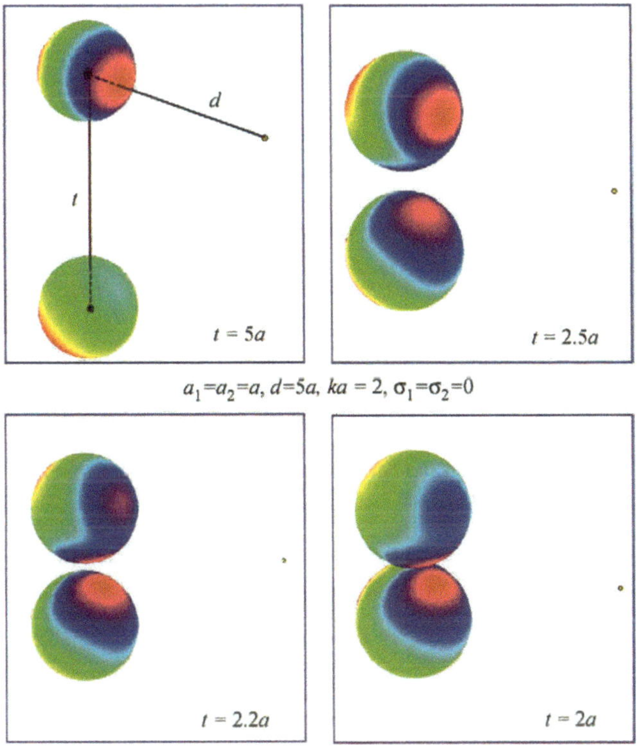

Plate E: Surface distributions of the modulus of the potential of the Helmholtz equation for two spheres of equal radius exposed to the field of monopole source as the spheres approach each other.

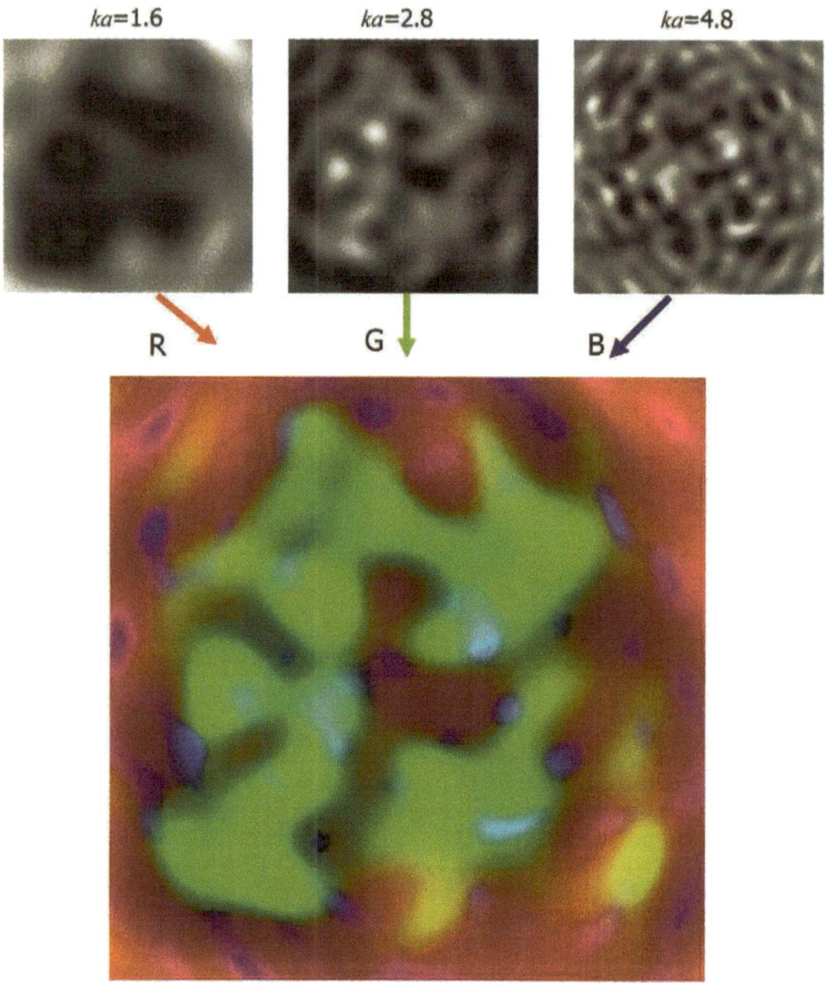

Plate F: Speckle patterns on the imaging plane computed using the FMM for scattering of a plane wave from 100 scatterers (the geometry is shown in Fig. 10.3 on the left). Computations are made with three different *ka* to produce grayscale images, which further put into the RGB channels to produce a polychromatic image.

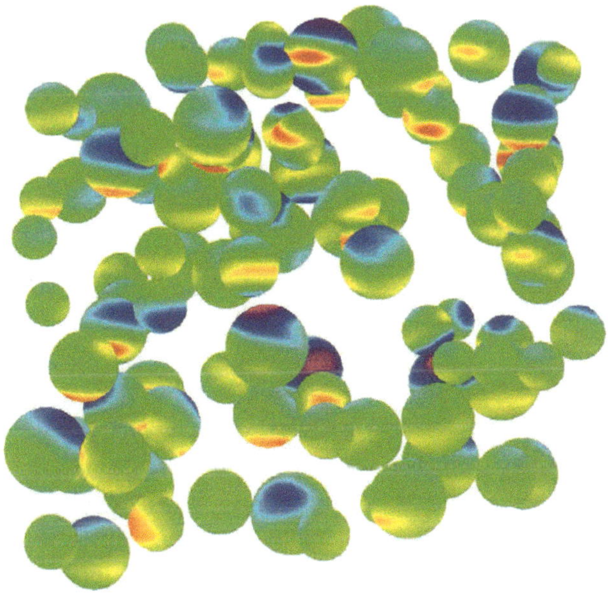

Plate G: An example of surface imaging for scattering from a set of 100 spheres ($ka = 1$). The numerical solution is obtained with the FMM. A colormap used to represent the amplitude of the surface potential.

Bibliography

[AB79] J.B. Allen and D.A. Berkley. Method for simulating small-room acoustics. *J. Acoust. Soc. Am.*, 65:943–950, 1979.

[ADD02] V.R. Algazi, R.O. Duda, R. Duraiswami, N.A. Gumerov, and Z. Tang. Approximating the head-related transfer function using simple geometric models of the head and torso. *J. Acoust. Soc. Am.*, 112: 2053–2701, 2002.

[AR91] B.K. Alpert and V. Rokhlin. A fast algorithm for the evaluation of Legendre expansions. *SIAM J. Sci. Stat. Comput.*, 12:158–179, 1991.

[AS64] M. Abramowitz and I.A. Stegun. *Handbook of Mathematical Functions.* National Bureau of Standards, Washington, DC, 1964.

[Atk49] F.V. Atkinson. On Sommerfield's radiation condition. *Phil. Mag. Ser. 7*, 40:645–651, 1949.

[CGR99] H. Cheng, L. Greengard, and V. Rokhlin. A fast adaptive multipole algorithm in three dimensions. *J. Comput. Phys.*, 155:468–498, 1999.

[Che90] W.C. Chew. *Waves and Fields in Inhomogeneous Media.* Van Nostrand Reinhold, New York, 1990.

[Che92] W.C. Chew. Recurrence relations for three-dimensional scalar addition theorem. *J. Electromagn. Waves Appl.*, 6(2):133–142, 1992.

[CIGR99] C.H. Choi, J. Ivanic, M.S. Gordon, and K. Ruedenberg. Rapid and stable determination of rotation matrices between spherical harmonics by direct recursion. *J. Chem. Phys.*, 111:8825–8831, 1999.

[CK98] D. Colton and R. Kress. *Inverse Acoustic and Electromagnetic Scattering Theory*, 2nd ed. Springer, Berlin, 1998.

[CT65] J.W. Cooley and J.W. Tukey. An algorithm for the machine calculation of complex Fourier series. *Math. Comput.*, 19:297–301, 1965.

[CLR90] T.H. Cormen, C.E. Leiserson, and R.L. Rivest. *Introduction to Algorithms.* MIT Press, Cambridge, MA, 1990.

[CMSJ01] W.C. Chew, E. Michielssen, J.M. Song, and J.M. Jin. *Fast and Efficient Algorithms in Computational Electromagnetics.* Artech House, Boston, MA, 2001.

[CRW93] R. Coifman, V. Rokhlin, and S. Wandzura. The fast multipole method: A pedestrian prescription. *IEEE Trans. Antennas Propagat.*, 35:7–12, 1993.

[Dar00] E. Darve. The fast multipole method I: Error analysis and asymptotic complexity. *SIAM J. Numer. Anal.*, 38:98–128, 2000.

[DS00] J.J. Dongarra and F. Sullivan. The top 10 algorithms. *Comput. Sci. Eng.*, 2:22–23, 2000.

[DGR96] A. Dutt, M. Gu, and V. Rokhlin. Fast algorithms for polynomial interpolation, integration and differentiation. *SIAM J. Numer. Anal.*, 33: 1689–1711, 1996.

[DGZD01] R. Duraiswami, N.A. Gumerov, D.N. Zotkin, and L.S. Davis. Efficient evaluation of reverberant sound fields. In *Proceedings of the IEEE Workshop on Applications of Signal Processing to Audio and Acoustics*, pp. 203–206, 2001.

[DH94] J.R. Driscoll and D.M. Healy. Computing Fourier transforms and convolution on the sphere. *Adv. Appl. Math.*, 15:202–250, 1994.

[ED95] M.A. Epton and B. Dembart. Multipole translation theory for the three-dimensional Laplace and Helmholtz equations. *SIAM J. Sci. Comput.*, 16:865–897, 1995.

[EJB96] W.D. Elliott and J.A. Board, Jr. Fast Fourier transform accelerated fast multipole algorithm. *SIAM J. Sci. Comput.*, 17:398–415, 1996.

[EW31] G.E. Wigner. *Gruppentheorie und ihre Anwendung auf die Quantenmechanik der Atomspektren*. Vieweg, Wiesbaden, Germany, 1931.

[FHLB99] P. Filippi, D. Habault, J.P. Lefevre, and A. Bergassoli. *Acoustics, basic physics, theory and methods*. Academic Press, London, 1999.

[GR87] L. Greengard and V. Rokhlin. A fast algorithm for particle simulations. *J. Comput. Phys.*, 73:325–348, 1987, 135:280–292, 1997.

[GD01a] N.A. Gumerov and R. Duraiswami. Fast, exact, and stable computation of multipole translation and rotation coefficients for the 3-D Helmholtz equation. Technical Report 2001-44, University of Maryland, Institute for Advanced Computer Studies, 2001.

[GD01b] N.A. Gumerov and R. Duraiswami. Modeling the effect of a nearby boundary on the HRTF. In *Proceedings. (ICASSP'01). IEEE International Conference on Acoustics, Speech, and Signal Processing*, Vol. 5, pp. 3337–3340, 2001.

[GD02] N.A. Gumerov and R. Duraiswami. Computation of scattering from n spheres using multipole reexpansion. *J. Acoust. Soc. Am.*, 112: 2688–2701, 2002.

[GDB03] N.A. Gumerov, R. Duraiswami, and Y.A. Borovikov. Data structures, optimal choice of parameters, and complexity results for generalized fast multipole methods in d dimensions. Technical Report CS-TR-4458, University of Maryland Institute for Advanced Computer Studies, 2003. Version available as University of Maryland UMIACS-TR-2003-28, CS-TR-4458.

[GDT02] N.A. Gumerov, R. Duraiswami, and Z. Tang. Numerical study of the influence of the torso on the HRTF. In *Proceedings. (ICASSP '02). IEEE International Conference on Acoustics, Speech, and Signal Processing*, Vol. 2, pp. 1965–1968, 2002.

[GL89] G. Golun and C. Van Loan. *Matrix Computations*, 2nd ed. Johns Hopkins University Press, Baltimore, MD, 1989.

[Han81] S. Hanish. *A Treatise on Acoustical Radiation*. Naval Research Laboratory, Washington, DC, 1981.

[IR96] J. Ivanic and K. Ruedenberg. Rotation matrices for real spherical harmonics. Direct determination by recursion. *J. Chem. Phys.*, 100: 6342–6347, 1996.

[Jac75] J.D. Jackson. *Classical Electrodynamics*. Wiley, New York, 1975.

[JCA97] R. Jakob-Chien and B.K. Alpert. A fast spherical filter with uniform resolution. *J. Comput. Phys.*, 136:580–584, 1997.

[JN92] H.M. Jaeger and S.R. Nagel. Physics of granular states. *Science*, 255: 1524, 1992.

[KC98] S. Koc and W.C. Chew. Calculation of acoustical scattering from a cluster of scatterers. *J. Acoust. Soc. Am.*, 103:721–734, 1998.

[Kel95] C.T. Kelly. *Iterative Methods for Linear and Nonlinear Equations*. SIAM, Philadelphia, PA, 1995.

[KMHR00] P.J. Kostelec, D.K. Maslen, D.M. Healy, Jr., and D.N. Rockmore. Computational harmonic analysis for tensor fields on the two-sphere. *J. Comput. Phys.*, 162(2):514–535, 2000.

[KSC99] S. Koc, J. Song, and W.C. Chew. Error analysis for the numerical evaluation of the diagonal forms of the scalar spherical addition theorem. *SIAM J. Numer. Anal.*, 36:906–921, 1999.

[K99] R. Kurzweil. *The Age of Spiritual Machines: When Computers Exceed Human Intelligence*. Penguin, Baltimore, MD, 1999.

[LL60] L.D. Landau and E.M. Lifshitz. *Electrodynamics of Continuous Media*, volume 8 of *Course of Theoretical Physics*. Pergamon Press, New York, 1960.

[LL75] L.D. Landau and E.M. Lifshitz. *Fluid Mechanics. Course of Theoretical Physics*. Pergamon Press, New York, 1975.

[GM65] G. Moore. Cramming more components onto integrated circuits. *Electronics*, 38:8, 1965.

[MF53] P.M. Morse and H. Feshbach. *Methods of Theoretical Physics Vol. 1*. McGraw-Hill, New York, 1953.

[MI68] P.M. Morse and K.U. Ingard. *Theoretical Acoustics*. Princeton University Press, Princeton, NJ, 1968.

[ML78] C. Moler and C. Van Loan. Nineteen dubious ways to compute the exponential of a matrix. *SIAM Rev.*, 20:801–836, 1978.

[Moh99] M.J. Mohlenkamp. A fast transform for spherical harmonics. *J. Fourier Anal. Appl.*, 2:159–184, 1999.

[OM84] J.A. Orenstein and T.H. Merret. A class of data structures for associative searching. In *Proceedings of the Third ACM SIGAT-SIGMOD Symposium on Principles of Database Systems*, pp. 181–190, 1984.

[Pea90] G. Peano. Sur une courbe qui remplit toute une aire plaine. *Math. Annalen*, 36:157–160, 1890.

[Rah96] J. Rahola. Diagonal forms of the translation operators in the fast multipole algorithm for scattering problems. *BIT*, 36:333–358, 1996.

[Ray04] J.W. Strutt (Lord Rayleigh). On the acoustic shadow of a sphere. *Phil. Trans. R. Soc. Lond. Ser. A*, 203:87–89, 1904.

[Rok93] V. Rokhlin. Diagonal form of translation operators for the Helmholtz equation in three dimensions. *Appl. Comput. Harmon. Anal.*, 1:82–93, 1993.

[Saa92] Y. Saad. *Numerical Methods for Large Eigenvalue Problems*. Halstead Press, New York, 1992.

[Saa03] Y. Saad. *Iterative Methods for Sparse Linear Systems*, 2nd ed. SIAM, Philadelphia, PA, 2003.

[Saa93] Y. Saad. A flexible inner–outer preconditioned GMRES algorithm. *SIAM J. Sci. Comput.*, 14(2):461–469, 1993.

[Sam90] H. Samet. *Applications of Spatial Data Structures*. Addison-Wesley, Reading, MA, 1990.

[Sam94] H. Samet. *The Design and Analysis of Spatial Data Structures*. Addison-Wesley, Reading, MA, 1994.

[ST02] R. Suda and M. Takami. A fast spherical harmonics transform algorithm. *Math. Comput.*, 71(238):703–715, 2002.

[Ste61] S. Stein. Addition theorems for spherical wave functions. *Quart. Appl. Math.*, 19:15–24, 1961.

[Tan03] Z. Tang. *Fast transforms based on structured matrices with applications to the fast multipole method*. PhD Thesis, University of Maryland, College Park, MD, 2003.

[Tay95] M. Taylor. Cubature for the sphere and the discrete spherical harmonic transform. *SIAM J. Numer. Anal.*, 32:667–670, 1995.

[TDG04] Z. Tang, R. Duraiswami, and N.A. Gumerov. Fast algorithms to compute matrix vector products for Pascal matrices. Technical Report, University of Maryland, College Park, MD, 2004.

[TS99] T. Kailath and A.H. Sayed. *Fast Reliable Algorithms for Matrices with Structure*. SIAM, Philadelphia, PA, 1999.

[VMK88] D.A. Varsalovich, A.N. Moskalev, and V.K. Khersonskii. *Quantum Theory of Angular Momentum*. World Scientific, Singapore, 1988.

[VS01] V. Olshevsky and A. Shokrollahi. A superfast algorithm for confluent rational tangential interpolation problem via matrix-vector multiplication for confluent Cauchy-like matrices. In *Structured Matrices in Mathematics, Computer Science, and Engineering I*, Vol. 280, pp. 67–78, AMS, Providence, RI, 2001.

[Wat58] G.N. Watson. *A Treatise on the Theory of Bessel Functions*. Cambridge University Press, Cambridge, 1958.

[WI156] C.H. Wilcox. A generalization of theorems of Rellich and Atkinson. *Proc. Am. Math. Soc.*, 7:271–276, 1956.

[YR98] N. Yarvin and V. Rokhlin. A generalized one-dimensional fast multipole method with application to filtering of spherical harmonics. *J. Comput. Phys.*, 147(594):609, 1998.

Subject Index

Kolibri 360 GmbH
Ferdinand-Jühlke-Straße 7
99095 Erfurt, Deutschland
produktsicherheit@kolibri360.de